MUSEUM DIGITIZATION

www.royalcollins.com

MUSEUM DIGITIZATION

Edited by

Beijing Union University and Beijing Association
for the Popularization of Digital Science and Technology

RC

Books Beyond Boundaries

ROYAL COLLINS

Museum Digitization

Edited by Beijing Union University and Beijing Association for the Popularization of Digital Science and Technology

First published in 2021 by Royal Collins Publishing Group Inc.
Groupe Publication Royal Collins Inc.
BKM Royalcollins Publishers Private Limited

Headquarters: 550-555 boul. René-Lévesque O Montréal (Québec) H2Z1B1 Canada
India office: 805 Hemkunt House, 8th Floor, Rajendra Place, New Delhi 110 008

ISBN: 978-1-4878-0446-6

To find out more about our publications, please visit www.royalcollins.com

We are grateful for the financial assistance of B&R Book Program in the publication of this book.

Editorial Board

Preface

In the rapidly developing "Internet+" era, the application of digital technology in science and technology museums is becoming increasingly mature, especially in the implementation of strategies that support culture through science and technology, bringing new opportunities to the construction and development of science and technology museums. Science and technology museums of all levels throughout the world not only actively promote the application of information technology, but also make use of the Internet, mobile communication, Big Data, and other technology to make museums more open, inclusive, and dynamic, and to serve as public cultural and educational institutions serving the all-round development of human beings. A number of digital display projects are being launched, such as the immersive digital audio-visual program "Stepping into the Riverside scene at Qingming Festival," the large-scale virtual reality show "The Forbidden City – Juanqinzhai and Turret Tower," "Digital Yuanmingyuan," "Digital Three Mountains Five Gardens," and "Roaming the Dajue Temple." The application of WiFi technology to experience the micro Palace Museum, micro navigation, and micro video has been implemented in many science and technology museums. Facing a new situation of deepening reform and accelerating the prosperity of science, technology, and culture in China, what will the future development direction of museum digitization be? How can science and technology museums make full use of the advantages of new technology to showcase culture and services? How can they tell a positive story? How can they bring cultural relics to life? How can they increase, strengthen, and accelerate the dissemination of information in science and technology museums and improve the level of public services with the help of new media? How can they make it easier for the public to access and enjoy museums? These and other important topics created the need for the Beijing Digital Museum seminar.

Under the guidance of the Beijing Association for Science and Technology, the Beijing Municipal Bureau of Cultural Relics, and the Beijing Municipal Bureau of Economy and Information Technology, on June 14–15, 2015, the 2015 Beijing Digital Museum seminar took place at Beijing Union University and the Software Park of the Chinese Academy of

Sciences. It was jointly hosted by the Beijing Association for the Popularization of Digital Science and Technology, the Capital Museum Alliance, the Beijing Museum Society, the Museum Digitization Committee of the Chinese Museum Association, the Cultural Relics Photography Committee of the China Cultural Relics Academy, Beijing Union University, and the Computer Network Information Center of the Chinese Academy of Sciences.

The theme of the symposium was Promoting Public Cultural Services — Museum Informatization. The five topics in this book, ("Letting Cultural Relics Live — The Creative Digitization of the Museum," "Remembering Nostalgia — The Digitization and Conservation of Cultural Heritage," "Sharing and Utilization as the New Status Quo — Digital Development and the Sharing of Museum Resources," "Mobile Navigation for the Thousands — How Emerging Media can Boost Communications and Services for Museums," and "Entering the Museum — Excavating and Utilizing the Cultural Tourism Value of the Museum") lead this exploration of the digitization of museums.

The seminar received widespread attention. It was attended by than 300 professionals and scholars in fields such as technology, culture and museology, the popularization of science, education, design, art and the Internet came from the Beijing Museum (Science and Technology department), the University Museum, the Chinese Academy of Sciences Museum, the science popularization base, colleges and universities, scientific research institutes, and Zhongguancun science and technology innovation enterprises in Shanghai, Zhejiang, Yunnan, Shandong, Fujian, Hubei, and other areas. More than 90 papers were received, and nearly 50 people gave speeches.

This seminar introduced the most forward-looking technological developments and the most innovative demonstration cases, promoting the most pragmatic approaches and serving as a good reference for the present and future of the digitization of science and technology museums. The organizing committee set up an editorial board, which collated and collected 70 papers, and published them based on the five topics discussed in the seminar. These papers focus on the museum's realization of its own social functions and their responsibility in social and cultural development, and also discuss the important role that science and technology museums and other regional cultural resources can play in the public service system. They analyze development opportunities and application cases brought by new media to museums, and also discuss museum data processing and the management and standardized operation of digital resource utilization of cultural relics. Moreover, they expound the reform and innovation of the museum in adapting to new situations, and the planning and vision for the construction of future museums.

This book is the fifth in a series of papers on the development of Beijing's digital museums since 2005. It is suitable for those who are engaged in public cultural services in facilities such as science and technology museums and libraries, and personnel involved in cultural creativity, exhibition, science popularization, and research and development. It can also be used as a reference for people in the field of digitization or science communications, and for enthusiasts who are interested in the development of science and technology museums. Readers are welcome to contact us by e-mail (bjszkpxh@163.com) to exchange experience, offer their opinions, and make suggestions. If you encounter any problems with the compilation, please comment with corrections.

In the process of editing this book, Zhao Hong, Li Fengyang and their colleagues from the Beijing Association for the Popularization of Digital Science and Technology have done much work in the collection and arrangement of manuscripts.

Thank you.

The Editorial Board

Contents

⇨ CHAPTER 1

Letting Cultural Relics Live

– The Creative Digitization of the Museum

Preliminary Exploration into the Construction of a Digital Museum

Qian Wenzhong
(China Agriculture Museum Information Center, Beijing 100026, China)

Abstract: With the development of digital museums, increasing numbers of Chinese and foreign museums have begun research and exploration into digitization, making achievements in both theory and practice. The following paper introduces the current status quo of museum digitization, and the relationship between physical museums and digital museum applications. It discusses prevalent issues that must be attended to in the development of the digital museum, engaging with construction practices, and using the China Agriculture Museum as a key example.

Keywords: digital museum; network; construction

Introduction

Since their introduction in the 1990s, digital museums have continued to develop over more than two decades, with increasing numbers of museums both in China and abroad engaging in relevant research and exploration. Early proponents such as the Digital Palace, the International Dunhuang Project, and China University Digital Museum Construction set the precedent. Since then, the digital museum has seen continuing research and booming construction, with guidance from the Ministry of Education, the State Administration of Cultural Heritage, and other relevant government departments. It has also received support from academic organizations such as the Association for the Digitization of Chinese Museums, the Informatization Committee of the China Society of Cultural Relics, and the Beijing Association for the Popularization of Digital Science and Technology, as well as positive responses from a variety of museums and active participation from several professional IT companies.

1. Outline of the Digital Museum

1.1 The concept of the digital museum

There are many concepts involved in the digital museum, but generally speaking, they emphasize the use of digital network technology, relying on digital methods to fully present the substance of a physical museum, including an on-site digital display system for the physical exhibition hall, a digital service management system using network technology, and a network platform system.

1.2 Current status of digital museums

In the late 1990s, digital museums began attracting the attention cultural and artistic circles in China, sparking a surge of research. In 1997, Nanjing Museum built the first online museum, and hundreds of museums in China followed suit.

In June 2000, the Palace Museum put forward the "Digital Palace" concept, establishing a website the following year. This was the birth of China's first genuine digital museum. Since then, the Palace Museum's website has released a time-walking section, allowing users to explore a complete rendering of the Palace online.

In 2001, the Ministry of Education launched a project called "Modern Distance Education Using Online Public Resources — The Construction of a Digital University Museum". As of the start of 2007, the project had built 18 digital university museums, providing a wealth of digital resources for teaching, scientific research, and popular science.

In September 2003, the State Administration of Cultural Heritage formally began research into digital museum engineering in China.

In 2002, the National Palace Museum in Taipei implemented a digitized collection plan. At present, its website has four components: selected collections, topical websites, 3D appreciation of cultural artefacts, and a collection database system. Aside from providing introductions to various cultural artefacts, it also offers an enquiry function.

Furthermore, the Capital Museum and the Nanjing Museum have also developed relatively complete and valuable object informatization databases and digital museums.

1.3 The relationship between digital museums and physical museums

1) Both digital and physical museums have strengths, and are mutually beneficial

Physical museums allow visitors to be fully immersed in the historical and cultural environment of a meticulously arranged museum hall. They can view actual valuable cultural artefacts, listening to and enjoying them in close proximity, and gaining knowledge and experience. By contrast, digital museums break the restrictions of physical space, offering visitors an online experience anytime and anywhere. They use digital resources to reorganize, consolidate, and improve collections, providing lifelike, vivid, and moving displays. They also use expert, network-based remote education to offer more convenient, enriched, and systematic knowledge. They utilize forums, message boards, public mailboxes, and various other interactive forms to evaluate individual feedback.

2) Physical museums are the foundation and support for digital museums
Digital collections are at the core of digital museums, and these collections are entirely reliant on the artefacts of a physical museum. Physical museums revolve around the collection, protection, research, and dissemination of cultural artefacts, all of which are necessary for the development of the digital museum. Each artefact within a physical museum is an irreplaceable symbol of a specific moment in history, with its own life and character. Without these artefacts, the physical museum would not exist, and there would be no source for digital artefacts.

3) Digital museums are the supplement and extension of physical museums
Digital museums can reduce the contradiction between the collection and preservation and research and display of cultural artefacts, allowing them to escape quantitative confines, and expanding their scope and spread. They are an opportunity for physical museums to attract potential visitors, and a crucial channel for the spread of their influence. Digital museums can enhance displays and maximize the hidden depths of exhibitions. They can also satisfy the need to protect and safeguard the legacy of intangible culture, meeting the demand for personal experience in a mutual interaction between visitors and displays. Moreover, they offer an effective means by which to improve the management of the physical museum.

1.4 Digital museum applications

Digital museums integrate the digitization, management, and interactive display of cultural collections, working to provide services for their preservation, management, and utilization. Most importantly, they provide the following applications:

1) Digitization of collections, including the collection and processing of text, 2D images, and 3D scan data regarding basic information, management information, and research information.
2) Business information management, i.e. collecting statistics, inquiries, and information regarding collections, accumulating conservation research and management information to form a "life archive" for the collection.
3) Digital information display, including an on-site digital display system in the physical museum's display hall, and network platform display system.

2. Crucial Focal Points for the Construction of Digital Museums

The development of digital museums is a huge and highly complex systematic project, requiring large capital investments, and involving a large amount of content, a long process, and complex techniques. As such, it appears crucial to clarify the theory, system architecture, technical requirements, and target visitors of digital museum development.

2.1 Users

Cultural artefacts are at the center of the physical museum, which acts as a means for their collection, preservation, research, and spread, achieving the purposes of education, research, and appreciation, with user seeing only what is exhibited. Alternatively, digital museums display "cultural artefacts" to users via the Internet, using digital technology, network technology, multimedia technology, and virtual reality to create lifelike, three-dimensional "virtual artefacts". With service at its core, user demands are studied so that they may see whatever they choose.

2.2 System architecture

The development of the digital museum contains both an on-site digitized display system of the physical museum exhibition hall and a network platform display system, as well as a basic network management system. These systems each have their own storage, processing, and display platforms, formed of huge and complex systems that require distinct and logical design, sensible architecture, and a clear hierarchy. They must distinguish among basic platforms, production and processing platforms, and application management platforms, determining the logical relationship and data integration between each, on a foundation of construction, safeguarding, and expansion.

2.3 Digital techniques

Within the development of the digital museum, many key technologies have arisen to promote rapid growth. Choosing appropriate digital techniques can improve visitation, benefitting user experience, increasing the sense of immersion, and offering users necessary knowledge while they take part in learning and entertainment.

These key digital techniques include: multimedia technology, used for the processing, production, and display of text, images, and audiovisual components of artefacts; virtual reality, used to recreate the environments of ancient artefacts, and conjure 3D, panoramic forms, and virtual roaming; 3D scanning, used to digitize artefacts, obtaining 3D data on cultural objects; 3D display, used to display artefacts in solid, 3D form, using methods of 3D modeling and automatic 3D imaging systems; QR codes, used in intelligent mobile terminals to obtain information efficiently, offering internet access, messaging, calls, material exchange, and automatic text import; mobile applications, used for location-based information push, and information browsing, and including mobile websites and mobile apps. Furthermore, advanced technologies, such as augmented reality, are also used in digital museums. Given the characteristics of these technologies and the requirements of displaying collections, with the appropriate selection, twice the results can be achieved with half the effort.

2.4 Standards for organizing information

Due to differences in organizational nature and professional categories, domestic museums have various classifications, recording regulations, structural systems, and management records. In the development of a digital museum, the organization of information is not standardized, meaning that digital museums become "information islands", which are not conducive to the sharing, dissemination, and utilization of data in collections.

1) Digitization of collections and database design
Formulating the digitization and database design specifications for a collection requires reference to the relevant national standards, other specifications such as the Information Index System Specifications for Museums (Trial), and the Information Index Specifications for Museum Collections as promulgated by the State Administration of Cultural Heritage. This is necessary to maintain consistency in the digitization of collections, the design of databases, and the development of information systems.

Collections should be comprehensive, complete, and standardized, and should describe their content accurately, supplemented by multimedia information including images, audiovisual content, 3D content, and animation from multiple perspectives. They should be well-rounded and deeply interpreted, all the while ensuring compatibility with system data.

2) Websites must abide by web standards
The design and production of a digital museum's webpage should abide by web standards, abandoning the traditional form layout and adopting an XML (extensible markup language) and CSS (cascading style sheet) layout. Webpage content and form should be created completely independently, with simplified site access and maintenance, and improved speeds for file download and page display. The appearance and feel of the page must be consistent, providing easily attainable information and greater usability.

2.5 Target visitors

Digital museum visitors fall into different groups, the positioning of which relates to the depth and breadth of digital collection resources, as well as the choice of interface style and expression. Specific visitors have different levels of knowledge and cognitive ability, and different preferences, meaning that the target positioning, function construction, overall content, style and form of a digital museum must be specifically determined to suit them. It is also possible to target multiple visitors with different columns.

3. Construction of the digital China Agricultural Museum

The development of the digital China Agricultural Museum entailed a unified framework and standards, focusing on user service, with an on-site digital exhibition display system, business management system, and network platform display system as the main construction content. Various technologies were applied – such as multimedia, virtual reality, and 3D display – to build a network-based, integrated information system with business, office, and public information service functions. The following section explains the network platform's display system.

3.1 Construction objectives

The digital China Agricultural Museum is based on its physical counterparts. With information resources at the core, information standards as the basis, information technology as the means, information networks as the carrier, and information application as a guide, it: digitally preserves and archives agricultural artefacts, sites, documents, historical figures, and customs; digitally simulates agricultural activities; digitally displays agricultural cultural artefacts and cultural heritage, creating a public service multimedia display platform that disseminates agricultural knowledge, publicizes agricultural conditions, and develops and expands agricultural civilization. The use of techniques such as information virtualization, digitization of information resources, and diversification of information display offers another window into the physical museum's content. The digital China Agricultural Museum focuses on agriculture, rural life, and peasantry, and has user service at its core. Its design ideas, content, and expression meet a variety of needs in terms of knowledge and entertainment for many different visitors.

3.2 Functional framework

The digital museum system involves five functions: gathering information about the collection, processing and production, display, publication, and application support.

1) A comprehensive platform for information about the collection
The integrated information platform is the foundation of the entire system. By digitizing the collection, it is transformed into digital information that can be accessed via computer, including sounds, graphics, and moving images, involving 2D and 3D forms. Currently, 2D photography and 3D scanning are largely used to gather this information in lieu of a highly automated, computer-controlled, high-definition photography and 3D information scanning system.

2) A multimedia processing and production platform
A key component of the overall system, the multimedia processing and production platform processes basic information attained by the integrated information collection platform, selecting various complete materials to be displayed on the network platform. Using computer system applications (such as Adobe Photoshop, Flash, Authorware, Director, 3D panorama, virtual roaming, 3dsMax, Web3D, 3D automatic imaging systems, and other professional software) the platform processes 2D and 3D information, creating 360° panoramic roaming, a virtual exhibition, 3D models, online virtual reality experiences, and other crucial content for the digital museum.

3) A display platform
Focusing on public web services, this platform configures various application servers to offer services via the Internet. Content and information displayed here is not limited to text and images, but also includes videos, panoramas, animations, puzzles and games, virtual reality (VR), 3D displays, and augmented reality. As such, the display platform extends the depth and breadth of information dissemination from the physical museum via storytelling, full-scale multimedia stereoscopic display, mass data, and interactive functions, conveying a rich web of information to visitors.

4) Platforms for the publication of information and application support
At the core of the system, the publication of information is achieved using a free framework system function and category settings, independently establishing the management permissions of each category, and publishing and offering on-demand features including text, images, panoramas, and 2D/ virtual 3D displays. The application support platform hosts various operating environments, including VR, 3D displays, and panoramic displays, providing a convenient background for management across a variety of systems and platforms. This platform is the support framework for the entire overall system.

5) The design of the system's background database
The system's background database offers support to the whole system, and must be built in conjunction with the information publication and application support platforms. Construction of the database must conform to standards for the organization of information, and must be built with the requirements of the overall system in mind. It must be unified and connected to the collection's database system, designed with a library and record format that is aligned with the existing collection database, with unified metadata, input, and output, so that data can be transplanted quickly and efficiently. Meanwhile, the interfaces of other digitally-networked business management systems for museums must be taken into account.

3.3 The content of construction

1) Digitization of collections: Basic information about collections, including photos, videos and 3D scanning information, is obtained through the comprehensive information collection platform, and then entered into the standard database and processed by the multimedia processing and production platform into integrated display materials.
2) Web portal: Multimedia and database technologies are used to build a full-text search system; this includes animations of agricultural history, memorabilia, agricultural figures, and sites, as well as sharing information from a variety of research publications and electronic magazines. Furthermore, expert blogs, question and answer systems, and the like must be created.
3) Pocket museums: This involves the development and utilization of mobile apps and the WeChat public platform, using network positioning systems, QR codes, RFID, and other such technologies to identify visitors by their identities and locations, then pushing customized guides and collections based on specific interests and hobbies. It also involves offering information services such as explanations, location services, and data downloads.
4) Virtual museums: This involves constructing a web-based 3D panoramic view of the exhibition hall and a virtual 3D exhibition using digital simulation technology (Virtual Reality/VR) and exploring the restoration of monuments and the protection of cultural heritage (augmented reality/AR); enabling visitors to enter a virtual world via high-precision 3D display, immersing them in "real" ancient scenes, and allowing for intimate contact with historical artefacts.

4. Conclusion

The digital museum offers an escape from the restrictions of time and space, allowing the public to enjoy cultural heritage and global cultural artefacts without leaving their homes. However, the development of digital museums is a complex project, involving interdisciplinary integration of the cultural and IT industries, and requiring support from experts, capital, and technology. Only with in-depth research, continuing exploration, and the gradual mastery of the theory and regulation of digital museum development will it be possible to build public digital museums as new windows to culture.

References

[1] Chen *Gang. Analysis of the concept, characteristics and development model of the digital museum.* National Museum of China, 2007, 3: 88–93.

[2] Zhang Haoda. 'Lessons and reflections on the recent construction of digital museums', *Digital Library Forum*. Beijing: Editorial Office of the Digital Library Forum, 2010 (1, 2).

[3] Qi Xiande, Ma Qi, Han Jun. *Preliminary study on digital museum solutions and current trends in digital museum development*. Beijing: Communication University of China CUC Publishing House, 2014, pp. 28-32.

[4] *Thinking about digital museums*. Bobo Art Network: http://news.artxun.com/siyangfangzun-1467-7330289.shtml.

[5] The Digital Museum. Baidu Encyclopedia: http://baike.baidu.com/view/1528687.htm.

The Digital Palace Museum Project: A Digital Solution for Communicating Museum Culture

Wu Meng
(The Palace Museum, Beijing 100009, China)

Abstract: With the development of information technology, the application of new technologies in the development of digital exhibits has become an important trend in the museum sector. In order to offer public education and achieve effective dissemination of knowledge and culture, we must pay attention to interactive exhibit design when using new media to create digital museums. Through applying the concept of interactive design, digital technology can serve the goals of digital creation, opening cultural communication and effectively enhancing audience experience while conveying information in a constructive way. This paper discusses the evolution of certain digital projects at the Palace Museum, introducing the digital experience and forecasting the future developmental directions of this work.

Keywords: cultural display; interactive technology; interactive design; Palace Museum; digital exhibition hall

1. Digitization: the leading trend in museum development

As defined in the charter of the International Museum Association, the museum is a permanent mechanism to serve the development of society without profit, and is open to the public. Research, education, and appreciation are the main purposes for the collection, preservation, research, and spread of natural and man-made artefacts. This stresses the joint and collaborative social service functions of the museum, while also highlighting the need for museums to offer opportunities for appreciation, study, and education to the public. With revolutions in science and technology, and the rapid development of information technology, individuals can access educational resources more easily. Learning habits are also changing. In the past, people were accustomed to accessing knowledge in the classroom, but with new multimedia technology, interactive technology, and the popularization of mobile Internet technology, we are growing increasingly accustomed to using the Internet for active exploration, mobile learning, and more exciting forms of education. In continuing the pursuit of public education functions, museums have also begun trialing the use of new technology to improve knowledge dissemination. Traditionally, individuals in an exhibition hall have access to verbal explanations, but new technological means are now being tested to create digital guides with descriptions and explanations of objects of value, allowing visitors to gain a deeper understanding in the process of exploration. Examples of new technological methods include using platforms like Weibo and WeChat to publish information and make contact with the public, creating apps to showcase artefacts and other cultural spaces, setting up interactive screens for guidance and explanation, and constructing digital museum halls.

If differentiating by location, digital projects can be divided into online and offline projects. Online projects focus on accessing knowledge resources quickly and conveniently, anytime and anywhere, including interactions with museums and collections, as well as access to basic information regarding a museum or exhibit, such as addresses, directions, reviews, ticket booking, and visiting the online store. Offline projects, such as interactive museum equipment and mobile apps for exhibitions, combine digital technology with real objects for the purposes of display. Research has been conducted on the motivations of family groups visiting museums. The results* indicate that family groups that visit Chinese museums are mainly concerned with children's education, children's entertainment activities, some hands-on operations, parents' learning needs, children's social needs, and the attractiveness of the venue. In a nutshell, the core needs are education, entertainment, interactivity, social activities, and venue quality. Museum visitors want to learn as well as participate in hands-on interaction, and entertainment is essential for interactive activities. As such, the main goal of the museum in setting up offline digital projects should be to help visitors understand exhibits more effectively and efficiently, appreciate the collections in depth, collect and record useful information, and plan routes for their visits. Of course, attention must be paid to interest and knowledge gained in conjunction, which is impacted by psychological and cognitive characteristics of a given individual. Scientific research regarding human cognition demonstrates that people are largely selective in receiving information, and tend to focus on areas that

* 'Chinese Family Groups' Museum Visit Motivations: A Comparative Study of Beijing and Vancouver', *Curator*, Jan 2014, 57 (1).

are interesting or familiar. Therefore, improving interest levels with digital museum exhibitions proposes to effectively deepen the exhibition's impression on visitors, as well as improve the efficiency of exhibition information dissemination.

Generally speaking, given their differing positions, online museum projects more focused on depth, while offline projects are more focused on convenience. However, these two aspects complement one another. For example, in the currently popular O2O model, it is necessary to achieve a seamless connection between online and offline paradigms. Before they visit a museum, people are attracted by advertising, and want to access simple, basic information about the venue and exhibition online. People who are in the exhibition hall want convenient access to information about the exhibits and layouts that will assist them on their visit. They want to use mobile equipment of take photos and collect information to supplement and continue their learning. In achieving a seamless connection between online and offline paradigms, a museum may stimulate the content of an exhibition, effectively guiding the behavior of visitors and offering them a smoother and more enlivening viewing experience than would be possible within the physical space.

2. *The digitization of museums according to the principles of 'interaction design'*

As museums are focused on the general public, in order to improve visitor experience, the design and standards for digital projects must consider the application of 'interaction design theory', utilizing a variety of methods of interaction and uniting the unique applications of online and offline projects.

'Interaction design' is discipline concerned with interactive experiences. It was initially proposed in 1984 by Bill Moggridge, the founder of IDEO, at a design conference. From the user's point of view, it is a technique for making products easier, more effective, and more enjoyable to use. By designing the interactive interfacing and behavior of a product, an organic relationship can be established between the product and its users, so that a user's goals can be achieved effectively. In interactive design,* one must attend to the following:

1) Product availability, that is, products that are easy to master and use, with an effective system that provides user satisfaction;
2) Product visibility, referring to the visibility of functions;
3) Product limitation, referring to an effective interface design to prevent misuse;

* Baidu Encyclopedia, entry on Interactive Design: http://baike.baidu.com/link?url=KnylhmJfl-yo4ZWz3tA6WyxoVFQoAdtZo9KNKxYRzscrgjy Pr50GBcP RohOc_ABxVJq3TXXEfcQa11imKnjFSq

4) Product mapping, namely the relationship between accurately expressed controls and their effects;
5) Product consistency, that is, the performance of an operation must be consistent;
6) Product inspiration, meaning tips for fully accurate operation.

Amidst the creative practice of digital museum exhibits, it is necessary to pay attention to the functional design of a digital exhibition's interface. It should be easy to operate, as visitors are intent on acquiring knowledge. If too much energy must be spent on learning the means of accessing knowledge in the first place, there is the potential for blurred learning objectives and frustration.

Secondly, interactive projects must consider functional interface design, arranging important operation buttons in a way that aligns with the habits of the human eye. If operations require a gesture, it must conform to customary specifications and have a high degree of fault tolerance and error correction in user operation. Operation prompts must be direct, with an interaction design that is concise and consistent, in order to avoid overly complicated processes. Indeed, only by incorporating interaction design can a museum's digital exhibits provide positive experiences. Otherwise, they risk missing the point.

Interactive projects can present either strong or weak interaction. Although weak interaction projects display fewer interactive links, they are suitable for revealing certain content and explaining certain cultural themes. For example, specific short films or situational display projects science and education within an exhibition hall have relatively low interactivity, mainly offering functions of appreciation and knowledge communication. Strong interaction projects focus on human-computer interaction under the guidance of interactive programs. Because these projects focus on the timely feedback of human operations and procedures, the design requirements are much higher. Strong interaction projects intend to offer users a full understanding of all information, with an interaction process that must be minutely designed. Every aspect must be fascinating. The high sense of operation and participation in strong interaction projects is incomparable to that of weak interactive projects. They are suited to visitors who are willing to explore actively, forming a communication model that explores, discovers, and considers in persistent two-way interaction.

Strong interaction projects can be divided into linear and nonlinear interaction. Linear interaction is a simple structure in which the entire interaction process has a certain order. Generally, only one process entry is set, and must be completed step by step. The consistency of this style makes it more suited to online projects or single-action projects within the exhibition hall. User operation begins a process, and that process is interrupted only when a user finishes the operation.

The structure of nonlinear interaction is more complex, with a large number of layers, shifts, interspersions, and supplements being used to guide ideas. In terms of careful interaction design, all links in an interactive node can be traversed via a series flow in the interactive process. This interactive structure is better suited to exhibitions in the offline exhibition hall, as users may enter the interaction with more flexibility. Visitor requirements are not high. They can begin an interaction with the activity they perceive as most interesting, and may even pick up an activity where another visitor left off.

From an audience perspective, good interaction design comprehensively attends to the psychological and physiological characteristics of the visitor, offering the most informative and emotional experience. This piques audience enthusiasm and deepens the impact of the exhibition, which are the very goals of digital museums.

3. Creating interactive exhibition items for the digital Palace Museum hall

As the author has participated in the development of interactive exhibition items in the digital hall of the Palace Museum, we will now discuss the placement of digital exhibition components within our digital museum hall in relation to digital items within the museum. Since the digital showroom is a typical offline project, we will focus on offline projects, and highlight two interactive painting and calligraphy projects in which the author collaborated.

The Palace Museum is a historical and cultural facility located within an ancient structure, meaning that the visitor experience is different to that of an ordinary museum. From the visitor's perspective, the motivation to go to the Palace Museum instead of an ordinary museum involves the higher entertainment and educational value that is attached to the Forbidden City as a historical monument, due to a widely-held fascination with royal life and court culture. Therefore, in creating interactive exhibits, attention must be paid to attractiveness and enjoyment in the means of interaction. Exhibits must support varying levels of exploration, with gradations of interaction that become more distinct and increasingly abundant as opportunities for exploration. Interactions premised on appreciation and entertainment are at a shallower level, while knowledge-based interactions are deeper. Interactive projects should include multi-person projects, and project operation should attend to visitors' social needs.

In terms of interaction, the digital Palace Museum exhibition hall has a mixture of weak interactive projects (such as a digital sandbox and virtual reality theater projects) and strong interactive projects (such as paintings, calligraphy, utensils, embroidery, and other digitally displayed of artefacts). The two are interspersed in order to comprehensively interpret the history, architecture, and artefacts of the Palace Museum.

When it comes to the scale of interaction, there are both single and multi-person interaction projects, which are also planned with regard to the collection of cultural artefacts on display. For example, given the length of a calligraphy book, a full-volume display requires longer screen support. Therefore a digital version was designed that can accommodate multiple people interacting at the same time. The digital painting project uses smaller screens to showcase small-scale, highly-detailed paintings, guiding visitors to focus on the brushstrokes in close proximity. These screens are designed to be used by one visitor at a time. Given the relationship between linear and nonlinear interaction in terms of structure and scale, the digital book is designed as a nonlinear interactive project, while the digital painting is a linear interactive project.

In this exhibition, we chose Huang Quan's *Study of the Birds* as a digital cultural artefact for the interactive painting project. This was because it is the only authentic work to have been handed down through five generations. It depicts twenty-four vivid examples of daily life, with a stunning style and painting technique. Huang Quan and his son's *Huangjia fugui* style had a far-reaching impact on the highly-detailed paintings of later generations. However, this work is drawn on silk. In the world of calligraphy and painting, the saying goes that "paper is a thousand, while silk is five hundred", meaning that the latter is fragile and difficult to preserve. This poses issues for the long-term display of the work. Therefore, we used digital means (such as high-definition images showing the painter's brushstrokes, and multimedia techniques that bring the birds to life) to offer a vibrant and interaction experience. In addition, as this work is actually a "draft manuscript" (a blueprint for students to learn by copying), there is no composition or background. In developing this project, we imagined that the elements of the birds and insects would be framed separately, using backgrounds such as branches, flowers, and trees to bring the creatures to life in a natural environment, making the image even richer than the painting itself. In order to render the rare birds flawlessly, we hired avian experts from the Chinese Academy of Sciences to identify the species in the original painting and their dynamic characteristics. Then, an animation was designed for each creature, which the experts then checked. This project also created a link for the comparison of the birds in the painting with the pictures we collected, to highlight the accuracy of the painting.

Our digital interactive model for a calligraphy book involved *The Copy of the Preface to the Poems Composed at the Orchid Pavilion (Lantingji Xu)* by Feng Chengsu, the original of which was reputed to have contained the first running script in Chinese calligraphy. Because the piece in our collection has been extensively restored to its original condition, it attracts attention from around the world. However, as the work is written on low-quality mulberry paper, it is difficult to display in long-term, large-scale exhibitions. In comparison to paintings, calligraphic works are generally more abstract, with philosophical levels that are hard to understand. In particular, amidst the dots and strokes, calligraphy often reveals the author's temperament and psychological state, and even environmental factors at the time of the work's creation. The writing of *Lantingji Xu* occurred during the Lanting purification ceremony, so the author appears to have been in a good mood. Interestingly, the work begins with a more upright font. As the author's sentiments come to the fore, the font becomes increasingly relaxed, and the work develops a strong sense of rhythm. Thus, in developing the interactive program, we allowed participants experience the 'floating goblet on meandering waters' (*qu shui liu shang*). This not only expresses the background of the work, but also hints at the important literary practices of ancient times. Moreover, when viewing calligraphic works, visitors like to trace them with their fingers. In fact, this is the best way to deconstruct a piece of calligraphy and experience the artistic conception between the strokes. Therefore, using multimedia and interactive means, we can simulate '*qu shui liu shang*', allowing visitors to use a 'digital brush' and 'digital ink' to simulate writing, bringing them closer to the original means of artistic production. As well as this digital writing activity, the project also includes a display and evaluation session, with a "like" approach as in social media, offering visitors the opportunity to evaluate and interact with one another. Meanwhile, the project attempts to integrate each participant into the exhibition, improving the atmosphere and enhancing audience interest.

The interactive processes of these two projects also takes into account the use of social platforms for communication and publicity. With the rapid development of internet technology, social media is exploding. On social platforms, people tend to pay the closest attention to those who have common hobbies or visions, meaning that social media promotion is often the most accurate and effective means of positioning target groups. Therefore, these projects provide a sharing interface for the WeChat platform, allowing visitors to share their highlights. This element satisfies the public desire to use social media, express opinions, and communicate within social circles, while also achieving an element of promotion for the museum.

4. Conclusion

To summarize, in response to the rapid development of information and interactive technology, museums are using new media to publicize and interpret exhibitions and collections. This facilitates the invigoration of the museum atmosphere, and allows for greater flexibility in the viewing of exhibitions. In addition, the seamless integration of online and offline projects allows museums to integrate publicity, display, and follow-up services. Visitors will deepen their exploration and learning, acquiring new knowledge with enhanced participation and simulation, and an enriched experience. As such, people who come to the digital museum become part of the exhibitions, which promotes cultural communication and builds a direct interaction between the museum and its visitors.

In practice, we recognize that an important factor in the success or failure of digital interactive exhibition design is compliance with human habits of observation and operation. At present, with mobile devices in wide public use, observation and education via touch screens and interactive gestures are broadly accepted. A better experience can be achieved using similar interactive design ideas and gesture specifications, but we must also realize that is only a single stage in the development of interactive technology.

When discussing artificial intelligence technology, Myron Kruger proposed that "computers must adapt to people, not people to computers". In the design of human-machine interfaces, high-speed computer technology must adapt to human characteristics, which are relatively unchanging. This theory is also applied in interaction design. As technology develops, more new interactive technologies that operate close to inherent human habits of observation and operation (such as physical tags and somatosensory technologies) are emerging, making it possible for even more natural manipulation. This further reduces the learning cost of ordinary visitors in engaging with interactive operations and improving the viewing experience. Furthermore, with no impact on the safety of the collection, certain physical exhibits (such as 3D printed models) can offer direct interfaces for human-computer interaction. When people observe and touch these exhibits, they can directly interact via precision projection or AR technology, with the promise of more innovative technologies in the future, and deeper access to the cultural richness of the exhibits.

With all of this in mind, we have come to realize that digital exhibit design should not be led by technology, but should focus on the interaction between exhibits and visitors. The interactive process is not simply about human participation, but about the spiritual and emotional intercommunication between visitors and the exhibits. The cultural value of the exhibits must be transmitted to the audience, but the results of visitors' explorations, information feedback, and emotional stimulation are also key. The digital form creates a relaxed atmosphere for viewing, and enhances the visitor's experience while promoting the all-encompassing communication of culture. Furthermore, it involves a process of "equal communication" between museums and visitors, enhancing the former's cultural value and public service capacity.

The Use of Augmented Reality Technology in Museums

Qiu Ke
(Shanghai Shangte Culture Communication Co., Ltd., Shanghai 200235, China)

Abstract: Augmented reality technology has a unique real-time interactive function, which can fuse the environment and information with museum visitors, allowing people to experience real-world simulations in a virtual space. This enhances the enjoyment and interactive value of a tour, and offers a stronger feeling and better experience.

Keywords: augmented reality; virtual reality; real-time interaction; visitor experience

Preface

In recent years, the emergence of augmented reality (AR) technology has blurred the line between real and virtual computer environments, offering new opportunities to museums for interaction design and heightened visitor experience. The unique real-time interaction of AR, and its integration of the surrounding environment with information and audience feedback, are in line with the educational and psychological requirements of knowledge dissemination, inducing inspiring interest and curiosity with an organic combination of education and entertainment. The combination of 'virtual' and 'real' has increased participation and interest in museum exhibitions, adding a new highlight to museum curation. Furthermore, the cost of augmented reality is relatively low. Simply by using a camera to view information, the information-machine environment appears well integrated. With the combination of AR technology and the development of mobile terminals, a new era is beginning for museums. Indeed, the combination of virtual/real, highly interactive, and 3D positioning stands to greatly stimulate visitors' enthusiasm and participation, improving their perception of history and science, restoring the value of the Internet, and encouraging greater interactivity.

1. What is augmented reality technology?

Augmented reality refers to a virtual, computer generated environment that engineers realistic feelings of sight, hearing, force, touch, and movement. The real picture seen by a user, through a camera or similar equipment, involves real-time integration with computer-generated virtual screens, providing an enhanced image that contains a variety of additional information. Real environmental factors and virtual objects are superimposed in the same screen or space in real time. The audience sees a virtual scene superimposed on the real environment, presenting a heightened environment with real effects.

Real objects and virtual objects within augmented reality are seamlessly integrated with the user environment. There is also synchronous interaction between these objects in the virtual 3D space, providing a 3D connection between actual and virtual reality, offering a novel and interesting interactive experience overall.

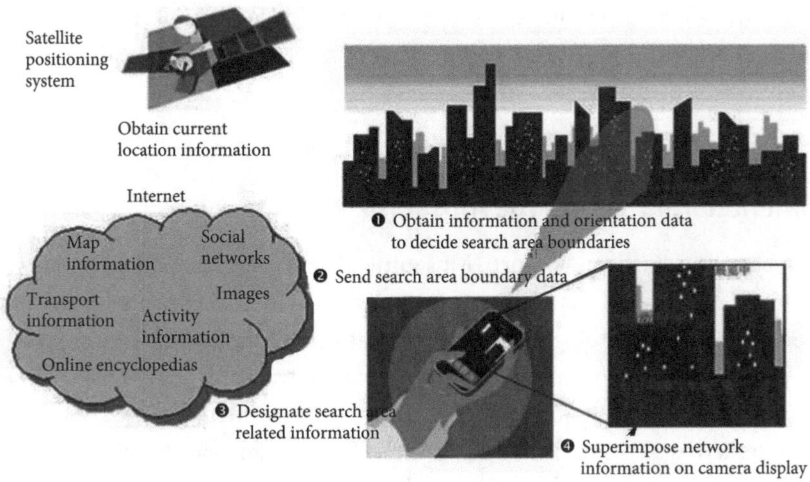

Figure 1 Schematic diagram of augmented reality

2. Augmented reality and virtual reality

Virtual reality (VR) refers to a virtual environment that uses computer technology to generate realistic virtual, auditory, and tactile interactions. With the appropriate equipment, users can interact with objects in the virtual world in a natural way, and can experience a mutual influence that engenders the feeling and experience of the real world.

According to varying levels of immersion and degrees of interactivity, virtual reality systems are divided into four kinds: immersive, desktop, augmented, and distributed. Augmented reality is a form of virtual reality, and can be understood as a new computer application and human-computer interaction technology developed on the basis of VR.

Virtual reality technology calculates scenes in real time via a computer, then provides information about the entire space to an audience according to the audience's own preferences. A viewer is able to walk along a self-guided route, and the computer works to generate corresponding scenes. Virtual reality emphasizes immersion in the virtual world, encouraging people to interact naturally with objects in it. Augmented reality systems are dedicated to the integration of a computer-generated virtual environment with the real environment, enhancing a viewer's understanding of the real environment without blocking any connection with the real world. Augmented reality has lower hardware requirements, higher registration accuracy, and a greater degree of realism. As AR has the capacity to enhance an audience's perception of a fused environment, it has obvious perceptual advantages in museum applications compared with virtual reality technology.

In the last two years, the augmented reality search volume has surpassed that of virtual reality, as people pay increasing amounts of attention to the concept of wrapping the information space. With augmented reality technology, museum visitors have experienced transformations in terms of information access, mutual knowledge, and ways of understanding. This can be the direction of development for the digital museums of the future.

3. *Augmented reality and the museum*

In recent years, the Horizon Report (Museum Edition), published by the New Media Consortium in the United States, has repeatedly reminded museums to pay attention to augmented reality technology, considering it as one of the technological applications with the most potential to change the development track of museums. Through the transmission of information and processing of real-time video footage, AR technology forms an interactive 3D image that offers a new and more realistic visitor experience, allowing museums to be more accessible and to provide more personalized services and perspectives. Key features of augmented reality include the ability to interact with and respond to users, as well as the creation of new layers of interpretation on top of past sensory experiences, providing far-reaching potential for the development of museum-based education. The interactive nature of augmented reality enhances museum visits and museum learning with interaction, curiosity, and points of interest, constructing new understandings and activating the potential of existing data. Dynamic processes, massive databases, and objects that are special and/or difficult to manipulate can be introduced proportionately into a visitor's personal space in an easy-to-understand, operational form.

3.1 Pre-show publicity

In celebration of the 100th anniversary of the founding of the Los Angeles Museum of Natural History, augmented reality technology was used for pre-show publicity, achieving remarkable results. In collaboration with technology company Blippar, the museum launched four different centenary posters and postcards, which were distributed in museums, major shopping centers, schools, and libraries. Using the mobile museum app to process images, visitors were able to participate in a 3D interactive experience, taking a photograph with the Tyrannosaurus rex and sharing it online. Meanwhile, the mobile app offered discounted museum tickets and a chance to enter a prize draw. Prizes included souvenirs and the opportunity to bring a family member to the museum for free. The grand prize was an overnight museum experience with VIP accommodation for the winner and three friends.

The traditional methods of disseminating information used in museum posters, bulletin boards, and brochures are only available in the physical environment, causing a lag in delivery and assimilation. Indeed, some information, such as curator interpretations, audience comments, and audio/video files prove difficult to disseminate in this mode. However, augmented reality information dissemination services make it possible to actively (rather than passively) transmit rich and diverse multimedia and exhibition information to potential visitors, overcoming the defects of traditional methods. Mobile augmented reality dissemination possesses the on-site quality of traditional dissemination methods, while also providing information that is inconvenient or impossible to provide on site, allowing for information provision that is rich, timely and effective. Meanwhile, mobile augmented reality push has become essentially a one-button operation. It is extremely easy to learn, and visitors can access information simply by opening the program.

3.2 Positioning guidance

With the rapid development of the Chinese economy, both old and new museums in China are growing ever larger, with tens of thousands of square meters of space for a single museum becoming common. Contemporary museums are designed largely in wide, open spaces, with scattered functional layouts and efficient guidance systems. The development of electronic maps and geographic location services have allowed some large museums to develop electronic map navigation systems. Combined with mobile augmented reality technology, this allows visitors to access real-time path orientation, offering an increasingly convenient experience. In the mobile environment, there is a correlation between the location of the viewer and an object. By obtaining the location data of both the viewer and the object, positional relationships can be calculated, and guidance information can be generated. Therefore, in mobile augmented reality, guidance information is superimposed over live video to provide practical museum guidance, as if a live attendant were offering assistance within a physical space. Guidance can be provided in either text form or in a 2D/3D format that directly displays target orientation. If a viewer does not specify a target, systems may customize route guidance towards several potential targets.

The Sunshine Aquarium and technology company HAKUHODO Tokyo have jointly developed an application that uses AR and the SLAM positioning function to create a virtual penguin that guides tour routes. Located in downtown Tokyo, the aquarium's area can often be confusing to tourists. It is now possible to simply open your phone, download the "Penguin Navi" app, and point the phone display to the real world in front of you, showing a group of penguins milling about on the sidewalk or zebra crossing. Combining technical satellite navigation with warm and familiar "lead-the-way" behavior, the app's operation is simple and effective. If users slow down, the penguin guide will stop and wait, beckoning for them to keep up.

Augmented reality is also helpful for parking. In the museum parking lot, a mobile app such as Car Finder employs software to record your car's location. Later, it guides you back using a camera, satellite navigation, a compass, a gyroscope, and other sensors.

3.3 Exhibition descriptions

Museum visitors often spend a significant amount of time taking photos on their mobile phones, and sending text messages and photos to friends. Therefore, the use of AR technology to turn a mobile phone into a storytelling tool offers new vitality to static exhibits, enriching the transfer of knowledge on a variety of levels. Museum visitors can use AR technology to view locations and routes within the museum, and can also access specific data and read further information of interest, offering a new dimension for them to understand and interact with museum collections and the world around them.

Upon entering a museum, visitors generally begin by reading the brochure. The first page usually offers photographs and statements from the curators. Now, with a swipe of their screen, visitors can also access a video of a curator giving a speech. Through their mobile phone cameras, users can see interpretations of key exhibits outlined in the brochure, explaining certain curatorial decisions and offering comments from famous critics. This full integration of media and interactivity allows for an easier understanding of the museum's content.

In recent years, many foreign museums have begun to use augmented reality technology to turn static images into dynamic videos or 3D animations, such as the Ultimate Dinosaurs at the Royal Ontario Museum, the Life of Art at the Getty Museum and the Laguna Art Museum at California, the Rijksmuseum van Oudheden (Dutch National Museum of Antiquities) and its use of the MovableScreen augmented reality system to display Ancient Roman remains, as well as the McNay Art Museum, which has trialed its own Reality Check app. Harvard University Art Museum has used Layar's augmented reality recognition technology for "In Harmony: The Norma Jean Calderwood Collection of Islamic Art", demonstrating the various functions of smart phone and tablet cameras in practice, as well as the impact of low museum lighting on image recognition.

At the Hong Kong International Art Fair, the City University of Hong Kong held an exhibition entitled "The Pure Land of Man: Entering the Dunhuang Mogao Grottoes", which showcased murals that are inaccessible to the public. When a user aligned a connected iPad screen with a wall surface that displayed the simple outline of a mural, a more detailed version appeared on the screen.

San Francisco's Asian Art Museum exhibited China's Terracotta Warriors using augmented reality technology to display further educational materials for certain segments of the exhibition, enhancing audience interaction. Using an app, visitors could scan six different cards to display 3D content and videos in "real space", seeing virtual items from a variety of perspectives in real-world situations, and taking photos to share on social media.

Augmented reality is a convenient and effective way of adding data to real-world objects, facilities, and processes, to promote the deeper understanding of objects of interest. Currently, most of these tools propose an enhancement of the user experience with real-time control functions. For example, curators can use augmented reality to extend a venue into specific virtual areas. For example, artist Will Pappenheimer designed an exhibition filled with gaps at the Los Angeles Museum of Art using Layar AR software. He then flooded the museum courtyards and corridors with colorful digital objects and speech videos, allowing visitors to use their smart phones to view scenes that do not exist in the real world.

At the American Museum of Natural History, augmented reality has been applied to an exhibit entitled "Beyond the Planet — The Future of Space Exploration". Whenever a viewer finds any of the eleven augmented reality icons displayed throughout the exhibition, they can use the camera on their device to activate 3D animation.

Information within AR target objects is either derived from an existing "point of interest" database, or from a curator's input. Information can be linked to museum buildings or featureless spaces such as squares and gardens, and effects can be improved with image analysis and accurate overlapping of information on the actual screen. Visitors can take a photo of a cultural artefact in the exhibition hall, for example, and access information about its source and original owner. They can view an image of the original collection, and then a simulation of how it would look after being repaired or restored. They can even view an item using an "X-ray function", looking at the inner layers and structures of objects, or interact with 3D models, videos, and digital descriptions.

Some museums also use speech recognition and image synthesis to incorporate visitors and computer interpretations into a single scene, offering a live Q&A function between an interpreter and an audience. Like a live-action introduction video, such operations are very attractive to visitors. The "Making the Modern World" show at the Science Museum in London has developed an augmented reality application to complement the exhibition. BBC presenter James May, best known for his work on Top Gear, acts as a virtual guide, explaining the exhibition and showcase. The magic of AR has greatly improved interactive performances, as the virtual guide can attend directly to visitors through mobile terminal screens, explaining famous collections, providing tour routes, interpreting exhibits, and offering a variety of information services.

3.4 Reproducing History

In regard to historical and cultural artefacts, the "Augmented Walking Tour" mobile AR application has broad prospects. It allows visitors to access photographs, historic documents, recordings, and videos directly from a Cloud database while they visit historical and cultural sites. As a collection of historical and cultural heritage resources, a museum's value depends on the maximization of resources. Mobile AR technology provides a new path for the development and utilization of these resources, as it enhances physical objects within the collection.

The Stedelijk Museum Amsterdam has launched an enhanced travel event called ARTours, in cooperation with the national tourism board. At famous historical sites in the Netherlands, visitors can use their mobile phones to view historical photos, paintings, and related information provided by the museum in real time. Standing in front of ancient architecture, visitors can use simulation technology to view the original appearance buildings that have since been repaired, as well as relevant historical information. Meanwhile, messaging and annotation features enhance interactivity, as visitors can view comments and external resources in a network, communicating with one another for educational purposes. Augmented reality methods combining virtual reality and authentic life can trigger an enhanced emotional, which is conducive to remembering and understanding information.

The London Street Museum's mobile app was another early success in the field, allowing users to enjoy the city's art and photography alongside a temporal and spatial experience. Upon starting the program, a visitor's current location is determined before relevant surrounding attractions are displayed. Selecting an attraction brings up a small dialog box showing its specific location, its maker, and the time of creation. Clicking on a righthand blue arrow will display an image of the actual work. Within the browsing interface, users can also see detailed information about the work, or zoom into the image to see details. Interestingly, the program supports a 3D browsing effect. When standing in an art-related location, enabling 3D browsing allows the program to automatically identify the location and overlay work of art with the actual scene based on buildings and the environment, meaning that users can engage with various historical events, people, stories, and images through their mobile phones. Walking between virtual 3D history and real scenery, houses, streets, high-rises, and lanes, visitors can experience a huge historical and cultural museum.

Sydney offers yet another interesting example, namely an augmented reality mobile app developed by the Powerhouse Museum. When visiting Sydney, users can use the app to see sights superimposed with their hundred year-old historical appearance. Even if not in Sydney, users may input the exact address of a location with latitude and longitude coordinates to see a monument from a century ago. Further improvements are in the works, intending to achieve superimposition of different modern photographic works at a single address, and even introduce gaming elements for entertainment.

The City of York Hologram Tour, supported by the York City Council in the UK, also provides entertainment and education to visitors using smart phones. After installing the program, aiming the phone at a specific attraction will bring up a costumed actor on the screen to perform a comedic history over a minute and a half. Users can also take photos with the avatar. The app then directs visitors to the York Museum, downloading full application resources over a free wireless network so as not to waste visitors' data allowance. Using this app creates a digital experience for visitors to the historical city, especially members of the younger generation who were born in the digital age. Sound control technology is also being developed for acoustic holographic images. For example, outside the Chocolate Workshop at the York Cocoa House, a holographic image of a chocolate tycoon has been generated to remind visitors to use the app for further information about the attraction.

3.5 Improving Quality and Enhancing Interaction

Augmented reality technology enhances a visitor's learning experience by overlaying digital images, online learning materials, and museum-related academic information over the real-life collection. AR makes a museum visit more interactive, establishing an inquiry-based learning model in which visitors can read relevant metadata, images, and videos to deepen their understanding of a collection.

As a trigger for augmented content, AR applications not only recognize graphic features but also the shape of objects. For example, items such as vases or plaster casts in the exhibition hall can be used to trigger content. Viewers can also measure length with mobile camera functions, using autofocus detection for resolution and scaling. With unit conversion support, they can measure the length, width, angle, and area of any item, anywhere and at any time, without needing a ruler.

In cases where the narration is in an unfamiliar language or text, words may also be translated with augmented reality technology. At the end of 2010, the American company Quest Visual developed an optical character recognition (OCR) translation software called Word Lens using AR technology. A viewer needs simply to point a mobile phone camera to the text requiring translation (for example, in a book or magazine, or on a menu or road sign), and the translated text will appear on the screen almost immediately. Word

Lens involves recognition, translation, and substitution functions, and it is extremely easy to use. It employs OCR technology to recognize text, then searches and translates via a dictionary database. It then uses AR technology to overlay translated text on original text, meaning that aside from the text, the camera image remains unchanged. Currently, this software supports translation among only a few languages, including English, Spanish, French, and German. If an augmented reality program were to be developed to convert simplified and traditional Chinese characters (even ancient texts like bronze or oracle bone inscriptions), it would be of great convenience to younger generations who cannot understand these scripts. Furthermore, in addition to the translation function, academic resources recommended by the museum can be displayed in real time, allowing visitors to browse and save collection resources on their mobile phones, and also to send push results to designated email addresses or social networking platforms. Sharing content assists understanding, and expands learning resources in an efficient way.

In fact, international scholars have begun to consider the use of augmented reality in the study of ancient literature. While protecting ancient books, access is provided to the original literature, offering a rich network of materials for improved study, such as ancient documents or special editions. Frequent use of these resources does not compromise the protection of ancient literature. The John Rylands Library at the University of Manchester has established a project for research and learning around special collections using AR technology. Damage can be caused by examining medieval manuscripts, rare editions, and archives. However, in displaying digital pictures, text, online learning materials, and related information collected by museum libraries and other institutions, technology can preserve the original manuscripts. They even allow readers to "turn the pages" of the digital versions, zoom in on details that would otherwise go unnoticed, listen to medieval English readings, find metadata, read other literature, and compare materials. These technologies provide examples for the rest of the world.

3.6 Improving the Reading Experience

Suppose you are reading a book purchased in a museum. Simply scanning your phone camera against the cover and pages inside will reveal visual and interactive elements on the phone screen, with the stories within the book coming alive in front of your eyes. A flat image becomes a mobile 3D model; a 2D map becomes a rotating globe; a dinosaur stands up and walks around; a racing car is under your control. A book may seem ordinary, but on the screen of a mobile phone it can come alive in three dimensions. This innovation will change traditional reading practices, combining AR scenes and 3D models with the real world. For museum visitors, particularly children, this offers an immersive supplementary experience.

Augmented reality technology that integrates text, images, 3D animation, sound, video, interactive games, and other information provides a crucial carrier for digital publishing and museum education. It enhances participation in textual information, enabling knowledge and information to be more effectively accepted and absorbed by the audience. In these applications, a reader becomes an active participant in a world of sensory elements.

Museums can develop AR books for specific collections and promotions. For example, visitors may see a virtual guide appear in a book on painting, then listen as the guide tells related stories. Using mobile phones to identify the markers on books, three dimensional scenes will appear, integrating the virtual world into reality for a 3D reading experience.

Multimedia books based on augmented reality technology break the limitations of the content and form of original books. With the help of terminal devices such as mobile phones, they can offer a variety of manifestations, providing users with real-time access to relevant digital resources. With the lowering of production costs and the maturity of technology, the application of AR books in museum education has bright prospects for development.

3.7 The Development and Promotion of Cultural and Creative Products

Augmented reality technology also has the capacity to promote entertainment related to cultural and creative museum products, making ordinary, flat pictures move, and turning tickets, posters, brochures, and even transport cards into fun animations related to promotional themes via mobile phone recognition technology.

Applying the same principle to cultural and creative products also culminates in a comprehensive sensory, audiovisual experience. With a mobile camera, flat images (such as toys, posters, or magazines) can be captured to present an interactive, 3D animation; a simple picture jumps beyond its planar limits and comes alive in three dimensions, with movements, sounds, and interactions that produce a multi-sensory space integrating vision, touch, and hearing. Technology like this is highly attractive to children.

AR subtly combines virtual objects or characters (such as famous scientists) with photo capture in order to improve the museum experience and provide entertainment. Sharing content can attract the attention of wider visitors. Viewers can rotate, resize, and correct objects appearing in the image through the touch screen on their mobile device, meaning that virtual characters can engage with tangible cultural and creative products. Viewers can also save AR images, and share them with friends via social networks and email, or create e-cards or AR albums to keep and share.

Zapper has developed a t-shirt featuring Disney animations, on which Mickey Mouse appears to be dancing in the mobile app. It also created two baseball caps printed with either "I am Obama" or "I am Romney". When a phone is aimed at someone wearing one of these hats, a virtual cartoon of Obama or Romney appears, allowing users to imitate the candidates and share the experience with friends. In this way, consumers can buy attractive, practical, and tangible cultural and creative products, and also engage in the virtual world. This could be the future trend for AR technology and museum products.

3.8 Sharing and Statistical Functions

In real life, people want to communicate at any time, in any place, to express opinions and share emotions with friends, hoping to find recognition and identification. With augmented reality, online communication can be established with mobile phones and location services. Users are free to express thoughts, share information, build friendships, and spread their ideas to a greater number of people.

AR technology also makes it possible to collect mobile phone behaviors and habits, collecting large amounts of sensor data and performing automated statistical archiving. This data can be used for secondary updates of interactive content as well as scientific analysis and filtering of usage effects and reader experience. This can be used to create a visitor database for museums, to link the real world with the vast amount of online information that is required for data support in subsequent planning and projecting.

4. Conclusion

Our lives are becoming ever more digital, and the information around us is becoming easier and faster to obtain. In light of this, museums are facing ever-increasing amounts of information, and are looking to mobile technology to provide a more intuitive and convenient user experience.

With the development of AR and other technologies, the interaction between museums, visitors, and the digital world will become more comprehensive and flexible, enhancing audience engagement with technology. For young visitors especially, this will improve learning autonomy, and change the way they relate to education. Interactive learning interfaces with vivid and innovative models immerse children in a learning environment that is rich in knowledge and discovery, stimulating their enthusiasm, and allowing them to learn through play. Passive ways of learning have given way to completely sensory active modes that mobilize the imagination, encourage independent understanding, and improve education overall.

Through the camera on a mobile device, AR users can see virtual objects up close. A strong sense of immersion is provided by the perfect combination of the virtual and the real. In this process, museum visits will become even more interesting, and learning outcomes will be achieved with less effort.

Augmented reality is more than simply commercialism; it is an aggregated link to a variety of resources. Successful production of AR technology requires continuous innovation of both hardware and software, the results of which will inspire significant change in the interaction between visitors and mobile terminals. Through sound, gesture, and other input methods combined with novel, seamless information browsing methods, visitors in the real world will experience vivid virtual simulations, enhancing entertainment and interaction, and allowing museums to offer a long-lasting emotional experience.

References

[1] Chen Jing, Wang Yongtian, Guo Junwei, Liu Wei. 'Applications of Augmented Reality in Smart Phones', *Journal of the University of Electronic Science and Technology of China*, April 2010, 39.

[2] Cai Su, Song Qian, Tang Yao. 'Architecture and Practices of Augmented Reality in Education Environments', *China Educational Technology*, August 2011 (295).

[3] Chang Yong, Xue Liming, Jiang Shufang. 'Research on Key Augmented Reality Smartphone Technologies', *Computer Engineering and Applications*, April 2011, 69.

[4] Li Qing, Zhang Liaodong. 'Empirical research on mobile learning based on augmented reality', *China Educational Technology*, January 2013 (312).

[5] Wang Ge, Wang Xiaoning, Xu Shunqian. 'Augmented reality technology and its military applications', *Gansu Science and Technology*, November 2013, 29 (22).

[6] Shi Xiaolin, Bian Naizheng. 'Research and Implementation of Mobile Augmented Reality Indoor Guidance', *Computer Applications and Software*, February 2013, 30 (2).

[7] Yu Riji, Tang Cunchen, Hu Shushan. 'Research and development of cultural tourism innovation based on AR technology', *Art 100*, 2014, 133 (4).

[8] Wang Pu. 'Research into Mobile Augmented Reality Technology Application in Museums', *Library & Information*, 2014, 1.

[9] Kats, Rimma. 'Natural History Museum of Los Angeles County taps augmented reality to drive interaction', compiled by Chen Lizi, *New Museum Technology*, 2014, 2 (1).

[10] Yeung, Ken. 'The New Asian Art Museum in San Francisco unveils new augmented reality applications for its "China's Terracotta Warriors" exhibition', compiled by Chen Kexu, *New Museum Technology*, 2014, 2 (1).

[11] Marsden, Rhodri. 'The "City of York Holographic Guide" Application', compiled by Zhang Xin, translated by Chen Lizi, *New Museum Technology*, 2014, 3 (2).

[12] Houlier, Pascal and Jean Laneurit, Isabelle Marshall, Eric Marchand, Fabien Servant, and Anne Bationo Tillion, 'A day at the museum: An augmented reality fine-art exhibit', 2010 IEEE International Symposium on Mixed and Augmented Reality, Arts, Media, and Humanities Proceedings, Seoul, Korea.

[13] Yurkov, Kirill, Maugry: 'Augmented Reality Guide for Museums, From Proof of Concept to the Museum as a Service,' International Journal *Information Theory and Applications*, 2012, 19 (2) WSCG (18th International Conference on Computer Graphics, Visualization, and Computer Vision).

[14] Hseih, Min-Chai and Hao-Chiang Koong Lin, 'A Conceptual Study of an Augmented Reality E-Learning System based on Usability Evaluation', *CISME (Communications in Information Science and Management Engineering)*, World Academic Publishing, 2011, 1 (8).

Research into the Application of 3D Digital Technology in Museums

Shi Limin
(Beijing Intelligent Unlimited Technology Co., Ltd., Beijing 100085, China)

Abstract: Moving from two dimensions to three dimensions requires the irreversible development of computerized vision technology. This is an important direction for the future of the digital museum. Based on the three-dimensional model, this paper explains a series of completely independent core technologies such as: high-precision 3D modeling of movable cultural artefacts with common digital images; automatic 3D real-time modeling of non-movable cultural artefacts with aerial oblique photography; 3D data management of large-scale cultural artefacts and an Internet-based interactive 3D model for browsing artefacts; panoramic roaming 3D data and 3D terminal display of mobile artefacts, using this core technology as a basis upon which to propose an overall structure and series of product plans for the creation of 3D digital museums, providing useful exploration and experimentation in informatization.

Keywords: 3D modeling of cultural artefacts; 3D model management; 3D Internet model display; 3D panoramic roaming data; 3D digital museum products

1. Current Research

3D digital technology has an important role in the permanent digital storage of cultural artefacts, as well as academic research, virtualized restoration conservation, autonomous interactive appreciation, remote stereoscopic display, and characteristic product development. Therefore, an in-depth study in this technical field will have far-reaching implications.

With the increasing pace of development of digital museums, major museums across the country have built basic computer hardware platforms to support the operation of information systems. 2D digital technologies including virtual simulation, augmented reality, video imaging, and panoramic roaming have become popular, and museums have carried out many other useful explorations into the 3D digital conservation of cultural artefacts and the development of cultural and creative products. However, restrictions in core technology required for data production mean that bottlenecks still exist for the development of 3D digital museums. They are noticeable in four aspects, namely that construction costs are too high, production times are too long, data standards are too complex, and professional thresholds are too high. Core technological breakthroughs are required.

With more than 30 years of operation in computer vision, image-based 3D reconstruction technology has been widely used in fields such as robot navigation, film and television entertainment, and virtual reality. This technology enables a rectification of the relationship between the camera and the scene, restoring 3D structures in a scene using images captured at different angles by using feature detection, matching, and camera calibration. Compared with data generation methods based on structured light and laser scanning, image-based 3D reconstruction technology has various advantages within the digital production field, including convenient data collection, low hardware costs, and high automation. The field has also attracted early attention and formed a solid application scale. However, due to a high technical threshold and algorithm difficulties, the corresponding software has yet to be produced in China. At present, the main software, such as Autodesk 123D, Smart Acute 3D, and PhotoScan, is foreign. Although this software has been able to meet data production requirements thus far, a lack of self-determination means that data formats are singular and unable to be customized, making them unfit for a variety of applications.

In addition, in the field of digitizing 3D cultural artefacts, digital production and its applications are largely separate. Because data processing software is basically a foreign monopoly, there is no necessary connection between digital products and their applications. Existing digital products cannot intrinsically meet application requirements, meaning that producers must utilize a variety of digital means and make repeated investments.

The development of 3D digital production software with completely independent core technology is key to closing the loop of 3D data production to application. The development of independent 3D digital software is necessary to meet future requirements. Museums must be able to generate digital products suitable for a variety of applications, and realize flexible data customization for these applications. Potential application directions must be explored to find solutions for data production, data management, and 3D digitization, which will reduce input costs.

2. *Core Technological Breakthroughs*

2.1 High-precision 3D modeling technology for cultural artefacts based on ordinary digital images

The basic process of image-based 3D reconstruction technology is shown in Figure 1.

1) Camera calibration. A multi-view self-calibration camera is used to reconcile the relative positional relationship between camera, scene, and imaging model, using feature points of the same point in the scene or object in the image.
2) Dense point Cloud reconstruction. Based on an imaging model obtained through camera calibration, multi-view feature point matching, diffusion, filtering, and other such strategies are utilized for dense point Cloud reconstruction on the surface of a scene (or object).
3) Grid and automatic texture mapping. A mesh surface of the scene (or object) is created via the recreated point Cloud, and automatic texture mapping is implemented via occlusion analysis.

For the respective parts in the above process, we have developed a robust self-calibration camera algorithm based on multi-feature detection and matching, a dense reconstruction algorithm based on feature diffusion and depth map fusion, and a point Cloud network and texture algorithm based on the Poisson equation, integrating each module to form WXJ-X – the first image-based 3D modeling application software in China. See Figure 2.

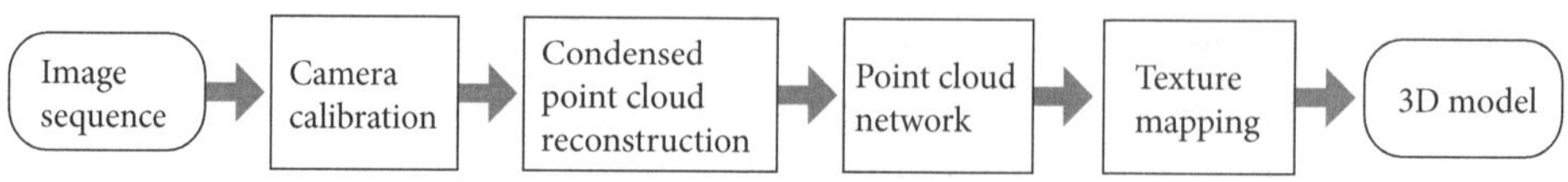

Figure 1 The basic process of image-based 3D reconstruction technology

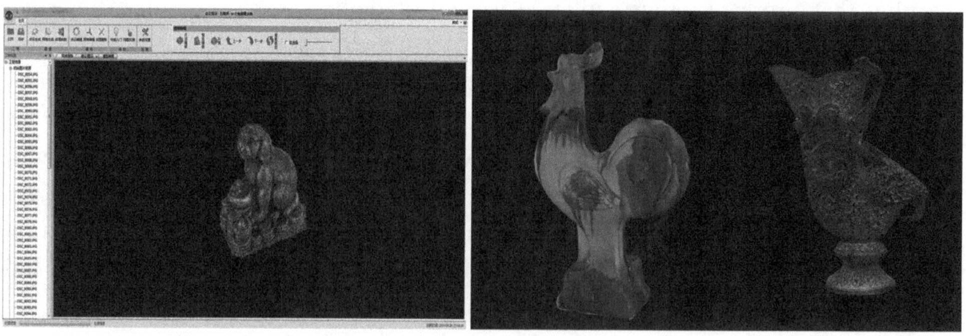

Figure 2 Software interface and data processing results from WXJ-X (rooster, wine goblet)

As a completely independent core technology, this software has the following for 3D digital modeling:

- A wide range of modeling objects. High-precision and fast 3D modeling is possible for movable cultural artefacts (such as stoneware, pottery, bronze, gold and silverware, porcelain, lacquerware, jade, handicrafts) with rich textures and relatively simple structures.
- Low cost. WJX-X reduces costs to less than one-fifth compared to domestic uses of laser scanning for 3D modeling of cultural artefacts.
- High quality and high fidelity. Using WJX-X professional software in 3D modeling provides a geometric accuracy of ≤ 0.2 mm (comparable to the accuracy of standard laser scanning equipment); texture accuracy and picture height are consistent (fully automatic texture mapping is better than laser scanning), and the effect is in no way inferior to foreign software such as Autodesk 123D, Smart Acute 3D, or PhotoScan.
- Multiple format support. The software can obtain 3D model data for cultural artefacts in any standard file format such as obj/stl/vrml/ply/ive. For data security, a specific WXJ encryption format can also be provided.
- Meets individual needs. High-precision 3D modeling of cultural artefacts is the basic foundation for the development and utilization of creative and cultural products; because WXJ-X technology is completely independent, the company may provide required source codes according user needs, as well as collaborate in research and development for major creative and cultural products.

2.2 Fully automatic 3D real-time modeling technology using aerial oblique photography

The core process of real-world modeling technology based on aerial photography is largely similar to the reconstruction of smaller objects, with the biggest difficulty occurring in

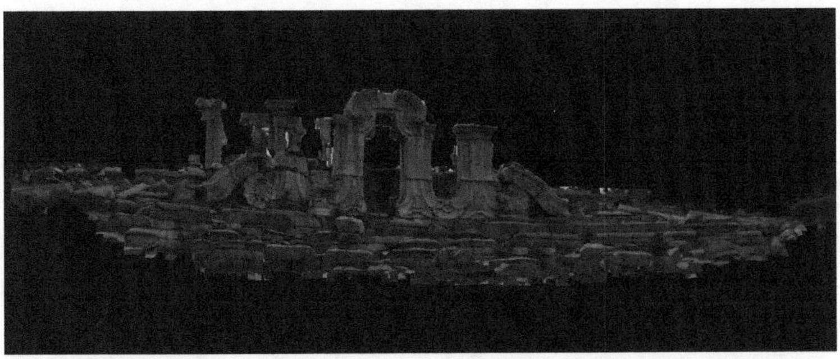

Figure 3 Software interface and data processing results from WXJ-D (the Old Summer Palace)

the processing and display of Big Data alongside structural maintenance of a larger scene. In response to these problems, technological breakthroughs have been achieved in the formulation of WXJ-D, the first domestic application software that uses aerial oblique photography for the development of real-time 3D geographic information, as shown in Figure 3.

1) Data partitioning and multi-CPU + GPU parallel strategy
Aerial oblique image data is generally above 1,000, with an image resolution higher than 30 million pixels and an image data size up to several hundred GB; such large data processing requires careful design and parallel processing. Parallel processing with multiple CPU + GPUs is implemented in key processes such as feature point detection, matching, and dense reconstruction, significantly improving processing power and efficiency.

2) Aerotriangulation based on oblique photography
Oblique photography usually collects data with five simultaneous lens groups, with long-ranging aerotriangulation imaging at a high accuracy. In the case of aerotriangulation solutions, epipolar constraints, single constraints, and internal lens constraints are adopted, which provide good error filtration, guaranteeing the stability and accuracy of aerotriangulation.

3) Structural reconstruction of buildings
Maintaining the detailed geometry of buildings is widely recognized as an issue in the 3D reconstruction of urban settings. Based on traditional point Cloud reconstruction, linear matching and straight line reconstruction are combined to support fine-edge contour geometry and preserve the geometric structure in building models.

4) OSG-based PLOD data processing a display technology
Based on OSG, hierarchical processing of large data occurs, allowing for the smooth loading and preview of large-scene PagedLOD (PLOD) data.

Technical features:

- For most immovable cultural artefacts (such as cave temples, stone carvings, ancient sites, and ancient tombs) fast 3D modeling is possible.
- One-time processing of around 3,000 to 5,000 images in a 30 km^2 range, with an accuracy that meets surveying and mapping requirements.
- WXJ-D reconstruction is at an equal leading level applications in the international market such as Leica LPS, Pix4d, Smart3d, Street View Factory, and Inpho. WXJ-D is the only Chinese company with completely independent technology.

3. *Product Development for Applications*

Based on the above two image-based 3D reconstruction software programs, full consideration has been given to potential applications and media platforms for digital cultural artefacts, after research and development into supporting applications. This has formed a closed loop of digital 3D cultural products from production to management to application.

3.1 A 3D data management system for large-scale cultural artefacts

The main components of this system include a 3D model database, a basic information database, a data management system, and a 3D display system for cultural artefacts, with the ability to store, display, dispatch, retrieve, measure, and manage information related to large-scale objects in three dimensions. Stored data involves professional grade 3D models of cultural artefacts, and is mainly used to establish a permanent digital 3D artefact archive and to perform information management.

3.2 A digital network 3D display system for cultural artefacts

This system is composed of three main parts: a database of 3D models of cultural artefacts, a basic information database, a data management system, and a 3D network display system. Using this platform, 3D models of single objects can be displayed smoothly via the Internet, meaning that users can interact with them at 720° on the network without needing to install a plugin. The mobile display has been adapted to perform 3D display of cultural artefacts via the Internet.

3.3 A system for panoramic roaming and interaction in the 3D display of cultural artefacts

This system is composed of a panoramic roaming database, a database for 3D models of cultural artefacts, a basic information database, and a hotspot loading system, a data management system, and a 3D display system. It offers efficient voice, text, and image loading via unique 3D model hotspot loading technology, realizing simultaneous remote panoramic display and visualization of cultural artefacts. It gives users the impression that they are actually present in the museum, for a fully immersive experience.

3.4 A mobile terminal system for the 3D display of cultural artefacts

This system is composed of a 3D model database, a basic information database, and a 3D mobile display system for cultural artefacts. It offers synchronous 3D display data from the existing museum network on tablets and mobile phone terminals.

4. Application Solutions

Technology and products are integrated organically on the basis of research and development, offering a complete and comprehensive solution for the digital display of cultural artefacts involving hardware and software to cover digital production and application.

4.1 Basic design ideas

Data interface standardization: Ensuring the versatility and applicability of production data to all application layers; also conducive to effective data sharing and dissemination.

Diversified application platforms: Supporting a variety of media platforms, focusing on mainstream and emerging media, so that cultural artefacts are in public reach and can be enjoyed everywhere, facilitating cultural diffusion.

Original base compatibility: The digitization of cultural artefacts as advocated by the State has been occurring for years, and most museums have a digital foundation. Future digitization must make full use of these existing foundations to avoid repeated investments.

4.2 Planning the overall framework

Under the existing frameworks for standards and security of information, the existing system can be divided into four levels: an infrastructure layer, a data layer, an application layer, and a terminal display layer. The functions and composition of each layer are shown in Figure 4.

1) The infrastructure layer provides basic hardware support for 3D digital museums, including computing and network hardware such as servers, network equipment, and storage devices.
2) The data layer includes a 3D database of artefact display models, a basic 3D information database, and a detailed cultural artefact model database, offering data storage and basic data for upper-level applications.
3) The application layer is composed of application systems with a variety of characteristics, providing customized services for diverse user needs, including an infinite 3D artefact reconstruction system, a 3D file management system, a panoramic roaming system, a digital 3D desktop display system, a 3D digital network display system, a 3D mobile terminal display system, a 3D holographic display system, and an on-site interactive stereoscopic 3D display system.
4) The terminal display layer is mainly composed of various applicable data media platforms, including websites, PCs, holograms, 3D TVs, mobile phones, and tablet computers, with the capability to integrate and install software products on different platforms for the multi-channel display of digital cultural artefacts.

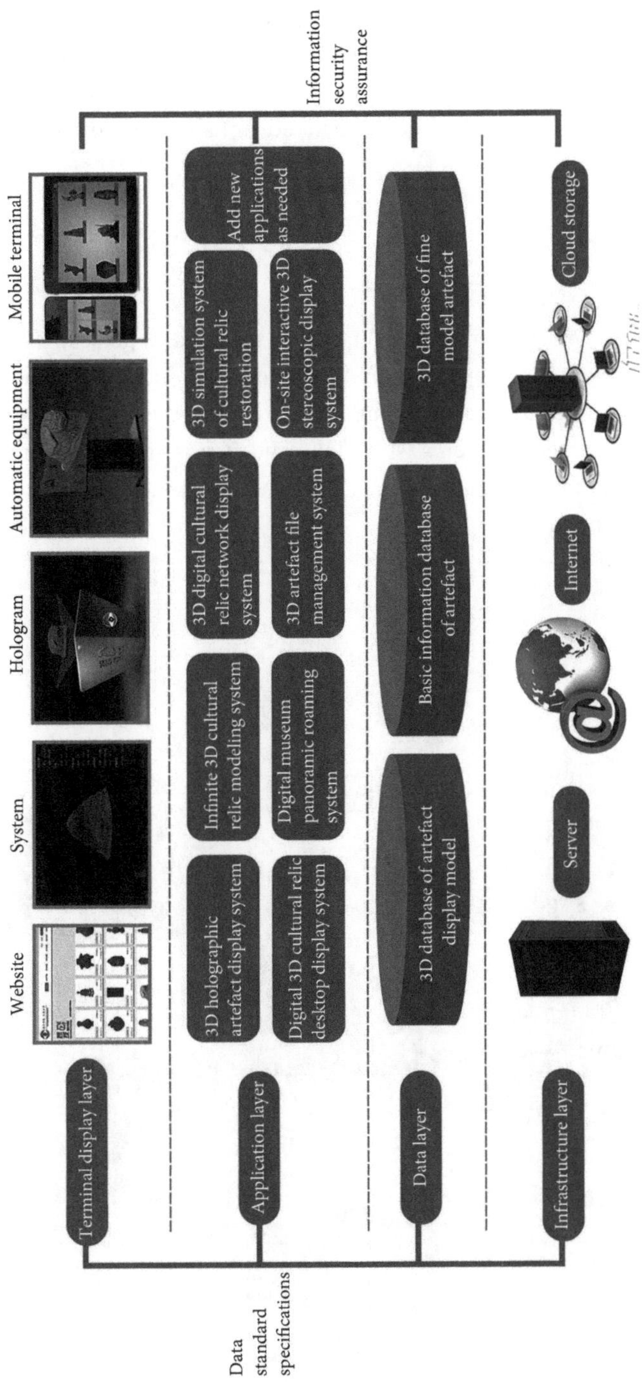

Figure 4 The overall structure of the 3D digital solution for cultural artefacts

5. *Case Study*

Digital practices have been carried out at the Chengdu Chuanshang Sculpture Art Museum in accordance with the above solutions, as shown in Figure 5. Based on the overall architecture and the WJX-X 3D modeling software, a series of application products have been formed, showing both the technical advantages and cost improvements of a one-stop data service.

6. *Summary*

The 3D digitization of cultural artefacts is a systematic project involving digital production, management, and application. With independent research and development, we have effectively explored and experimented with 3D digitization. Meanwhile, due to internationally recognized problems with 3D image reconstruction technology, we have yet to overcome residual difficulties to provide complete domestic technological support for digital 3D museums.

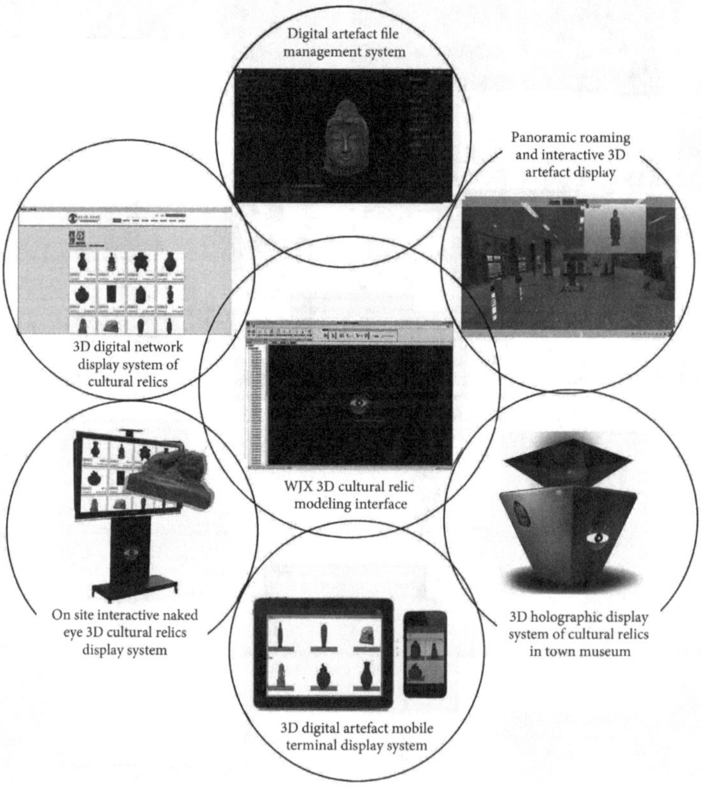

Figure 5 Chengdu Chuanshang Sculpture Art Museum's digital solution application

The Crucial Role of Video Image Storage at the Palace Museum

Zhao Simeng
(Ministry of Information at the Palace Museum, Beijing 100009, China)

Abstract: In the process of modernization, the ancient construction techniques that have been passed down from generation to generation by masters and apprentices are now facing serious threats from neglect and even complete loss. One of our top priorities is using modern multimedia technology involving text, images, and video to preserve ancient building repair processes in an accurate, systematic, and comprehensive way. A major issue in the cultural realm today is how to use modern imaging technology to save, preserve, spread, and develop intangible cultural heritage via digital museum platforms.

Keywords: video footage acquisition; storage; important function

Preface

At present, the research and application of digital information technology in the field of material cultural heritage is on the right track, but for intangible cultural heritage it is only just beginning. Considering the preservation of traditional techniques and intangible cultural heritage, experts are now facing the issue of how to use digital imaging technology firstly to preserve it, and then to explain and disseminate it in a fresh way. The repairing, preservation, and inheritance of intangible cultural heritage is key, and the emergence of digital imaging technology has provided new ideas and technical support. Therefore, in regards to the protection and inheritance of intangible heritage, the application of digital images in museums has both theoretical and practical value.

1. The Origin of Images

Early man-made images take the form of ancient patterns on cliffs, rocks, and oracle bones, and are regarded by many scholars as carriers of visual information. Vision is the primary source of human consciousness, and the most basic medium for human thought. Photographic images are characterized by authenticity, directness, presence, timeliness, imagism, symbolism, stillness, and virtuality. Digital images are simply those that are acquired, stored, and transmitted digitally.

2. Video Image Storage Formats

Commonly used video formats include WMV, ASF, RM, RMVB, MOV, AVI, DAT, MPG, and MPEG. As the saying goes, birds of a feather flock together, and video files are no exception. As a result, video files can be divided into two categories: image files (such as common DVD files) and streaming video files (formats that have arisen with the development of the Internet). The latter forms of video, such as online live broadcasts, are built on video streaming technology.

3. Methods of Transmitting Video Footage

1) From "authoritative release" to "grassroots production"
With the large-scale popularization of video shooting tools such as DV, DC, and camera phones, with just a computer connected to the Internet, a video camera, and perhaps some simple video editing, videos can be shared easily. Compared with the "authoritative release" of radio and television media, in addition to content such as movies, TV dramas, TV news, variety shows, advertisements, and music, users can also upload personalized entertainment content. The democratic nature of this "grassroots production" emphasizes audience subjectivity, and visitors have a certain amount of sway on the content that is produced.

2) From "content is king" to "entertainment wins"
If "content is king" in traditional terms, then online media purports that "entertainment wins". The US-based video-sharing site YouTube benefits from this every day, providing easy-to-follow classified videos on all manner of topics including art, comedy, music, travel, animals, and cars. It is the entertainment factor that is the site's trump card.

3) From "push-to-play" to anytime "on-demand downloads"
The "push-to-play" playback mode of commercial television has gradually reduced in attractiveness, due to the advantages of anytime online on-demand video download and

playback, and its diverse interaction methods. Visitors are no longer passive, one-way acceptors of information, but are fully invested in initiative and interactivity.

4) From "linear propagation" to "interactive" communication
Online video advertising allows consumers to "experience" products, services, and brands in person, as well as book, trade, and pay online. A combination of text, images, sound, and video provides a different experience of goods and services. Strong interactivity is another advantage of network-based video communication, with users able to reply to and comment on the video and its publisher.

5) From "ratings survey" to "statistical and retrievable"
Traditional TV determines the effect of advertising through a sample survey of ratings. On the Internet, through authoritative and impartial statistics about visitor traffic, one can accurately calculate browsing volume, click-through rate, and other indicators, as well as the time and geographical distribution of user access. This allows for an evaluation of the advertising effect and a verification of advertising strategies. In addition, users can actively retrieve content and watch it repeatedly.

4. Collecting and Storing Video Images in the Palace Museum

We established a video acquisition and processing division in 2003, and it is mainly responsible for shooting in a project called "Maintenance of Ancient Buildings in the Imperial Palace". Since 2003, the department has operated within each ancient construction, photographing the repair and post-repair processes, and storing tens of thousands of precious items. With regard to image recording for a building, it is necessary to mobilize our understanding of its function and era. The clearer our understanding of this background, the more accurate the record. When shooting, we must firstly consider authenticity, as incorrect image data lowers the worth of heritage and has no use or value for preservation. Secondly, we must be able to present and explain historical heritage when shooting. Recorded content must be rigorous and clear, accurately conveying its communicative intentions. Finally, we must consider artistry – not simply recording times, places, scenes, and characters, which can be historically accurate, but lack soul. Indeed, since Ming and Qing Dynasty buildings have unique and artistic characteristics, our collectors must be aware of how to distinguish between and see similarities among ancient buildings, magnifying unique features.

Meanwhile, we have developed specific shooting requirements and plans for special features of ancient buildings under repair. In 2006, our institute collaborated with the Dayang Company to launch a video and audio image utilization system to suit the unique characteristics of the Imperial Palace. The system involves several links,

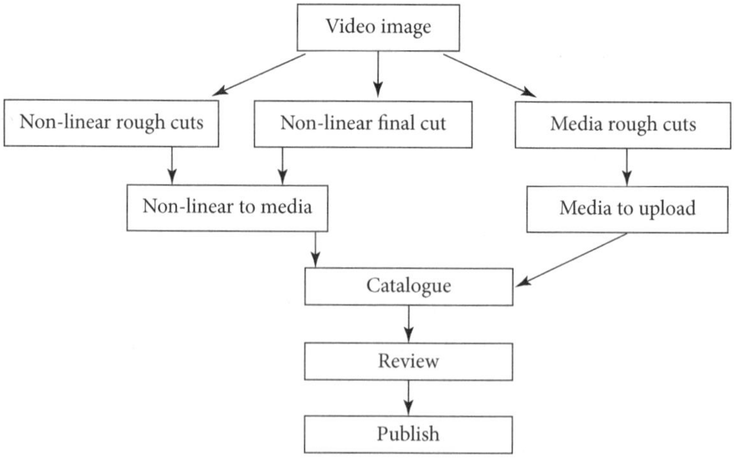

Figure 1 Flowchart of the video and audio imaging system

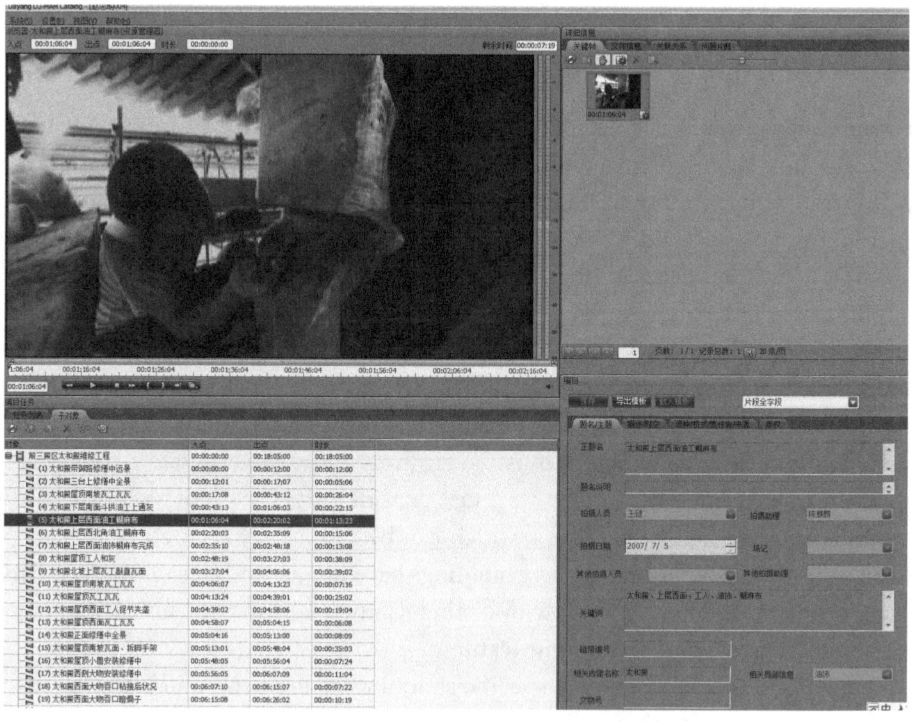

Figure 2 The audio-visual system catalog

including the uploading of material, cataloging, review, retrieval, and download (see Figure 1). It is capable of storing digital videos and audio resources, as well as functions such as the migration, cataloging, archiving, storing, retrieving, and invoking of video and audio data. It is complete video asset storage management system for audiovisual resources. Digital storage of such resources is based on standard-definition SD 8/19 bit uncompressed DV25, DV50, MPEG-21, and IBP formats, and high-definition HD 8/10 bit uncompressed, MPEG-21, DVCPROHD, and HDV formats. It can provide data streams in MPEG-4, Real, or WMV encoding formats, and can be used to query, browse, and download programs through a computer local area network. High and low bit rate materials are collected automatically in the acquisition process, with high code stream material used for archive storage, and low code stream material used for cataloging, searching, and browsing. According to the content, the multi-level cataloging method divides among the clip layer, the scene layer, and the lens layer (see Figure 2). Photographic content and related files are stored in corresponding areas, with key frames extracted to facilitate future retrieval and use. Stored video and audio resource files are arranged in a hierarchical structure for online storage, nearline storage, and offline storage for security, high availability, and high cost performance. The retrieval system can also provide audiovisual browsing services and support multiple combinations of full text, keywords, photographers, and photographic time searches, providing a convenient and fast search method by which to retrieve program materials. Required materials can be located via various matching and full-text searches.

At the start of 2003, the newfound division began to use the latest Panasonic AG-DVC200 HD camera (hereinafter referred to as Panasonic DVC200), with 48 megapixels, DV tape storage, optional shock-proof digital gain, and shock-proof white balance tracking modes. Since it weighed more than 5kg in total, the shortage of manpower at the time meant that team members were faced with difficulties in shooting as they completed the recording process for a large number of construction projects including Taihe Temple and Taihe Gate. After several years of hard work, hundreds of pieces of high-definition video footage of the repair of ancient buildings was recorded. However, with the continuous development of technology, the DV HD camera eventually became unfit for purpose.

In 2011, we purchased a Panasonic P2HD professional HD camera (hereinafter referred to as Panasonic P2). Using a memory card for recording, and at just half the weight of the Panasonic DVC200, it was much more flexible, and better suited to recording the meetings, lectures, sports events, and leadership inspections that comprised the fundamental shooting process.

In 2012, imaging technology entered another new era. While many people were still impressed by new iPad screens, image technology with twice the resolution of the new iPad entered the field, i.e. 4K technology. At this point, we purchased the CineAlta series 4K high-definition digital camera F55 (hereinafter referred to as the F55). It had a high-definition recording screen and a full-area electronic shutter that allowed the F55 CMOS to output images frame by frame like a CCD device, completely eliminating CMOS jello and flash band effects.

In terms of storage, the Panasonic DVC200 and F55 are generally suited to shooting maintenance projects for ancient constructions. Before 2013, we used the DVC200 to track and record these projects. According to various vintages in different construction areas, tape numbers were recorded, and field records were input into an Excel table. In 2014, in order to save these tapes in higher quality and for greater longevity, we purchased a number of professional anti-magnetic libraries for tapes of various categories. After recording, we used the Dayang Media Asset Management System for uploading, cataloging, reviewing, and archiving. With this method, the tapes and the system act as double backups of the recorded materials, so that material captured on the F55 will be backed up first on the network disk, and then uploaded to the media asset system as well. Field records must account for shooting content, document numbers, and data duration from start to finish. The Panasonic P2 is now mainly used to shoot daily event meetings, using a standard definition format to shoot. Recorded material is copied to the network disk for backup. Meanwhile, the meeting is edited by the Dayang non-linear editing system, and exported to the Media Asset System for cataloguing and review. Finally, it outputs a DVD file for disk backup. The department has established a complete and smooth working model that is used to this day.

5. The Crucial Role of Video Imaging in the Construction of Museums

There is a persistent issue with intangible cultural heritage in digital museums, namely low numbers of video footage. Video imaging can be used not only as a means of communicating this heritage, but also as a navigational tool. Provided that video footage are of high enough quality and accessible file sizes for Internet dissemination, it is feasible to invoke a large number of video applications, which ultimately offers significant advantages. Video images allow visitors to perceive real objects and experience physical materials, evoking the spirit that is at the core of intangible cultural heritage. As the pace of human life accelerates, and people begin to consume content in different ways, visitors need intuitive, clear, elemental representation. Video images fully satisfy this need, meaning that they have distinct advantages in the exhibiting of intangible culture in today's museums.

The Forbidden City is the largest, most intact ancient building complex in China. Completed in the 18th year of the Ming Emperor Yongle, the complex has a history of more than 560 years. The Palace Museum was opening in October 1925, standing as a key national site for cultural preservation. Existing for thousands of years, ancient Chinese architecture provides a great testimony to the longevity of the nation's civilization. Prime examples include the many large-scale palaces that were built in the Ming and Qing Dynasties, grand in terms of both quantity and quality.

Under the purview of the Central People's Government, the State allocated special funds for the repair of ancient buildings in 1949. In the early 1950s, the Palace Museum established a department for the management of ancient constructions, in which a professional construction team was responsible for all construction projects. At that time, investigations were undertaken to ascertain the degree of fundamental damage to buildings, and maintenance and reconstruction plans were formulated.

Nowadays, with the continuous development of science and technology, it is no longer a challenge to record the maintenance process with high-definition video equipment. As time goes on, people are paying more attention to the protection of traditional buildings, yet the phenomena of "destruction in preservation and loss in construction" is undeniable, meaning that the protection of ancient buildings is still unsatisfactory. As such, we must work to preserve original elements with modern technology, which will have profound significance for future reference and historical research. If living culture disappears, it cannot be regenerated. Therefore, protecting intangible cultural heritage does not simply entail preserving "old as old", as with tangible culture. It inherits development and develops with inheritance, raising a question about how traditional culture is directed in modernization. In the process of answering this question, emergent digital imaging technology has provided new ideas and technical support for the safeguarding, preservation, and inheritance of intangible culture. Digital imaging technology allows for more convenient and low-cost preservation, and utilizes the ubiquitous Internet to inspire public interest in order to carry ancient wisdom forward. Meanwhile, the establishment of a video imaging database assists personnel who manage museum information resources to deal with images more efficiently and conveniently, while also enabling easy access to resources. Tens of thousands of minutes of video footage already exist in our department's database, including the maintenance projects of the main Forbidden City area, which are important for current and future research.

Fewer and fewer people now have the skills required to use core technology for the repair of ancient buildings. It takes at least 10 years to train a mature technician. Since this sort of repair is a key element of the preservation of cultural artefacts, the status of skilled individuals is somewhat indicative of the future of our cultural heritage. With the passage of time, large-scale ancient buildings must be repaired. At present, four to five projects are begun each year. In order to record the repair process more comprehensively, in recent years our department has hired a professional shooting team to track each project, amassing a detailed and comprehensive record of the original state before repair, the construction site stage, and finally the site post-renovation.

We currently require further interviews with senior repair experts, because the Forbidden City opened in the 1950s without a construction team. Expert Shan Shiyuan invited ten famous craftsmen known for their skills in the "eight masterpieces", together with their apprentices from the construction department. They were the first and second generation of artisans at the new Palace Museum. All of the old experts have now retired or passed away, and members of the third generation will soon be retiring. If we do not address the scarcity of skilled craftsmanship, we risk losing the transmission of traditional culture.

With the development of multimedia and video footage, the rise of virtual reality technology, and rapid network development, the cause of cultural heritage protection has come to the fore. Heritage can now be saved using high-precision digital images. We must record exceptional craftsmanship with video footage and similar means in order to preserve ancient techniques for construction and repair. Video footage is worth a thousand words, as it is able to present reality directly in a series of unique images.

References

[1] Xiao Yefei. *Advantages and Prospects of Video Communication*, Sept 2008.
[2] Interview with reporters. 'Beijing's Forbidden City carries out routine repair on four ancient buildings this year'. *Beijing Morning Post*, August 2014.
[3] Wang Jing. *Image application for intangible cultural heritage in digital museums*, Dec 2007.

A Case Study of the Cultural Position of Museums and Digital Applications in the Context of New Media

Ye Yipei
(The Palace Museum, Beijing 100009, China)

Abstract: Technological advances have spurred immense changes in public lifestyles. With the rapid transition from print to broadcast, television and network communications have increased the information capacity of culture itself. The mass publication of books and newspapers, the development of audiobooks and video resources, and the explosion of information in the Internet age all demonstrate that culture is undergoing an informational evolution under the influence of the media. This paper analyzes the technical trends, contemporary characteristics, audience characteristics, and aesthetic tastes in the context of new media, considering the cultural position of the museum as a carrier of traditional culture in the new media era.

Keywords: new media; era of image interpretation; audience; aesthetics; museum

Professor Xiong Chengyu, director of the New Media Research Center at Tsinghua University, believes that "new media is a relative concept. Today's new media refers mainly to media forms that are generated and influenced on the basis of computer information processing technologies, including online media and other digital offline media forms." In *The Interpretation of Key Terms of Communication* (Huang Xiaozhong, Yang Xiaohong, Feng Gang. Sichuan University Press, 2005-8), the authors translate the English term "new media" and define it as communication carried by core computer technology, referring to fiber-optic cables, large-scale computer database communication systems, communication satellites, live satellite TV systems, and high-definition TVs, as well as the Internet and other multimedia practices that have grown rapidly since the 1990s. The Internet is a mainstay.

1. The Technical Background and Characteristics of New Media

Internet technology has developed rapidly since its inception. In the 1960s, the Advanced Research Projects Agency (ARPA) at the US Department of Defense created an experimental network, the ARPA Network, during the Cold War, which was considered to be the first prototype of the Internet. In the 1990s, Englishman Tim Berners Lee invented Web technology whilst working in a large research institution in the European Community. The World Wide Web, also coined by Lee, integrates information from various computers connected to the network, and allows for data transfer and retrieval transfer from one Web server to another via Hypertext Transfer Protocol (HTTP). Lee's invention transformed difficult character commands into intuitive graphic information. Over the past 50 years, Web technology has become one of the most important network application technologies, developing rapidly into an information ecosystem that supports search, aggregation, and interaction of data supported on the Web. In the meantime, Web technology has also made three qualitative. Web 1.0 began in the 1990s, featuring an extensive use of static HTML pages upon which to post information. It was able to reintegrate all kinds of scattered and fragmented information into a more organized bundle to align with user reading habits. However, information dissemination in Web 1.0 was still dominated by unidirectionality, and did not reflect the subjective status of the audience. It did not offer users the faculty to participate in the collaborative development of information.

Therefore, Web 2.0 was launched, in which users were invited to give feedback. The Internet has evolved from a series of websites to a mature service platform for providing network applications for end users, emphasizing user participation, online network collaboration, data storage networking, social network, RSS applications and file sharing. These elements have become the main support for the development of Web 2.0. Users are no longer passive receivers of information, but are participants, promoters, and producers of it.

Further developments in Web technology have spawned Web 3.0. Although its characteristics are controversial in the industry, given overall development trends, it will use the Internet for database conversion, the evolution of artificial intelligence, semantic networking, and the evolution of service-oriented architecture (SOA). Based on Web 2.0, it continues to split jumbled micro-content into minimum units, at the same time standardizing and structuring the meaning of words, and realizing interactions between micro-information, and semantic-based links between micro-content. Web 3.0 digs deeper into information, interoperating directly from an underlying database, aggregating and connecting various data points scattered across the Internet based on user needs, and adding metadata to webpages to enable machines to make sense of web content. It thus provides semantic-based retrieval and matching, offering users more personalized, precise, and intelligent information retrieval. In short, users can participate in the behavior patterns offered by Web 2.0, with the addition of easier access to content of interest.

Due to the trend for user personalization, new media also involves significant contemporary features. First, from a technical point of view, it involves digital elements. Wither text, images, or video footage, digital components can be woven into binary code and recognized by a computer, converted into a pulse signal, and finally pushed out as a recognizable symbol to users. In network-based communities, social relationships are based on the compilation, storage, transmission, exchange, and control of pieces of digitized information that are reflected in a series of basic digital interaction processes. Digital media has changed the outlook of media information that was strictly controlled in the past, allowing for freer dissemination of information and use the Internet, especially to garner huge amounts of information through a variety of powerful functions. Digital media has thus altered regional media communications, with characteristics that allow extension across the world – a powerful factor in the drive of globalization.

Secondly, from a communication perspective, new media is interactive. On the one hand, "media" is a carrier of art. On the other, it is an experience received by an audience. Interactivity is largely reflected in the direct interaction between a work and its audience. For example, a viewer can experience the function of an object, or appreciate the beauty of an work of art with a few simple clicks, touches, sounds, and other operative behaviors. Interaction between users and digital media devices are manifested in two-way human-machine communication founded on input and output. In technical terms, computer and digital technologies have developed popular tools such as keyboards, mice, microphones, and stereos for this purpose. With the continuous advancement of technology, tools such as Kinect and Siri have also come into use, featuring gesture and speech recognition technology that liberates a user's hands.

Thirdly, communication occurs across media. Thanks to the development of the mobile internet and information technology, new media has accelerated media convergence. Not simply addition or subtraction, this process involves the mutual penetration and blending of two or more new media at multiple levels, in multiple fields and multiple dimensions. Meanwhile, new media and traditional media (such as newspapers, television, and radio) can also be effectively integrated. Such integration reduces the cost of information collection while expanding the reach of media dissemination.

Fourthly, in terms of artistic expression, new media is virtual. It stems from reality, yet is different to reality. It is another version of the real world that involves virtualizing scenarios that are currently not possible in reality, providing an experience that appears real to the human eye, but is virtual reality.

2. Audience Analysis

2.1 Demographics in the new media era

Due to the wide-ranging and significant differences between new and traditional media, their respective visitors show various discrepancies. The first is diversity. Each link and chain in the new media industry has a different audience. Due to a higher number of use functions, smart phones prove relatively complicated to operate, and are therefore more suited to young or middle-aged individuals. Meanwhile, for the elderly, phones are largely a tool for making calls, so there is less demand for smart phones. Furthermore, among young and middle-aged users, each brand of mobile phone requires a specific operating system – iOS for Apple, Android for Samsung, and BlackBerry systems for BlackBerry handsets.

Second is the issue of identity complexity. With the current orientation of popular culture and social entertainment, new media breaks traditional text and knowledge barriers, with widely used images and videos that enable people of any identity, age, education, or class to enjoy convenient new media services.

The third discrepancy is liquidity. Diversified information and diverse communication methods make audience distraction relatively simple. In an environment of fierce media competition, without high-quality information and convenient, diverse channels of communication, visitors will simply eliminate the service.

The fourth issue is uncontrollability. Identity complexity and the relative independence of new media visitors are crucial, uncontrollable factors in new media. Meanwhile, high-tech involvement means that the formation and dissemination of public opinion are no longer limited to word of mouth or correspondence, nor to singular institutions or mainstream media channels.

2.2 Aesthetic Trends in New Media Audiences

2.2.1 Visual experience

American scholar Nicholas Mirzoeff has divided the communication of human information communication into three historical stages. First came the ancient image stage (1650–1820), in which images were authentic and appeared only in the real world. Second was the modern image (1820–1975), in which the emergence of photography and film meant that images were created as a visual relationship between present visitors and an image from a past time and space. Third was the contemporary (or postmodern) stage (1975–present), in which the image is contradictory or virtual. In the era of digital media, images have long been the main channel for accessing information. As such, this period has been dubbed the "image reading era".

Professor Jin Yuanpu from the Chinese Department at Renmin University believes that "the era of reading pictures is related to great changes in modern society. In terms of communicative methods, original print media has developed into electronic and digital media, with a mass of visual images and other phenomena being produced. In the past, words may have been the dominant mode of knowledge acquisition, but now visual images have revolutionized both our social lives and ways of thinking."

Alvin Toffler, in his near futuristic book *The Third Wave*, points out that: "human society is gestating three forms of cultural illiteracy – textual illiteracy, computer illiteracy, and visual illiteracy. The latter two are products of an industrialized society, namely a post-industrial society."

These two views demonstrate that social change is closely related to the advancement of technology.

From the perspective of the museum industry, physical exhibits are largely used to spread culture and play a role in public education, and include a high proportion of image information. Therefore, for the interpretation of artefacts, a large number of images and animations is necessary. In recent years, the maturating of augmented reality technology has provided new ideas for the interpretation of artefacts within museums. At the Cleveland Museum of Art, the "ArtLens" iPad app offers a rich array of video, audio, text, and image content, with a "scanning" feature that applies online AR technology combining the iPad camera with Qualcomm Vuforia image recognition software. When a user scans a work of art featuring the ArtLens logo, the app automatically recognizes the image and provides interpretive information. It is worth mentioning that the interpretations here do not interrupt the viewer's appreciation of the overall original image, instead using the camera to analyze the actual image and presenting line segments that guide users to information about partial details (see Figure 1).

However, the trend for "reading images" has been met with much concern and criticism from scholars. From a traditional point of view, language is representative of rationality and advancement, while visual content offers a subjective interpretation of ideas, and is therefore less advanced. Scholar Zhao Yiheng believes that fears about globalization in current cultural development have arisen due to "dumbing down", that is, cultural

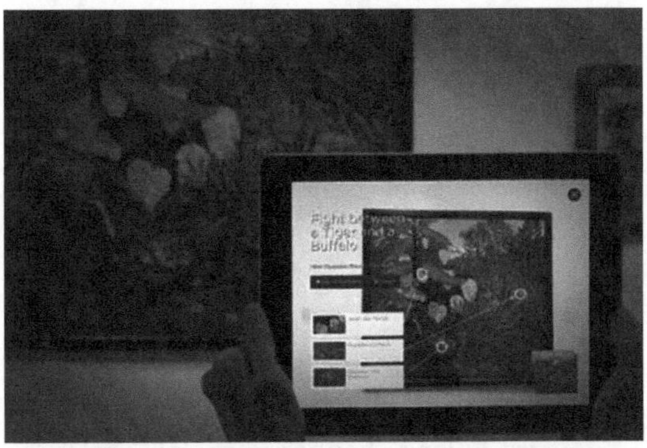

Figure 1 The "scanning" function in the ArtLens app

activities rejecting wisdom and speculation, as cultural products move toward fulfilling the needs of the public. Many individuals who rarely participated in cultural circles in the past have now become the masters of "view count", "play count", and "click count". Meanwhile, the media (especially film, television, and network culture) increasingly highlights the negative aspects of commercialization, imaging, and oversimplification. As the hardware of production culture rises, software is constantly being "dumbed down".

This view is certainly worthy of attention, but from a museum perspective, cultural artefacts are largely unrelated to public daily life. It is challenging to resonate directly with a theory that involves secondary aesthetic trends.

2.2.2 Perceptual entertainment

In the era of traditional media, aesthetic beauty stood as a classic, perceptual, and philosophical subject. Art was respected by the public, with a willing engagement in metaphysical thought and an overwhelming pursuit of rationalism. Conversely, the new media era is carried by a cultural ethic that emphasizes interaction, pursues equal rights, returns to the true self, and advocates for freedom. It also highlights grassroots, resists elitism, elevates sensibility, and respects individuality. Visitors have transitioned away from the relative "eternal" and compassionate psychology of traditional aesthetics, moving toward the instant, popular, and hedonistic aesthetics of consumerism. From the museum perspective, the new media era invokes developed technology to enable visitors to complete a second interpretation of cultural artefact collections, with the assistance of all-encompassing 3D media tools. Visitors actively use new media to empower personal speech, based on the second creation of the original work, presenting a scenario not dissimilar to the famous post-structuralist theory proposed by Roland Barthes in *The Death of the Author*. Barthes suggests here that when an author composes a work, at the moment of its completion, his or her relationship with the text ends, as it is now in the hands of an interpretative reader. Suggesting that "the text be liberated from the authority of the author", this literary theory has been widely applied to other fields, including cultural and artistic practice and visual art.

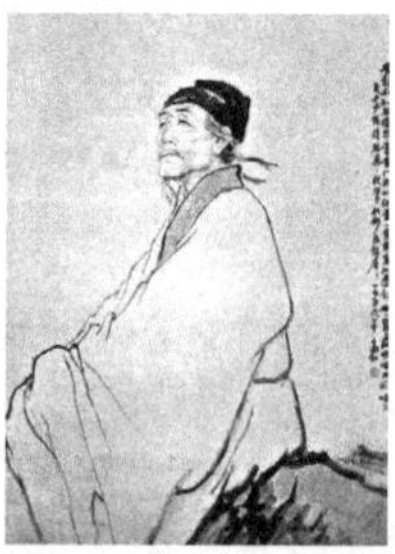

Figure 2 Jiang Zhaohe's *Portrait of Du Fu*

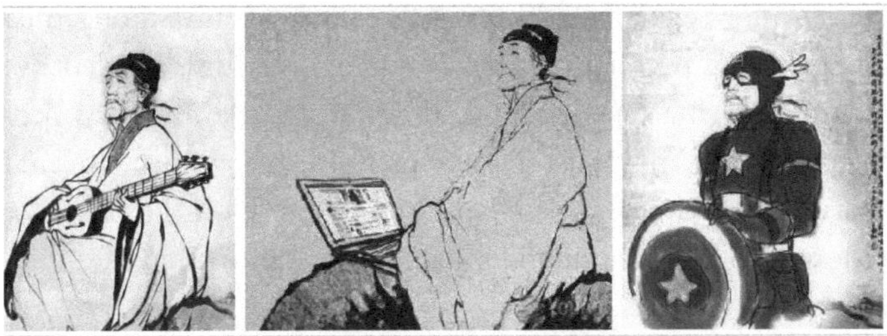

Figure 3: Spoofs by web users

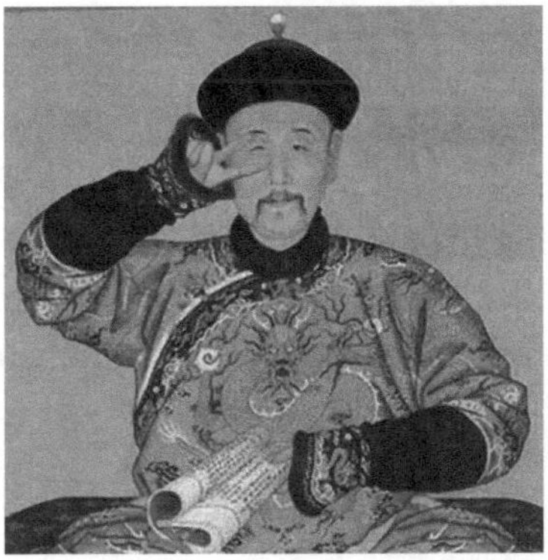

Figure 4: "Yong Zheng: Feeling Adorable" - one of the pictures in the series

One example is a series of Weibo graffiti entitled "Du Fu is Busy", which was initiated by high school students in Mainland China in 2012. Jiang Zhaohe's original picture was created in 1959, and can be seen at the Museum of Chinese History. The image is presented as an illustration of Du Fu's text *Climbing* in the Chinese textbook *Language (Compulsory 3)* (see Figure 2). It has been spoofed by Weibo users (see Figures 3a, 3b, and 3c) and forwarded extensively across the platform.

Banter, irony, absurdity, and an anti-authority stance are some the means by which people publicize individuality and freedom, eliminating orthodoxy and sublimity, and welcoming a form of aesthetic content characterized by splicing, spoofing, and vocalization. A similar phenomenon has appeared in the promotion of museums. In 2014, a group of pictures created by the staff of the Palace Museum, entitled "Yong Zheng: Feeling Adorable" (see Figure 4), caught the attention of web users with its dynamic images, and was forwarded more than 800,000 times.

Digital media offers convenience, and is also guiding a change in our ways of thinking. In the digital environment, the aesthetics of graphic media are created by both the external environment and the requirements of internal visitors. Even with the refinement of museums, the public continues to experience a certain degree of distance, so there is a demand for secular media alongside consumption. Lofty museum culture collides with unrestrained popular culture, requiring museums to identify with market demands. Museums catering to popular aesthetic tastes are met with mixed attitudes, as some are suspicious of media as a sales method, believing it to entail a level of frivolity and disrespect. However, from a museum standpoint, using digital means to promote collections and increase public interest in traditional culture may simply be a contradiction of modern society.

The new technologies of the future are unpredictable, but given the current environment, a general cultural trend towards individual emphasis may encourage technological developments such as Big Data collection, analysis, and the mobilization of new media. Promoting the popularization of history and culture is a key function of the museum. In facing an increasingly young audience, the museum's first task is to cultivate interest. At this stage, it is necessary to reduce the challenges of text, and pay attention to the language used to express information, in order not to bombard visitors with too much textual information, matching the degree of technology and content for the purpose of entertainment. Technology is changing by the day. The era of information is not only an opportunity for museum development, but also a serious challenge, requiring museums to explore, experiment, and learn from one another.

References

[1] Li Linrong. *Introduction to New Media*. Beijing: Law Press China, 2015, pp. 17–27, 56–61.

[2] Li Yi. *Media Aesthetic Culture in the Context of Digital Media*, Beijing: Beijing United Press, 2014, p. 51.

[3] Han Han. *The Order of Discourse: A Critique of Issues in Cultural Modernity*, Taiwan: Taiwan Xiuwei Publishing House, 2010, p. 24.

[4] Gong Chengbo, Tian Xu, Liang Peipei. *Introduction to Digital Media Art*, Beijing: China Radio and Television Publishing House, 2014, p. 20.

[5] Li Jingxiu. 'Research into the aesthetic psychology of the audience from an overall media perspective'. PhD thesis at Jilin University, 2013.

[6] Gong Chengbo. 'Analysis of the Spirit of New Media Culture', [J] *Shandong Social Sciences*, 2010, 5.

[7] Gong Chengbo, Tian Yuan. 'Changes in audience ecology in the new media era', [J] *Young Reporters*, 2014, 3.

An Analysis of Fixed Augmented Reality Technology Applications in the China Science and Technology Museum

Chu Kaili
(School of Humanities and Social Sciences, Beihang University, Beijing 100191)

Abstract: Augmented reality is an emerging field of research in virtual reality technology. This paper analyzes two examples of augmented reality technology at the China Science and Technology Museum, namely *The Exploitation of the Works of Nature* and "Our Motherland" exhibits. It discusses the effects of fixed AR technology applications, proposing its enhancement with several suggestions for the effective combination of real technology and science museum exhibition.

Keywords: augmented reality; virtual and real fusion; Science and Technology Museum; case analysis

Preface

Augmented reality (AR) is premised on virtual reality (VR), featuring forms, technologies, and implications that have spawned new developments and extensions. Augmented reality refers to the superimposition of virtual computer-generated information with the real world, enhancing real-world information and enabling users to gain new insight and interact with the virtual world.

The basic formation of AR was proposed in the 1960s. In the years that followed, the continuous development of computer technology saw consistent improvements to human-computer interaction equipment. Over the past two decades, AR technology has become a reality, attracting a huge amount of attention, and becoming one of the most influential emerging technologies. Outside of China, AR is used widely in advertising, medicine, machine assembly and maintenance, navigation systems, archaeological and cultural activities, art, entertainment, gaming, and education. In China, however, AR technology had a later start, with applications occurring mainly in the advertising industry, entertainment, and gaming.

Alongside the rapid advancement of digital technology and the recognition of human-computer interaction, children enjoy increased interaction with multimedia for entertainment, discovery, and learning. However, many of the scientific and cultural education products that use AR technology still lack the richness of contemporaneous digital entertainment products on the market.[1]

The China Science and Technology Museum is a location that popularizes information and educates the public, especially young people. Applying AR technology has been revolutionary for such institutions, making it possible to display large, complex, and fragile exhibits. Augmented reality visually presents invisible or abstract concepts to the viewer, helping visitors to understand and interact with immersive, virtual objects and scenes. Using digital technology in the design of exhibitions can ensure better dissemination of knowledge and offer visitors an interactive experience. It has therefore become a primary task for science and technology museums.

1. An Overview of Augmented Reality Technology

1.1 A basic definition

In 1997, research pioneer Ronald Azuma described AR as an environment that includes both a virtual and a real world. Users see not only the real world, but also a computer-generated "enhancement" created with virtual information.[2] Augmented reality systems have three characteristics, namely the fusion of virtual and real, real-time interactivity, and three-dimensional registration.

Virtual reality is a completely engineered artificial digital environment that uses computer hardware and software to simulate the real world, providing users with an immersive experience in virtual scenes accompanied by visual, tactile, and auditory senses.[3] Augmented reality integrates digital information into the real environment, providing users with an interface to experience and interact with this integration. In addition to three-dimensionality and immersive virtual reality, the interactivity and integration of virtual reality in augmented reality systems enhances a user's perception of the real environment.[4] Due to these various characteristics, augmented reality technology can provide its users with a unique experience, and also entails broad application prospects.

1.2 Augmented reality technology

AR technology has arisen from information technology at a certain extent of development, and research into its application systems is extensive. The AR system first constructs a virtual 3D object via optoelectrical display technology, interactive technology, computer graphics, and visualization technology. When interacting with an AR system, users are recognized by sensing technology that places virtual objects given specific identifiers. In the real environment, users look through a display device and see virtual objects integrated with the real environment, producing a sensory effect that "convinces" them that the virtual object is part of the real surrounding environment.[5] In this scheme, the hardware includes computers or mobile devices, cameras, tracking and sensing systems, displays, computer networks, and identifiers. The software includes applications, web services, and content servers.[3]

AR systems can be divided into two categories according to their ability to move, namely fixed systems and mobile systems (also called indoor and outdoor systems). Most science and technology museums use fixed AR systems for displaying exhibits, while mobile systems are used mostly for tablets and smart phones, which can be carried around with a greater degree of flexibility. With regard to tracking methods, there is a marker method in which a camera captures a specific identifier that causes software to retrieve corresponding information. This identifier is generally a QR code, commonly using a black-and-white geometric pattern. Unmarked object tracking is a different method, and is typically applied to intelligent mobile terminals using geo-base stations or GPS data. Regardless of these differences, high-speed data networks and efficient processors are required for the efficient processing of relevant data that allows for more natural integration of reality and virtuality, as well as more cooperative human-computer interaction.

2. Fixed Augmented Reality: A Case Study

2.1 Introduction

This section describes two cases of the popularization of science using fixed augmented reality technology at the China Science and Technology Museum, namely *The Exploitation of the Works of Nature* show in the China Light exhibition hall, and *Our Motherland* in the Science, Technology, and Life exhibition hall.

The Exploitation of the Works of Nature is the title of the world's first comprehensive project on agricultural and handicraft production, including written descriptions as well as clear and simple illustrations. The project had a strong guiding role, and was called the "Encyclopedia of Chinese Processes in the 17th Century" by European scholars. Systematically summarizing various techniques in Ancient China, the work outlines a complete scientific and technological system. Many of the production techniques described in the book are still being used today. However, strictly speaking, *The Exploitation of the Works of Nature* is not a technical guide, but "a popular science book for noblemen who cannot tell the difference between grains, and Confucian scholars who are out of touch with reality".[6] The objects depicted in the book are simple in construction, with clear illustrations. Based on this physical book, augmented reality was used to reproduce certain scenes and content, allowing visitors to flip through a paper volume on a desk while objects emerge from the pages (see Figure 1).

The *Our Motherland* exhibition also applies fixed augmented reality technology, involving a Chinese map puzzle game that integrates geography and human knowledge. Visitors place puzzle pieces corresponding to certain provinces and cities under the camera, and the natural or manmade landscapes of the corresponding provinces and cities are displayed on the screen, with folk songs and regional features appearing on the right hand side. Visitors are also invited to place the puzzle pieces together on the mapped floor to create a complete map of China (see Figure 2). The exhibition is highly interactive and stimulating.

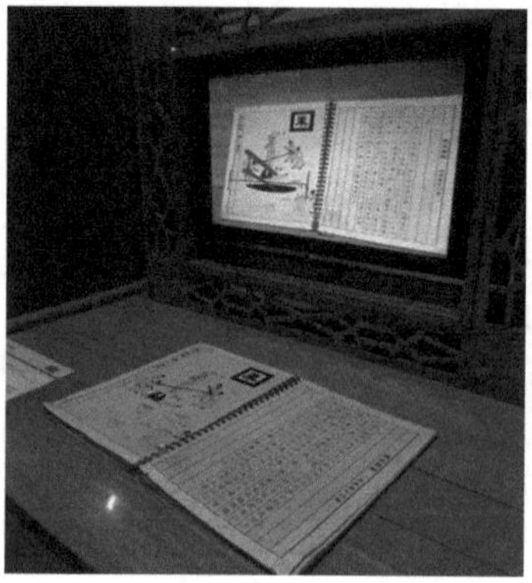

Figure 1 ***The Exploitation of the Works of Nature***

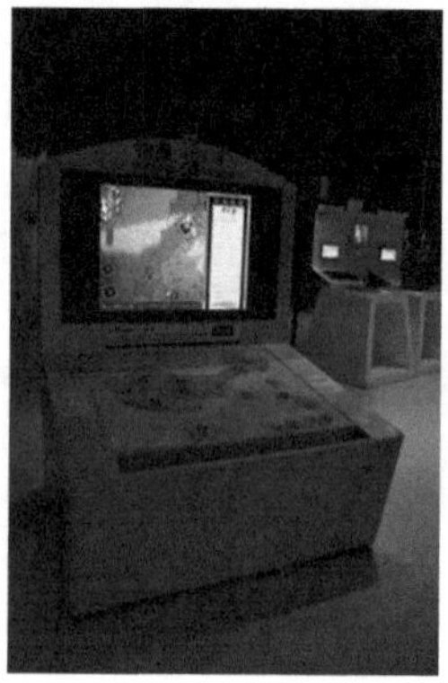

Figure 2 ***Our Motherland***

2.2 Analysis of Design Concepts

Both cases mentioned above utilize fixed augmented reality technology, yet there are nuances in the exhibitions as a whole and their respective design philosophies.

In *The Exploitation of the Works of Nature*, augmented reality is used to convert text descriptions and 2D images into stereoscopic images, which are presented on the pages through a screen, as if opening a 3D book. This allows visitors to understand the structure and principle of ancient production techniques. When turning to a different page, corresponding artefacts appear on the screen, allowing visitors to browse according to their own interests.

Meanwhile, in *Our Motherland*, augmented reality technology is combined with a map puzzle game. Only when puzzle pieces are placed under a camera do corresponding text information and stereoscopic landscape images appear on the screen, so that visitors can put the puzzle together accordingly. During the game, visitors learn about folk culture and characteristics of various regions in China, while also experiencing a form of patriotic entertainment.

Taken together, the design concepts of the two exhibitions include the following:

1) Firstly, visitors' interest is aroused by combining augmented reality technology, real-time interaction, and 3D graphics, leading to an intuitive appreciation of augmented reality.
2) The conversion of text descriptions and 2D images into 3D images allows visitors to combine text and 3D graphics, combine specific observations with abstract symbols, and adapt to the cognitive laws of understanding, enabling access to a wealth of information. The "cone of experience" theory divides learning experience into three levels. The bottom level is the most specific; with more experience, it becomes more abstract. Education and teaching should begin with the most intuitive and specific experience, then slowly rise to the abstraction of conceptual understanding. The three levels of experience are not isolated from one another in pedagogical activities, but are interconnected and interactive. In augmented reality, designing virtual 3D objects allows learners to understand abstract language symbols via solid experience.
3) By identifying various cards, visitors can actively select objects of interest for observation, reflecting control over the display in a real-time interaction with the exhibit. During the process of interaction, the visitor gains instant feedback, making it easy to generate an immersive experience and prolong participation time.

Human-computer interaction is one of the most important components in the field of computer science, setting up a dialogue and communication between humans and computers using input and output devices. Computer technology has boomed in recent years, with people demanding increasing interactivity in human-computer relations. Natural, convenient, and cooperative human-computer interaction has become the public goal.[4] In order to stimulate visitors' interest and disseminate scientific knowledge, science and technology museums tend to focus on the audience, designing exhibitions that provide strongly interactive experiences.[7] The real-time interaction of augmented reality technology demands that displays present efficient, natural, and cohesive human-computer interaction.

2.3 *Suggestions for Improvement*

2.3.1 Drawbacks

In these two exhibitions, fixed augmented reality largely caters to scientific knowledge, but some shortcomings still exist regarding specific details.

1) At the China Science and Technology Museum, the fixed AR system is easy to implement and maintain, and involves only low network service requirements, as well as being a good display aid. However, this technology is only used by temporary exhibitions. As it is not within an exhibition hall, it is like an island, and is thus slightly inferior to mobile AR devices that can be used more freely.
2) Before attending an exhibition, visitors could be distracted by the AR technology portal, interested in the fusion of virtual and real space to an extent that neglects other intellectual content, resulting in an intuitive understanding only of the AR scenario. In *The Exploitation of the Works of Nature*, for example, people can simply look at objects on the on-screen pages, observing how they appear with the flip of each page, enjoying the entertainment without paying attention to the words. In this way, visitors simply observe 3D objects without learning about more abstract concepts, meaning that the full process of learning can be interrupted. This not ideal for popular science. Secondly, visitors are invited to choose viewing content freely in order to reflect real-time human-computer interaction, yet this engagement alone is low-level, involving a simple and repetitive process by which visitors simply choose markers and witness a screen display response, without necessarily engaging with the information provided. Following a primary engagement with AR technology, if exhibitions have no other attractive features, sustained interest is sure to be reduced.
3) The ideal goal of AR technology is to mobilize multiple viewer senses, yet these two exhibitions primarily engage vision, that is, visitors are offered virtual 3D objects superimposed in real space. Although experiments have confirmed that vision engages more than 80% of perceived information, it is not ideal to stimulate such a broad sensory experience in a quest to build audience knowledge.

2.3.2 Planned improvements

1) In order to popularize AR technology and its applications, a themed exhibition can be designed, introducing the developmental history, principles, and application of the technology. This can satisfy visitors' curiosity regarding advancements, while potentially stimulating ideas for future AR applications, so that AR can better serve the public. If AR technology is approached as an aid to enhancing and expanding the real world, we must work to engage visitors' attention in the themes of exhibitions rather than simply devices. For example, in *The Exploitation of the Works of Nature*,

markers could be added that allow interaction with ancient artefacts, or visual design principles could be created for operative demonstration, attracting visitors' attention away from AR technology and towards the exhibition's theme. This would avoid the risk of visitors moving to the next display without obtaining any new knowledge.

2) These two exhibitions involve a simple superimposition of virtual graphics over markers. However, if virtual objects were laid upon other real objects, and these real objects were thus promoted and enhanced, the effect of mutual fusion would be greatly enhanced. However, there are two main challenges here. First, the market must be attached to the real object, or the real object itself must be the marker. Second, there must be a seamless connection between virtual and real objects so that visitors do not feel any sense of violation.
3) Making full use of AR technology visualization characteristics, we intend to create a spatial learning environment that uses 3D effects to attract visitors' attention and participation. "Enhanced Chemical Reactions" is an augmented reality tool created to assist in the teaching of basic chemistry. It mimics molecules and molecular reactions, and is a good example of how AR can be used to present abstract structures that students cannot see. Compared with real molecular models, this technological example does not cause damage or require a specific environment. Another example is the use of superimposition to display magnetic field distribution, visualizing virtual magnetic fields over a real magnet, with virtual 3D graphics appearing in the real world. In the interactive process, learners are first curious about the phenomena, and then engaged by the entertaining models. They experience psychological change and are encouraged to interact with the virtual world in order to gain an understanding of certain phenomena.

3. Summary and Outlook

This paper has described two exhibitions at the China Science and Technology Museum that utilize fixed augmented reality technology applications, analyzing the design concepts of these exhibitions and their weaknesses, as well as putting forward suggestions for improvements. AR technology is a concept that integrates virtual and real space, seamlessly connecting the two worlds with certain advantages over virtual reality technology in enhanced human perception. When applied at the Science and Technology Museum, there should be a good integration of knowledge, exhibits, and technology, not merely the mechanical application of technology. Of course, some technical issues still exist, such as target recognition, single human-computer interaction, and small display size. However, given the continued development of computer technology, AR is sure to progress towards perfection.

References

[1] Li Ziyang. 'The Application of Interactive Media Technology in Childhood Education—French Science and Technology Festival Interactive Augmented Reality Design as an Example', [J] *Art, Science and Technology*, 2013, 10: 381–382.

[2] Reinhold, Behringer and Ronald Azuma, Yohan Baillot, et al. 'Recent Advances in Augmented Reality', *IEEE Computer Graphics and Applications*, 2001, 21 (6): 34–47.

[3] Kipper, Gregory and Joseph Rampolla. *Introduction to Augmented Reality Technology*. Translated by Zheng Yi. Beijing: National Defence Industry Press, 2014, pp.18–19, 25.

[4] Du Fengyi. *Augmented reality applications in cultural artefact display* [D]. University of Electronic Science and Technology of China, 2009.

[5] Chong Tian, Liu Yue, Hu Xiaoming. 'Research on Key Technologies and Applications of Outdoor Augmented Reality Systems', [J] *Journal of System Simulation*, 2003, 4: 329:337.

[6] Zuo Jianjun, Liu Shiqing. 'Analysis of the Graphic Context Relationships in "Tiangong Kaiwu" for Multimedia Education Purposes', [J] *Journal of Nanjing Xiaozhuang University*, 2010, 03: 117–120.

[7] Liao Xiayan. *Design and Integration of Human Science and Technology—Human-Computer Interaction Design and Application in Science and Technology Museums* [A]. Chinese Digital Creative Annual 2012 [C]. Chongqing: Chongqing University Press, 2013, pp. 310–311.

Creating a Virtual Experience and Innovating the Display Model: The Construction of the Guihai Beilin Museum's Virtual Landscape Tour

Wei Weineng
(Guilin Museum, Guilin, Guangxi 541001, China)

Abstract: Digital virtual technology has become an important mode for museum display, enhanced services, and the extension of cultural knowledge. The Guihai Beilin Museum's virtual landscape tour uses high technology (3D digital technology, multimedia, and virtual reality) to display the excavation of the stone carvings hidden in Guihai's beautiful landscape, increasing interactivity and offering visitors a unique and entertaining cultural experience. By exploring the museum's cultural displays and service methods, increasing numbers of people will be attracted to the museum and engage with museum culture.

Keywords: Guilin stone inscription; landscape tour; virtual experience

Virtual reality (VR) technology is a new and comprehensive branch of information technology that emerged at the end of the 20th century. It uses computer technology to generate realistic three-dimensional visual and aural sensations, enabling users to experience and interact with virtual space and time naturally using appropriate devices and passwords. The technology integrates the latest developments in computer graphic technology (CG), computer simulation technology, artificial intelligence, sensory technology, display technology, and network-based parallel processing. It is a high-tech simulation system that completely generated by computer technology.[1] In a nutshell, virtual reality is a way for individuals to visualize and interact with complex data via computers, and a technical mode by which to display entities and tangible bodies.

What is an experience? Pine & Gilmore, the co-founders of US company Strategic Horizons LLP, believe that "an experience is a beautiful sensation that occurs in a person's consciousness when one is engaged at a certain emotional, physical, intellectual, or even spiritual level".[2] In short, people who join the virtual environment via passwords and settings are invited to have a real physical and mental experience.

Nowadays, virtual technology is used widely in museum exhibits and interactive experience projects as a way to extend the cultural implications attached to museum collections. It allows visitors to engage and participate in experiences that deepen their understanding of artefacts and their own cultures.

Josh Basseches, former deputy director of the Peabody Essex Museum in the United States, states: "Good building facilities plus good collections equal a good museum. This is a common rule of the museum community. However, if you want to turn a good museum into a great museum, then you must also add strong visitor experience to the equation".[3] In regard to collections and cultural experiences, visitors' needs present new concepts, new challenges, and higher requirements for museum exhibitions and services.

1. Virtual display of a natural landscape

There are more than 2,000 stone carvings in Guilin, including cliff carvings, statues, and inscriptions. The cliff carvings are mainly distributed throughout the urban area in more than 30 famous cave dwelling areas such as the Longyin Caves, Fupo Hill, Duxiu (Solitary Beauty) Peak, and Diecai (Piled Silk) Hill. They include a wide variety of inscriptions that are rich in content and huge in size. They are the most characteristic features of Guilin's landscape, recording its development and history, as well as the aesthetics of mountains and rivers. The inscriptions contain paintings, poems, and calligraphy, which are important aspects of the city of Guilin thanks to their historical, scientific, and artistic value.

The Guilin Beilin Museum is an important window into this stone carving culture. Given the wide distribution of the carvings and their inseparability from the natural environment, it is difficult for people to fully engage with and understand them. A physical exhibition hall cannot accommodate them, so an important part of the museum's work is figuring out how display and share this culture.

Over the years, the Guilin Beilin Museum has explored new ways and means of displaying the stone carvings in full, in order to maximize the scientific utility of its cultural resources. In order to spread this precious cultural heritage more widely, and to promote the integration of culture and technology that is necessary for modern museum culture and service, Guilin Beilin Museum has created a "virtual landscape tour" experience hall, with auxiliary display projects such as an in-situ display of the Longyin

Cave cliff inscriptions, a small stone carving park, and *Sounds of the Landscape — An Exhibition of Guilin Stone Carving.*

Focusing on creating an interactive experience for the appreciation of ancient culture, high technology (3D digital technology, multimedia, and virtual technology) is used in a series of interactive activities, including simulated tours, games, compositions, music, drinks (tea and wine), inscriptions, engraving, and monuments, allowing visitors to experience ancient Chinese rituals. Interactivity allows for participation and the experiencing of certain aspects in person. Visitors can touch ancient inscriptions on the landscape, gaining cultural enjoyment and entertainment. Here, the Guilin Beilin Museum's goal is to promote education, the inheritance of culture, and patriotic leisure activities. To this end, it integrates conservation, research, exhibitions, tourism, culture, and leisure.

2. *Initial plans for the "Virtual Landscape Tour"*

Stone carving in Guilin stems from a love of natural landscapes. Through development, exposure, and appraisal by the Tang and Song literati, a deeper humanistic meaning was built, forming a "world class" appreciation of natural landscapes. As such, the origins and development of the Guilin stone carvings are inseparable from the culture of landscape appreciation of the ancient literati. These aspects are mutually dependent and symbiotic. Thus, the content of the experience hall is displayed in three parts: 'Beautiful Landscapes of Guilin', 'Virtual Dragon Call in a Deserted Valley', and the 'Ancient Literati Landscape Appreciation Tour'. All three use high-tech means to display the activities of ancient people and the origins of stone carving, raising audience participation through integrating high-tech products for lighting, sound effects, and visuals.

As its name suggests, 'Beautiful Landscapes of Guilin' largely displays attractive landscapes. As its textual content suggests, "the beautiful scenery of Guilin's mountains and the majesty of its caves were carriers for the emotions of the literati, inspiring scholars and bureaucrats in developing and expressing their appraisals, adding a humanistic element." Guilin's stone inscriptions have made it famous throughout China, as affirmed by the sentiments "Guilin's landscape tops the world", and "Guilin's name tops the list in the Tang and Song Dynasties inscriptions".

This project involves four sets of projection screens and 3D model scenes (landscapes and scenery), offering interpretations of the Guilin urban landscape in spring, summer, autumn, and winter respectively. There are simulations of misty spring rain, summer thunder and lightning, autumn leaves, and winter snow, as well as images of fisherman fishing and literati touring rivers, with sound effects of birds, *guqin*, wind and rain. This combination of 3D images and landscapes offers an unobstructed recreation of the Guilin landscape. As a virtual simulation tour, it unites visitors with the ancient people of Guilin with a strong atmospheric effect, beautiful images, and 3D visuals (see Figure 1).

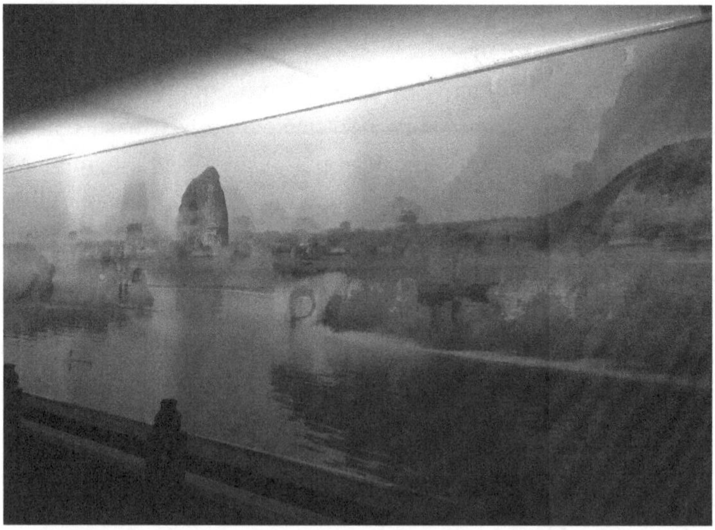

Figure 1 'Beautiful Landscapes of Guilin'

The second component is the 'Virtual Dragon Call in a Deserted Valley', featuring content based on the legend of Longyin (Hidden Dragon) Cave. This is a perfect combination of natural and human landscapes. In the Song Dynasty, Zhang Tian wrote in his *Longyindong Poem*: "Challenge not the dragon hiding in the dark, for his spring captures one by surprise. Kong Ming's long seclusion builds character; Lao Zi's sudden flight takes him to the sky".[4] Ancient literati were inspired by the reclusiveness of the dragon, believing that it was cultivating its soul while in seclusion, and waiting for opportunities to take flight, so as to realize its aspirations. A person with talent was often referred to as a dragon.

3D projection screens are used to explain the origins and legends of the Longyin Cave, complete with a virtual dragon. When summoned by visitors, the dragon emerges from the cave. It smashes through the wall before flying overhead and disappearing into the third scene (a landscape appreciation by the ancient literati). The scene includes interactive content: when visitors shout into the cave at a certain decibel, the dragon wakes. It spirals of the cave into the sky, and finally disappears into the audience.

The third segment is the 'Ancient Literati Landscape Appreciation Tour'. People of the Song Dynasty upheld acts of exposition, such as the scenes reflected in Li Gonglin's *Elegant Gathering in the Western Garden*, and the inscriptions left in large numbers of Guilin stone carvings. These elements hint at the rich and elegant lives of Song scholars, like the *Longyinyan Trip Inscription of Zhao Shaoyin and his Four Friends*, where details like the day's itinerary, activities, moods, and scenes were recorded. Shaoyin had dressed for the feast he had prepared for his guests Gongbi, Jingyuan, Mingfu and Shiju. The meal was followed by a boat ride to Longyin Rock, where they drank spring water from the rocks and cooked aquatic plants. In a casual mood, they stayed until sunset, and went home the same way they came. It was dated the fifth year of Shaoxing, on the 10th day of the third month.[5] These records not only portrayed the social life of their day, but also acted as a way for the literati to express artistic conceptions, thoughts, and emotions about the natural landscape in the form of stone carvings. A large number of landscape carvings were thus born.

This exhibit involves a simulated scene on a 7.4m-wide and 3.8m-high arc screen. Based on Li Gonglin's *Elegant Gathering in the Western Garden*, the 3D projection technology displays interactive ancient literati activities, including smelling scents, tasting tea, drinking, painting, playing stringed instruments, and carving wall inscriptions. The interactive session requires participants to continue the line "looking at mountains is like looking at a painting" from Chen Yuanlong's poem, *A Trip to the Longyin Cave*, from the Qing Dynasty. The correct answer is "a trip to the mountains is like reading history". With speech recognition and interactive technology, smoke appears following the correct answer, accompanied by the scent of aromatic incense and the sound of the *guqin* (see Figure 2).

Through the virtual display of the three scenes, visitors can experience ancient tours and the composition of poetry in Guilin, as well as sentimental attachments to its aesthetics, further deepening their understanding of stone carving.

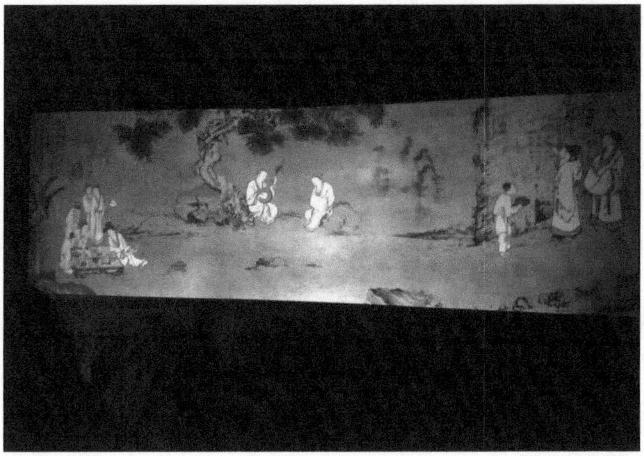

Figure 2

3. *Significance and Development*

The Guihai Beilin Museum is surrounded by mountains and rivers, with beautiful scenery and environments that are rich in culture. The stone carvings in this area are unique in terms of characteristics, with the historic Longyin Cave inscriptions closely integrated with the natural environment,. The value of the landscape is a huge draw for tourists, making this area the most historically and culturally rich attraction in Guilin. Meanwhile, the museum shoulders the responsibility of education, so communications and display are important methods by which it can exert its social functions. Traditional display methods are irreplaceable to a certain extent, but with constraints of time and space, they can also hinder cultural communication. A variety of computer-, network-, and multimedia-based technologies can be used in digital displays to create vivid images and exceed the space and time boundaries. This way, more people will be attracted to history and culture.

The creation of the 'Ancient Literati Landscape Appreciation Tour' demonstrates an attempt to advance the means and level of museum display technology. The goal is to display Guilin's stone carvings while providing a high-grade leisure experience to visitors. The tourism sector is conducive to the establishment of a historical and cultural brand for Guilin. It must meet growing needs for cultural display and information dissemination, and be suitable to today's rapid development of cultural tourism. This is of great significance for the promotion of tourism in Guilin.

The stone carving viewing garden and virtual experience offers an innovative, fresh, and attractive arena for students and the public to participate in patriotism, local history, culture, education, and art. It is an important window for the display of Guilin's history and culture, making a major contribution to these fields.

Cultural services and exhibitions are important components of the digital museum. High-tech multimedia equipment (such as virtual 3D display technology) provides visitors with a more comprehensive, detailed, and vivid experience of cultural artefacts and displays, as well as new ways to learn and appreciate culture. These rich, stimulating, and attractive displays can inspire more individuals to visit cultural sites.

Just as Jack Ma, Chair of the Board of Directors for the Alibaba Group, predicted when he suggested that "physical stores will never die, but online stores represent the trend of today and tomorrow", the broad development prospects for digital virtual display represent the future for museums, bringing forth social development in the presentation of richer cultural information as the main force of social education.

References

[1] http://baike.haosou.com/doc/6982730-7205460.html.

[2] http://www.ectime.com.cn/Emag.aspx?titleid=658.

[3] http://shizheng.xilu.com/20140709/1000150002561180.html.

[4] Guilin Cultural Relic Management Committee. Guilin Stone Carving (I). Internal Information, 1977: 110.

[5] Guilin Cultural Relic Management Committee. Guilin Stone Carving (I). Internal Information, 1977: 127.

Using a VRay Renderer to Enhance Light and Shadow Effects in the "Juan Qin Zhai" VR Program

Dong Gangying
(Ministry of Data and Information, National Palace Museum, Beijing 100009, China)

Abstract: "Juan Qin Zhai" (Lodge of Retirement) – a virtual reality program released by the Palace Museum in 2010 – entails an initial attempt at a three-dimensional reconstruction of two-dimensional data in a *trompe l'oeil* reconstruction. In 2014, the Lodge of Retirement VR improvement project worked to edit and sort through original scene data, using VRay rendering technology to render a new light and shadow scene map, enhancing the original. The visual effect of the VR program has brought cultural artefacts to life, realizing the digital 3D display of 2D cultural forms.

Keywords: light and shadow enhancement; VRay renderer; lightmap texture

1. Purpose of the Project

"The Forbidden City — Lodge of Retirement" (hereinafter referred to as the Lodge of Retirement) was a virtual reality program released by the Institute for the Research of Digital Asset Application of the Palace Museum in 2010, recording the status of repairs to the eponymous Lodge of Retirement. The work is bold and innovative in concept, staging the virtual 3Dization of 2D data from a landscape painting, while displaying an accurate, panoramic view of the interior decor of the lodge. Virtual reality technology is used to reproduce Emperor Qianlong's fantasy vision of a Jiangnan retirement lodge, using a digital creation program to "enliven" ancient structures.

Limited by the technology and equipment of the time, the model data was compiled by CG producers. The relationships between the models were manually drawn, with room for further improvement in their completion, and the coordination of their light-shadow relation.

After four years, updated equipment and improvements in light and shadow technology allowed the Institute for the Research of Digital Asset Application of the Palace Museum to establish a VR light and shadow improvement project of the Lodge of Retirement in 2014. Based on data from the 3D scene program, the VRay advanced light and shadow renderer plug-in from the 3dsMax software has been used to render light and shadow operations on the two 3D model scenes, enhancing the overall visual effect of the program.

2. *Assessment of Existing Data Resources*

Operations for the rendering of light and shadow require a significant amount of manpower and time. Once rendering work is begun, it is almost impossible to rework. As such, it is necessary to make sufficient preliminary preparations, sorting existing model data and classifying and evaluating content and model accuracy so as to ensure the effective execution of subsequent light and shadow rendering.

2.1 Analysis of scene data

The Lodge of Retirement VR program consists of four 3D scene modes. Within the data manager, these scenes are labelled Jqz01, Jqz02, Jqz03, and KK25_jqz. Jqz01 is the main core scene in the program, including indoor and outdoor models of the lodge structure as well as exterior models of the Fuwang Pavillion, Zhuxiang Hall, and the courtyard environment that can be seen from inside. These are A-grade, ultra-high precision 3D models.

Jqz02 entails scenes of the internal structure, excluding the roof, which is in two parts: the Eastern Hall and Western Hall. These scenes are used to depict the structure of the lodge, and are simplified versions of the Jqz01 indoor models, standing as B-grade high-precision models.

Jqz03 is a hypothetical scene of three panorama landscape paintings within the Western Hall of the Lodge of Retirement. The Western Hall is divided into three parts: the northern wall, the western wall, and the roof. The scenery in the paintings is combined with interior decor, blending surrounding walls to present a full springtime landscape. Here, the program uses VR technology to render 3D buildings, flowers, birds, mountains, and streams from the landscape paintings in a logical layout based on perspective and distance, presenting a courtyard in the full bloom of spring. All the models within this scene are A-grade ultra-high precision models.

KK25_jqz is a large scene model of the Forbidden City. The scene involves a snow effect, used in the opening overview. The models here are composed of three-level A-, B-, and C-grade precision models.

2.2 Data volume analysis

Table 1 shows the data amounts, the number of Polygon surfaces, map data amounts, and number of maps within the four modes of the VR Lodge of Retirement program.

Table 1 Data volume, Polygon surface number, mapping data volume, and number of mapping sheets of the four major model scenarios in the VR Lodge of Retirement program

Scene name	*Data volume of scenarios*	*Polygon surface number*	*Mapping data volume*	*Number of mapping sheets*
Jqz01	177 MB	646,028	2.79 GB	885
Jqz02	98.6 MB	375,452	5.84 GB	471
Jqz03	268 MB	1,083,520	1.87 GB	650
KK25_jqz	339 MB	934,672	1.77 GB	1663

2.3 Analysis of the scene switching relationship

According to the plans and requirements of the Lodge of Retirement program script, viewpoint paths are switched and displayed in the scenes across all four modes.

The program begins with the large KK25_jqz Forbidden City scene. With Jqz01 at the core, the program then depicts the interior and exterior architecture of the Lodge of Retirement in depth, interspersed with Jqz02 scenes featuring a 360° rotation showing the indoor structure, providing a deep analysis of the architecture of the Eastern and Western Halls. Then, to the interior decor of the Eastern Hall and the stage of the Western Hall, and on the hidden stage in Jqz01, the full scene of the three panoramic paintings is displayed. This leads to Jqz03 – a fantasy exploration of these three hypothetical landscape paintings, restoring Emperor Qianlong's vision. Finally, returning to the Jqz01 scene, the program exits the Lodge of Retirement and reaches its conclusion.

Figure 1 shows the scene switching sequence.

The Jqz01 architectural scenario is applied repeatedly as a core scene within the program, with an A-grade degree of accuracy. Jqz02 scenes are used for the interior

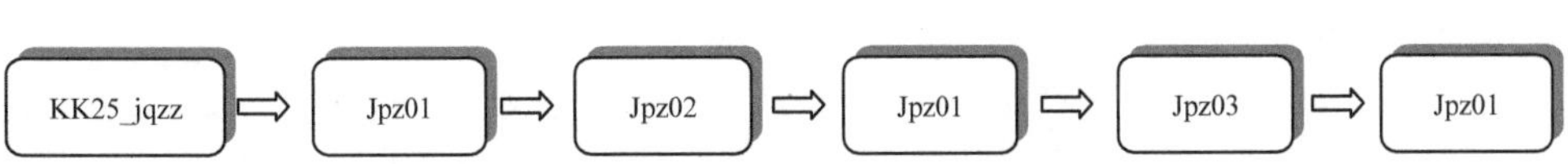

Figure 1 The scene switching sequence

building environment, with B-grade accuracy. The imaginary Jqz03 scene is a core highlight of the program, with A-grade accuracy. KK25_jqz features background information for the program, with data accuracy inter-mixes A-, B-, and C-grade scenes. Following careful data analysis regarding the rendering of light and shadow, enhancement work was applied largely on Jpz01 and Jpzo3 scenes.

3. VRay renderer

To complete light and shadow rendering work, we compared four rendering programs: VRay, Brazil, Lightscape, and Renderman. Each of these renderers have strengths, but VRay is more efficient, flexible, and easy to use. As such, the project selected the 3ds Max VRay plugin.

The VRay software is produced by Chaos Group and ASGVIS. In China, a high-quality rendering program is promoted by Manheng. VRay is currently the most popular rendering engine in the industry, offering a special material called VRayMtl, which can be used to achieve more accurate physical illumination, faster rendering, and more straightforward reflection adjustment and refraction parameters within scenes. VRayMtl has the capacity to apply different texture maps, control reflection and refraction, add transition and displacement maps, make direct global illumination calculations, and select bidirectional reflection distribution functions.

VRay combines ray tracing and radiosity, with a real light calculation that produces professional lighting effects to accurately meet the various lighting needs of Jqz01 scenes featuring the indoor lobby and Western Hall of the Lodge of Retirement. VRay also has a reputation as the 'king of caustics', with the capacity to perfectly represent the true outdoor effects of the Sun, mountain streams, and courtyard atmosphere within the Jqz03 vista.

Lightmaps generated by the VRay renderer are inversely attached to existing VR scenes, making it possible to improve light and shadow effects within the VR program.

4. Aims of the Project

This project intends to generate a new version of the data model with modifications to the original 3D models in the VR Lodge of Retirement program. To this end, it aims to remove hand-painted light and shadow, creating clean and organized data with realistic, soft, and unobtrusive light and shadow effects. Specific project quality requirements are shown in Table 2.

Table 2 Specific project quality requirements

Name of scenario	*Light and shadow effect requirements*
Jpz01 East Hall	Modify the model, remove hand-painted light and shadow, interior light is authentic, clean and soft, highlight steps and key cultural relics by filling light into the area
Jpz01 West Hall	Modify the model, remove hand-painted light and shadow, interior light is authentic, clean and soft, highlight key cultural relics such as landscape paintings and stage by filling light
Jpz03 Hypothetical scenario	Modify the model, remove hand-painted light and shadow, realize immersive effect of outdoor sunlight

5. The Production Process

Light and shadow rendering work must be carried out in stages. First, existing data is sorted, modifying any discrepancies in models and textures. Once the data is complete, scene lighting is set, object sensitivity parameters are adjusted, and the rendering work is performed. Finally, according to the requirements of the program, the scenes are reorganized, splitting old and new versions of data for later debugging. The specifics of this process are as follows:

5.1 Modifying the original data

1) Modifying the model
Original 2010 data was almost perfect within the contemporaneous production environment. In order to ensure high precision in the map and model, the architectural and cultural artefacts included in the scene were first assembled by CG production staff, with large-scale elements such as walls and floors inserted separately in surface units. Although the data was sharp, the integrity was poor.

The VRay renderer has high requirements for the integrity of 3D models, demanding that building models be structurally intact and free of gaps, with the ground and walls integrated without redundant intersections. The structure of cultural artefacts must be complete, featuring full artefacts without unfinished tops and bottoms, and fully opaque.

The modification of the models is the top priority in the rendering of light and shadows. Only by ensuring that all models are complete, seamless, and free of light can correct Lightmap textures be obtained after rendering.

2) Modifying the map
Original light and shadow maps between objects and buildings within the scenes were hand-drawn by CG producers, with light and shadow information directly applied to the texture map of the object's surface.

Ultimately, the goal of VRay rendering is to generate Lightmap textures of light and shadow based on the environment around the object, then map these textures onto the original object, and finally superimpose and generate objects with better light and shadow information. As such, the texture map of the object must be a blank canvas for light and shadow information. It is necessary to remove hand-painted light and shadow information on the original texture before rendering work begins, restoring the texture mapping of objects.

5.2 Light settings

Lights in the Eastern and Western Halls of the Jqz01 scene are set for global illumination. In each compartment, different local light source points are set according to the position of cultural artefacts on the *kang*, with key cultural artefacts highlighted with light fitting methods.

In the Jqz03 scenes, an imitation of the Sun and spring light in the imaginary scene is achieved by adjusting lighting, sunlight, and shadow parameters. The sunlight parameters are set to the blooming of the wisteria flower in Jiangnan at 11am on May 1st.

5.3 Object sensing parameter settings

Each object in the scene must have unique sensitivity properties, set one by one according to the various materials of the cultural artefact and the building.

1) Key object parameter settings for Jzq01
In the Jqz01 scene, the most common objects are ornamental doors, bifacial embroidery, *kang* seating, *kang* tables, wallpaper, and gold bricks. Following careful classification, an object's attributes are analyzed, and several experiments take place to determine parameter settings.

- Ornamental doors are mostly made of a matte wood material, so the diffuse reflection is set to wood grain material, with a reflectivity of 40, a highlight glossiness (HGlossiness) of 0.65, and reflection glossiness (RGlossiness) between 0.7 and 0.8, with bump of 15% for the wood grain texture.
- Bifacial embroidery involves cloth, a surface with minimal roughness, low reflectivity, both smoothness and roughness, diffuse reflection set to cloth texture, a reflection degree of 10, locked HGlossiness, and RGlossiness set to 1, with associated diffuse roughness maps.
- Most of the *kang* tables are composed of glossy, varnished wood. The diffuse reflection is set to wood grain with a 55 reflectivity, a HGlossiness of -0.8, and an RGlossiness of 1.
- *Kang* seating and seat cushions are largely made of silk, so diffuse reflection is set to cloth material, color is a default value, reflection is 20, HGlossiness is 7, the

surface roughness is 8, and the brightness is set to 0, 5, and 100.
- Wallpaper is made of paper, so the diffuse reflection is set to wallpaper material, with a reflection of 20, locked HGlossiness, RGlossiness of 0.5, maximum depth of 1, and no light tracking.
- Gold bricks are mirrored stones, with a smooth and reflective surface and low highlights. The diffuse reflection is set to stone texture, with a reflection of 30, a HGlossiness of 0.8, and an RGlossiness value of 1.

2) Key object parameter settings for Jqz03
In the Jqz03 scene, the most common materials are bamboo fencing, building tiles, rockeries, vegetation, and water. The following key parameters have been determined for these objects:
- Bamboo fencing is made of glossy, varnished wood with a smooth surface and a low level of reflection. The diffuse reflection is set to wood grain with a reflection of 55, a HGlossiness of –0.8, and an RGlossiness of 1.
- Glazed tiles made of porcelain have a smooth surface and bright highlights. The diffuse reflection is set to porcelain materials with attenuated reflection and HGlossiness at 0.9.
- Vegetation and green space are standard plant material, with diffuse reflections set to plant color, and a glossiness of 50, fine-tuned according to different forms of vegetation.
- Rocks have smooth surfaces, with some irregularities and low levels of highlighting. The diffuse reflection is set to stone texture, with a reflection value of 30 and an RGlossiness of 1.
- The waterfall, formed of water, is set to a black diffusion, with a reflection of 255, attenuated Fresnel reflections, a refraction of 255, and a refractive index of 1.33. The fog color is light cyan and the fog multiplier is at 0.01. The bump map is noisy with a bump of 25, fine-tuned according to the fluctuation of the water surface.

5.4 Inverse Lightmap mapping

Upon completion of preliminary preparations comes the final rendering stage. Formal light and shadow rendering work with the VRay renderer is now begun, generating a Lightmap light and shadow map of the object. These Lightmap textures are inversely applied to the modified scene data, ultimately obtaining scene data with real light and shadow effects.

5.5 Scene reorganization

Following this rendering work, the new data versions are inserted into the old program platform's debugging system to ensure seamless integration of old and new data. In this program, the stage area of the Western Hall within the Jqz01 scene involves a candlelit illumination effect, and object models following VRay rendering were not affected by this light. In order to ensure the integrity of the program, old and new scene models were thus reorganized, dividing the Eastern and Western Halls into two sections at the nine-minute demarcation point in the program, so that the Eastern Hall could take up new data while the Western Hall retained original data (without Lightmap), then merging and exporting the two versions – new east with new west scene data, and new east with old west scene data. Finally, the program was debugged to generate a new finished version of the VR program.

Figure 1 and Figure 2 are examples of data rendering comparisons in which before and after scene light effects have been improved within the program.

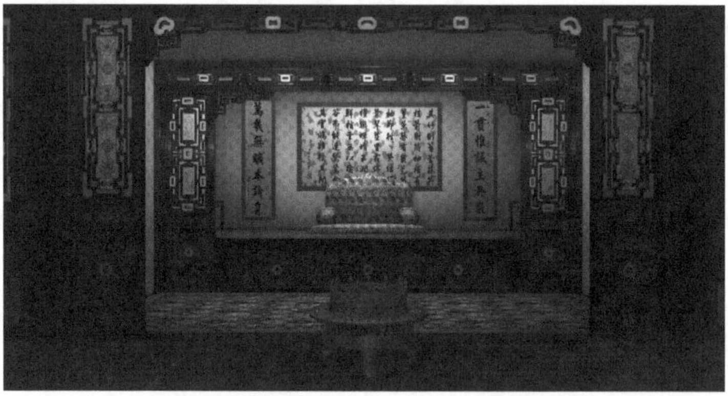

Figure 1 Original data from the Eastern Hall

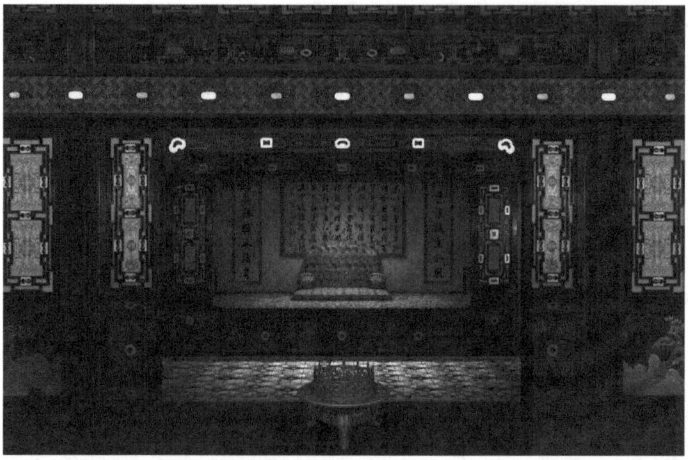

Figure 2 Light and shadow data from the Eastern Hall after upgrading

6. *Summary of VRay rendering VRay*

Given the actual manifestation of the light and shadow improvement project for the Lodge of Retirement VR Program, this section sums up the lessons learned and the notes taken about the production of Lightmap texture, which can be used as a reference point for similar projects.

6.1 Ensuring model integrity

Before rendering begins, models in the scene must be examined carefully so as to ensure the integrity of "baked" objects. In order to control the amount of data involved in existing VR programs, some objects beyond the view path (as well as some B-grade and C-grade artefacts) can be replaced by low-precision models or Polygon figures. Before "baking" objects, one must first assess the object scene, replacing low-precision models and Polygon elements with solid models. As light is highly penetrating, "baked" objects must comprise only complete models with no gaps. Building models must also be intact, with no gaps within the building, or between the building and the ground. This will guarantee that there is no light transmission during rendering.

6.2 Setting the required light source

The types and properties of light sources must be designed in advance so as to avoid reworking.

According to the needs of specific VR programs, given the varying lighting effects in different scenes, outdoor scenes can be taken directly in Beijing according to the corresponding dates and seasons within the 12 months, selecting a specific time of day to determine the direction of sunlight. Parallel light is generally used for interiors, with light source points provided to achieve partial lighting of focus objects.

When using the VRay renderer to apply a light and shadow map, model objects will affect one another, and projected shadows will be superimposed on one another in the final Lightmap of the scene.

6.3 Steps in VRay light and shadow rendering

Following the early stages, the VRay renderer is used to begin actual rendering work. The specific steps are as follows:

Step 1: Use the 3dsMax plugin to add new UV coordinates (UV2) to scene models.

Step 2: Divide the scene according to the size of the area (the specific area depends on specific needs, where objects requiring special display can be grouped separately); each small scene is grouped separately, and the Script plugin is used to combine elements and form single objects. Split scenes are saved separately for light and shadow work, keep raw data area separated.

Step 3: Use the Script plugin to turn the object material into dedicated VRay material.

Step 4: Use the VRay plugin to change the model export file format to .exr.

Step 5: Run the "baking" job and obtain Lightmap texture.

Step 6: Inversely apply the Lightmap texture in .exr format to the original scene model to create the final rendering model.

References

[1] Working Group for the Protection of the Lodge of Retirement at *the Palace Museum*. 'Report on the Protection of the Lodge of Retirement — Landscape painting portion', The Palace Museum, 2004, 1.

[2] Li Daming. *Southern Affection of Past Qing Emperors*. Forbidden City, 2006 (11, 12).

[3] Ouyang Hong. *The virtual reproduction of the Lodge of Retirement — Creative Technology Boosts the Digital Museum*. Communication University of China (CUC) Publishing House, 2012 (1st edition), pp. 297–301.

[4] Wang Yumei, Zhang Bo. *3ds Max + VRay renderings, from entry to mastery*. Posts & Telecom Press Co., Ltd., 2010.

[5] Tang Lianxi. *Super Rendering King: 3ds Max/VRay material light and shadow rendering performance techniques*. Posts & Telecom Press Co., Ltd., 2014.

[6] VRay material parameter references. Baidu Literature, http://wenku.baidu.com/view/e47f3b7602768e9951e73872.html

[7] VRay encyclopedia entry. Baidu Literature, http://baike.baidu.com/view/750173.htm.

Discussing the Digital Display of the Paintings and Calligraphy Collection at the China Agricultural Museum

Ning Gang (mtzl@ciae.com.cn), **Chen Ying, and Li Shuangjiang**
(China Agricultural Museum Information Center, Beijing 100026, China)

Abstract: The digital display of collections is an important innovation in the business, technology, and service of museums. This article details the content and progress of the digital display of paintings and calligraphy at the China Agricultural Museum.

Keywords: painting and calligraphy collection; digitization; display; virtual exhibition

1. Digital Display and the Function of a Museum

A museum is a non-profit, permanent public institution that serves society and its development. Its chief functions include the collection, conservation, research, dissemination, and display of material and intangible artefacts and environments for the purposes of education, study, and appreciation. The museum is not only a place in which to collect and preserve cultural artefacts, but also has the important mission of promoting social development in the service of the public. The implementation of these features relies heavily on museum collections themselves. In order to serve the public more effectively, it is necessary to display collections as efficiently as possible for greater engagement with visitors. The China Agricultural Museum attaches great importance to the technological development and application of digital museum collections.

Traditional museums, also known as "physical museums", are limited by size, meaning that often only a small portion of their collection can be exhibited at any one time. The majority of museum artefacts are kept in storage, which is not an effective use of resources. A physical exhibition has many advantages, such as allowing visitors to appreciate cultural artefacts closely and intuitively on site. However, there are also obvious shortcomings. Besides a lack of space, opening hours also restrict freedom and the overall effect of the exhibition. Physical displays generally last for a prolonged period with the contents unchanged, meaning that visitors may see the same exhibits numerous times and therefore lose interest. Even with the opportunity to get close to a variety of precious cultural artefacts, physical exhibitions often lack interactivity and entertainment value. Meanwhile, from the perspective of safety, a physical exhibition is not conducive to preservation.

With the development and application of computer and network-based technology, digital museum display is becoming more prevalent. By means of various information technologies, and a lack of limits on time, region, and space, digital display can solve many of the issues encountered by traditional physical displays. Indeed, digital displays can be shared freely through the Internet at any time, and have a wide scope of regional dissemination, making it possible to display a much greater number of artefacts than a physical exhibition hall. Almost entirely free of space constraints, a greater number of cultural artefacts can be freed from collections. For the vast majority of people, the purpose of a museum visit is to appreciate cultural collections and be educated. Both of these needs can be fully met by digital displays. Not only does a digital display increase freedom of visitation, it also plays a role in the preservation of cultural artefacts, reducing potential damage. All in all, it is a useful supplement to tradition museum practices.

2. Features and Forms of Display for Painting and Calligraphy Collections

2.1 Characteristics of painting and calligraphy collections

Museums generally house a wide variety of cultural artefacts formed of different materials, in which paintings and calligraphy are an important category. In terms of materials, paintings and calligraphy are generally paper-based, and are therefore much more susceptible to damage than other collections. Slight negligence may lead to damage from pollution, mildew, and insects, and long-term preservation and multiple relocations are difficult. Paintings and calligraphy are two dimensional works with their content largely reflected in images. Therefore, unless being viewed for specific research purposes, it is usually only necessary to display the image itself to the audience. Current digital imaging technology has the capacity to closely reflect original images, appearing almost exactly like the real work. These characteristics suggest that paintings and calligraphy can be digitized to protect the physical originals, and are also well-suited to digital display.

2.2 *Forms of display*

The characteristics of painting and calligraphy collections determine a certain form of display without the need for complex digital techniques and equipment. The digitization of such collections, that is, the convergence of digital technology with paintings and calligraphy, involves the collection of visual information via photography and scanning. This information is input into computers in image form, and the processing of these images creates a virtual collection that can be displayed across various media. The China Agricultural Museum has a large collection of peasant paintings, agricultural posters, and New Year folk paintings. In recent years, it has carried out extensive exploration and practice in digital display, accumulating a certain level of experience. A collection of this kind is displayed in two main forms. The first is the standard Internet form of displaying a "photo + text description" online. The second is in a professionally designed virtual exhibition, which simulates the form of a traditional exhibition hall, in which a series of work is displayed.

The conventional "photo + text description" method is simple and clear, and the content of such exhibitions is relatively random. Professional processing is performed on the photo and a text description is added. Meanwhile, works can be displayed online with certain online effects, so that visitors can appreciate each piece at close range. The disadvantages of this display method are its relative blandness and a lack of flexibility. Without the addition of special effects and multimedia elements, visitors can become bored.

The virtual exhibition is a departure from conventional methods, combining computer technology with the collection itself so as to redesign and focus on a specific kind of work. It does not simply display work independently, but considers the beauty of design and artistry, as well as the logic of presentation and the integrity of the work. Displaying artefacts with embedded music and animation effects can be a novelty for visitors. This new exploration into digital practices has been embraced by the China Agricultural Museum in recent years. The following section entails a detailed description of the virtual exhibition.

3. *The Development of the Virtual Exhibition Hall*

The virtual exhibition is a comprehensive method of display that offers visitors a more flexible and vivid experience in comparison to the display of a single collection. Over the years, the China Agricultural Museum has carried out the design and production of two kinds of virtual exhibitions in the practice of digital display. The first virtual exhibition has been developed with a reliance on existing exhibition entities, that is, the physical exhibition is saved in digital form, or designed on the basis of the physical exhibition, then developed by relevant technical personnel. The second does not rely on physical exhibitions, but operates entirely on design and production by relevant personnel. Displays that do not rely on the physical exhibition are built on the basis of cultural collections, with no physical point of reference, making the process much more challenging and time-consuming. The creation of a purely virtual exhibition requires topic selection, script writing, interface design, digitization of the collection, and a series of post-production activities. The following section is about a hall of this nature.

3.1 The technical process

1) Topic selection
Based on the museum's collection and exhibitions, researchers select the theme and range of the virtual display.

2) Script writing
Without a corresponding physical exhibition for virtual display, this stage addresses the form in which the collection will be displayed, which collections will be displayed, and according to what logic. The script for the virtual exhibition is written by a display designer, including the specifics of the theme, collection, textual material, and multimedia elements.

3) Digitization of the collection and material processing
Naturally, virtual display is inseparable from digitization. For paintings and calligraphy, digitization largely involves photography, with professional photographers working to document and process images of the work.

4) Design
Good exhibition design highlights a theme, giving visitors a systematic sense of knowledge alongside aesthetic enjoyment. The design of a virtual exhibition includes interface design and exhibit display, acting as a blueprint for the production of the virtual display. It includes the design of a welcome page as well as the framework and structure of the display. This stage involves static interface design.

5) The integration of various design materials
Based on preliminary preparations, this stage involves processes of instantiating scripts, designing dynamics, and vivid visualization by professional technicians, including the production of buttons, adding links, and the inlay of content such as text, images, and music.

6) Performance display following modification, improvements, and debugging
The production of virtual exhibitions requires cooperation between various departments and professionals. Museum researchers, special designers, and professional production personnel must work together for maximum effect.

3.2 Means of developing a virtual exhibition

The virtual exhibition at the China Agricultural Museum has an online version and a local touch-screen version. In practice, the technical means for the development and application of these two forms are different, using mainly Flash and Authorware software.

Authorware is an interpretive, process-based graphic programming language used to develop interactive programs that integrate sound, text, graphics, simple animation, and digital cinema. It is an icon-oriented multimedia production tool, and does not require traditional computer language programming. The program can be edited by adjusting the icon sequence in a flow chart control model, bringing together various forms of multimedia data to achieve multimedia presentation. In practice, Authorware is largely used for local touch screens in virtual exhibitions.

Flash is an authoring software that combines the creation of animations with app development. Flash provides full authoring and editing features for the creation of digital animations, interactive web sites, and desktop and mobile app development. It can encompass simple animations, video content, complex applications, and anything in between. Rich media Flash applications can be built by adding images, sounds, videos, and special effects.

Extensible Markup Language (XML) is a language used to mark electronic files as structural elements. It can be used to tag data and define data types, and is a source language that allows users to define a unique markup language, making it ideal for web communications.

By implementing XML, Flash can quickly and easily obtain external text and images, turn external albums into dynamic elements with specific design effects, and realize interactive functions. A multimedia system produced by the combination of Flash and XML features small animation size, small file size, fast online delivery, and high interactivity.

4. Case Study: The Painting and Calligraphy Collection at the China Agricultural Museum

4.1 "Image + text" online display

The "image + text" display method is best suited to painting and calligraphy. Using it, the China Agricultural Museum has selected hundreds of peasant paintings, agricultural posters, and high-density New Year pictures for its official website. It includes webpages with simply processed photos in which the original paintings are presented to the viewer as closely as possible, alongside webpage effects and auxiliary text to display basic information, as shown in Figures 1 and 2. The works can be sorted by region or content. Clicking on the corresponding thumbnail to zoom in has the effect of showing a movie.

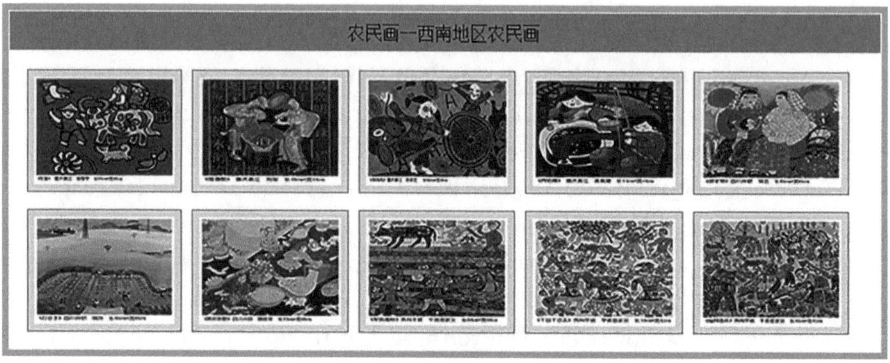

Figure 1 Peasant paintings from Southwest China

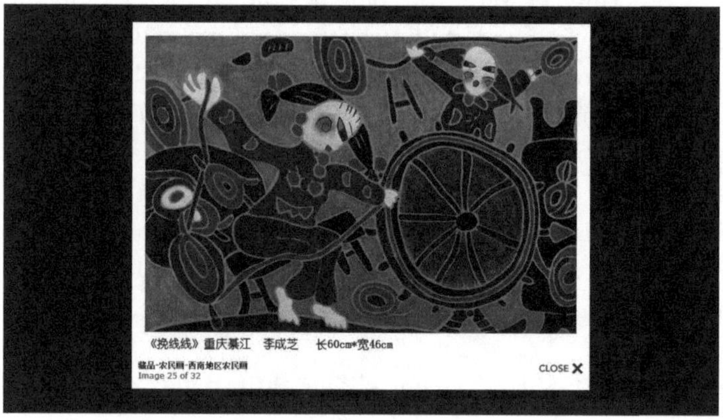

Figure 2 Yarn spinning

4.2 The virtual exhibition

Since 1998, the China Agricultural Museum has been experimenting with the digital display of collections and exhibitions. More than ten virtual exhibitions have since been produced based on physical exhibitions, and are linked on the official website in order to facilitate visits and expand the display forms of the museum. For over a decade, the China Agricultural Museum has been accumulating a wealth of experience, and continues to explore new forms of digital display. With special regard to painting collections, two exhibitions called *Memory Village – A Collection of Peasant Paintings* and *Unforgettable Memory – A Collection of Agricultural Posters* were launched in 2014 in conjunction with International Museum Day on May 18th. These two virtual exhibitions had no physical counterpart, and were realized through the use of computer technology.

Memory Village – A Collection of Peasant Paintings was the China Agricultural Museum's first virtual exhibition to be launched without also holding a physical exhibition. It displayed peasant paintings against the backdrop of rural walls. Content was categorized

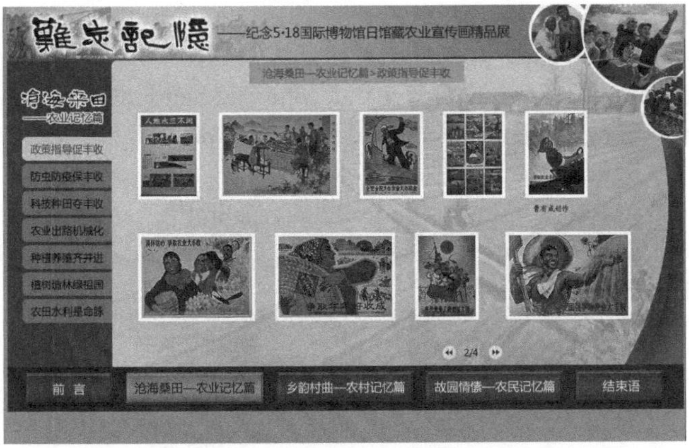

Figure 3 The online version of *Unforgettable Memory – A Collection of Agricultural Posters* (2014)

by region, creating a virtual showroom with both touch screens and online versions. The touch-screen version could be accessed anywhere. With only a small occupancy rate but a wealth of content, it was especially suited to the limited exhibition area and the resulting inability to display more pieces. In this virtual showroom, visitors were able to access a larger number of items using only a few touch screens. Meanwhile, the online version provided a more convenient way for visitors to browse and appreciate peasant paintings, increasing their interest.

The digital exhibition of agricultural posters was developed using Flash technology, involving a selection of more than 360 distinctive pieces from 1949 onwards, classified according to various themes. Visitors could choose to view any given poster, selecting from a multi-picture display containing more than 900 images. After clicking on a particularly work, they could view the details of any section, gaining even closer contact with the poster than would be possible on site.

From a structural point of view, a virtual exhibition can be divided into three parts: the preface, the main body, and the conclusion, with further subdivisions within the main body. The menu appears on the left at the bottom, with generally no more than two layers for the sake of brevity. The entire exhibition consists of multiple smaller exhibitions, categorized in secondary menus that are easy to develop and produce.

5. Summary

The digitization of museum collections is important for the protection of cultural artefacts. Moreover, the actual digital resources garnered in the digitization process also stand as important materials for digital display. Developing new forms of digital display allows for expansion of a museum's visitor base and service scope, promoting higher functionality. A collection must be displayed in a digital form that is suited to its distinct characteristics. It is necessary to allow collections to "live". This way, they will become more vivid and therefore easier to understand, appreciate, and interact with. Thanks to the rapid development of information technology, the means of digital museum display will become increasingly abundant, and technology will mature, having a profound impact on the traditional museum industry. As professionals in the field, we must maintain the current rate of progress, using all possible means to present stored collections to the world in a variety of new forms.

References

[1] Chen Gang, Zhu Kongqiang. 'Analysis of Digital Museum Related Issues', [J] *Intelligent Construction and Urban Information*, 2004, 9.

[2] Wang Yuchang. 'Issues in the construction of digital museums', [J] *Gansu Science and Technology*, 2006, 3.

[3] Wang Yu, Zhang Xu, Shi Minfeng. 'On the Planning and Construction of Museum Networks', [J] *Journal of the Beijing Institute of Graphic Communications*, 2008, 2.

[4] Zheng Jinchun. 'Thoughts on Digital Museum Construction', [J] *Tianjin Science and Technology*, 2006, 6.

[5] Qi Yue, Shen Xukun. *Management and display of digital resources in museums* [M]. Shanghai: Science and Technology Press, 2008.

Applied Research on Augmented Reality in Museum Education

Yang Xin, Liu Shuhan, and Zhang Tianmeng
(China Aviation Museum, Beijing 102211, China)

Abstract: With an increase in free admission to museums, social and public requirements are increasing. Using modern technology as a tool to enhance exhibitions and stimulate visitors' interest has become a key concern for contemporary museums, especially in the field of science and technology. Augmented reality is a new technology that blurs the boundaries between the real world and virtual computer space by enriching sounds, sights, and feelings. The application of augmented reality technology in museum exhibitions brings static exhibits to life and enhances "interactivity" and "participation", while also offering entertainment value through education. This paper first offers a brief overview of AR technology and museum education, then focuses on the correlation between the two, finally outlining the application of augmented reality technology at the China Aviation Museum.

Keywords: augmented reality; museum education; interactive experience

Since free museum admission was first launched in China in early 2008, most museums and memorials throughout the country are now free to visit. However, social and public requirements are increasing exponentially. Static exhibits remain the mainstay in most venues, where information is transmitted between visitors and exhibits via captions or commentary – a method that no longer satisfies diverse learning needs. Therefore, the use of modern technology to enhance attraction and stimulate visitors' interest has become common in contemporary museums, especially those in the field of science and technology.

With the development of computers and information technology, augmented reality has come to the fore. This new technology is capable of blurring the lines between the real world and the virtual computer world by enriching sounds, sights, and feelings, offering wide range of applications in the military, medical, industrial, and cultural heritage industries. In the current mobile Internet era, increasing numbers of apps and virtual applications are being developed for mobile devices such as tablets and smart phones, bringing AR technology within reach. As such, this paper focuses on methods for the application of augmented reality to museum exhibitions and the enhancement of the interactive experience.

1. Overview of AR technology

In the early 1990s, Tom Caudell and his colleagues at Boeing proposed the idea of "augmented reality". In the system they designed, information such as route paths and text in the form of simple lines were superimposed through a head-mounted apparatus onto a mechanic's high-tech eyewear in real time, assisting the mechanic in the step-by-step completion of disassembly. The goal of augmented reality is to generate virtual objects by means of optoelectronic displays, interactive technology, computer graphics, and visualization, accurately "placing" virtual objects within a real environment with registration technology. In this new hybrid environment, users have a genuine, interactive sensory experience that does not distinguish between the real and the virtual.

AR technology is an emerging multidisciplinary field of research. According to Ronald T. Azuma at the University of North Carolina, the augmented reality system contains three characteristics. The first is the integration of real- and virtual-world information, that is, the fusion of virtual and reality. The second is real-time interaction, and third is the positioning of virtual objects within three-dimensional space, i.e. 3D tracking registration. Amongst these features, real-time human-computer interaction is the most important, as it enhances immersion and perception.

AR technology is now more than 20 years old, and has transitioned from a gimmick to a tool with endless potential. In recent years, given the rapid development of mobile Internet technology and computer vision image recognition, there has been a rapid expansion from military to commercial to civilian applications. With significant support from IT giants such as Google, Apple, Intel, and Qualcomm, AR technology has been widely used in the fields of interactive advertising, exhibition display, wearable devices, digital publishing, and popular science education. From bar codes to QR codes, artificial rendering and physical recognition, augmented reality tools continue to evolve, and the use of AR technology is changing rapidly.

2. Museum Education

As a social institution, a museum makes scientific and cultural knowledge accessible, promoting national culture and patriotism while enhancing the public's aesthetic conception via planned educational activities. With the development of society in the early 21st century, the educational functions of museums have seen them expand into important classrooms for off-campus education, becoming another education system for society at large. Museum education has the characteristics of physicality, intuition, autonomy, socialization, and education. Methods of education generally include informative styles based on exhibitions, experiential modes based on dynamic interaction within exhibits, exploratory styles based on inquiry activities, and universal education through mainstream popular science.

Among these features, experiential education has been exploited widely in recent years, particularly by science and technology museums. Experiential museum education utilizes exhibits that offer immersive, hands-on, or interactive features to promote a greater understanding, interest, and experience of natural phenomena and scientific principles. Be it a simple demonstration project or a complex, large-scale simulation, experiential museum education emphasizes direct audience engagement. In comparison to conventional static exhibits, experiential education is far more interesting, and can achieve more specific educational goals.

The rapid development of modern science and technology has directly promoted innovation in experiential museum education, utilizing mechatronics, automation control, multimedia, simulation, and virtual reality technology. The resulting exhibits are more vivid and easier to understand, with higher attraction, appeal, and impact levels, stimulating various senses to provide a stronger psychological experience. As such, museums must actively introduce new technologies, methods, and techniques, innovating technological means to attract more visitors to learn, discover, practice, and be entertained through the educational process.

3. The Correlation between AR Technology and Museum Education

The key feature of AR technology is the ability to interact with users in real time, which aligns with the concept of experiential museum education. The application of AR technology in a museum allows exhibits to come to life in a fresh way, enhancing "interactivity" and "participation" while playing a role in education.

3.1 Augmented reality technology enables visitors to access more scientific information in a smaller area

In museums, AR technology is largely implemented through small screens (currently smart phones and tablets). For specific parts of an exhibition, museums develop an iOS/Android app that utilizes AR technology, allowing visitors to see stereoscopic multimedia educational materials on their smart phone or tablet. For example, the Asian Art Museum in the United States hosted a *Chinese Terracotta Warriors* exhibition. The site of the First Qin Emperor's Mausoleum in Xi'an is very large and complex, and is still being excavated. The exhibits at the Asian Art Museum showed only a small portion of the unearthed artefacts. By downloading an AR iOS app developed by the museum, and scanning the six signboards associated with the exhibits, visitors were provided with further 3D content and videos about unearthed artefacts in real space.

3.2 Augmented reality has unlimited potential for live learning experiences and exploration

AR technology offers vitality to static exhibits, and stands to enrich the transfer of knowledge at a variety of levels. As an example, AR allows virtual exhibits to revolve so that visitors can see them from all angles. Augmented reality can turn a fossilized dinosaur skeleton into a moving creature, and can "restore" the original appearance of Yuanmingyuan, allowing visitors to walk on the ruins and see spectacular views from more than a century ago. In an exhibition held at the Natural History Museum in London, visitors were able to install software on their iPhones and see BBC radio show host James Mayn give a virtual guided introduction to the museum. The *Towards the Modern World* exhibition includes nine signal-sensing cameras, which visitors can use to "summon" a virtual James Mayn who will tell historical stories about the invention of the steam engine and the first home computer. AR technology's combination of virtual and real displays offers a fresh view and provides visitors with a more immersive experience.

3.3 Augmented reality technology is bundled with location services, becoming an important tool in breaking down the limitations of time and space within the museum

Location tracking provides services based on changes in user location. In particular, the precise position of buildings and interior spaces allows users to obtain personalized information tailored to their indoor location, offering detailed information in geospatial and three-dimensional space. In the museum, location services can help visitors locate specific areas and exhibits within a huge exhibition space, and can recommend other areas according to users' preferences. Positioning services are bundled into AR technology, so visitors can enjoy a variety of sensory stimulation and experience a completely personalized tour.

4. The Application of AR Technology at the China Aviation Museum

The China Aviation Museum began to develop a digital version in 2013, and has reached its preliminary goals. Augmented reality applications are one of the highlights. In order to support AR application, the Aviation Museum has implemented the several facilities. First, WiFi coverage is offered throughout the museum to provide access to the mobile Internet portal on mobile devices such as smart phones, tablets, and laptops. The second facility is a customized smart navigation terminal, offering richer audiovisual AR resources with a large capacity and a large screen. This enables self-guided tours, location services and navigation, AR experiences, and a unique Aviation Museum gaming experience. All of this greatly enhances the provision of knowledge, entertainment, and interactivity during on-site visits. There are two main types of augmented reality applications in the Aviation Museum, as detailed below.

4.1 Exhibits

The application of AR technology enables static exhibits to display vivid content in three dimensions. Viewers are not limited to what they see, but are provided with high-impact AR experiences within real space, offering a new tour experience that includes 'Working Principles of the Aircraft Turbine Engine', the 'Wreckage Recovery Collection', 'Ejection Seat Projection', 'The Wang Hai Brigade Air Show on November 18th, 1951' and 'The Return of Air Refueling'. As shown in Figure 1, there are five buttons along the lower half of the page. By clicking the corresponding button, the page displays an introduction to the content. Visitors can click the "start" button to scan a physical object or QR code and begin the AR experience.

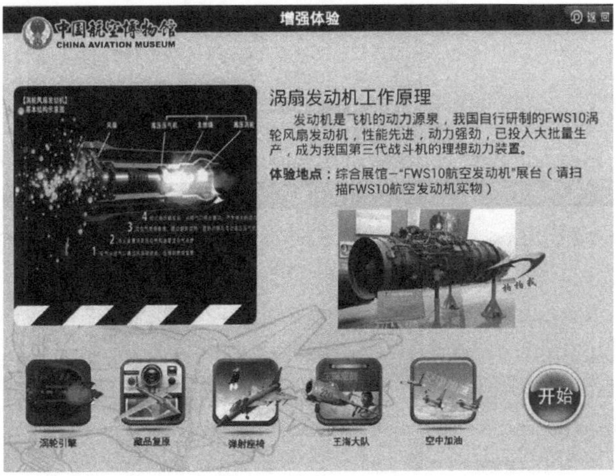

Figure 1 The "AR experience" interface

4.2 The children's playroom

Augmented reality and other technologies have been applied to the design of games for children, introducing them to simple knowledge related to aircraft in a colorful and entertaining way. The playroom includes a variety of interactive experience games such as the Nine Star Challenge, 3D Aircraft Disassembly and Assembly, Guess the Aircraft, and Air Refueling Flight Simulation, as shown in Figure 2. In the Nine Star Challenge, nine aircraft are randomly generated within the exhibition hall. Successfully scanning the aircraft's QR code illuminates a five-pointed star, completing a challenge with each successful scan.

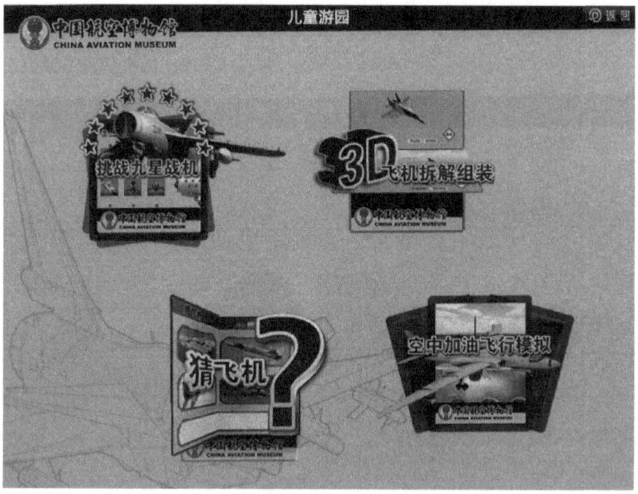

Figure 2 Children's playroom interface

5. Conclusion

One of the most important functions of a museum is to provide a space for the public to learn about science and culture. Methods by which to promote public education and cultivate self-exploration are at the root of innovations in museum display. Augmented reality technology is a high-tech field that full of potential and possibilities. In our current era of creative information, AR technology is being applied to museum exhibitions, utilizing a combination of virtual and real spaces to transform passive visits into interactive, multi-sensory experiences, while increasing the entertainment value. This more effective dissemination of information and knowledge helps museums to exert their role in social education.

References

[1] Ren Fujun, Li Zhaohui. *China Science and Technology Infrastructure Development Report (2012–2013)* [M]. Beijing: Social Sciences Academic Press, 2013.

[2] Li Wenxia, Si Zhanjun, Gu Chong. 'Research on the Application of Augmented Reality Technology in Museums', [J] *Computer Knowledge and Technology*, 2014, 1: 160–162.

[3] Li Xinzhi, Cao Mo, Guo Ji et al. 'List of Alternative Technologies for the 2013 Horizon Report (Museum Edition)', [J] *New Museum Technology*, 2014, 1: 8–9.

[4] Teng Jipu. 'AR Technology: Enhance your perception of the world', [N] *Science and Technology Daily*, 2012-12-15 (6).

[5] Chen Shuangshuang. *Augmented Reality technology for the "glory" of museums* [EB/OL], accessed 2012-10-12 <http://www.hnmuseum.com/ hnmuseum/ generalIntro/introContent.jsp?infoid=013a52caf215402884833a4d5706000a>.

The Perceptual Experience of the Information Design of the Virtual Chinese Opera Museum

Chen Yan
(Department of New Media Art, China Academy of Opera, Beijing 100375, China)

Abstract: Art is not static, but is constantly in motion. Good design is the product of continuous advancement, drawing on traditional roots while promoting the consistent renewal, development, and improvement of design concepts. While enriching itself, it also provides convenience for users. The design of the modern digital Virtual Chinese Opera Museum is based on the fast dissemination of information and the rapid development of science and technology. Through the processes of conception and problem-solving, we have realized purposeful, value-creating human activities. We have also established the conception of movement in time and space, forming aesthetic ideologies for design work, and ultimately having a powerful visual impact and offering personalized, interactive experiences to users via technologically advanced media platforms.

Keywords: information; insight; experience; creativity

Information design indicates a designer's perception of life, created as kind of transcendence, an internal shock, or a Gestalt experience. When we design, we not only work with the expression of external form, but explore the inner world of the heart and soul. As we often say, this kind of work moves with the self. Its magnetic field spreads to the surrounding area, affecting visitors and making them remember. The purpose of any information design is to convey specific information to the audience in a clear, accessible, and understandable way.

1. The Perception of Information

The accelerated pace of life and the rapid development of science and technology have created a mass of extraneous information, whether useful or lacking in value, which constantly flows into human sight, and at the same time is mechanically stored in the memory. When we lose useful information, we sigh and reflect, asking "Why is memory so short-lived?" when in fact all information must be personally promoted. This has two components: its own practicality, and an additional artificial visual expression. Our brains are often stimulated, our senses recognizing external environmental stimuli and temporarily storing them in perceptual memory. At this point, the brain transfers new information to memory, and when visual stimuli appear, people retain images through perceptual memory such as "image memory". It takes only a few milliseconds for this information to be transferred to short-term memory storage, so when external stimulation causes psychological satisfaction and pleasure, information will be sealed in the depths of the mind.

Modern design is inseparable from the ever-changing transformations in society and the rapid development of science and technology. Due to the continuous improvement of technology, the design process proposes the adoption of an integrated approach for information dissemination between people and between people and things. For example, the design of the Virtual Chinese Opera Museum integrates various forms of information such as pictures, text, audio, and video (see Figure 1), to provide a completely free information platform from multiple angles. Participation stimulates visitors' sensory nerves, allowing them to memorize information and slowly recollect it later.

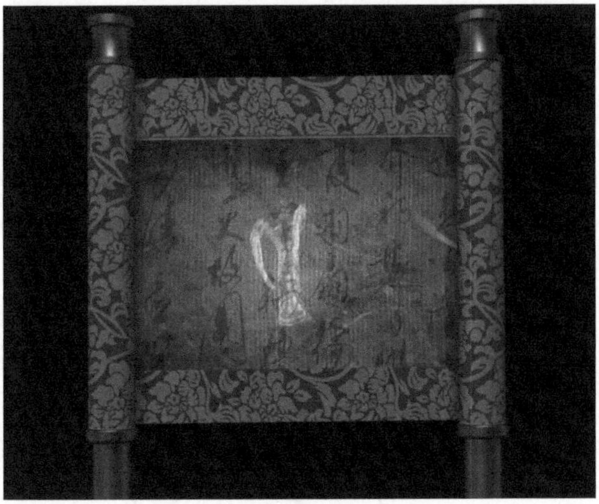

Figure 1

2. The Naturalization of Information

Information design is different to traditional design, using modern means to create new contexts in which users can communicate and interact with one another in real time. In a sense, it changes the passive appreciation and reception of information of traditional art, offering a chance to feel a special artistic aesthetic and "free artistic experience" while perceiving the content of a work. People thus communicate with their senses through both virtual computer space and real space.

We first come into contact with visual images via the brain, storing perceived objects in the memory and reproducing them later. Consciousness forms a sensory perception of image-borne information, and thus imagery produces an attached sense of tangibility. Experiments have shown that 83% of human access to information comes from sight, 11% from hearing, 3.5% from smell, 1.5% from touch, and 1% from taste, demonstrating that we access information through a variety of natural pathways. As we perceive information through their eyes, ears, nose, tongue, and skin, we can also collect information through facial features. Natural elements of environmental life that enable access to all kinds of information through the sensory organs, including light, sound, and taste, can be transformed into information we need, providing us with a broad space of association, and a vast world for design inspiration.

Modern design is most satisfying when it involves natural communication, the transmission of information, and continuous access to information via sound, color, and interaction with the natural world. To attain information from signals within the natural world, we must communicate with the environment. Communication is thus one of the key ways to naturalize information. Gestures, eye contact, and expressions can all inadvertently convey information, enabling us to obtain it without deliberate attention or by being informed that information exists. When we smell food, hunger arises. This is a natural transmission of sensory information. When the TV is too loud and its decibels exceed a certain limit, noise is the most easily registered result of signal feedback. This can also be understood as invisible communication – a natural signal that people can get used to. A designer will follow this easily-manipulable pattern, working in a way that encourages people to understand and control design intentions more naturally. For example, when we see a "smiley face" symbol in the design of a camera, we press it to capture the smile of a subject.

When visiting a museum, we are eager to know the background, age, name, and other general information about an exhibit. Therefore, voice explanation is the most natural form of information exchange. When designing the Virtual Chinese Opera Museum, "Cue" was one of the design points regarding natural information interaction within the project. Mainly reflected in prompts for visitors to navigate information, it takes the role of a live commentator, offering timely and accurate information. The venue guide takes the cartoon form of an opera character (see Figure 2), and provides basic spoken information to visitors upon their entry. The voice control icon to the bottom right of

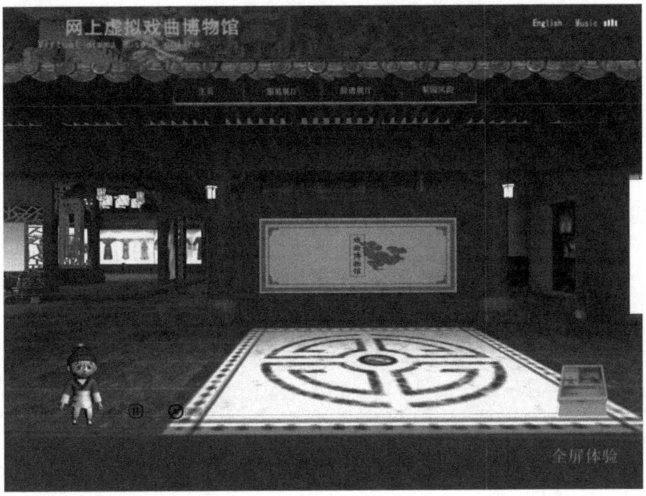

Figure 2

the character can be clicked to turn the voice on and off, and to choose whether voice prompts are required. Voice prompts are designed to offer visitors targeted reminders, and any delay in the information may cause an interruption to the train of thought, an unwelcome distraction, or even negative psychological reactions. Rich imagery, dynamic effects, colorized emotions, humanized information prompts, and upbeat language are all used to mobilize visitors' emotions, retain their attention, stimulate their interest, and enable the maintenance of the highest quality tour for a prolonged period.

3. Requirements and insights

Designers' goals are to fulfill users' needs and create works that can impress the community and achieve independent value. This requires them to consider practicality. They must be able to describe a design in words, and also to understand it on a fundamental level in order to meet the needs of the potential user.

The Gestalt psychologist Wolfgang Kohler undertook experimental research on chimpanzees in Tenerife. Caged chimpanzees like to eat bananas. Placing a long bamboo pole near to the cage, and a short bamboo pole inside the cage, the chimps were able to access the fruit. This proved that even when an animal is confined in a geographical space, and does not have the necessary behavioral environment, it may exceed our expectations with insightful thinking. These experiments showed that animals must be aware of all factors in a problematic situation before engaging in real psychological problem solving. If an animal does not know that food exists behind an obstacle, no behavioral environment exists, and therefore no problem exists. Thus, an epiphany due to desire or demand will not occur. On the contrary, if an animal knows that food is behind the obstacle but has no desire or demand for it, the problem does not exist, and again an epiphany due to demand cannot be obtained.

Design is the same. The desire or demand of a user group promotes the direction of product design. A single design element cannot satisfy human curiosity and desire. Designers must experience and empathize with user group perspectives, going for the "banana" in the immediate vicinity in order to achieve Gestalt design. Of course, this process is slow, and requires repeated testing.

Solving life's problems is a challenge. Solving them in an easy and even enjoyable way is an even greater challenger. This can inspire a talented designer realize new design concepts and exploit good product design. For example, the problem of obesity is affecting a huge number of people, and requires better regulation of dietary intake. However, the science of nutrition is complex and inaccessible to the general public. Designer Jennifer Panepinto came up with a solution. In the process of measuring food with a cup, she

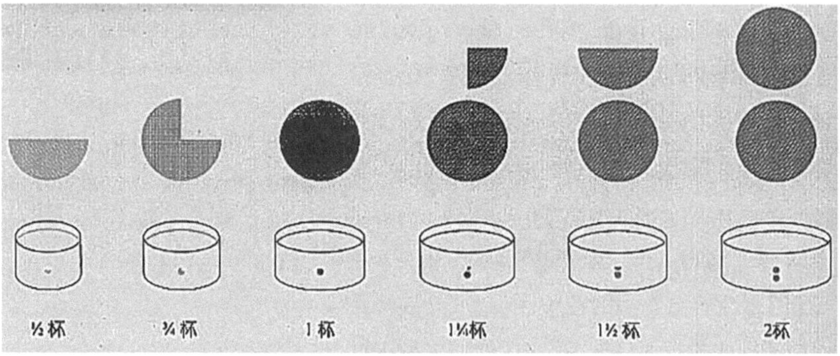

Figure 3

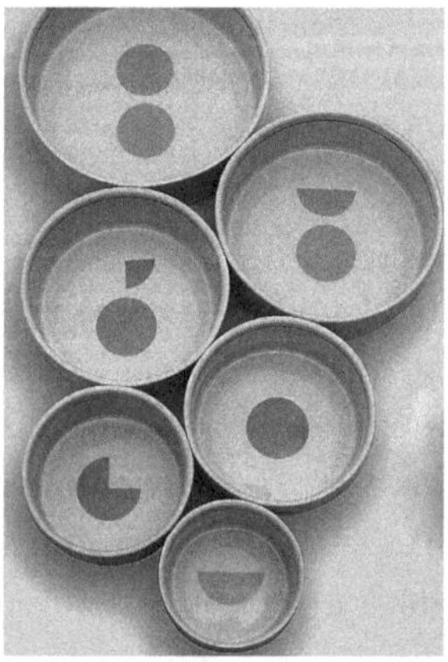

Figure 4

realized that people need a tool that offers easy food control, rather than complexity and irritation. So, she designed a series of high-end tableware called Mesü (see Figures 3 and 4). Painted icons indicate capacity, allowing people to quickly and easily identify the size and proportion of measurements required. Thanks to Panepinto's empathy with the dieting mindset, this conveniently designed identification scheme helps people who need to control food portions.

The Virtual Chinese Opera Museum is a design concept that we hope will spread information about opera more widely. However, for complex and large-scale exhibition halls or museums, production is complicated, the costs are relatively high, the storage of exhibits is cumbersome, and numerous financial and material resources are required. There are also risks such as transporting exhibits as well as wear and tear. However, with a special storage and management mode, the "virtual museum" (see Figure 5) format can be used to display digital information. This design point allows for the wide dissemination of low-cost information, so that users can visit the Opera Museum from the comfort of their homes. Therefore, designers are heavily guided by the desires and needs of user groups in their work.

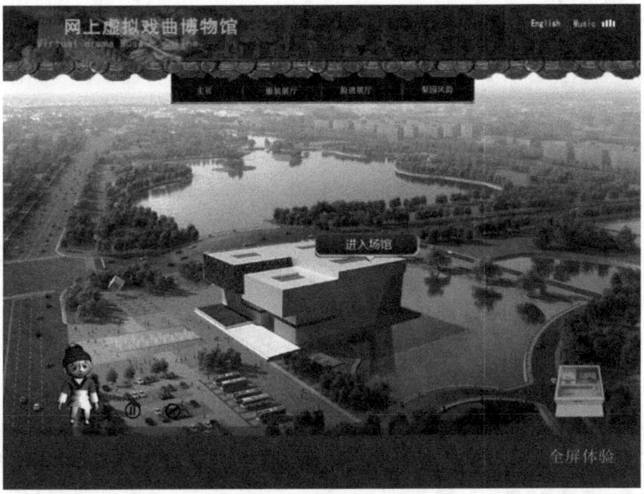

Figure 5

4. *Creative Design*

Designers do not simply work to their own preferences; they are inspired by direct encounters with users' needs, both current and historic.

Creativity in information design requires recognition and resonance between designers and user groups, that is, the analysis of perspectives. Blind copying or plagiarism is out of the question. True creativity requires perception, memory, thinking, association, and understanding. Design concepts must be comprehensive, exploratory and innovative, with a mind to quality and the accumulation of knowledge. The process of creative design is inseparable from cognitive abilities such as reasoning, imagination, association, and intuition. Creative thinking is actually a pioneering activity, opening up new cognitive areas and new achievements for human understanding.

Creative design requires us to fully understanding the structure of design content, and we must regard each element as part of an organic whole. Any creativity must be understood and developed with regard to the unity of content structure, seeking an appropriate Gestalt development direction. Creative thinking is not a purely intellectual activity. It is restricted by factors such as human motivation, emotion, and accumulated knowledge. Designers must fully understand the issues involved in design content, seeing discrepancies and constraints in structure, and using discrete structural points to sort the aspects of the problem into a clear hierarchy, or a relational structure of reorganized problems.

In the case of the Virtual Chinese Opera Museum, n the preliminary design phase involved a consideration of existing problems. For example, with so many knowledge points, how could they be connected in an educationally conducive manner? How to display information about traditional opera in a virtual museum was the first design concept to be determined. We decided to simulate a real museum space to create a virtual, three-dimensional area. This 3D space is divided into several display areas, and visitors can roam and browse freely (see Figure 6) via a mouse and keyboard. All of the exhibits in the virtual museum (see Figure 7) can also be viewed at 360° from all directions, enabling interactive activities such as close-up viewing, rotation, and zooming (see Figure 8), while text, pictures, videos and other materials can be viewed at any time (Figure 9). As a result, no matter where a visitor is, as long as the virtual multimedia museum is displayed on a computer desktop, visitors can experience the true essence of traditional Chinese opera.

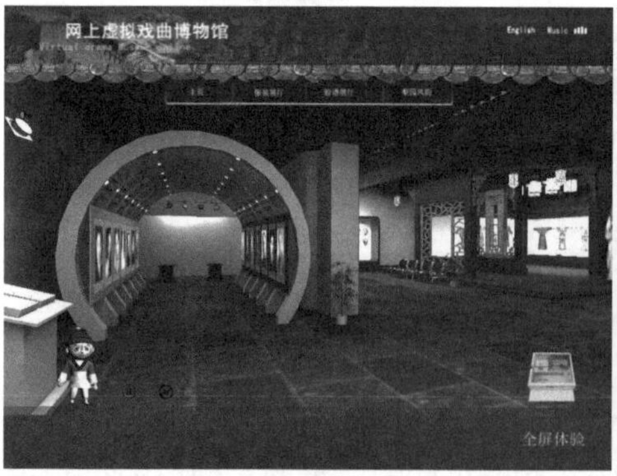

Figure 6

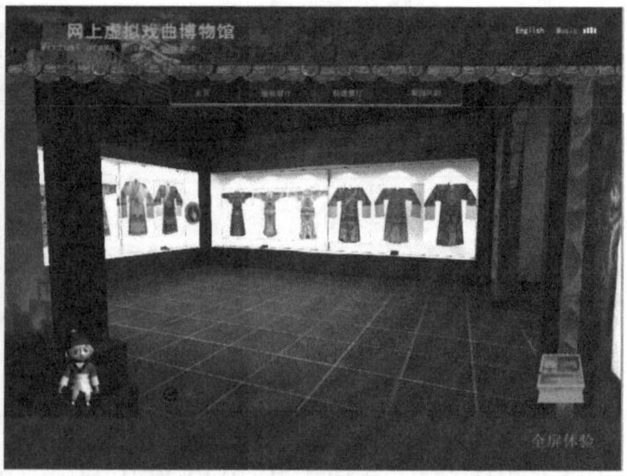

Figure 7

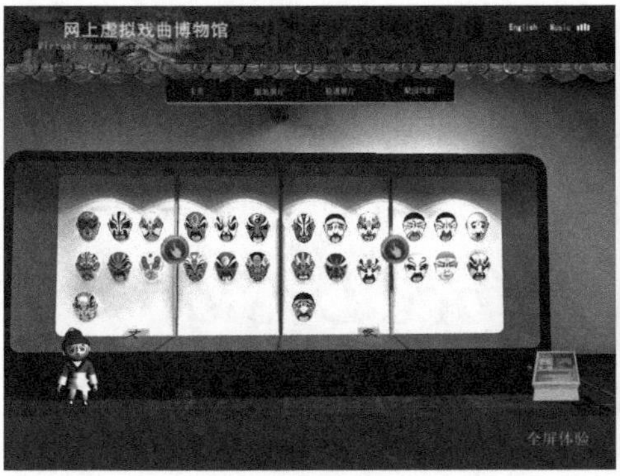

Figure 8

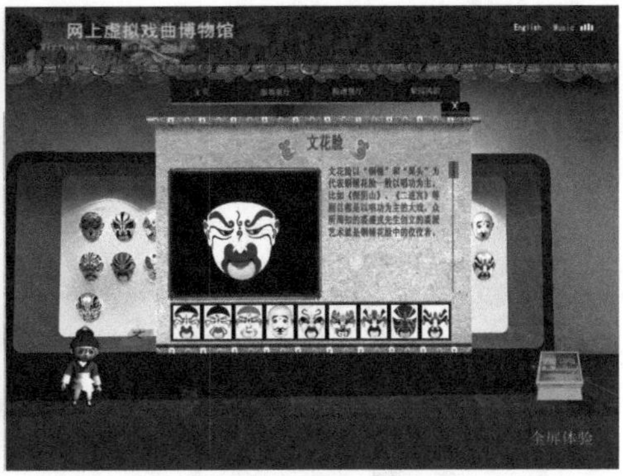

Figure 9

Information is a crucial element of the creativity behind the design of the Virtual Chinese Opera Museum. A chaotic and disorderly design will only confuse visitors, leading to low engagement and retention of information. Keen intuition, careful observation, and a self-aware evaluation of visual performance, as well as audience reactions, are part of good design methodology. Design can be understood as a process, and audience participation can be understood as a mode of re-creation for modern design. Creativity in design is reflected in the understanding of a work's essence and the Gestalt experience of visual perception. Design is a materialization of behavior controlled by consciousness – a kind of consciousness that is the source of design ideas. Designers embody the language of design in their own artistic practice, judging the transmission of design works through user experience and cognition.

Research into the Interface Design of the Virtual Exhibition at the China Agricultural Museum

Chen Ying
(China Agricultural Museum, Beijing 100026, China)

Abstract: With the rise of Internet technology, traditional museums have built their own virtual exhibitions. The virtual exhibition generally uses a computer or multimedia touch screen as a display carrier, and its interface is particularly important as a medium for information exchange between people and machines. Excellent interface design is of great significance in exploiting the advantages and realizing the functions of the virtual exhibition. Interface design has its own unique components, and is topic worthy of attention and discussion.

Keywords: virtual exhibition; interface design; design elements; interaction

Preface

As a form of social education, museums play a huge role in the dissemination of cultural information. However, in recent years, the rise of Internet technology has challenged the traditional museum, and both exhibit content and display methods have been unable to accommodate increasing cultural demands from the public. Given these conditions, major museums in China and abroad have begun to use new technology, breaking traditional display patterns and building virtual exhibitions. These virtual exhibitions present an opportunity to expand the physical museum, changing the static model of traditional display in the physical museum, and acquiring greater means for artistic expression and possibilities for cultural creativity.

As China's number-one specialized agricultural museum, the China Agricultural Museum has carried out explorations and practice in this field. From relying on physical exhibitions to creating independent virtual exhibitions based on physical displays, the institution has accumulated rich experience. In recent years, every May 18th (International Museum Day) the China Agricultural Museum has launched a series of virtual exhibitions with different themes, providing visitors with new scientific and artistic experiences.

1. The Significance of Interface Design in the Construction of the Virtual Exhibition

An interface is the medium for information exchange between people and machines, and virtual exhibitions generally use computers or multimedia touch screens as display carriers. The term 'interface' here generally refers to everything that a viewer sees on the screen when browsing the virtual exhibition, including text, graphics, and windows. For visitors, the interface represents the entire virtual exhibition. Excellent interface design provides a sound structure, rich content, and clear navigation, optimizing interaction between users and the information system. It also enables visitors to engage in a positive psychological experience when visiting virtual exhibitions by improving public access to information, increasing the efficiency of resources, mobilizing the advantages of virtual exhibitions, and realizing their full functionality.

2. Content of Interface Design for Virtual Exhibitions

In general, the content of interface design for virtual exhibitions at the China Agricultural Museum includes a welcome page, a primary interface, and a secondary interface. Depending on the specific situation there can be a deeper interface, but this never goes below three levels, so as not to cause confusion.

The welcome page (or title page) is the first interface that the visitor sees, and is the "façade" of the virtual exhibition. In its design, the title appears in large font in a prominent position on the screen, highlighting the overall theme. Meanwhile, the color, text, and layout design of the welcome page is a systematic display of the exhibition, indicating the style of the interface design as a whole.

By clicking the "Welcome to Browse" button on the welcome page, visitors enter the first interface level, i.e. the navigation interface, featuring a color and style that continues throughout the design. This functions to display the structure and logic of the virtual exhibition, with clear navigation. The design of the secondary interface is based on this first level, functioning to display detailed information with many words and pictures in order to convey information efficiently and accurately.

3. Components of Interface Design

3.1 Analysis

Analysis is the first step in the design of an interface for virtual exhibitions, including subject content analysis and audience analysis.

Subject content analysis uses the content and characteristics of the virtual exhibition to determine the artistic conception and style of the interface. Point, line, and plane composition, the use of color, and the layout of components strive to highlight the subject and reflect specific aesthetic and artistic qualities. For example, a virtual exhibition themed around "Chinese Tea" should utilize traditional Chinese patterns and curves to create a lasting artistic effect.

Audience analysis refers to the analysis of visitors to the virtual exhibition. Factors such as age, cultural background, and computer literacy directly or indirectly affect the final interface form.

Virtual exhibitions produced by the China Agricultural Museum are mainly focused on agriculture-related subjects, such as farming, agricultural policy, peasant art, and agricultural folklore, with forms of expression that are largely popular, simple, and mainstream. From the perspective of audience analysis, visitors include young people, lovers of peasant art, agricultural experts, and scholars. All of these groups have their own orientations and preferences, which must be considered in the design of the interface.

3.2 Text design

In terms of composition, text is not limited to the communication of information. It is also a form of artistic expression. Its attributes include font, font size, kerning, and line spacing, with each variety having its own unique visual characteristics.

With font layout, attention must be paid to the composition of points, lines, and planes, with text characters arranged in lines, and lines arranged on a plane. In layout design, it is important to consider font case, font size, alignment, blank space, color, and character spacing in order to create a fresh, natural, simple, and pleasing layout.

Title fonts must be conspicuous, and are generally large in size, mostly featuring artistic or handwritten fonts. For example, the title of *Unforgettable Memory — A Collection of Agricultural Posters* appears in a handwritten calligraphy font with substantial cultural and artistic resonance. While text fonts on traditional media platforms can be various, in general, the choice of text fonts in the interface should not exceed three, as the appearance of too many font types in one interface may lead to serious visual confusion.

3.3 Color

Color in interface design often has an advantage over text in providing a first impression. The attributes of color are reflected in three aspects of hue, brightness and clarity, and their cohesion can produce ever-changing visual effects.

The semantics of color are indispensable in design, as different colors prompt different psychological reactions. For example, an interface with yellow or orange as the main color provokes a feeling of vitality, while a blue or green color invokes a cool and quiet mood.

In interface design, a color scheme should be chosen according to the specific project, considering the emotional and semantic performance of color. For example, in the China Agricultural Museum's virtual exhibition, *Our Fields — Contemporary Chinese Agriculture; Rural and Peasant Images* has a green color base, in line with the theme of "fields", offering a sense of abundance and vibrancy. The *Chinese Farming Culture* exhibition has a dark brown base – the color of the land, indicating both fertile soil and a sense of culture.

3.4 Graphics and images

Graphics and images belong to a category of figures, and are essential elements in the interface – the most important in terms of visual impact.

Graphics generally refer to man-made images. Graphic design brings a new visual perception to interface design, with strong abstract capability and ample space for artistic imagination.

Images refer to the reproduction of reality without artificial action, i.e. faithful reflections of the objective existence of things. In the process of design, images are applied largely for the purpose of interpreting content.

Graphics and images also accompany text for visual communication, facilitate understanding, and make a layout three-dimensional real. Image positioning is directly related to layout composition. Left, right, top, bottom, and the four sides are all focal points in the line of sight. A picture sits well at any of these points, clearly revealing the visual impact of the layout. Picture size is also directly related to the visual effects and emotional communication of the layout. In general, important images that attract attention are magnified, while subordinate images are reduced in order to create a pattern of distinction between primary and secondary fields. These are the basic principles of the layout.

3.5 Multimedia design

Multimedia elements in a virtual exhibition interface generally include audio, video, and animation. Audio involves input/output voice, and video has both sound and image, while animation increases the vividness of an image. Multimedia such as audio, video, or animation can be used to complement plain text or 2D graphics, accompanying the visual design of virtual exhibitions. For example, in *Unforgettable Memory — A Collection of Agricultural Posters*, a variety of sound and animation effects have been added to showcase the posters. Visitors are invited to experience them in virtual form on the Internet, increasing interest and overcoming the monotony of physical viewing.

3.6 Interaction design

Interaction design involves a viewer directly interacting with the computer or multimedia touch screen while browsing the virtual exhibition. The most straightforward example of this is when a user clicks on a button in the interface or touch screen to access information, an interaction takes place in this process. Good interaction design often refers to accessible understanding and efficient icon design.

Dynamic effects in interaction are effects generated when switching between pages and information modules, such as moving to the next page of a novel in a natural and interesting way by deconstructing, shifting, or adding scenes. When the mouse slides over an icon and clicks, different dynamic effects occur, increasing the entertainment value of the interaction process.

4. Case Study – Memory Village – A Collection of Peasant Paintings

Memory Village — A Collection of Peasant Paintings was the first virtual exhibition to be produced by the China Agricultural Museum without a physical basis. The goal of its interface design is to evoke enthusiasm for the artistry of peasant paintings, and at the same time provide interactivity that is effective and convenient for visitors.

According to the structure of the exhibition, the display interface can be divided into four levels: the welcome interface, the primary interface (introduction), the secondary interface (exhibition hall), and the tertiary interface (exhibits).

The welcome interface (see Figure 1) presents the theme of "memory village" in large font in a central screen position. In the layout, there is a close-up image of eaves. A Jiangnan water village is arranged in the middle followed by mountains in the distance. The color scheme is warm, reflecting the theme of "memory", and the animations of boats, birds, and falling petals evoke the tranquil atmosphere of the countryside.

The primary interface (see Figure 2) is divided into the title area, content display area, and directory area. The title area is decorated with artistic graphics of peasants and

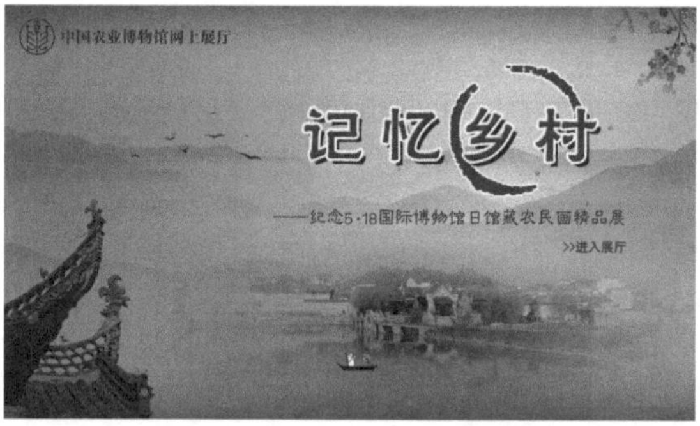

Figure 1 The welcome interface

the countryside. In the content display area, the text is arranged vertically. The visual effects are more prominent, with peasant paintings from different regions designed with local characteristics to highlight regionality. The directory area takes the form of buttons, showing a clear exhibition structure, while mouse strokes and button clicks have different artistic effects.

The secondary interface (see Figure 3) is unified with the form of the primary interface. As per the characteristics of peasant painting, imagery of a rural courtyard is used as the backdrop, with peasant paintings displayed on the courtyard wall.

The tertiary interface (see Figure 4) is the final display interface for peasant paintings. When a peasant painting on the wall in the secondary interface is selected, the background darkens, as if the lights have been dimmed, showing the painting in a full, independent display.

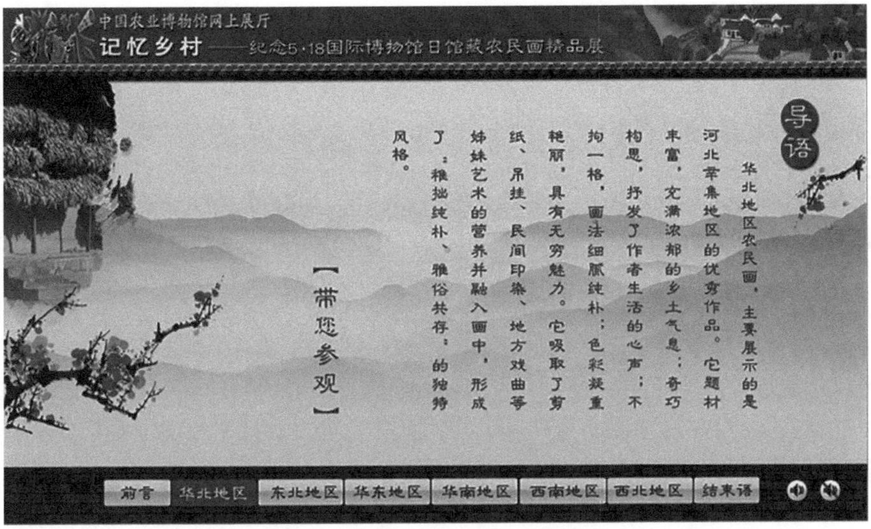

Figure 2 The primary interface

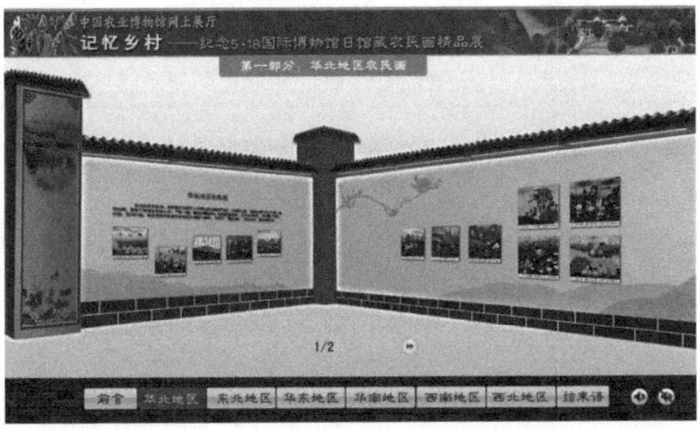

Figure 3 The secondary interface

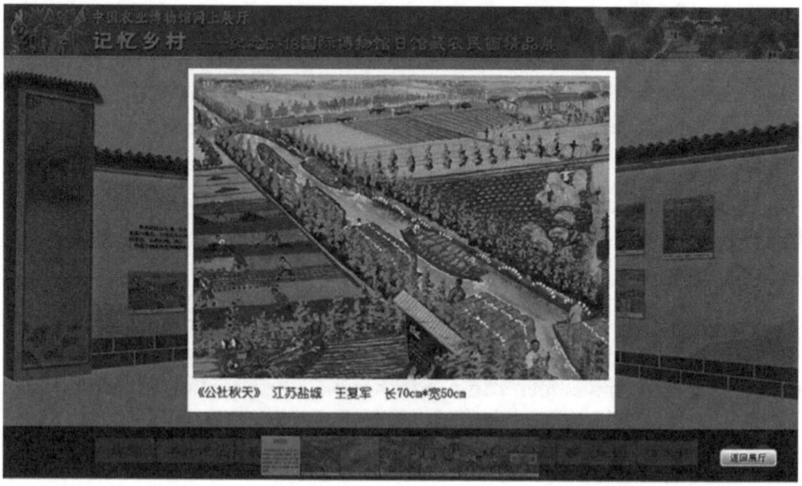

Figure 4 The tertiary interface

5. Conclusion

Virtual exhibitions at the China Agricultural Museum combine rural themes with modern virtual display technology. Depending on their themes, the virtual exhibitions have a variety of atmospheres and content, focusing on different expressions. Excellent interface design offers a positive viewing experience and acts as a comprehensive expression of a designer's skill and aesthetic experience. Contemporary designers must harness traits from various cultures and integrate them into their designs, filling the space with rich and diverse visual elements. They must skillfully blend technology and art while working to improve interface design for virtual exhibitions.

References

[1] Lu Xiaobo, Zhan Binghong. *Digital Graphic Interface Design*. Beijing: Tsinghua University Press, 2006, p. 49.

[2] Liao Hongyong. *Digital Interface Design*. Beijing: Beijing Normal University Press, 2010, p. 50.

[3] Yan Chen, Chai Chungang, Xu Na. *Multimedia interface design*. Beijing: Publishing House of Electronics Industry, 2011, p. 95.

[4] Han Zhongyuan. 'Application of Visual Communication in Multimedia Tourism Consulting Systems — Shaanxi Red Tourism multimedia visual design research', Master's Thesis, Full-text Database, p. 12.

CHAPTER 2

Remembering Nostalgia

– The Digitization and Conservation of Cultural Heritage

The Application of 4K Image Technology in the Acquisition of Movable Cultural Archive Images

Si Bing
(National Palace Museum, Department of Information, Beijing 100009, China)

Abstract: Image technology for the acquisition of cultural artefacts is in a constant state of innovation alongside advances in photographic technology. The collection of cultural archives is the basis of a museum's informatization, and provides the most important data resources for cultural display in the creation of digital museums. The completion of comprehensive and efficient image collection for movable cultural archives is a consistent issue in basic data collection. This paper looks at the use of new digital imaging technology to capture artefacts. 4K video image technology is used for image collection of movable 3D cultural archives, to obtain both a video and photo in one shot. This provides a comparative analysis based on complete image data, exploring a digital technology that is more suited to the daily work of cultural artefact image acquisition, making the collection of cultural images much more efficient.

Keywords: Movable cultural artefacts, image archives of cultural artefacts, 4K, Slog-Gamma Curve

Preface

In recent years, museum informatization and digitization have been encouraged and improved with new technologies, and new technological applications have consistently been explored. Within museum informatization, the digitization of image archives of cultural artefacts persists as the basis of informatization. Images of cultural artefacts are at the core of museum archives, and are essential information for management and display. In the creation of information systems for museums, digital resources will become the main element of practice. The collection of digital images will be the most integral component, and the most important resource data in the development of digital museums. The collecting of digital cultural information involves the conservation, reproduction, and transmission of images and documents to facilitate use. Its advantages are mainly seen in collection management within museums, digital museum exhibition displays, academic research, and cultural diffusion.

Image acquisition technology used for cultural artefacts continues to innovate with the transformation of photographic technology. Before the popularization of the digital camera, the film camera existed as an essential shooting device for image collection. However, the later stages of film processing, such as developing, saving, and copying, proved complicated and time consuming, requiring a specific environment. Therefore, with the development of digital technology, digital imaging immediately replaced traditional image acquisition methods, becoming the main acquisition technology in the collection of both movable and immovable cultural artefacts. Image record carriers have since grown from a few million pixels to tens of millions, and the continuous innovation of digital technology provides more choices for the collection of cultural artefacts.

1. Image Collection for Movable 3D Cultural Artefacts

Images in museum collections are basically divided into two kinds according to their end use, namely archival images and display images. Archival images pursue fidelity. The style, color, and proportion of each part of the artefact should be as accurate as possible. These images are not to be used in the publication of traditional exhibition catalogues or in digital museum collections, as they may interfere with visitors' observation of the collection due to excessive detail.

In collecting information about cultural artefacts, the most important factor is to record every detail of an artefact, allowing images to preserve historical value. In the first census of movable cultural artefacts, the State Administration of Cultural Heritage emphasized the use of modern information as a means to investigate, identify, and register movable cultural artefacts, master basic information such as the overall state of movable cultural artefacts, and realize the integration and utilization of national information resources. Therefore, the collection of image data for movable cultural artefacts is crucial in the archiving of cultural information.

Movable cultural artefacts can be divided into two-dimensional and three-dimensional forms. 2D cultural artefacts include calligraphy and painting, rare and ancient books, weaving, and embroidery. For these kinds of artefacts, collection has been carried out using large-scale, flat digital scanners, while the collection of movable 3D artefacts largely involves digital cameras. Image data for both is stored in a file format that meets basic recording requirements. However, issues still exist with these collection methods:

1) The information collected is not comprehensive, and cannot guarantee a complete, 360° angle record of the image;
2) Because collection standards for archival images are relatively low, acquired images cannot be widely used;
3) Incomplete collection of cultural artefacts results in repeated extraction, which may incur safety hazards;
4) Longer periods of archival image collection (such as at the Palace Museum, where movable 3D cultural artefacts make up 80% of its 1.8 million artefacts) raise the question of how to achieve comprehensive and detailed image collection in a short time span.

2. *The Application of 4K Ultra-high Definition Video Technology*

Thanks to the continuing developments and innovations in digital technology, 4K ultra-high definition video image technology has become a possibility. 4K is a standard for image resolution, referring to 4,096 × 2,160 pixel resolution (four times the resolution of 2K projections and HDTV), belonging to the realm of ultra-high definition. 4K video is a motion picture. When shooting at 25 frames per second in PAL, 25 digital images at 4,096 × 2,160 resolution can be obtained for each second of video material. When using a 4K digital camera to shoot movable 3D artefacts, shooting a piece of art at 360° in 60 seconds amounts to 1,500 digital images at 4,096 × 2,160 pixels. The fastest continuous shooting speed of digital SLR cameras is 14 fps (Canon 1DX), taking much longer than one minute to complete 1,500 shots with guaranteed image clarity.

Following technical demonstration and equipment selection, we tested a 4K high-definition camera on our collection of 3D cultural artefacts. Our aim was to avoid potential safety hazards caused by multiple interactions with cultural artefacts in the process of image collection, to maximize conservation while completing our archival resources. We used the Sony F55 4K camera to capture video images. After the video files were processed, both 4K video image files and high-pixel (4,096 × 2,160) photo files were obtained, using acquisition techniques to obtain two or more archival materials in a single shot.

2.1 Selecting cultural heritage

Regional and common handicrafts were chosen as the cultural artefacts for this experiment (see Figure 1). During the shoot, the acquisition equipment had to maintain a safe distance from the artefacts (the subject), and had to meet protection requirements in terms of stability (firmness, heating attributes, mobile properties) based on the premise safety.

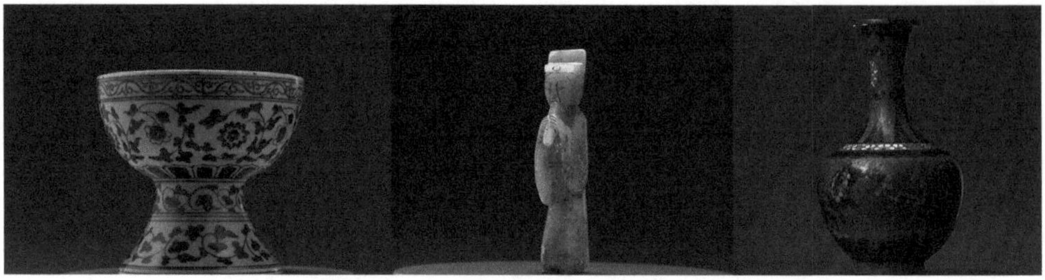

Figure 1 The handicrafts selected for the experiment

2.2 Attributes of 4K video data

The 4K image data obtained had be compatible with mainstream post-production hardware and software. It needed to be versatile, and had have expandability potential for the long-term preservation and utilization process, including video format setting, color gamut range selection, and single frame (digital photo) format output.

2.3 Pre-shooting preparation

Location: the Cultural Relic Studio at the Palace Museum's Information Technology Department.

Acquisition equipment: 4K – Sony PMW-F55 digital cinema camera, Sony AXS-R5 16bit RAW module; continuous light source – Kino FLO (5600K daylight color temperature); post-production device – Apple MacPro workstation; final output software – Premiere Pro CC, Photoshop CC, DaVinci Resolve 11.

Shooting parameter settings: 4K 16bit RAW lossless video file format, (gamut and Gamma) CineEl/Slog3, 5600K color temperature.

Other shooting devices: electric constant-speed still life turntable (30 cm diameter, 60 s/360° speed), color swatch.

2.4 The acquisition process

Following pre-shoot preparations, the shooting of movable 3D artefacts was carried out. Firstly, a subject (a substitute artefact) was placed on a professional digital still-life turntable to determine image composition, with lighting effects set according to the size, shape, and characteristics of the artefact (see Figure 2). Because this style of shooting uses a continuous light source, it is different to flash photography. The continuous light source allows a photographer to be more intuitive with the light and shape of artefacts and the contrast between light and dark. Meanwhile, for safety reasons, the kinoFLO

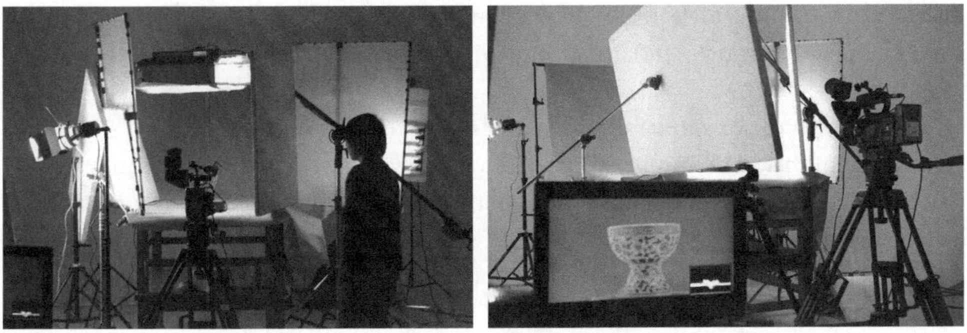

Figure 2 Setting up lighting effects

low-temperature high-color lighting device has stable source output, with a cooler light source than flash. If major adjustments have to be made to shooting equipment, or if anything occurs that may threaten the artefact, it could be moved to a safe area.

Three kinds of utensils were selected for this shoot (jade, ceramic, and enamel) with heights of around 20cm. For the sake of efficiency, they were classified according to type and size, with specific standards for light distribution. After setting up the lighting, confirming the image composition, and placing the color swatch in front of the table, the table was set to a constant rotation speed and turned on, completing a 360° video image capture in 60 seconds. During the filming process, specialist personnel were available to protect the artefacts. If a safety hazard was noted, shooting was immediately halted.

2.5 Post-production

The captured video files were in the XAVC format. First, the original video file was used to output a single frame image using DaVinci or Premiere video editing software. Image formats such as TIFF or PNG were selected. The extracted TIFF digital files were then imported into Photoshop for retouching and color management. Brightness and color correction were performed on completed video images in DaVinci, then the files were saved into the AV asset database, which could be turned into an AVI as per usage requirements. Mainstream digital files, such as MOV, MP4, GIF, TIFF, JPG, and PNG, can also be used to provide image material.

2.6 Analysis of the results

As per the image acquisition process described above, data acquisition attributes are an important aim for these test shots. It turns out that a 4K digital camera can be used to collect video images of artefacts in RAW mode, and to extract and store static images (digital photos) from the video information. The photos collected are clear and rich in layers, reaching 4,096 × 2,160 pixels in size. This is equivalent to the image quality of a 10 million pixel digital camera, and meets the requirements of file storage. Meanwhile, two kinds of file information are obtained in one shot, namely video and photo images, broadening utilization potential and improving the efficiency of image collection.

1) Feasibility study for similar filming modes from this shooting

The nature of this work can be summarized as video image acquisition for movable 3D artefacts, using a video camera as the shooting device, and simultaneously performing two forms of video and photo shooting, i.e. extracting photo images from video images. So, is it possible to shoot both video and photo at the same time with a shooting device such as a digital camera?

This test is based on a technical survey in the professional field.

Table 1 Comparison of the Main Technical Parameters of Mainstream Video Shooting Equipment

	CANON EOS-1D X	*CANON EOS-1D C*	*Sony PMW-F55*	*RED Epic-Dragon*
Sensor type	CMOS	CMOS	CMOS	CMOS
Measurement	36 mm × 24 mm	36 mm × 24 mm	24 mm × 12.5 mm	30 mm × 15 mm
Pixel	18.1 million	18.1 million	11.6 million	19 million
Highest resolution	5,184 × 3,456	5,184 × 3,456	4,096 × 2,160	6,144 × 3,160
Image type	JPEG + RAW	JPEG + RAW	TIFF + RAW (output single frame)	TIFF + R3D (output single frame)
Video type	MOV (MPEG-4 AVC/ H.264) (1920×1080) (1280×720) (640×480)	MOV (MPEG-4 AVC/ H.264) (4096×2160) (1920×1080) (1280×720)	RAW (4K/2K 16bit) XAVC (4K/2K 4:2:2 10bit) MPEG-4 MPEG-2 ProRes DNxHD	RAW (6K/5K/4K/2K 16bit)

The following conclusions can be drawn from comparisons in Table 1.

Mainstream photography equipment in professional fields cannot meet the mainstream technical requirements of video images when shooting video. The reasons for this are:

- Mainstream equipment cannot shoot RAW video images and adjust the Slog-Gamma curve;
- Due to equipment size and technical limitations, video images remain mostly at 2K standards;
- The limited lens selection when shooting video on digital cameras is not compatible with cinematic photographic lenses.

Therefore, it is not possible to shoot RAW video format, basically negating the possibility for a reverse of this test (non-RAW or Slog video formats are too small for future use and expansion of video image files, so they do not have long-term retention value).

In terms of photographic images, digital cameras can take RAW format stills, with pixels slightly higher than photo files obtained through video extraction. This still has irreplaceable value in certain fields, especially when publishing is required.

The photo files extracted from the video files taken in this experiment firstly meet the format requirement (RAW uncompressed TIFF), and secondly reach a size of 4,096 × 2,160 pixel resolution. Therefore, they can be used as general image files and in other network multimedia applications, such as web pages, mobile device terminals, and multimedia digital displays.

Given the above analysis, it is possible to conclude that a video camera can shoot both video and photos, while a digital camera can only shoot photos and low-quality videos. In the collection of image and video information about cultural artefacts, the continuous and uninterrupted characteristics of video images enable the artefacts to be displayed more vividly and comprehensively, no longer just in a single "flat" photo. For example, characters or asymmetrically shaped artefacts can be filmed in 360° from all angles, which is ideal for digital museums and multimedia displays. One-time handling for two types of images is a more practical approach to the protection of cultural artefacts.

2) Afterthoughts on the use of acquisition equipment

The core pieces of equipment (the camera, lighting, and still life table) used in this process are all existing equipment, for maximized value. However, some of the accessories of the existing equipment require improvement, such as lighting stands and RAW modules. Meanwhile, due to lens resolution and the adjustment of the video Slog-Gamma curve, the image quality in this test also has plenty of room for improvement.

3. Exploring the development of ultra-high definition video image technology

At the time of this experiment, RED and Sony launched 6K and 8K video cameras, providing long-term development opportunities for the shooting mode. In both the United States and Japan, increasing numbers of photographers are using 4K video cameras for flat portraits and still life shots, because video cameras can more easily capture fleeting moments. They can also provide more post-production options, and are more flexible and vivid in terms of creation. Increasing numbers of institutions and individuals in China have begun using this shooting mode, with professional manufacturers continuing to develop their technology and increase capital investment. This means more sophisticated high-definition cameras with higher pixels and higher image quality.

With the rapid development of the mobile Internet, mobile phones, and iPads, as well as the growing popularity of mobile networks (4G), it is possible to release video image information from the constraints of space and time. At the end of the experiment, the WeChat client upload on Apple iOS and Google Android was used to enable the mutual transfer of video files on social apps and other clients.

4. Conclusion

The rapid development of digital technology has spurred the digitization of museums, and the collection of cultural artefacts is the basis of this work. The establishment of detailed and complete information files for each cultural artefact is conducive to safety and preservation, knowledge dissemination and cultural studies, fulfilling the ideals of every museum manager and cultural artefact curator. The film photography that dominated for nearly 80 years was instantly replaced by digital imaging technology, and the emergence of newer imaging modes is not too far off. Informatization requires the use of superior technical resources to improve information levels, ensure the safety of cultural artefacts, and permanently preserve comprehensive cultural information. The realization of digital technology that is better suited to cultural artefact image acquisition is the current challenge as we move towards more efficient and more standardized archival image collection.

Data Storage and Protection at the Palace Museum in the Era of Big Data

Peng Yan
(The Palace Museum, Beijing 100009, China)

Abstract: Emerging services such as Cloud computing, the Internet of Things, and social networking have promoted the growth of data forms and scales at an unprecedented rate, and the era of Big Data has officially arrived. Data has changed from a simple processing object to a basic resource, and better management and utilization of Big Data has become a topic of general concern. The scale effect of Big Data proposes great challenges to the storage, management, and analysis of data, and changes in management methods are afoot. This article looks at the Palace Museum, explaining how this institution is an industry frontrunner in the development of the digital museum, and suggesting ideas for how it could store and protect existing data more effectively, as well as manage it for future applications.

Keywords: Big Data; Forbidden City; storage and protection

1. Big Data and the Forbidden City

1.1 What is Big Data?

With the emergence of new kinds of information distribution methods such as blogs, social networks, and location-based services (LBS), and the rise of technologies such as Cloud computing and the Internet of Things, data is growing and accumulating at an unprecedented rate, and the age of Big Data has arrived. Academia, industry, and even government agencies have begun to pay close attention to Big Data issues, which have generated strong interest.

Firstly, it is necessary to define Big Data. While its literal meaning is data on a huge scale, it is a relatively abstract concept. However, it is impossible to see a difference between the concept of Big Data and previous conceptions of "mass data" and "ultra-largescale data". There is one universal definition of Big Data, with different definitions based on a generalization of its unique characteristics. Amongst these definitions, the most representative is the 3V version, i.e. that Big Data satisfies three characteristics: scale, diversity, and high speed. International data companies believe that Big Data has to be valuable, and this value is often characterized by sparsity. IBM believes that Big Data has to be authentic. Wikipedia's definition is straightforward: Big Data is a collection of information that uses common software tools to capture, manage and process data in excess of tolerable time.

From databases to Big Data, there has been an increase not only in the volume of data, but in the variety of data types, the relationship between patterns and data, and data processing objects and tools. All are undergoing fundamental changes. In the Big Data era, data is transformed into a basic resource, and must be used to solve problems collaboratively across many fields. Google's chief economist once said that data is widely available, but what is lacking is the ability to extract knowledge from it. The fundamental purpose of data collection is to extract useful knowledge from data and apply it to specific areas. Big Data applications have a variety of characteristics depending on the field in which they are used.

1.2 Big Data in the Forbidden City

How can the variety of mastered data be adapted to the fast pace of the current era at the Palace Museum?

Let us first look at the existing data types available there. After years of informatization, the data collected by the Palace Museum on a daily basis includes images of cultural artefacts, data on ancient constructions, cultural heritage monitoring data, ticketing information, and web browsing information. Although the amount of data is increasing exponentially year by year, we must consider how to analyze and understand it, and how to apply it to daily museum maintenance and future development.

So that the variety of the data collected over the years can be applied in future, data preservation is the key issue of the day.

2. The status of data at the Palace Museum

2.1 Data storage

In view of the existing storage situation, the Palace Museum has protected its data in terms of software and hardware.

Software storage and protection entails the following:

1) Storage devices use the RAID-5 file storage format.
2) Permission settings imposed in Windows file sharing to restrict user access to important data. Access is only open to a small number of users who frequently use certain data, while most users have read-only access, and are unable to modify or delete raw stored data.
3) Anti-virus software is used to perform overall virus prevention on server systems and files.

Hardware storage and protection entails the following:

1) An online spare disk exists in most disk arrays. When a disk array fails, the spare disk will work to replace it.
2) Offline backup is performed for almost all important data, burned onto either a CD or DVD for offline backup protection, then stored in the CD-ROM library.
3) Portions of frequently used data use multiple sets of the same data at the same time to achieve load balancing and data backup effects.
4) Tape backup of important image and video data is achieved with a large tape library device.

In short, the Palace Museum uses a variety of media and multiple backup methods to protect its data, aiming to have at least one backup of each piece of data in order to reduce the potential for corruption or loss.

2.2 Potential threats to data

Given the current storage status of data, potential threats mainly arise from hacker attacks, viruses, illegal operations, destruction or tampering by internal personnel, misuse, hard drive destruction, and natural disasters.

1) Hacking: Since important data is generally stored in the office network, if the internal network is not protected by physical disconnection from the Internet, it is highly vulnerable to external hacking.
2) Viruses: As above, if not physically disconnected from the Internet, the office network is likely to be infected by viruses. With the popularization of mobile storage devices within the daily office network, viruses are increasing. Those that target system execution may have little effect on data format files, but can easily cause system crashes.
3) Deliberate deconstruction, tampering, and misuse by internal personnel: Generally, strict access control is set for important data, restricting access and applying strict rules and regulations. However, human factors are the hardest to control, and it is

most difficult to prevent intentional destruction or unintentional mis-operation by insiders.

4) Hardware failure of data storage devices: This is the highest threat to data security. The average user adopts the RAID 5 file storage format, meaning that if a disk in the array suddenly suffers a hardware failure, the backup disk will automatically replenish the data. However, if two disks are damaged at the same time, or if there is a damaged disk when a backup disk is added to the entire disk array, data loss may occur. Meanwhile, disks also have service lives. Generally, a storage device is self-initiated. When used for about three years, some disks can become faulty, resulting in data loss.
5) Natural disasters: In disasters such as earthquakes, tsunamis, typhoons, and fires, all forms of local data security protection are futile.

3. Local storage protection

In order to protect data more effectively, we propose the following security solutions as a response to the previous analysis of the current data storage status and potential threats.

1) Make full use of existing tape library equipment, backing up all important data and saving it offline. Tape disks have the advantages of a large capacity and fast recovery in comparison to optical disks.
2) Blu-ray disks are now in use, with the advantages of larger storage capacity compared to DVDs, and longer-term storage compared to tapes. However, they are also relatively expensive at present. Currently, the maximum capacity available for purchase is 50 GB, though higher capacity disks are under development. As such, it is recommended to use Blu-ray disks instead of DVDs to backup important fine image data offline, saving it to the data processing library.
3) SATA hard drives are now widely used, with a large capacity, low price, fast access speed, and long-term preservation capabilities. SATA hard disks can be used to back up data offline and create an offline hard disk library.
4) It is recommended to replace earlier disk array storage devices (which use small capacity disks that are no longer available for purchase and are unable to support existing large-capacity disks) with current larger-capacity server hard disks. Small capacity disk array libraries can be used for temporary file storage so that critical data is not lost in the event of failure.
5) Regarding storage devices purchased in the last few years, certain newer models should be purchased as replacements to reduce the risk of disk failure.
6) Develop a solid working protocol, checking various storage devices regularly, performing frequent data verification and disk checks, and identifying problems as quickly as possible.

7) For storage array libraries holding critical data, the hard disk should be replaced every three years to ensure security and real-time availability of key data.

4. *Remote storage protection*

The various methods listed above are based on the premise of a single data storage space. As the amount of data grows and its importance increases, the establishment of a single data space in a museum cannot fully guarantee data security. Thus, we must consider establishing a disaster recovery backup center for data collection, eventually forming a disaster recovery method for the two or three centers.

4.1 Objectives

According to the current disaster recovery plan at the Palace Museum, the project is divided in two parts:

1) Intra-city disaster tolerance planning

- The Palace Museum's central server room: the main data center, which is involved in business production and operation, is the centralized monitoring center.
- Beijing's intra-city disaster tolerance server room: the intra-city backup data center is used in the event of short-term unresolved failure or non-regional disasters in the central data center, switching part of the services or all of the application systems to the intra-city backup data center.

2) Remote planning for disaster recovery
Remote disaster recovery center: the remote backup data center is used for short-term critical business application system recovery in the event of a regional disaster in Beijing.

4.2 The Palace Museum's intra-city disaster solution

When composing a disaster recovery plan, we must first define the goals of disaster recovery based on specific business requirements. There are unique technological implementations for different disaster recovery goals, as well as different investment scales and operating costs. The MirrorView synchronous mirroring of storage devices (see Figure 1) implements intra-city disaster recovery. Implementation and hosting, and the transparency the operating system and database do not occupy the host, operating system, or database. Different hosts, operating systems, and databases can be recovered through a set of MirrorViews. MirrorView supports two-way disaster recovery and mutual backup, and is used to replicate data based on SAN storage networks. In the process of disaster recovery planning in different places, it can be implemented step by step according to the actual conditions.

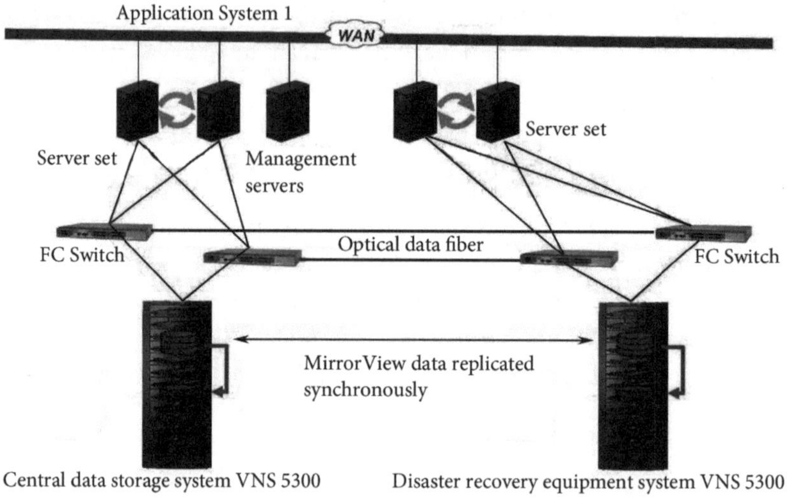

Figure 1 The MirrorView solution

For the Palace Museum, a storage array-based MirrorView solution is recommended for central disaster tolerance. In the Beijing production center and the intra-city disaster recovery center, a storage array-based MirrorView-synced disaster recovery solution with zero data loss is recommended. This scheme proposes zero data loss under the premise of ensuring normal operation of the production system without affecting performance.

4.3 The Palace Museum's remote disaster recovery plan

In order to improve data security and protection, the Palace Museum has established a central disaster recovery plan for three centers in two locations, with the aims of achieving urban disaster recovery and preventing disasters at the regional level. According to the Informatization Development Plan for the Palace Museum, the Beijing intra-city disaster recovery center is the basis for the development of a remote data recovery center. It will eventually forming a complete disaster recovery framework for three centers in two locations (see Figure 2).

The three-center disaster recovery plan for an intra-city and remote disaster recovery outlines that the production data center serves as a central production center, running most business systems and acting as a centralized monitoring center. The intra-city backup center implements application-level disaster recovery, ensuring the continuous operation of RTO and RPO functions, and defending against non-regional disasters. The remote data disaster recovery center implements data-level backup, and is generally used for testing systems to defend against major regional disasters.

A disk array-based Mirror sync data replication system has been established between the Beijing production center, the intra-city disaster recovery center, and the remote

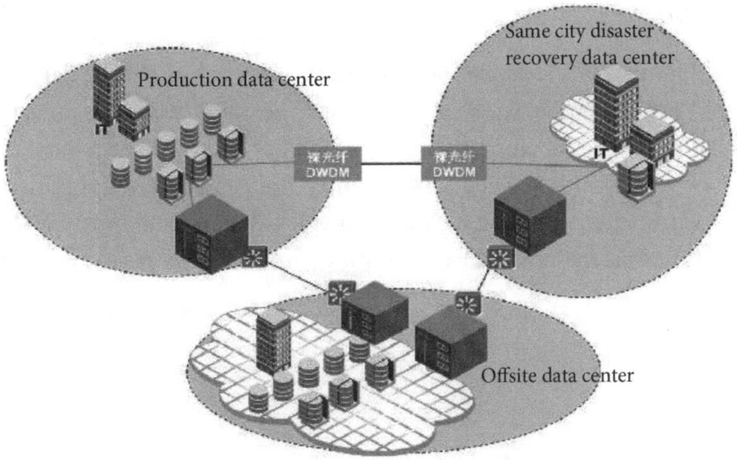

Figure 2 Disaster relief with two locations and three centers

disaster recovery center, in order to realize remote data backup with zero data loss and without limitations of distance.

- The disk arrays for the production center and the intra-city disaster recovery center are synchronized.
- The disk arrays of the intra-city disaster recovery center and the remote backup center are asynchronous.

In normal operation, the storage system at the production center is set to the source disk. The storage system of the intra-city disaster recovery center is the target disk, so data in the source disk is mirrored to the target disk via the data backup channel managed by MirrorView, with the Mirror synchronization mode ensuring that data written to the disk is never lost.

When a catastrophic failure occurs within the production system, Mirror backup data will have been synchronized with the intra-city disaster recovery center. The backup machine in the city disaster recovery center can be used to restore production data to a completely consistent state.

Meanwhile, the MirrorView of the intra-city disaster recovery center is transmitted to the remote disaster recovery center in real time, ensuring the integrity and automation of the city's disaster recovery system and remote backup system.

A disaster backup system has been established that can quickly restore system data following a disaster, so that business systems can resume operations as quickly as possible. This requires strong organization, management, and various protection procedures. Proper organization and well-trained personnel can greatly reduce recovery time, and organizational management is an important aspect of disaster recovery.

5. Conclusion

With the development of Cloud computing and the Internet of Things, data is all around us. The era of Big Data has arrived, and its correct use offers great convenience to public life. However, it also poses major challenges to traditional data management methods. In the face of new trends and challenges in this area, the Palace Museum must not neglect foundational work when applying new technologies. Saving data must be seen as a top priority to ensure that the museum will continue to showcase its renewed vision to the rest of the world.

References

[1] Meng Xiaofeng, Ci Xiang. *Big Data Management: Concepts, technology, and challenges.* School of Information, Renmin University of China.

Research into the Creation of the Beijing Museum of Natural History's Digital Collection

Huang Yanjun and Liu Jing
(Beijing Natural History Museum, Beijing 100050, China)

Abstract: The rapid progress of information technology provides new opportunities for the development of the museum industry. Its application in digital museum collections and conservation is pervasive. In 2014, the Beijing Museum of Natural History used advanced digital technology and multimedia to carry out digital processing of 15,200 of its 270,000+ artefacts, establishing an initial core collection. This paper elaborates on the institution's research, and experience in terms of the needs, users, business objectives, technical goals, and content requirements of creating a digital collection.

Keywords: museum; collection digitization

Preface

The rapid progress of information technology provides new opportunities for museum development, and its application in digital museum collections and conservation is pervasive. On entering the 21st century, the Beijing Museum of Natural History (built in 1958) began to see the importance of the development and application of information technology in the digitization and conservation of collections. Combined with functional characteristics, the museum has developed a digital protection scheme suitable for its own collections, with a guiding ideology to use digital technology for the collection and storage of related information in an all-encompassing manner. In this way, it provides services to the public through virtual museums, network communication, video broadcasting, and information inquiry, so that visitors can learn more about natural history. In 2014, the museum used digital and multimedia technology to carry out centralized digital processing of more than 15,200 out of its 270,000+ artefacts, forming a core data reserve for the whole collection, and taking a solid step towards digitization. This article will focus on the creation of the Beijing Museum of Natural History's digital collection.

1. Analyzing demand for digitization

To digitally collect and process 15,200 artefacts in a year, it is necessary to carefully analyze the demands and requirements of the creation of digitized collection. First of all, we clearly defined how to digitize the 15,200 items. All of the artefacts required 2D data collection and processing, and 80 also required 3D data collection and processing; 220 items required video data collection and processing. For this reason, we arranged for staff from the specimen department, scientific research department, science department, and exhibition department to analyze and formulate the operation flow for collecting the data, and to determine its future users.

1.1 Operation flow

The business requirements of digital collection processing for the Beijing Museum of Natural History cover operations such as collection, processing, and the creation of a database. A flow chart of these operations is shown in Figure 1.

1.2 Users of the digital data collection

The eventual users of the digital collection and processing data at the Beijing Natural History Museum are visitors inside the museum, and users outside the museum.

1) Eventual users within the museum:
   - The collection management department, i.e. staff from the specimen department, who carry out digital collection and collection management on a daily basis for the purpose of conservation.
   - Professional researchers: including researchers in the scientific research department at the museum. These users' work is closely related to the collection itself, and requires information about the collection. Unlike other users, they need more detailed data, such as utilization status and the actual effect of various conservation measures, as well as the characteristics, quantity, and type of local collections. On the basis of this knowledge, better solutions can be proposed for protection measures, and deeper scientific research can be carried out regarding the collection. This will be conducive to conservation and research.

- Popular science and exhibition users: including the science department and exhibition department staff from the museum. These users require information about classification, quantity, and preservation status, and also need high-quality images and video information for use in exhibitions and popular science.

2) Eventual users outside the museum

Users outside the museum are mainly members of the public. They require a variety of data from the collection. Digital engineering involves a platform for public communication, a window through which the public can understand collections, and a source of information

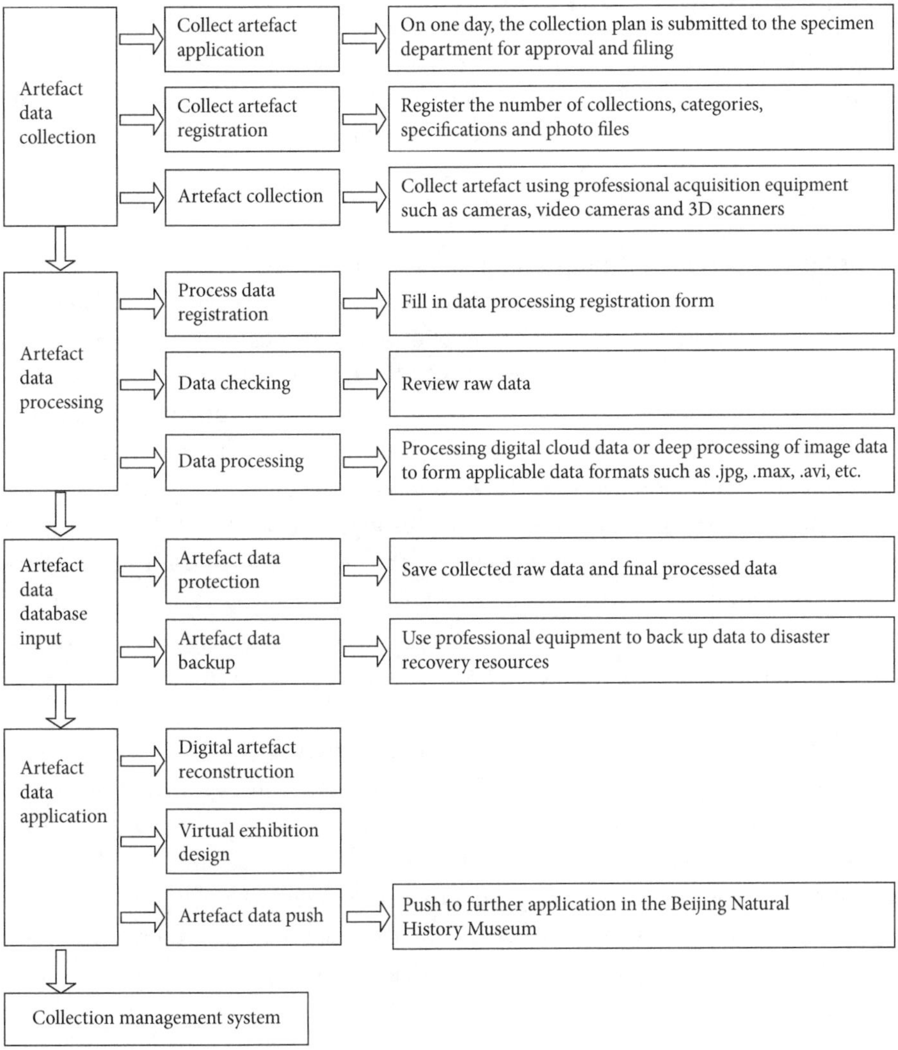

Figure 1 Digital collection and processing at the Beijing Museum of Natural History

by which the museum can obtain social cachet. Public users do not require so much professional information, but often need to obtain data on laws and regulations from the museum's public system. They also seek education from virtual exhibitions, expand their knowledge, and strengthen their understanding of natural history. Users outside of the library tend to obtain collection information through relevant channels of the Beijing Natural History Museum's website, which generally involves basic catalog information and image information, wherein image information comes in the form of a thumbnail image and a low-resolution image that can be published on the external network.

2. *Setting goals for digitization*

Clear goals are important in guaranteeing the digitization of 15,200 artefacts in the museum. After carefully analyzing the business process and various possible future applications of the data, business objectives and technical goals have been clarified.

2.1 Business objectives

1) Focusing on and standardizing existing business processes
In the creation of the digital collection, the museum must first comply with national unified standards, namely the Standards and Operational Guidelines of the National Cultural Relic Bureau issued by the State Administration of Cultural Heritage. These guidelines integrate the characteristics of natural science museums to customize standards and specifications for the processing and cataloguing system for the Beijing Natural History Museum's digital collection. Meanwhile, according to the business characteristics of the museum, digital collection and collection processing is the center, data management is the basis, and business processes are at the core. The result is coordinated, comprehensive acquisition and processing based on data sharing and data linkage, effectively improving standardization and efficiency of digitization while supporting and optimizing the overall standardization of internal business processes.

2) Objectives for business improvements within the museum
The fundamental goal of creating a digital collection is to use modern computer technology for focused digital acquisition and processing such as 3D scanning and 2D shooting, improving the efficiency and depth of our professional departments, and sharing the results more widely amongst internal employees and the public. At the core is a formation of digital resources in the collection to meet the developmental needs of the museum, ensuring ease of use, reliability, maintainability, adjustability, and scalability of data resources as well as the traceability of data.

2.2 Technical objectives

The creation of this digital collection has realized data collection and processing in the overall process of library digitization, such as data storage and applications. As such, technical objectives largely focus on the technical parameters of acquisition equipment and the format requirements of data collection.

The technical parameters of the 2D and 3D data acquisition equipment are shown in Table 1.

Table 1 Requirements for the Technical Parameters of 2D and 3D Equipment

Equipment type	*Equipment brand*	*Model*	*Specifications*
3D acquisition devices	Faro	120	Measurement accuracy: 0.07 mm Measurement speed: 122,000/244,000/488,000/976,000 point/s (adjustable)
	Handy Scan	VIUScan	Measurement: 18,000 /s Spatial resolution: 0.1 mm (0.004 in) Accuracy: Up to 50 μm (0.002 in) Content accuracy: 20 μm + 0.2 L/1000
2D acquisition devices	Hasselblad	H4D60	Effective pixels: 60 million, digital
	Cannon	EOS-1Ds Mark III	Effective pixels: 21.1 million CMOS sensors, 19 cross-focus points
		Cannon 5D Mark II	Effective pixels: 21.1 million
	Elinchrom	Light and rapid shadow-building system (lights, tripods, etc.)	Reflector, light barrier, beam hood, softbox, flash system for outdoor shooting, etc.

Format requirements for data collection: 2D data in line with the Digital Resource Processing Standard Specification and Operation Guide, with a resolution above 300dpi; 3D data in line with the use of 3D scanners for collection of scientific value, using 3D and CAD software processing, supporting resolutions above 2000dpi.

3. Digitization

The digitization of the collection involves the digital collection of artefacts and the digital processing of collections, as shown in Figure 2. We have thus far completed all major forms of 2D digitization for 15,200 artefacts, and 3D digitization on 80 valuable items.

Meanwhile, we have improved process specifications and management mechanisms for digitization resources, providing further guidance and reference for future work.

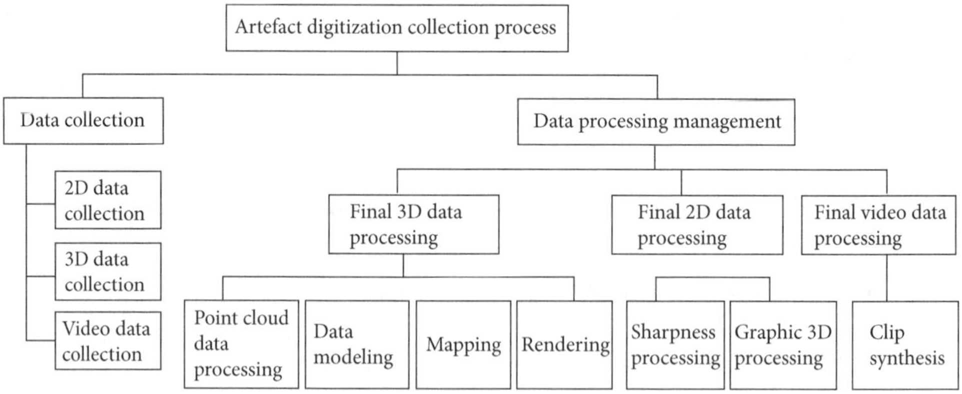

Figure 2 Development of a digital collection

3.1 Data collection

This entails recording 15,200 specimens according to normative 2D, 3D, and video collection standards. For example, for selected specimens like dinosaurs and ancient mammals, 3D data is collected with a 3D scanner. For living organisms, such as insects, fish, and ancient creatures, 2D data is collected with professional photographic equipment. Video data is collected on biological specimens using professional imaging equipment.

3.2 Data processing

Data processing refers to the specialized post-processing of acquired 2D, 3D, and video data. Amongst the data, original 3D information collected by scans is in the form of point Cloud data, which must be processed first by professional software such as PointCloud, then CAD, UG, and 3DMAX. Finally, data files are formed in a variety of formats that satisfy the museum's requirements for multiple tasks like scientific research and popular science. For 2D data, post-processing is performed by software tools such as Photoshop. It is also necessary to display the 3D state of the specimen using 2D image processing technology to reconstruct artefacts that cannot be scanned into 3D display data as per requirements. Video data is processed by professional non-linear editing platforms to output video with audiovisual synthesis.

Key digitization content and related quantities are shown in Table 2.

The 15,200 digitized artefacts are categorized in Table 3 according to the classification order.

Table 2 The Main Content and Quantity of Digitization

Type	*Content*	*Target output quantity*	*Task requirements*
2D artefact data collection and processing	2D data information collection	15,200 pieces	
	2D information processing	57,950 sheets (1 to 6 standard size photos per piece)	At least 3 to 5 shots per side are taken as an alternative and saved, totaling 173,850 to 289,750
3D artefact data collection and processing	3D data collection	80 pieces	Most specimens (series) have 50 to 300 bone parts (components) per body
	Point cloud data processing	80 pieces	As above
	Deep data processing and modeling	80 pieces	As above
	Model processing and management	80 pieces	As above
	Texture mapping	80 pieces	As above
	Model animation rendering	80 pieces	80 main pieces needed for overall production
Comprehensive visual effect production	Sample image data collection and editing	200 pieces	
	Comprehensive production and processing of late data of key specimens	20 pieces	

Table 3 15,200 Digital Artefacts Classified by Type

Collection classification	*Quantity*	*Collection classification*	*Quantity*
Insects	4,244 pieces	Ancient Humans	320 pieces
Fossils	2,723 pieces	Plants	4,006 pieces
Quality goods	2,045 pieces	Clay	155 pieces
Shells	1,650 pieces	Others	57 pieces
Total	15,200 pieces		

3.3 Overall content framework

The content of the Beijing Museum of Natural History's collection is based on ontology, with the overall structure following a logical hierarchy of selection, digitization, digital processing, and digital application of the collection from bottom to top, as shown in Figure 3.

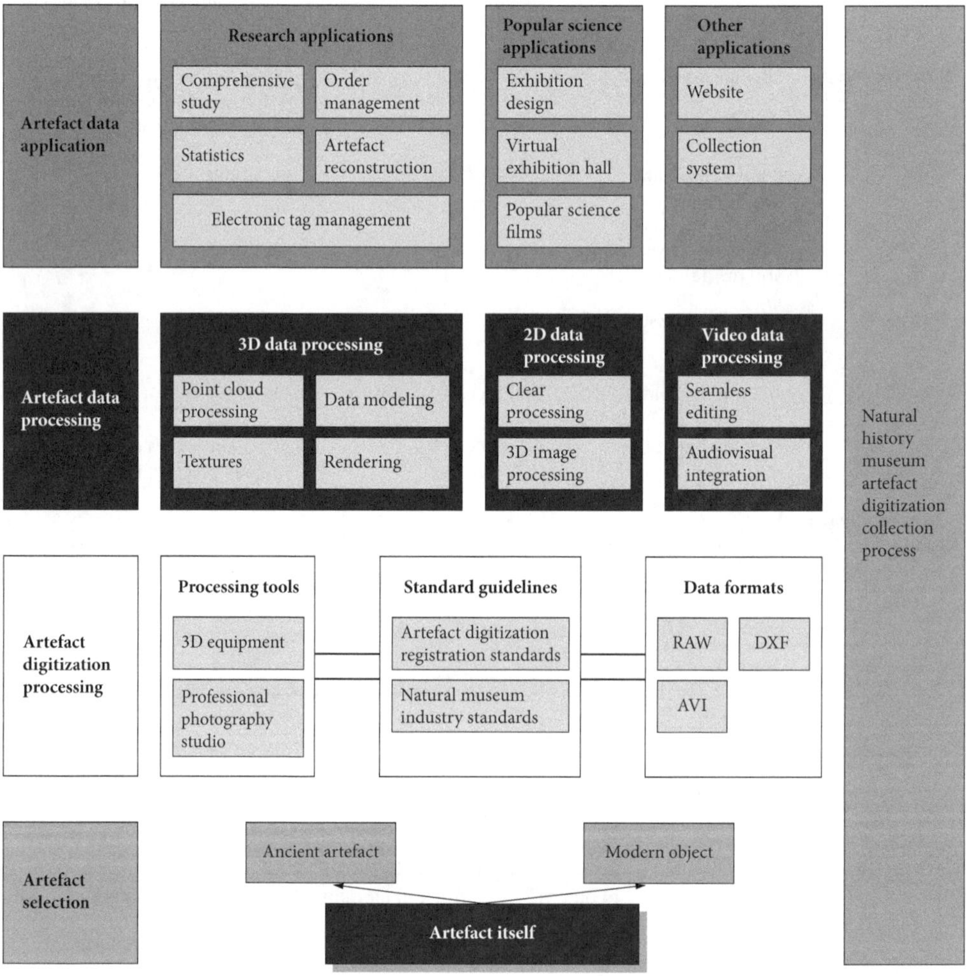

Figure 3 Digital development of the Beijing Museum of Natural History's collection

4. Lessons learned

The digitization of the collection at the Beijing Museum of Natural History has just begun. The digitization of 15,200 artefacts that has taken place so far accounts for only 5.6% of the 270,000+ items in the physical collection. We understand that there are still complex tasks to be completed. We have also realized that data collection is at the core of informatization in museums, as without digital data, there is no informatization to speak of, let alone to use in applications. Furthermore, we feel that the creation of a digital collection has great significance. Considering the need to strengthen natural heritage conservation, the creation of digital collections can effectively reduce the use of physical collections and therefore the probability of damage and destruction, whilst providing technical support for scientific research and popular science. From the perspective of cultural creativity, development, and prosperity, the results of recent digitization will offer a useful exploration into the integration of science, technology, and culture at the Beijing Museum of Natural History. It will also attempt to take a leading scientific and cultural position in China in the future.

Finding Something New in the Forbidden City: A Case Study on the Digital Aesthetics of the Jadeite Cabbage

Shen Shuqi
(Jiageng College, Xiamen University, Zhangzhou, Fujian 363105, China)

Abstract: In the 21st century, new and mobile media are the main communication trends. This article explores how and why the Jadeite Cabbage – one of the highlights of the National Palace Museum in Taiwan – creates and reproduces digital aesthetics and new visuality, examining the relationship between commodity aesthetics and creative cultural production.

Keywords: Jadeite cabbage; new media; digital aesthetics; visuality

Preface: From material orientation to people orientation

In the 1960s, the United States delivered the concept of new media, but domestic and foreign scholars still differ in their opinions about it. For example, there are claims that new media has an online basis, but it can also be built on computer information processing technology. However, what is unanimous is the emphasis of new media on "interactivity" and "technicality". Compared with traditional media, new media has the characteristics of immediacy, openness, personalization and differentiation, mass information, low-cost global communication, convenient retrieval, and integration [1]. In 1999, Paul Levinson's book, *Digital McLuhan: A Guide to the Information Millennium* predicted the important role online media would play in the era of electronic media. In 2001, Lev Manovich defined new media by its material and its logical organization in *The Language of New*

Media [2]. Doubtlessly, the 21st century is a digital age of ever-changing information [3], and the use of new media is a new direction for communication tools [4].

Currently, Chinese academic circles are still focused on the following aspects of digital museum research:

1) Education and teaching [5];
2) Collection design and display [6];
3) Historical development and development strategies [7];
4) New media applications [8].

In this new era, the traditional museum moves from being material-oriented to people-oriented. There is a change in emphasis from the "object" (artefact) to the "person" (visitor), and from traditional, static collection management, preservation, display design and history, and elitism based on academic research, to social development, lifelong education, the maximal use of high-technology to engage visitors in a large-scale dynamic way. Given the trends of new high-tech and visitor-oriented museums, how do the aesthetics of the collections themselves reflect both real and imagined digital aesthetics through the museum's official website in conjunction with new media applications? [9]

1. Digital aesthetics of the Jadeite Cabbage

In *The Work of Art in the Age of Technological Reproducibility*, literary critic Walter Benjamin (1892–1940) discusses art in the context of technical reproduction, examining how it can be expanded outdoors through indoor appreciation and performance due to mass reproduction of emerging technologies, with constant dissemination, publication, and distribution that copies the aura of the original artwork, diluting the mystery of the original. However, Benjamin was optimistic that the advent of the age of mechanical reproduction would be a positive change. For example, the emergence of photography minimized the distance between people and art, while also opening up a space for art in democratic politics. With this in mind, when entering new museums, how can the aura of art and collections be altered with continuous popularization and dissemination of new and mobile media? [10]

The era of Big Data brings public of information. Joseph Turow has pointed out that in the digital world, the Internet allows materials from the past to become the focus of attention, whether or not they are worthy of it. However, these connections do not exist in a vacuum; they are set within the context of social economy. If museums only display "objects" that can be seen at a glance as the center of their exhibits, then the extensive use of official websites offers the chance to include graphical images of artefacts. Museums will then be able to create a modern "object" through disassembling, restructuring, expanding, and piecing together. This leads to the existence of digital museums [12]. In a way, this mode transcends the category of antiquity. In another way, it also breaks the geographical boundaries of physical museums. Therefore it creates a new visual aesthetic and experience for collections.

The National Palace Museum in Taipei is a good example. In 2001, the museum began work on a five-year digitization plan for its collection. By 2012, it had completed 2D digitization of 400 pieces, producing 2,000 high-definition digital videos, and shooting digital images for 40,000 pages in rare books [13]. The virtual display of the Jadeite Cabbage – one of the most popular and best-known artefacts – was accompanied by the following textual narrative [14]:

"This piece of work, which is almost exactly the same as a real cabbage, is made from jade. Its warm theme, white body, and green leaves give a sense of familiarity and closeness. Look out for the two insects on the leaves: a katydid and a locust that symbolize many children and grandchildren…" [15]

Through the Internet, images of the cabbage are presented in the following three styles:

1) Panoramic image of the real environment: an image of the Jadeite Cabbage placed on a wooden frame at a 45° angle [16];
2) Close-up of the real environment: with the Jadeite Cabbage rotating at 360°, it is partially enlarged to show a close-up of the katydid and locust that symbolize "many children and grandchildren" [17];
3) Historical imagining: through image processing, we return to 1925, before the establishment of the Palace Museum, showing cabbages in small enamel pots [18].

As well as being constituted in the digital displays of the Jadeite Cabbage, carvings of the katydid and locust are become documents (object-lessons) in the exhibition hall, appearing in historical evidence of the real thing, and bearing witness to the historical relevance of the past in the present [19].

Thanks to developments in the science and technology of digital imaging, the Jadeite Cabbage is no longer simply a 18.7 × 9.1 × 5.07 cm collectible. It has been transformed into a display that can be viewed at any time – partially enlarged, moved by angles, displayed from different angles, in digital images that can be continually reproduced and transmitted. The work spans time and space, allowing public access to comprehensive information. In the virtual museum, information such as images, graphics, sounds, and words are combined organically using multimedia technology, offering visitors a deeper understanding of the historical and cultural information reflected by the objects. The question is whether the virtual aesthetics that exist in the online world carry the authenticity of artistic collections, or present the right to interpretation and the materiality of modernity [20]. The following is a discussion of the text and video aesthetics of the museum collection.

2. *The Digital Aesthetics of Text and Video in Museum Collections – A Case Study of the Jadeite Cabbage*

Using computer technology, digital imaging technology, and network technology, the digital museum takes digital images of exhibits from various angles and details, and then sorts them and classifies them with detailed text explanations. This data is then used to produces web pages on the Internet. The Palace Museum's official website brings its collection to life, and creates new value for the institution [21]. It contains the following text:

"We have extracted unique characteristics from world-famous artefacts, such as the Jadeite Cabbage, the Ivory Ball, the Mao Gong vessel, the Revolving Vase, and the Carved Olive Boat, through 3D imaging technology and film production, allowing the artefacts to display unexpected new expressions, reproducing skillful craftsmanship, and offering a new horizon full of fresh imagination." [22]

The digital display of the Jadeite Cabbage is enhanced from a flat catalogue menu, text introduction, and static images to a dynamic, diverse, and imaginative new exhibit. Through a 40-second video, we see vivid insects and a fresh green cabbage. As the lens draws closer, the insects fly around. Then, after a black screen, the carved cabbage appears in a field of Chinese cabbages. The two insects attempt to fly onto it, but are caught by a pair of hands. The Jadeite Cabbage remains in the background, finally entering the cabinet in the Palace Museum's exhibition hall. At this point, a single flying insect quickly turns 360° towards the cabbage, and merges into the carved jadeite. The video ends with a display of the whole jade carving on the left, and part of the enlarged image of the insect on the right, alongside the slogan "Find something fresh and new at the Palace Museum" [23]. Interestingly, this is a replacement for the heavily historic anecdote from the Qing Dynasty that used to accompany the piece:

"...this piece was originally placed in the imperial resting place of Emperor Guangxu and his concubine Jin at Yonghe Palace in the Forbidden City, so some have speculated that it was a dowry item symbolizing purity and the desire for lots of children and grandchildren..."[25]. Instead, the short film is light-hearted and entertaining, not limited to its original historical background. The Jadeite Cabbage is thus drawn from its original socioeconomic and cultural backdrop into the high-tech world of the Internet. Virtual time and space give birth to new digital image aesthetics. The cabbage can be rotated 360°, and its existence in space and time offers a new voice for digital aesthetics. If the digital museum is created using virtual reality technology, the final result of digitization is virtualization, which alters human activity and proposes the most revolutionary form of survival, i.e. being virtual [24].

In addition to the Jadeite Cabbage and its virtual electronic images, how else does the museum preserve ancient things and brand them to produce derivatives of cultural creativity?

3. Antiquities in the Present: Commodity Aesthetics of Cultural Products Derived from the Jadeite Cabbage

The National Palace Museum in Taipei displays a wide range of cultural products, and its cultural and creative output is estimated to exceed 200 million RMB [26]. Culture needs brands, and all brands have an ingrained cultural background to support. Brand experience is inseparable from cultural experience. Therefore, the creation and exploration of a cultural brand is the essence of cultural creativity, with cultural positioning at the core [27]. Today's museum is no longer a storeroom for goods, but a business that must attract visitors with the promise of a journey of discovery, rather than simply a corridor of dusty displays [28].

As the Jadeite Cabbage is defined as one of the most popular, easily accessible, and highly recognized artefacts within the National Palace Museum, it has spawned a variety of accompanying products, including umbrellas and compact umbrellas, vases, earrings, pendants, tableware, salt and pepper shakers [30], fruit forks, stationary, pens and erasers, and mooncakes.

The Jadeite Cabbage is reproduced on many cultural goods. For example, the folding umbrella looks like a Chinese cabbage surrounded by layers, with the edges imitating fresh leaves shaded with subtle color, and a white main body decorated to look like a lotus leaf. The vase transforms the original jadeite into high-grade porcelain with soft light and dark green lines. The original lush green id recreated with corn-like bright yellow-orange bugs running up and down the surface.

In the Internet age, advancements in digital reproduction technology have brought unprecedented capabilities for imitation. The Internet has become a visiting space that is accessible, readable, interactive, visually enriched and intensive – a replacement for the real world and the practice of "remote existence" [31]. An electronic image is not a substitute for a real object, nor is it another form. In fact, it is not material at all, but a metaphor for reality. We have come to accept that electronic bytes can represent atoms, and have learned to naturally welcome this metaphor [32]. New media and the Internet present the Taipei National Palace Museum with the possibility of "finding something new at the Forbidden City", rather like putting old wine in new bottles.

The representation of the Jadeite Cabbage encourages us to consider how the visualization and interactive dissemination of electronic images can display the underlying richness of Chinese culture within artefacts in collections, and how virtual images are used to transcend the boundaries of cultural geography and create virtual aesthetics and brand positioning within the online world in the new century.

References

[1] Kuang Wenbo, 'What is New Media Ultimately?', *News and Writing*, July 2012: 24–25.

[2] Levinson, Paul. *Digital McLuhan: A Guide to the Information Millennium*. London and New York: Routledge, 1999. Levinson, Paul, Song Wei, *Digital McLuhan*. Taipei: Owl Press, 2000.

[3] Lev Manovich, *The Language of New Media*. Cambridge, Ma: MIT Press, 2001 ,p.11.

[4] Joseph Turow, "Introduction: On Not Taking the Hyperlink for Granted," in *The Hyperlinked Society: Questioning Connections in the Digital Age*. Ann Arbor: University of Michigan Press, University of Michigan Library, 2008, pp. 1–18.

[5] Guan Xin, 'Digital Museums and Public Education', *Museum Research*, 2006, 93: 3–7.

[6] Guo Liping, Wei Kai, 'Research on the New Model of Cultural Relic Display', *Digital Museums, Art and Design (Theory)*, December 2009: 64–66.

[7] Chen Gang, 'Analysis of the Concept, Characteristics, and Development Models of Digital Museums', *Chinese Museum*, March 2007: 88–93.

[8] Chen Guoliang, Tong Yin, Hu Jiang, 'Research on Digital Museum Applications', *Computer Applications and Software*, Vol. 21, No. 5, May 2004: 50–52, 65.

[9] Zhen Shuonan, 'What is New Museum Science?', *Chinese Museum*, 2001, 1: 25–32.

[10] Benjamin, Walter, translated by Xu Qiling, *The Work of Art in the Age of Mechanical Reproduction*, Taipei: Taiwan Photo Studio, 1999, p. 70.

[11] Joseph Turow, *Hyperlinked Society: Questioning Connections in the Digital Age*. Ann Arbor: University of Michigan Press, University of Michigan Library, 2008, p.21.

The Application of Wireless Sensing Technology for the Monitoring and Control of Indoor Environments

Niu Riuyan and Li Yuan
(Institute of Technology, Beijing Union University, Beijing 100101, China)

Abstract: This paper discusses the way in which the indoor environment can be maintained according to the relay and control of received indoor monitoring data. This is accomplished by using indoor sensors for temperature, humidity, and illuminance to collect monitoring data for indoor environments, and then using wireless networks to transmit data, and wireless receiving systems to receive it.

Keywords: temperature sensor; humidity sensor; illuminance sensor; wireless sensor network

Preface

With the rapid development of wireless sensing technology and IoT applications, wireless sensor network technology has been used widely in the fields of industry, agriculture, transport, medicine, and defense. Thanks to improved modern living standards, environmental requirements are increasing. Therefore, a comprehensive indoor environmental monitoring system for the improvement of public life and work has been created, with broad prospects for application.

Traditional indoor environmental monitoring and control generally requires manual environmental detection. This not only wastes manpower, but also has certain limitations as a monitoring and control system, as there is no self-regulation function. At present, most indoor environmental parameter monitoring systems are wired, requiring a large quantity of wiring that affects interior aesthetics, and is inconvenient for system maintenance and management.

Wireless sensor network technology is mainly designed for short-distance, low-power, low-speed data transmission. Data nodes are connected and forwarded through a unique wireless transmission chip, forming a wide range of coverage to accommodate a large number of nodes. The network between sensor nodes can be composed freely and intelligently, and is self-organizing, meaning that its nodes can form smart network connections, and these connections can adopt different topologies depending on requirements. The network also has a self-maintenance feature, meaning that when a node experiences issues, the data transmission of other sensor nodes in the network is not affected. Due to these advanced and flexible features, sensor network equipment is easy to install and maintain, allowing for large-scale deployment without increased the cost of a single node.

With wireless sensor network technology, the indoor environment monitoring and control system is designed to achieve real-time monitoring of indoor environmental parameters such as temperature, humidity, and illuminance, as well as to organize monitoring data for the establishment of an environmental parameter database. Wireless sensor technology is able to form a wireless monitoring network through sensor nodes, collecting required data and transmitting it to the monitoring terminal in real time through the wireless network. According to the data collected, the system is ultimately able to collect data regarding temperature, humidity, illuminance, and smoke. It automatically controls relays that are connected to the device in relation to pre-set ideal indoor environment parameters, so that the indoor environment is always maintained.

1. Building the Wireless Sensor Network

Wireless nodes are used to transmit and receive wireless data, forward data, and form various network topologies of the wireless local area network. They reform a traditional sensor module socket to form a wireless sensor module, and forward data collected by the sensor module, thereby building a wireless sensor network. The wireless transceiver module uses CC2530 as the main controller. CC2530 has multiple operating modes, making it especially suitable for systems with ultra-low power requirements. Short transition times between operating modes further ensures low energy consumption. Sensor data management utilizes STM32F107 as the main controller.

First, one wireless node is selected as a coordinator, downloading the corresponding coordination program. Second, the router and data acquisition programs are downloaded to the corresponding wireless sensor module, with all wireless node channels set to one program channel. Once all wireless nodes are switched on, a blue LED flashes. The coordinator displays the channel number, and the router displays the corresponding address, indicating that all wireless nodes are successfully working on the network, as shown in Figure 1.

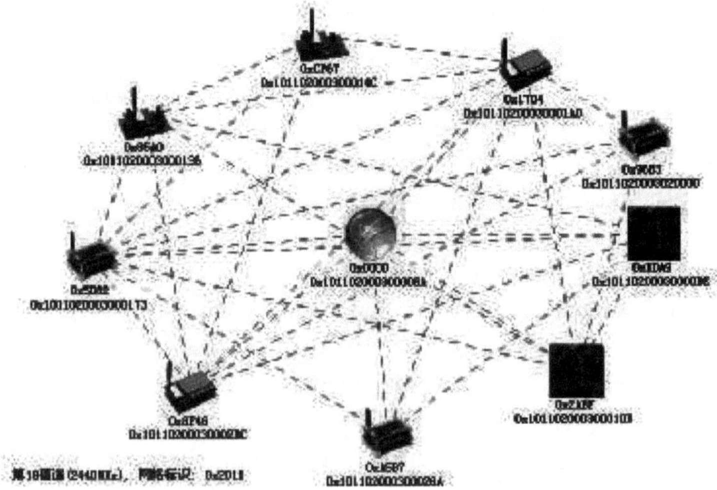

Figure 1 The topology of the wireless sensor network

2. Structure of the System's Hardware

The wireless sensor modules are an important part of the monitoring system, as they are responsible for the collection of relevant data within the space. The terminal node generally installs different sensors depending on the placement and type of monitoring data. The relay module uses high and low levels of CC2530 output as control signals for the drive circuit, controlling the closing and opening of followers, lights, humidifiers, alarm systems, gas switches, air conditioning units, and curtain circuits from outside the equipment.

The coordinator node is connected to the computer via a USB cable, with other sensor nodes arranged according to the actual conditions of the indoor environment, with hardware connections as shown in Figure 2.

3. Results of Actual Operation

By installing indoor environmental monitoring software on the computer, downloading corresponding programs to all wireless nodes, connecting system hardware according to Figure 2 and turning its power on, and configuring the corresponding serial port data, sensor nodes collect indoor environment parameters online, transmitting data to

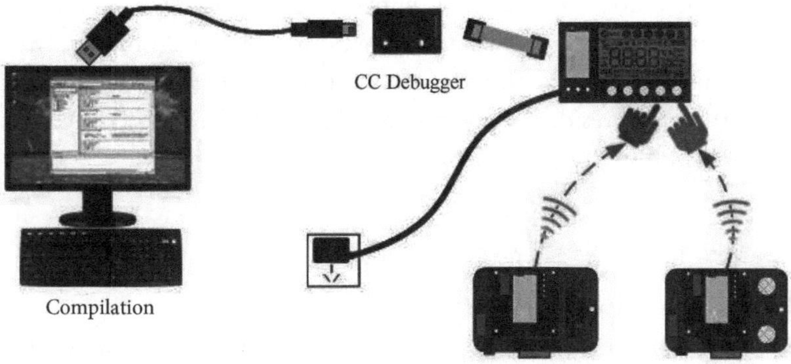

Figure 2 System hardware connection

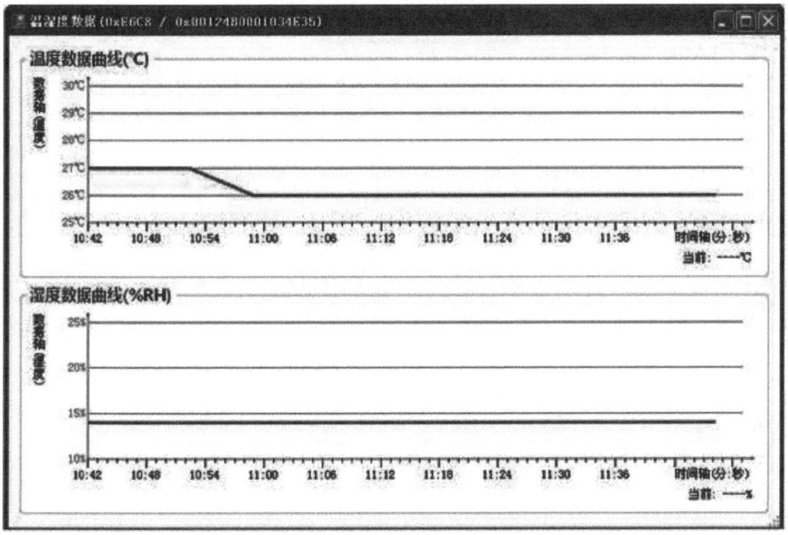

Figure 3 Curve of temperature and humidity detection data

the coordinator over a wireless network. Upon receiving data, the coordinator uses the serial port to collect and display it. According to indoor environment detection data, the monitoring system then issues corresponding instructions, controlling the opening and closing of corresponding devices wirelessly, and thereby realizing control over the indoor environment. The result of this operation is shown in Figure 3.

4. Conclusion

Wireless network sensing technology is used in the design of a monitoring and control system for indoor environmental parameters. The system realizes the real-time collection and processing of indoor temperature, humidity, illuminance, and smoke, with the ability to monitor the various signals in the environment in real time at any time, while effectively solving many drawbacks of the traditional monitoring system. This measuring and control system features definite development prospects thanks to its convenient use, low power consumption, and high reliability.

References

[1] Zhang Dehong. 'Research on Internet of Things technology for coal mine safety detection systems', [J] *Coal Technology*, 2011, 12.

[2] Xian Xiaodong, Chang Chao, Hu Ying. 'Design of an indoor environmental monitoring and early warning system based on WSNs and GSM', [J] *Sensors and Microsystems*, 2011, 6.

[3] Xu Yun. 'Design of an indoor environmental monitoring system based on the Zigbee wireless network', [J] *Electronic Production*, 2014, 24.

Digital Communication: A New Carrier for Nostalgia

Du Jianfeng
(College of Applied Arts and Sciences, Beijing United University, Beijing 100191, China)

Abstract: Nostalgia is a common human emotion, relying on a deep cultural matrix that is closely linked to the long history and culture of the nation. All nostalgia, in its final analysis, is cultural nostalgia, engaged with the profound cultural memory of a nation. With the progress of modernization, globalization, and urbanization, traditional nostalgia has been broken down, and a new modern nostalgia must be recreated. In preserving various carriers such as nature, society, and art, the interactive presentation and dissemination of information in digital form is regarded as a new carrier for nostalgia.

Key words: nostalgia; cultural implications; cultural memory; fracture reconstruction; carrier

Preface

Nostalgia is a common, eternal, and deep-seated human emotion, with close links to history and national culture. All nostalgia, in the end, is cultural nostalgia and national memory. How can a city pass on nostalgia? How are traditional folk memories to be transmitted? How are history and culture to be protected? How can we live in harmony with nature? How can we find new carriers for nostalgia? These questions have triggered new ideas about the inheritance and conservation of urban history and culture.

1. The Cultural Implications of Nostalgia

"Nostalgia" is a natural psychological mechanism and a universal, deep-seated human emotion. It is a pervasive desire to recapture lost times, particularly among people who have left their homeland. This feeling of homesickness "lies within the geography of nostalgia, or is based on the geography of nostalgia, but beyond region, blood, people, and even clan, deep nostalgia also injects deep cultural understanding… the farther away, the longer one is gone, the less it is used, the stronger the feeling, the more one must explain, the more out of hand it becomes" [1].

Nostalgia exists on three levels. The first level is thoughts about relatives, friends, and fellow citizens. The second is the nostalgia for one's homeland – its mountains and rivers, and scenes of old. The third level is the deepest: a profound affection for the history and culture that lies at the foundation of life. The first two levels are sub-levels, and the latter is a meta-level. Nostalgia is both a concept of time and space, as well as a concept of culture. Firstly, it is a psychological concept – a universal psychological condition when traveling abroad. Secondly, nostalgia is a concept of geographical distance; the farther one is from home, the stronger it is. Thirdly, it is a concept of time – its contents and objects are often people and things in memories of times gone by. Finally, nostalgia is a cultural concept; sometimes one's "hometown" or home is not a birthplace or place of origin in a narrow sense, but also involves a spiritual homeland in a broader sense [2].

Nostalgia is also a kind of pain. Yu Guangzhong's poems about the tragic feelings of rootless people drifting in Taiwan in a particular historical period express a sense of reliance in his mother culture, reflected in his nine works about nostalgia. Yu writes: "Nostalgia is a kind of memory for me. If it is the nostalgia of fellowship, it is simple, but a literati – a nostalgic scholar – adds time, multiplies time, and takes on cultural memory. As people, we all have short memories, only lasting for decades. However, a nation has memories lasting for thousands of years – huge memories, and those memories become allusions, myths, history, which are brought wherever one goes, so that the spirits of the Qin, Han, Song, and Tang Dynasties all inhabit in one's body." [3] History can never wash away human memory or the pull of the homeland. The distance inherent in nostalgia is not only physical, but also spiritual.

2. Nostalgia and Cultural Memory

History and culture are the fundamental spiritual carriers of any nation. The "homeland" is also a place built by culture, and guardianship and attachment to it constitute the traditional and modern ideas of "nostalgia". Like life, death, and love, nostalgia it a common theme in art and literature, expressed in countless works in ancient and modern China and across the world. Since ancient times, nostalgia has been interpreted as an eternal theme of wandering nomads. The poet Wang Wan of the Tang Dynasty wrote:

"Who will send my letter home without delay? I see no northward-flying wild geese here" (*Passing by the Northern Mountains*), while Zhang Ji wrote: "Outside Gu Su city lies Cold Mountain Temple; at midnight the sound of bells reach the ferries" (*Maple Bridge Night Mooring*). Wang Wei's hometown poem contains the lines: "a stranger alone in a foreign land" (*Thinking of My Brothers on Mountain-Climbing Day*), and Wang Changling writes that "men came from thousands of miles, never to return again" (*To the Frontier*). In *Meeting a Messenger Going to the Capital*, Cen Can writes "to the East, a long long road home appears", and Fan Zhongyan writes "a cup of wine does not drown nostalgia for far away home" in *Pride of the Fisherman*. This proves that the expression of nostalgia is a deep cultural memory, leaving significant spiritual wealth for future generations.

In *Jujube Stones*, Xiao Qian tells the story of an old classmate who yearned for his motherland and hometown while living abroad. Before his visit, the author received a letter from the old man who had settled in the United States many years ago, asking him to bring a few jujube cores with him. The author found this odd. Only when he saw the elderly man did he realize that, although he had lived abroad for many years, he still missed his motherland and hometown. Every year at Christmas, he thought of years gone by, the sound of firecrackers, and yearned to plant a lush green jujube tree in his laneway. He even built a simulated "North Sea" in his own garden, and tried to plant the jujube seeds in his American home. When it took root, sprouted, and grew, it served to dispel his endless nostalgia.

The writer Jia Pingwa is a major figure in the Shaanxi literary community. His novels are imbued with the deep cultural and spiritual heritage of Sanqin. The Shangzhou series of novels are seen as the cornerstone of Jia Pingwa's literary world. Shangzhou, like Shen Congwen's Xiangxi and Sun Li's Lotus Lake, has become the carrier of a distinct culture and an accumulation of national psychology. Shangzhou is Jia Pingwa's hometown – the place where he lived from a very young age, and he is extremely attached to it. He says, "comfort for this peaceful soul, for twenty long years, was the overlapping rock and round moon outside the front door. These were the two books I read at the time, and many wonderful things were obtained there… I think that my life, my written life, is that mountain stream" [4]. His early works were written in Shangzhou, and eulogize its mountains, water, moon, and people. It is the author's spiritual utopia.

The film adaptation of Taiwanese writer Lin Haiyin's autobiographical novel, *My Memories of Old Beijing*, directed by Wu Yizhu, has touched countless visitors with its "faint mourning and deep lovesickness". The stories, characters, spaces, and scenes are vividly attractive, demonstrating the emptiness and quaintness of the old capital through scenes that are full of life. In the afternoon, amidst the sound of buzzing, an old white-haired man has a chesty cough and feels drowsy. In the winter sun, children play and laugh. There are camels in the *hutong* (alley), bleak screams in the wild, and the clinking waterwheel as water flows through the wells. This constitutes the background of the film, forming a unique artistic atmosphere. The film relates the past from the perspective of a young girl named Xiao Yingzi. Through the presentation of the tragic fates of the characters, the helplessness of separation, and the impatience of life, we are offered a colorful folk painting of Beijing. Amidst the rustic feeling, there is faint mourning and sorrow, and also a remnant of the author's deep nostalgia for her childhood. The film presents three stories with no causal relationship, similar to the natural cross-sections of life, and eventually ends with "parting" emotions. Through Xiao Yingzi's acquaintance with the people around her, she experiences a range of emotions until the end, when "my dear father is gone, and my childhood is over". These ambiguous thoughts and relationships imply an attachment and nostalgia for the home, the *hutong*, the courtyard, environment, and relatives and friends from the past. They are also a recollection and memory of lost years and a vanished culture.

The above-mentioned works, such as poetry, novels, and movies, offer a clear representation of the deep nostalgia of their creators. Profound attachment and eternal feelings for one's homeland, loved ones, family, nation, history, and culture are carried in a variety of forms, creating cultural memory.

3. Traditional Carriers of Nostalgia

As a profound and lingering feeling, nostalgia relies to some extent on certain carriers. This was the case with the creation of ancient Chinese nostalgia-themed poetry. In terms of motifs, ancient Chinese nostalgia involves rich and complicated dimensions including the hometown and homeland, geography and culture, and emotional sustenance, forming several patterns of artistic expression that extends to modern poetry. In the writing process, several literary models have been formulated such as nostalgia from height, moon-related nostalgia, visual nostalgia, festival nostalgia, chromatic nostalgia, childhood nostalgia, distant dreams of angelica, and nostalgia for autumn winds at dusk [5]. As such, mountains, bright moons, objects, solar terms, sounds, dreams, heights, and sunsets have become carriers of nostalgia. For the old man in *Jujube Stones*, the jujube seeds he asked a friend to bring from his hometown were a way to transplant the deep-rooted jujube tree from Beijing and his hometown laneway into a foreign country. The overlapping rocks and the bright moon in front of Jia Pingwa's house were a comfort for his soul and a source of enlightenment in his life. In Yu Dafu's story, *Autumn in Peiping*, his cultural nostalgia appears in a visual psychological image. When he lived in the south, and arrived in Beijing in the autumn, he would always remember the reeds of Tao Ranting (Leisure Pavillion Park), the shadow of willows in Diaoyutai, the insects in Xishan, the night moon in Yuquan, and the bell of Tanzhe Temple. In his view, everything about Beijing in autumn was beautiful, be it a morning glory flower growing on a wall, lines on the ground after the sweep of a broom, sophora flowers scattered on the ground, or faint humming and chirping. He had a desire to "compress two-thirds of life, in exchange for a one-third fraction" to spend autumn in the North.

In *Yinzi's Nostalgia*, Lin Haiyin compiles 56 recollective essays about old Beijing, Taiwan, and teachers and friends. Amongst them, *Bitterness of Beiping* describes the author's interpretation of Beiping, including deep memories and infinite nostalgia for the four seasons: "Beiping can't be forgotten. I lived there for so long, like a tree rooted to the ground. Childhood, girlhood, and womanhood – half a life was spend there. Happiness and sorrow, laughter and tears, the ancient city once poured out all my feelings. Spring and autumn… How do I become acquainted with the seasons there?" [6] With its bright and beautiful springs, summer twilights, dull and desolate autumns, and early winters, when the old *hutong* have long disappeared, the hometown of Xiao Yinzi's memory is still alive. A word, a sentence, a line, and a sheet of paper carrying human expression represent those who have remained in Beiping – those who share a common mother culture.

In an era of rapid change, records are a cultural responsibility. The retention of images will allow us and our descendants to remember moments from the past, particularly through documentaries. The TV documentary *A Bite of China* shows that under the guise of modern civilization, Chinese tradition is being gradually lost. Old men stay in the countryside while young people pour into the city, leaving the fields barren, with virtually nobody engaging in traditional farming and labor. Small workshop-style production methods exist only in images, with almost no one taking over, as the younger generation is less willing to continue the cyclical lifestyle of their forebears. In this context, people who move to cities use food to dispel their homesickness. In *A Bite of China*, food acts as a familiar idyllic pastoral song, comforting modern people who are far away from home. As such, the traditional characters of nostalgia are now cultural artefacts such as poems, prose, novels, and sounds.

4. Fractures and Reconstructions

The village is an major carrier of traditional cultural nostalgia, and an important image of rural China. However, with the progress of modernization, globalization, and urbanization, the increasingly dire ecological situation has caused villages to stagnate, resulting in a barren wasteland. This runs contrary to the nostalgic image of quiet rural homesteads. Those who live on the land, especially the new generation of young people, are no longer attached to pastoral modes, choosing to view the village as a station on the way to the city. However, cities offer no roots, so migrants often feel a sense of transience. Meanwhile, people who grew up in cities have also lost aspects of themselves in the modern attack on tradition.

What are the consequences of this degradation of tradition? The ancient city of Beijing no longer exists. The checkerboard streets that once impressed Marco Polo have become thicker lines, with *hutong* alleys converted to rubble. The spacious and bright courtyards that were commonplace have become rare. Only a handful of *hutong* are yet to be demolished, and many of the courtyards have been turned into small shops. The buzz of traditional life is no longer apparent. The former atmosphere of the Beijing courtyard has been lost, and *hutong* housing has been razed to the ground.

Chen Kaige's 10-minute short film *100 Flowers Hidden Deep* depicts a scene in which the traditional culture corresponding to the material home (*hutong* courtyard) is destroyed during demolition. Mr. Feng's madness symbolizes the serious consequences of this destruction. The workers are sadly indifferent and numb in the face of disaster. The film's audio is particularly effective, with virtual moving sounds that are both interesting and relevant to the subjective perspective of the madman. When a non-existent "period vase" is broken by a slip of the hand, a sense of absurdity comes through in a long, sad sound, juxtaposed with the slow, forward-moving lens until Mr. Feng's mournful face appears and the tone of the film transitions from absurd to heavy. When Mr. Feng shakes his bell and runs to his "new home", an animated quadrangle appears, accompanied by a combination of wind chimes, screams of hawking, and traditional instrumental music. These sound effects take the audience back to the "good times of the past", powerfully setting the theme and pushing the film towards its climax.

Digital technology has changed ways of life and discourse. In today's digital communications, nostalgia must be transmitted through new carriers such as digital representation, the digitization of historical and cultural resources, re-presenting disappearing objects (such as hometowns, villages, landscapes, and homes), and transforming physical exhibits into interactive displays in digital form. In one way, this embodies the concept of respecting nature, conforming to it, and integrating with it. It enables residents to see mountains and water, remember their hometowns, protect and carry forward traditional culture, and continue the historical context of the city. In another way, through digital communication, we can realize a sense of reality and power in the process of information transmission. This can strengthen our identification with history and culture, and realize effective communication with nature, society, and the self through its interactive characteristics. Modern people can then experience the rich civilization and history of their ancestors and feel closer to their homeland, ensuring the survival of cultural memory.

References

[1] Yu Daxiang. 'Discussing cultural nostalgia', [J] *Journal of Hunan Normal University (Social Science Edition)*, 2003, 11.

[2] Zhong Haifeng. 'Cultural nostalgia in the context of globalization', [J] *Journal of Henan Normal University (Philosophy and Social Sciences)*, 2008, 2.

[3] Ma Yuyao. 'Looking at Taiwanese cultural consciousness in the modern nostalgic poetry of Yu Guangzhong — Research on modern Taiwanese nostalgic poetry', *Literature Research*, 2009, 5: 73.

[4] Li Zhaohong. 'Two cultural perspectives, two aesthetic tendencies', *Journal of Shangluo University*, 2007, 3.

[5] Lin Haiyin. *Yingzi's Nostalgia* [M]. Jiangsu People's Publishing House, 2014.

[6] Liu Yan. 'Historical writing from cultural nostalgia to homeland memories — Lihua and Yan Geling as examples', [J] *Journal of the Southwest Minzu University (Humanities and Social Sciences)*, 2010, 1.

[7] Du Jiangfeng, Sun Hongyun. 'China on the tip of the tongue: passionate praise and helpless elegy', [J] *Journalism Lover*, 2012, 12.

[8] Yang Jinglong. 'Traditional themes and modern writing of Chinese nostalgia poetry,' *Literary Review*, 2012, 5.

[9] Gao Ertai. *Looking for a home* [M]. Beijing October Arts & Literature Publishing House, 2011.

[10] Song Tao. 'The fracture and reconstruction of cultural nostalgia', [J] *Shandong Social Sciences, 2012*, 5.

The Display and Dissemination of Urban Cultural Heritage in the Context of New Media

Zhang Jing and Chen Yuexin
(Department of History and Culture, College of Applied Arts and Sciences, Beijing Union University, Beijing 100191, China)

Abstract: Urban cultural heritage is a unique symbol of urban identity, and a valuable basis for modern cities to trace their cultural origins, locate their positioning, restore urban identity, and gain public recognition. The in-depth study and excavation of urban cultural heritage, and the effective use and dissemination of urban cultural wealth are necessary measures in highlighting urban characteristics and promoting cultural development in cities. The application of new media display, dissemination, and means of service in the field of cultural heritage is conducive to the conservation of inheritance, and the innovative development of cultural heritage, thus promoting the sustainable development of urban culture and economy. This paper focuses largely on ideas and practices for the display and dissemination of urban cultural heritage in the context of new media.

Keywords: urban cultural heritage; new media; display; communication

Cities have long denoted a convergence in population, resources, public facilities, business, and cultural activities. The development of urban resources requires both the maintenance and promotion of the city by political and economic forces, as well as the agglomeration and inheritance of urban culture. Cultural heritage reflects the trajectory of a city – a concentrated expression and interpretation of urban characteristics, and the reality of urban memory [1]. A profound scientific understanding and exploration of cultural heritage to refine and enrich the identity and appropriately located in a city's characteristics are at the foundation of urban culture. Having an accurate grasp of urban cultural identity lays a good grounding for the acquisition of general public recognition and positioning for the direction of urban cultural development. The assuming of an "identity" is an important way to establish a positive cultural image, expand urban culture, create internal consensus among citizens, and inspire strong spiritual drive and intellectual support for urban economic and social development.

Combined with new concepts and technologies, new media's means of display and dissemination have broad and important application prospects for the development of cultural heritage. Using new media to explore, promote, and disseminate urban spirit allows cultural heritage to become active in the development of an urban identity, promoting the development and innovation of urban culture. It is thus a feasible way in which we promote sustainable urban development.

In recent years, as a new means of communication, new media has been applied to cultural dissemination in the art field both in China and abroad [2]. For example, the Cleveland Museum of Art has recently applied a large amount of new media art and technology, including large touch-screen walls and motion recognition matching (see Figure 1), offering a new look at contemporary museums. In China, the Nanjing Museum has built a digital experience hall (see Figure 2), and the new media approach has been applied centrally. New media art has attractive new entertainment features, with a strong audio-visual appeal and capacity for immersion. It is distinguished from other means of artistic communication by its opto-electronic characteristics. More importantly, new media art has brought about a new narrative style, with richness, vividness, interactivity, and non-linear characteristics that propose to change the traditional mode of museum discourse, providing a new way to experience urban culture.

The basis of cultural communication involves accurate analysis, understanding, and positioning of cultural resources. The display and dissemination of urban cultural heritage also refers to the refinement and expression of visual elements on the basis of an accurate grasp of a city's cultural heritage resources. For each city to understand its own specific value and locate its core spirit and traits, it must explore its historical trajectory

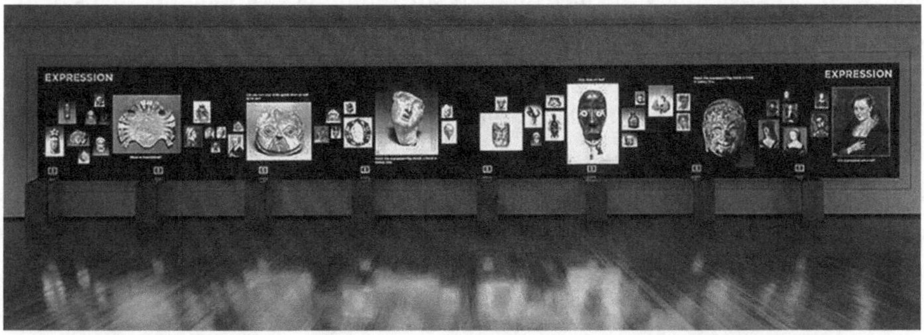

Figure 1 New media showcase at the Cleveland Museum of Art, USA

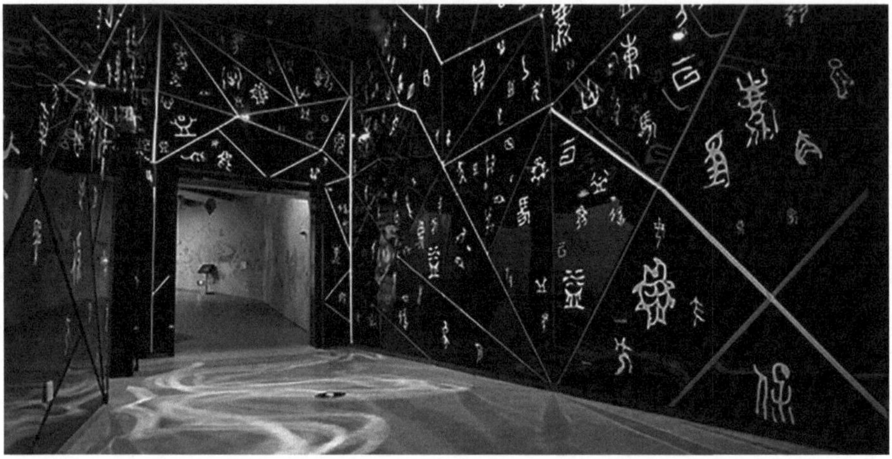

Figure 2 Digital experience hall at Nanjing Museum, China

from the aspects of urban roots, development context, and urban characteristics, thus grasping the lifeblood of sustainable development.

First, the city must pursue the "source" of urban culture from its origin and its geopolitical environment.

"Seeking roots" is a natural human compulsion, and the positioning of cultural heritage within urban identity is aligned with urban roots. China has a vast territory and obvious discrepancies in natural environment, and the causes of urban development are different. While the population gathers to survive and prosper in cities, this form of culture incorporates complex factors of corresponding natural humanities. The cultural starting point of diversity explains fundamental problems in urban existence, greatly influencing the direction of urban development.

Central and local political locations refer to capital cities and regions, such as Xi'an and Luoyang. Under the guarantees of political power, most of these cities are well-positioned and well-suited to long-term settlement. After a certain period of resource development and management, these areas form economic, transport, and cultural centers within a region, with a far-reaching influence on their surroundings. Although perhaps damaged or even ruined due to factors such as changes in dynasty or war-time disasters, cultural heritage is preserved within cities, temples, and large public buildings.

Military-based cities are located largely on borders and in inland areas, such as Jiuquan, Dunhuang, and the military stations and barricades along the Great Wall, as well as coastal areas established in the Ming Dynasty. Many of these cities hold the ruins of defense facilities such as fortresses and enemy barracks, and archeological discoveries have included military cultural artefacts such as weaponry.

Transport- and trade-oriented cities include land and waterway intersections, stations, towns, waterways, coastal terminals, and harbors. They rose up thanks to trade, transport, and business travel, and can be far from a political center, prospering with population growth. Archeological digs have unearthed carriages, ships, and related cultural artefacts.

Commercial, handicraft-based centers have developed into specialized cities based on single economic demand. After the Ming Dynasty, with the development of the commodity economy, many cities and towns emerged, such as Jingdezhen and Pingyao. The industrial cities that have materialized in the past 100 years record the footprint of national rejuvenation. Ruins of kilns, workshops, draft banks, shops, and other sites have been found in ancient cities, as well as old factories and oil wells preserved in more modern cities, offering visual illustrations of urban growth.

Some cities were planned artificially under special historical conditions, such as the modern colonial and reclaimed cities of Qingdao and Shanghai. The special economic zone city of Shenzhen was born out of the Reform and Opening. Such cities often have a unique cultural heritage due to integration with foreign cultures.

Obviously, the cultural heritage that originates from a city's urban formation is the most primitive and practical explanation for urban development, in direct contact with the foundation of a city's existence. It is the basis for understanding urban cultural characteristics.

In pursuing the "conception" of urban cultural development through historical transformation, the interpretation of urban identity via cultural heritage is also reflected in the inheritance of an urban cultural context. Thanks to the vast amount of historical material available, and a rich cultural heritage, the history of Chinese cities is usually accessible. With the political culture of dynasties as the backdrop, local culture has blossomed over thousands of years. Although cultural implications and extensions have undergone both minor and major shifts due to changing times, cities that have been fixed for long periods (such as Xi'an, Luoyang, Nanjing, and Beijing) have maintained a cultural mainline from generation to generation, and this is reflected in its existing cultural heritage. For example, Beijing was the capital during the Yuan Dynasty, and this was when the major planning of the city was carried out. Although the Ming and Qing Dynasties saw upgrades, basic plans have remained generally the same, such as the location of the Imperial Palace and the central axes of the city, through which we can still see traces of the 700-year-old Yuan capital. Layers of culture add density to a city's heritage, and show the various influences of different time periods.

Small and medium-sized cities have prospered at various points due to specific types, sizes, grades, or geographic locations. Handan, for example, was the capital of Zhao during the Warring States Period, and during the Cao Wei Period, when it was known as Yecheng. It was again established as the capital during the Northern Qi Dynasty, and was the center of China's politics, economy, culture and religion. After the Tang dynasty, the city became increasingly mediocre, famous only for the Cizhou Kiln. In modern times, it is a single-industry city with a basic economic form from coal mining. There are many examples in China of a famous metropolis becoming a prefecture-level city with no special features. To discover their history and highlight their individuality, one must look at their cultural heritage. Compared with single and limited industrial resources, internal sources of urban survival and development are present in the cultural context and inner spirit of long-lasting, generational lineage.

Therefore, from the cultural era of urban transformation, the weight and inheritance of those with a long history can be highlighted for survival in the present. By negotiating historical context, pursuing eras of prosperity and growth, cities can restore the most distinctive identities of their urban context, having far-reaching significance for future urban development.

The third consideration is pursuing the "features" of urban culture, such as special geographical environments, natural resources, well-known figures, major historical and social events, and even contemporary industrial markers.

Dunhuang is famous for its Mogao Grottoes. Jingdezhen is famous for porcelain making. Tai'an is named after Mount Tai. Zuoquan County in Shanxi is named after General Zuoquan. Characteristic resources, well-known figures, major historical/social events, and representative industries are labels of cultural heritage and urban identity. The attraction of certain cities is the witness of historical change and milestones in social development. For example, Chibi in Hubei was once the scene of the "Battle of the Red Cliffs", and was famous due to Wei, Shu, and Wu acting as "three pillars". Zunyi in Guizhou was the location of the CCP's Zunyi Conference during the Long March in 1935, which makes it a "famous city" in contemporary history. Today, there are still traces of the battlefield of Chibi, and the Zunyi Conference site is also well preserved within the city. These are valuable legacies marking the core of urban culture. Much like huge buildings and modern sculptures, these features have become the key markers of these cities, offering conspicuous and extraordinary significance.

The source, context, and characteristics of urban cultural heritage are prominent elements in the visualization of heritage resources. Systematic analysis and reconstruction is an effective way to deeply understand and carry forward urban cultural heritage, and highlight a city's identity. In the context of new media, these three aspects are visualized digitally. While existing artefacts are optimized for visualization, other non-material and intangible cultural heritage resources must be fully explored and analyzed so as to focus the display of the city's museum or other public media space (see Figure 3). The display and dissemination of new forms of urban cultural heritage is a powerful measure to establish a solid urban cultural image, expand a city's cultural brand, converge public consensus, and obtain strong spiritual power and intellectual support for urban economic and social development.

In the face of complex and diverse urban cultural heritage resources, the first thing that must be done is the integration and excavation of objects under the tripartite analysis model of source, context, and characteristics, forming a clear-cut theme, plot, and scene development in order to facilitate technical processing and the initial stages of artistic

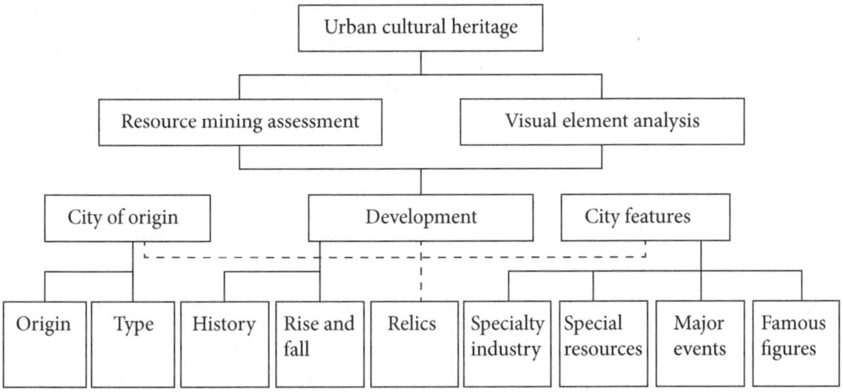

Figure 3 Guide for the analysis of urban cultural heritage resources

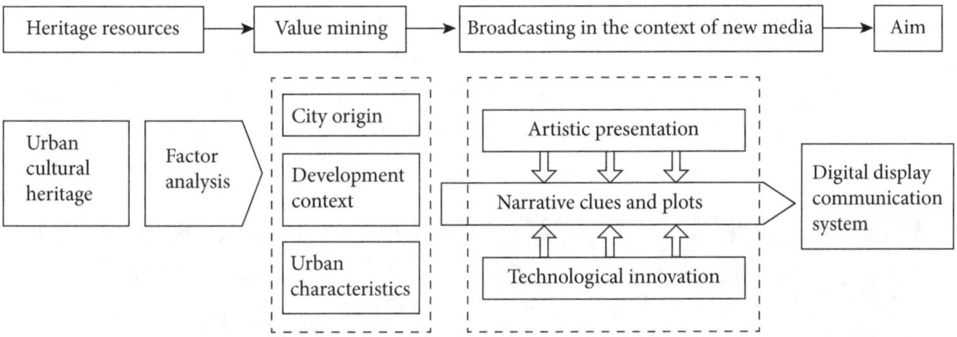

Figure 4 Establishing a new media digital display system for urban cultural heritage

presentation. Especially in the creation of new urban museums, designers, researchers, and experts from all areas work to combine the cultural contextual narrative of cities from the aspects of content integration and formal expression. They combine traditional culture and innovative technology, using the Internet and new media digital technical means to provide a key narrative plan for important core content, and finally establishing a complete digital display system that is suited to the specific cultural heritage of a city (see Figure 4).

Resource objects explaining "urban origin" may adopt the form of scene restoration, including scenes of ancestral life. Many museums in China employ wax figures, oil paintings, and landscape paintings, the effect of which is not outstanding, and can even be stereotypical. Dynamic presentations with 3D animation videos or large-screen projections are a better way of attracting visitors' interest. They are easy to modify, add to, and delete, and are rich in content and strong in relation. For typical cultural artefacts symbolizing the origin of a city and its emblem (such as Liangzhu jade, and the mask unearthed from the Sanxingdui site) can be expressed with sound and relief, that is, superimposing a projection on an embossed artwork, with careful programming of narrative scripts and precisely lit images, letting the traditional static relief "live" by presenting vivid and dynamic sound and light, which increased the expressiveness and mystery of artefacts.

Historical materials and recent research on cultural artefacts have gradually clarified the historical development of the city. Advanced spatial geographic information technology and the results of historical research can be fully integrated, i.e. expressing the context of urban development in units of geospatial time. Meanwhile as data has been collected and produced, the display of urban cultural heritage has generated corresponding materials and forms — urban 3D roaming based on virtual reality technology. In order to experience the geospatial characteristics of the ancient city and its temporal development, a digital 3D model is established. It is artistically rendered with natural scenes added, based on the fully verified urban layout and remains of the old city, integrating somatosensory Kinect equipment with simple instruments to realize virtual roaming in space and time (see Figure 5). In front of the big screen, visitors can choose the era and specific site of interest, simply gesturing to make basic movements. Be it galloping on horseback or walking through a workshop, one can travel through the city or town experiencing local customs.

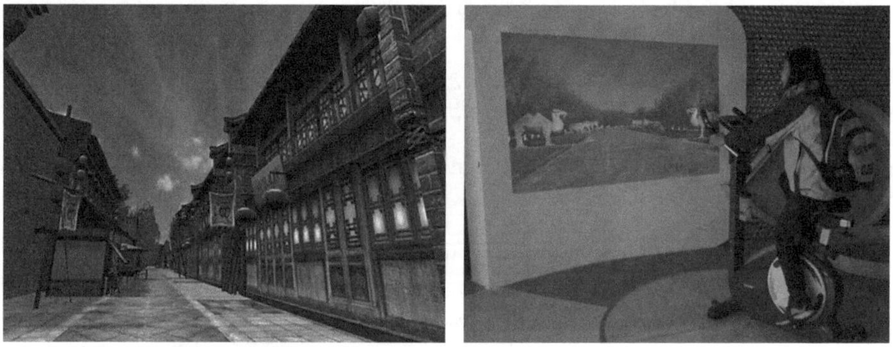

Figure 5 Virtual 3D scenes and physical urban roaming

For cultural artefacts, unique and appropriate display methods are set according to the characteristics of the artefacts themselves.

As for interactive new media artwork, designing interactive ancient chime music was an important trial. The designer adopted a traditional spatial design approach to landscape, using spatial modelling to create a semi-open independent space. On one side is a set of real unearthed chimes, and on the other is a fully digital interactive virtual chime. They are almost identical in shape, appearance, and scale. Visitors can understand the characteristics of this group of bells and their social and historical background through interactive images. They can also actually tap the digital bells. At this point, they hear the unique sound of real chimes while following the guidance on the screen to play music themselves. In this way, visitors are immersed in virtual and real space, with the novelty of music (see Figure 6). The display of experiential artefacts, such as tools and musical instruments, can be arranged in a similar way to engage the various senses.

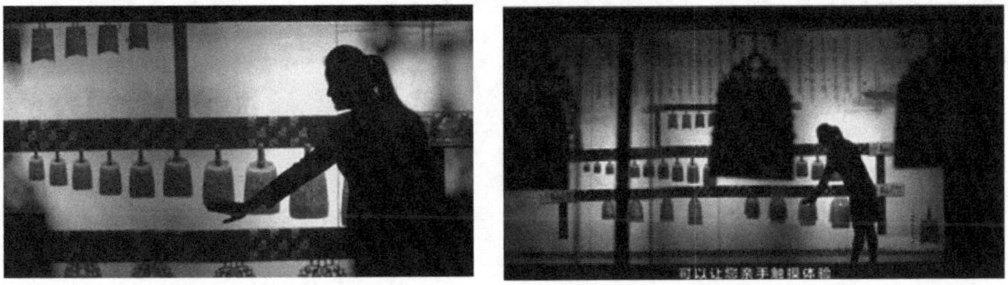

Figure 6 Interactive ancient chime music

Culture continues to evolve with the process of continuous innovation and transmission. Without innovation, culture loses its vitality. The cultural heritage that is presented to us today is precisely the result of constant innovation and unremitting ancestral exploration. To realize the inheritance and innovation of urban cultural heritage, we must examine and explore clues of historical evolution, carrying forward the wisdom and achievements contained therein. Meanwhile, advanced technology and artistic means must be combined to create innovative methods of display and dissemination that are timely, appealing, attractive, and infectious. In this way they will further highlight the cultural orientation and advantages of a city, giving residents a source of spiritual enlightenment and emotional engagement, and allowing cultural identity and innovation to become the spiritual driver and inexhaustible source of culture in the maintenance of sustainable urban development.

References

[1] Shan Jixiang. *Retaining the roots and soul of urban culture* [M]. Beijing: Science Press, 2010, p. 135.

[2] Marstine, Janet. *New museum theory and practice: An introduction*, Blackwell, 2006.

Digitization – An Effective Way to Protect Cultural Heritage

Li Jiayu
(School of Applied Arts and Sciences, Beijing Union University, Beijing 100101, China)

Zhao Lianwen
(Sanshan Wuyuan Institute, Beijing Union University, Beijing 100101, China)

Abstract: In recent years, the protection of cultural heritage has attracted increasing attention from all corners of society. This is because material and intangible cultural heritage are facing difficult conditions. While traditional methods of conservation have not been adequate to protect cultural heritage, digital technology can be a way to immortalize it. At present, the most common and simple means involve the use of scanners and digital cameras. Virtual reality technology has also become popular in recent years. The digitization of Chinese cultural heritage occurred relatively late, so it is now necessary to apply advanced technology in order to accelerate the pace of conservation. Thus far, China has developed some independent technologies, and many institutions such as Beijing Union University have also undertaken a lot of work in this field.

Keywords: digitization; cultural heritage; conservation method

What is cultural heritage? According to the UNESCO Convention Concerning the Protection of World Culture and Natural Heritage Nature and Culture, adopted in 1972 (UNESCO World Heritage Center), "cultural heritage" refers to the historical essence of the whole of human civilization. It is extremely rare, non-renewable, and non-replicable. It is not owned by the country in which it exists, but belongs to the entire human race. Cultural heritage can be divided into two categories: tangible and intangible. Human beings are obliged to recognize, protect, preserve, and display cultural heritage and natural heritage, as well as pass it on to future generations. Since the late 20th century, countries around the world have explored effective ways to conserve cultural heritage. In the 21st century, we have accelerated the pace of this exploration. Today, digitization is considered as the most effective method for the preservation of cultural heritage.

1. *Digitization – an inevitable requirement for the protection of cultural heritage*

The world's cultural heritage has endured for many years, and most of that which still exists faces challenges – a problem that is particularly prominent in developing countries. The living environment of China's cultural heritage is extremely fragile, and the damage caused by serious and unavoidable natural disasters such as volcanoes, earthquakes, fires and tsunamis is self-evident. For example, on March 14, 2013, the ancient city of Lijiang burned for three hours. At 1:37AM on January 11, 2014, a fire broke out in the ancient city of Dukezong in Shangri-La County, Diqing Prefecture, Yunnan Province, which took more an 10 hours to extinguish. The fire covered about a square kilometer, in an area of only 1.5 square kilometers, meaning that over two-thirds of the Dukezong were burned. At 2:49AM on January 3, 2015 in Dangshan in Yunnan's Dali Prefecture, a fire broke out in Gongchen Tower. Covering about 300 square meters, it destroyed a site that was more than 600 years old. More often than not, the disasters that occur are man-made. Especially in recent decades, with the rapid advancement of industrialization and urbanization, coupled with air pollution and over-exploitation due to the development of tourism, China's cultural heritage protection has been greatly affected. Many cultural heritage sites have been destroyed, and some sites are quickly disappearing. For example, the well-known Dunhuang Mogao Grottoes have been impacted due to the large number of tourists who visit each year.

The current status of intangible cultural heritage is even worse. Non-material items face the predicament of "death and destruction", while "non material" ancient culture experiences "modern traps". According to the Shanghai Morning Post, as reported on May 29, 2012, there are more than 1,000 items of intangible cultural heritage in Shanghai that are close to being lost as they are not being passed down, such as Chongming's *biandan* (traditional one-man puppetry on a stage supported by a shoulder pole), *luogushu* (a form of Chinese ballad singing with gongs and drums), bamboo carvings, Jiangnan *sizhu*, and the "wharf chant" – a song that workers once sang on the banks of the Huangpu River. However, as modern machinery and equipment are gradually being used on the docks, workers sing less and less. Seeking effective protection for cultural heritage is now crucial, and thus digital protection has emerged.

Practice has proven that traditional methods of protection can no longer conserve cultural heritage, and the development of digital technology stands to further some of the cultural heritage that is disappearing. For example, the Digital Dunhuang project intends to preserve the treasures of Dunhuang as much as possible, to reduce the damage to murals and sculptures caused when tourists visit the physical caves. It is also digitally regenerating some of the lost cultural heritage. The Old Summer Palace, known as Yuanmingyuan or the "Garden of Eternal Spring", is being recreated after its destruction over 100 years ago, with the help of digital technology. The site has been rebuilt as digital models, superimposed on the existing ruins in various optical displays, with the original scenes of the garden reproduced realistically with stereoscopic display technology, enabling visitors to roam and learn in a relaxed, enjoyable environment.

When it comes to intangible cultural heritage, digitization has advantages in the dissemination of cultural content, and also in the regeneration and preservation of lost intangible cultural heritage via digital technology.

2. The collection of digital resources and processing technology for cultural heritage

At present, the most commonly used and relatively simple means of collecting digital cultural heritage resources involves the use of scanners and digital cameras. Due to long-term natural or man-made damage, the patterns on many cultural heritage items are blurred and unclear. In terms of image acquisition, higher pixel devices are required, such as the world's first 1 billion pixel camera provided by Microsoft to the Dunhuang Research Institute. Certain intangible cultural heritage forms can be edited and converted into digital formats to be stored in computer hard disks and CD-ROMS, so that information can be better organized, collected, and recorded, to ensure more secure and accurate long-term preservation.

In the collection of digital cultural heritage resources, high-level 3D acquisition requires more specialized equipment. A 3D scanner can be used to obtain 3D cultural heritage data. For example, when the Taipei Palace Museum produced virtual animations of carved ivory balls, a large amount of raw data was collected using a 3D scanner. In terms of cultural heritage sites and ancient buildings, 3D laser scanning technology is the most important technical means for spatial data acquisition. If corresponding information about scene texture must be obtained to improve modeling realism, digital cameras must be used.

Virtual reality technology has become the most popular digital technology in recent years. Referred to as VR, and also known as spirit technology or artificial environment, it is a technology used to create a virtual world that is aligned with the real world. The application of virtual reality technology to cultural heritage preservation began much earlier in other countries than in China. The core of VR technology is the creation of virtual worlds as a development of film production, which has brought about tremendous changes in the spread of cultural heritage. Thanks to virtual restoration technology, we have seen lost buildings recreated in documentaries. With virtual reality technology, IBM developed the "Virtual Forbidden City" for the Palace Museum.

However, the high cost of virtual reality technology limits its use in cultural heritage protection, while augmented reality (AR) technology has a higher degree of realism and convenience. AR is the most promising technology at present, characterized by the combination of reality and virtual objects, like watching the real world through an interactive window. When the virtual restoration of ancient buildings and cultural artefacts is carried out, augmented reality technology can be utilized, complementing and displaying original sites and cultural artefacts. This avoids damage to sites and artefacts themselves, and also makes it possible to restore the original appearance of cultural artefacts. Therefore, AR technology is highly significant in the conservation of cultural heritage.

3. The status of digital cultural heritage protection

The United States has used digital technology to protect books and archives since 1990, referred to as "American memory". Due to the development of digital technology, especially the application of a series of 3D technologies under the umbrella of virtual reality, the status of cultural heritage protection has changed greatly.

Administrative bodies, from the government to the public sector, as well as many commercial companies, attach great importance to cultural heritage digitization, investing a lot of money, and obtaining advanced multimedia technology, virtual technology, or graphics. Many universities and research institutions have also carried out studies on the digital preservation of cultural heritage.

The United States strongly values the digital protection of cultural heritage, and its distinctive feature is commercial operation, meaning that the government allows commercial companies to participate in digital conservation. The "eHeritage" research program launched by Microsoft Research Asia provides technical and financial support for the Asian Cultural Heritage Digitization Project. For example, a 1 billion-pixel digital camera system was donated to the Dunhuang Research Institute in China, named "Feitian", providing ultra-high resolution images. It is specifically used to photograph the Dunhuang Mogao Grottoes and Buddhist altars. Japan also has a strong sense of cultural heritage protection, and has privately funded the digital processing of cultural heritage, such as in virtual reproductions of Ganjin's Toshodai Temple.

Due to technical issues and problems of understanding, the digital conservation of China's cultural heritage was late to begin. It is now necessary to make full use of advanced technology to accelerate the pace of development. Thus far, China has independently developed certain technologies, such as the research on digital protection technology for art in Dunhuang that was carried out by the National Key Laboratory of CAD & CG at Zhejiang University. The research proposed mural painting techniques and mural color gradient technology, developing the virtual tour system for the Mogao Grottoes and the auxiliary restoration system for the Dunhuang murals, completing the Dunhuang Mural Multimedia Recovery project, and virtual reality works such as The Forbidden City — Palace of the Son of Heaven. Professor Zhou Minquan at Beijing Normal University has done an excellent job in computer-assisted restoration of the Terracotta Warriors and Horses in the Qin Shihuang Mausoleum, realizing the virtual restoration of broken cultural artefacts. The creation of the digital museum in the Forbidden City has seen great success, and its image display system is number one in China.

The digital protection practices of these cultural heritage resources are only the beginning for China's large amount of cultural heritage resources, and there is still a lot of work to be done.

4. Digital practices in cultural heritage at Beijing Union University

In recent years, Beijing Union University has done a lot of work in the digitization of cultural heritage alongside teaching and research. In terms of cultural heritage digitization, virtual reality works have been developed, including a 3D Dajue Temple simulation navigation system project based on traditional virtual reality technology, and the Xu Beihong Memorial Museum network display system based on web 3D technology, as well as an interactive virtual tourism system based on mobile terminals, including the digitization of the Ming Tombs, and a virtual tour of Gaobeidian Lizhitang in Beijing. Based on virtual reality technology, the key technologies and industrial pathways for the development of Three Hills and Five Gardens Virtual Reality Project intends to provide remote access to the site through mobile Internet terminals such as Android smart phones and Apple mobile platforms, offering deep interaction through live, real-time camera images of specific landscapes, as well as virtual scenes and historical figures that are integrated into the live, real-time location on the camera screen. With cooperation from the Haidian District Party Committee's publicity department, the Cultural Promotion Center, Peony Group, and other units, the pedigree construction and visualization of the historical and cultural elements of Three Hills Five Gardens will use Cloud storage, network GIS, and virtual reality, adopting internationally-advanced 3D photography technology to build a panoramic open information platform for Three Hills Five Gardens, integrating virtual reality technology for the innovative reconstruction of historical and cultural landscapes in this scenic area. The digital display technology demonstrations at the Maiji Mountain Grottoes and the international cultural exchange applications use data mining, pattern recognition theory, integrated digital acquisition, content production, audio-visual, virtual reality, multimedia, and other suitable screen display technologies and equipment to restore the original appearance of sculptures and murals, revealing the unique position of the Grottoes (known as the "Oriental Sculpture Museum") in the context of Chinese Buddhist cave art and international sculpture.

References

[1] Huang Yonglin, Tan Guoxin. 'Research on digital conservation and development of Chinese intangible cultural heritage', *Journal of Huazhong Normal University (Humanities and Social Sciences Edition)*, 2012, 3.

[2] Wu Jianping. 'Technical construction of digital application platforms for cultural heritage purposes', *Computer Application Research*, 2006, 4.

[3] Wang Mingjie. 'The digital virtual space model for cultural heritage', *Youthful Years*, 2013, 5.

Action Performance in 3D Animation

Su Gaofeng
(College of Advertising, Beijing Union University, Beijing 100101, China)

Abstract: In society today, with the rapid development of media terminals, the ways and means of creating film and television are developing rapidly, and the digital museum shoulders the tremendous pressure of China's cultural heritage. Speaking of digital technology and computer technology, one method by which to reproduce Chinese culture is through animation, although most animation methods involve prohibitively crude and extensive production. In the following article, the author discussed some of the issues often encountered in the debugging of character action in terms of animation.

Keywords: animation; action debugging

Preface

Chinese cartoons have led the direction of world animation since their birth. Due to China's unique historical issues, modern animators have identified faults with traditional Chinese production concepts and techniques. One reason is that current animation production is constantly imitating Japan and the United States, and the core transformations and action performances of unique Chinese characters have been discarded. So, while current animation is not inferior, it lacks its own style. Another reason is that the excessive pursuit of economic interests results in gearing film production towards extensive development. Of course, there are many factors affecting the low standard of animation production. The author believes that improving the psychological transformation and motion performances of animated characters is a crucial aspect, the successful application of which can make a film lifelike, vivid, much more intriguing.

1. Psychological transformation – the lifeblood of an animated character

1.1 Psychological activity and external bodily movements are important in the formation of animated characters

So-called psychological activity is a reflection of the brain in the objective world. It is the action of ingesting, storing, encoding, and extracting information through neurophysiological processes of the brain. A person's psychological activities include cognition, emotion, and will. They are not three independent or juxtaposed psychological actions, but are three aspects that are connected and conflicting in the process of unified psychological activity.

Psychological activity is the thinking that a human or animal brain performs before actions such as language, behavior, and expression. There are many forms of psychological activity. In different environments, each person's psychological action is different, and emotional experience also has discrepancies in terms of time and space. Each person's psychological activities can be different, but may also have common starting points. No two people's psychological action are exactly the same; at most they are similar.

Physical movement is external motion, referring to visible movements that consume external body strength in performance. In the process of performance, bodily movements should be organic, purposeful, and logical. Meanwhile, any movement must be associated with psychological activity. Physical movements are external manifestations of psychological action, and because mental activities have unique characteristics, an individual's physical movements also have unique attributes.

1.2 The impact of psychological activity on physical movement

At present, many Chinese cartoons are being produced, but animators lack imagination, and characters move in a simple and mechanical manner. Each action occurs simply as a movement, without incorporating the emotional factors of each character, often displaying Character A with the same posture and the same way of speaking as Character B.

To create quality work, an animator or director cannot simply limit key poses to the ones found in animation reference books. These key frames are only a starting point. We must think about how to break down an action into parts so as to understand it more clearly. From this, we can create our own key frames to represent our own characters.

2. Physical action – the flesh of an animated character

2.1 Know what you want to do, and make plans in advance

The main aspect of planning is to understand every detail of a character, and to make detailed plans for each action in advance, so that each action can be grasped from an overall perspective, and the effect of each motion can be conceptualized. As each character's action has unique properties, different characters will produce different dynamic effects when performing the same general action.

With eating, for example, the movements of individual people in various situations are very different. Emotions expressed in the action shown in Figure 1 are hunger and a fondness for eating.

The action shown in Figure 2 is confusion about the food

The action shown in Figure 3 is eating often, not confused by the food

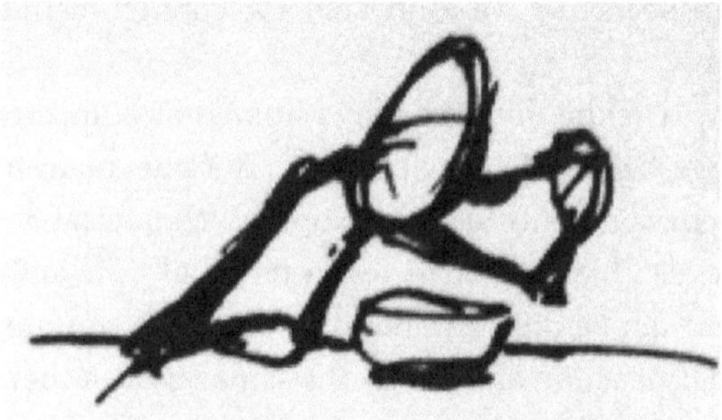

Figure 1 Eating action – enjoying eating

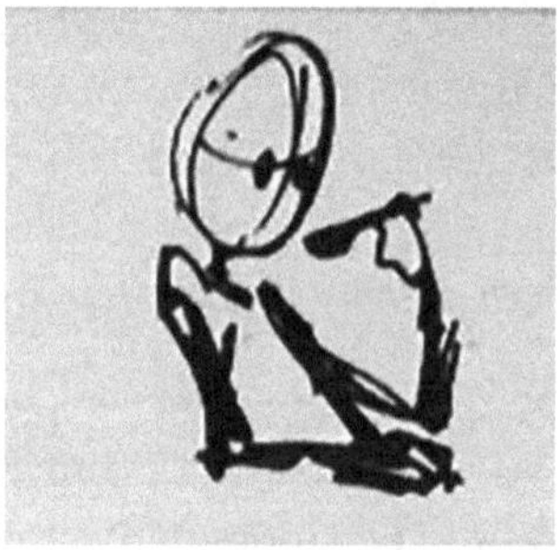

Figure 2 Eating action – confused by food

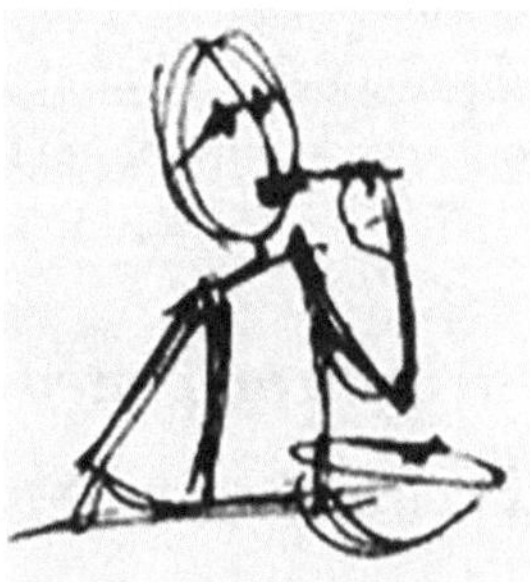

Figure 3 Eating action – eating often, not confused by food

The different eating states shown above involve emphasis on different physical movements based on different psychological activities. These are the qualities that we must carefully consider in the animation process.

2.2 *Considering emotions when creating actions*

When making adjustments to a character's action, there must be certainty in thought and attitude. Characters must be created in their own unique way, depending on the environment and mental state. They must not simply be performing an action for the sake of it, otherwise the resulting action will appear mechanized and without vitality.

The expression of the actions in Figure 4 varies according to changes in psychological state. There are different responses to the same situation. The first character shows a state of disdain; the second shows disdain, but with a desire to understand. The third shows surprise and curiosity, while the fourth is very curious and eager to understand. The fifth is also curious, but the degree of curiosity is not as deep as in the fourth.

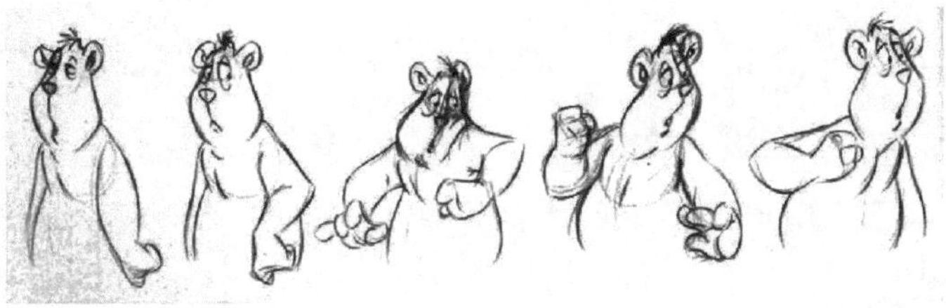

Figure 4 The expression of action changes according to different psychological states

3. Entering the mind of the character

How does one complete character action? This is a dilemma in many animation projects. Many animators have come up with a perfect pose for a character to perform as a specific action. For example, when surprised, they can be wide-eyed, or represented with an open mouth. Over time, these actions have been used and applied to multiple characters, making the movements of many characters increasingly stereotypical, lacking personality and difference. There are also a lot of animators who believe that small gestures such as blinking and swallowing are simply used to fill gaps, and ignore such movements altogether. In fact, if used correctly, blinking and other small movements can be more meaningful than movements of the whole body. For example, people blink for a number of reasons: to clear dust from their eyes, to express disbelief, to express understanding, to express boredom, frustration, and shock, or to indicate the beginning or end of the thought process. In a graduation assignment, the author designed two images of a man and a woman. The woman was ugly – a tough character. In order to highlight the image of a heroine, the author designed the male image as timid and weak. Although not a handsome man, in his bones he had the potential to be a good guy, generating comedy in the plot design. In creating motions for the male image, the author often used small actions to set off the character, such as changes in eyebrows, or eye expressions to match the actions, as well as minute motions to express specific characteristics.

Figure 5 shows the minor actions of a character before lifting the heroine's veil, using laughter to express the psychological activities of the lead actor. The male lead goes to the heroine and then turns around, his eyes turning with his head, followed by blinking, and finally the action of laughter. Many creators simply show laughter in the corner of a character's mouth, or with direct outward laughter, meaning that there is no show of emotional change in a character at all, or only a weak display. The author used this series of subtle changes to create an accumulation of emotional change before the final burst – the "outbreak" – demonstrating that characters have inner activity, while also creating a humorous rhythm.

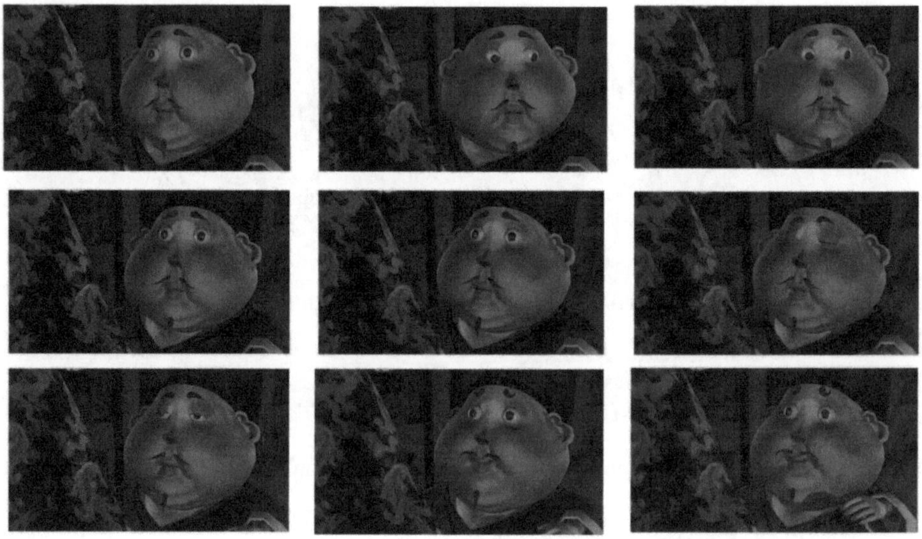

Figure 5 The male protagonist's minor action before lifting the heroine's veil

References

[1] Wu Zhenguo, Qi Xiaobei. *Film and television animation performance*. China Science and Technology Press, 2009.

[2] Zhang Hua, Yang Ruijiang. *Film and TV performing arts*. Zhejiang University Press, 2007.

[3] Zhao Qian. *Analysis of fine animation films*. Chongqing University Press, 2007.

[4] Liu Zhizi. *Methodology of film performance control*. China Film Press, 1999.

Skeleton Rigging in 3D Animation

Su Gaofeng
(College of Advertising, Beijing Union University, Beijing 100101, China)

Abstract: The development of digital museums is growing ever more diversified and complicated, with the capacity to realize virtual characters and scenes with digital control, and to reproduce culture and civilizations that have long since disappeared. For example, Yuanmingyuan and the Forbidden City have now obtained good digital representations, but the scene is only one aspect this. Another aspect is character rendering, which is mainly expressed in animation. At present, although China's domestic animation industry is developing, there is still considerable way to go to meet developed countries, such as processing in extensive forms, especially with the movement adjustment of characters and animals. This paper will discuss production issues, for example, how animators can find suitable methods to perform skeleton rigging.

Keywords: animation, skeleton rigging

Preface

With the development of society, film and television cartoons have also developed rapidly. Following the American film and television production of *Avatar*, which sold more than a million box office tickets in China, a series of 3D animations such as *Kung Fu Panda, The Adventures of Tintin, Frozen*, and *Wreck-It Ralph* were released in China, quickly opening up the market.

Many colleges and universities in China have begun to offer animation majors, training up a large number of animation producers. However, there is still a big gap between Chinese animation and animation from developed countries such as the United States. Research and investigation has led the author (a college professor) to identify many problems that affect the quality of Chinese cartoons. This paper offers a brief explanation of the issues with skeleton rigging of characters in 3D animation.

1. *Skeleton rigging*

In the process of 3D animation for general animation workers, the problem of skeleton rigging has always been a case of how to rig the bones of a character, and the conflict in rigging with several sets of bones. Some believe that using plugins provides an easier and less laborious way to perform skeleton rigging. However, the author believes that this method will place huge limitations on animation production, causing animated characters to move mechanically, and producing characters that have no soul or vitality. Obviously, this is not advisable in an industry that relies on vitality.

The problems with skeleton rigging systems are addressed in actual production practices and teaching processes that are more effective and easily grasped by students.

• Using three sets of bones to rig the arms more stably

In the skeleton rigging of the whole body, the author believes that rigging the arm is the most complicated, and rigging methods are different between individuals. Arm rigging is done in two ways: IK and FK. IK refers to inverse kinematics, that is, the rotation operation of a parent joint (arm joint) is controlled according to the positional movement of an end joint (wrist joint). FK refers to forward kinematics, controlling the position of child joints according to rotation of the parent joint (arm joint). More specifically, when an animation is adjusted, these two systems are applied to three sets of bones, then attached to the original bone, and the two power systems are switched by the original bone controller, which is more stable. If switching to IK when FK is adjusted to a suitable position, the limb returns to its initial state. The general adjustment method uses the Set Driven Key to follow FK through key frames, causing a lot of work in later animation adjustments. In the creation of a short film titled *Ugly Wedding*, the author summarized a set of practical rigging methods, which can be used to solve issues in the rigging process and greatly improve efficiency. The key is seamless transition between IK and FK in order to make animations more smooth and vivid in adjustment. Although switching modes is complicated in early preparation stages, it is relatively stable and highly adjustable. The author blends the modes perfectly with MALL, as shown below.

```
//IK TO FK (Reverse dynamics to forward dynamics)
float $L_jianbang_IKrx=`getAttr L_ jianbang_IK.rotateX`;float $L_jianbang_IKry=`getAttr L_jianbang_ IK.rotateY`;
float $L_jianbang_IKrz=`getAttr L_jianbang_IK.rotateZ`;
(The above are three rotational numerical sampling of the shoulder bones in the reverse dynamic direction.)
setAttr "L_jianbang_FK. rotateX" $L_jianbang_IKrx;
setAttr "L_jianbang_FK.rotateY" $L_jianbang_IKry;
setAttr "L_jianbang_ FK.rotateZ" $L_jianbang_IKrz;
```

```
(The above is the fusion switching setting of the two power systems of the shoulder bones in three directions)
float $L_zhou_IKry=`getAttr L_ zhou_IK.rotateY`;
(The above are three rotational numerical sampling of the elbow bones in the reverse dynamic direction)
setAttr "L_zhou_FK.rotateY" $L_ zhou_IKry;
(The above is the fusion switching setting of the two power systems of the elbow bones)
float $L_shouwan_IKrx=`getAttr Nv_23:L_ shouwan_IK.rotateX`;
float $L_shouwan_IKry=`getAttr Nv_23:L_shouwan_IK.rotateY`;
float $Nv_23:L_ shouwan_IKrz=`getAttr L_shouwan_IK.rotateZ`;
(The above are three rotational numerical sampling of the wrist bones in the reverse dynamic direction)

setAttr "L_shouwan_FK.rotateX" $L_shouwan_IKrx;
setAttr "L_shouwan_FK.rotateY" $L_shouwan_IKry;
setAttr "L_shouwan_FK.rotateZ" $L_shouwan_IKrz;
(The above is the fusion switch setting of the two dynamic systems for the wrist bone)
setAttr "IKFK.IKFK" 0;
Hiding of the controller
// FK TO IK(Forward dynamics to reverse dynamics)
float $FK_R_shouwan[]=`xform -q -ws -t FK_R_shouwan`;
xform -ws -t $FK_R_shouwan[0] $FK_R_shouwan[1] $FK_R_shouwan[2]
Numerical sampling of wrist bone dynamics variables
R_shouwan_kongzhi_IK;
float $FK_R_shouwan[]=`xform -q -ws -ro FK_R_shouwan`;
xform -ws -ro $FK_R_shouwan[0] $FK_R_shouwan[1] $FK_R_shouwan[2]
(Fusion switch setting of wrist bone dynamic system)
R_shouwan_kongzhi_IK;
float $R_zhou_PV_FK[]=`xform -q -ws -t R_zhou_PV_FK`;
xform -ws -t $R_zhou_PV_FK[0] $R_zhou_PV_FK[1] $R_zhou_PV_FK[2]
Numerical sampling of elbow bone dynamic variables
R_zhou_PV;
setAttr "R_shouwan|IKFK.RIKFK" 1;
Display of controller
```

Although three sets of bones can be more complicated in the early stages, they are more stable and reliable overall. Some animators use only one set of bones to rig, but the author's experiments have shown that is unstable to use one set with two rigging methods. This is because there are many controllers in a single joint, which easily becomes confused, as once a controller value appears there will be conflicts, affecting binding effects and causing problems for post-animation adjustments. If the switching controller is to be tracked seamlessly, it must be done twice on the same skeleton. With the controller and skeleton controlling each other, slight carelessness will lead to errors (see Figure 1 and Figure 2). These errors are very difficult to adjust, which will affect the rigging of the whole arm.

Using three sets of bones (see Figure 3), the relations between bone and bone, and controller and controller follows clearly. Individual bone and controller adjustments can be made independently without affecting overall arm control, the key being stability during the animation adjustment process. Stability is the premise of animation performance.

Before this, students often faced an important issue with production as, in general, the three joints of the arm must be in a straight line. If not, the locators and the IK set vector constraints will affect elbow control during IK, causing the wrist to turn (see Figure 4). If not attended to, this problem will cause major hazards, especially to animation adjustments.

Through creative and teaching practices, the author has come up with a solution to the problem that is more easily grasped by students (see Figure 5). Creating a plane level division to increase the height of three joints in a line can solve the issue of having three points on the same plans, so set vector constraints will not turn the wrist.

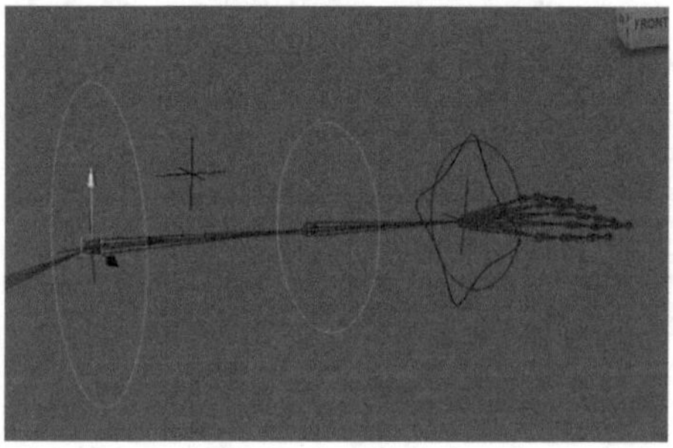

Figure 1 An example of a minor error due to carelessness

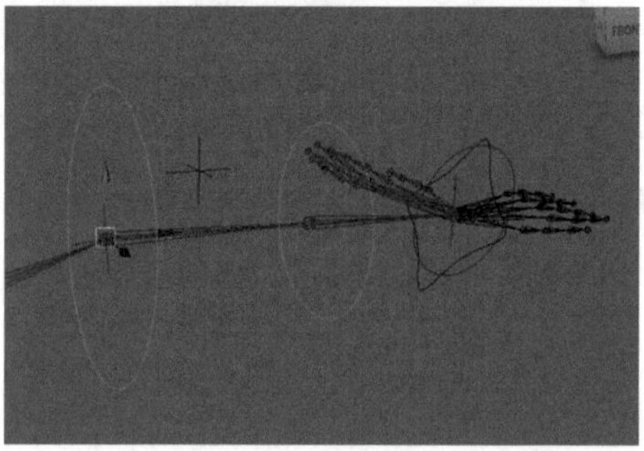

Figure 2 Another example of an error due to carelessness

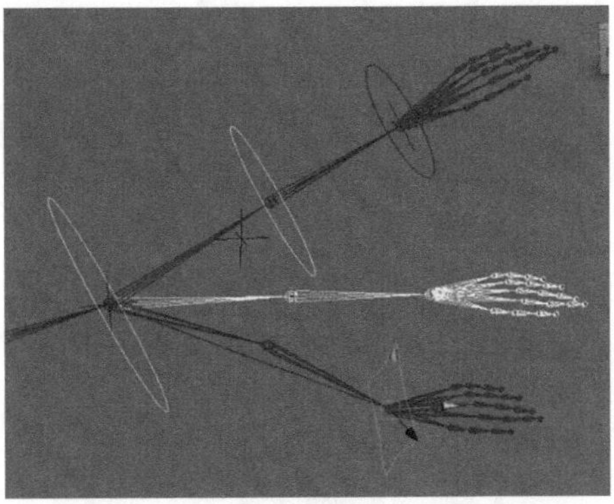

Figure 3 Using three sets of bones

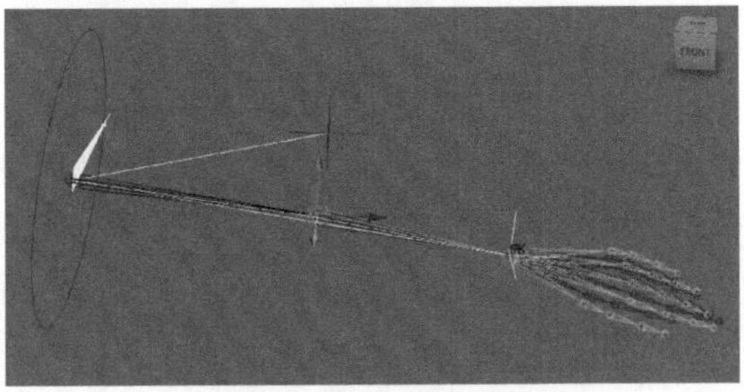

Figure 4 Wrist rotation

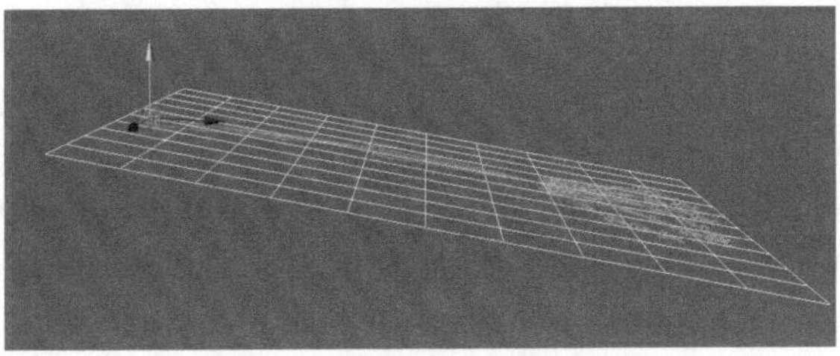

Figure 5 A solution to the error

2. Manual skeleton rigging is better than plugin skeleton rigging

2.1 Advantages of manual skeleton rigging

Usually, the performance of a cartoon character requires exaggeration to show humor and visual tension in the animation, and this requires comprehensive consideration when rigging, so as to lay the groundwork for later animation. For example, in the cartoon *Tom and Jerry*, the exaggerated performances of the two characters (such as the stretchability of the body and exaggerated expressions) leave a deep impression on the audience. For the two cases shown in Figures 6 and 7, while the skeletal system in Figure 1 has no issue in normal movement, it would be impossible to achieve major stretching actions as in *Tom and Jerry*.

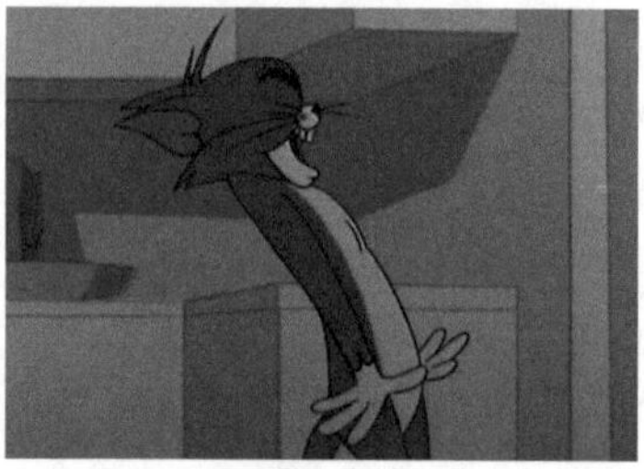

Figure 6 An example of action in *Tom and Jerry*

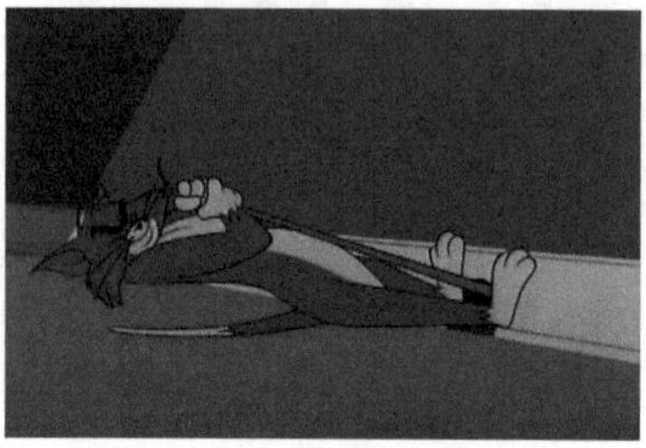

Figure 7 Another example of action in *Tom and Jerry*

Because manual skeleton rigging was used in the characters within the short animation *Ugly Wedding*, it was simple to add multiplication and division nodes so that bone stretching could be implemented as the controller stretches. This meant that more exaggerated movements were possible, and the action performance of the animated characters could play out to the fullest. If multiplication and division nodes are not added (see Figure 8), the controller cannot stretch with the control stretch, and is unable to satisfy exaggerated character performance (see Figure 9).

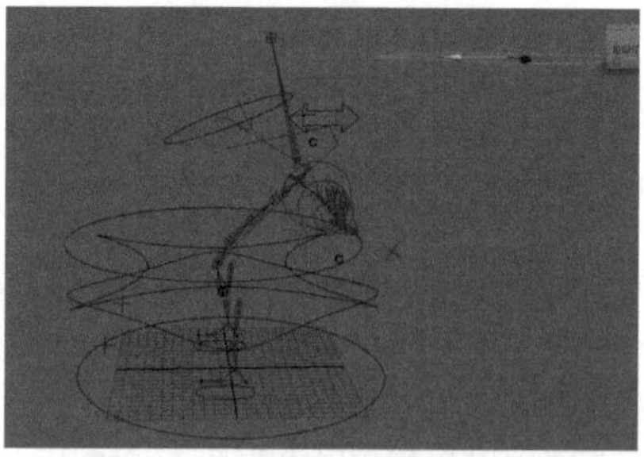

Figure 8 Without adding a multiplication and division node, the controller cannot stretch alongside the control stretch

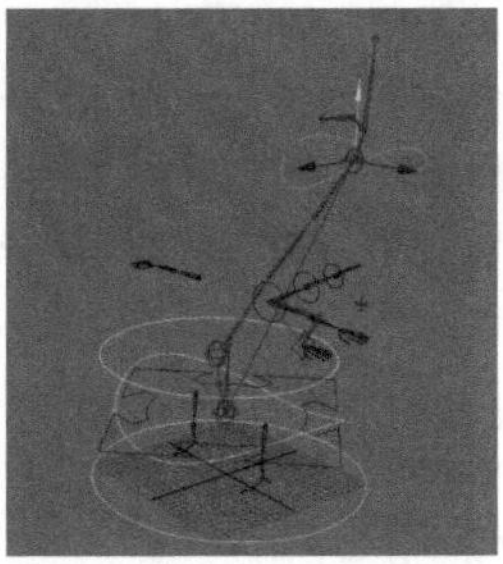

Figure 9 Failure to satisfy exaggerated character performance

2.2 Drawbacks of plugin skeleton rigging

Now, increasing numbers of animators are using plugins to bind roles to character bones. This seems to speed up the pace of skeleton rigging. The author has also attempted to bind using plugins during the creation of *Ugly Wedding*, but found several serious problems in the depths of production. Animator debugging actions become subject to the skeleton rigging system, and this is the reason why many modern cartoons seem to be more rigid. The main issues with the plugin system are as follows:

1) Plugin skeleton rigging cannot seamlessly switch between IK and FK with MALL language, as three sets of arm bones are based on a hand skeletal system set, resulting in errors when writing MALL language. If not using MALL language, seamless switching cannot be achieved, as when two systems switch, there is a return to the original coordinate point, imperceptibly adding a large amount of work for animators in the adjustment process.
2) A skeleton rig by plugin does not have full-body stretching and multiplication nodes, nor does it follow the design of the controller. It is unable to perform exaggerated stretching in action debugging, meaning that certain special actions cannot be displayed fully.

Given the combination of these factors, the author believes that is best to create special skeleton rigging systems based on individual roles and the performance of character action, in order to help animators enhance their work in the adjustment process.

References

[1] Wan Jianlong. *Maya Role Binding, Mars Classroom*. Posts & Telecommunications Press, 2012.

[2] Wu Zhenguo, Qi Xiaobei. *Film and television animation performance*. China Science and Technology Press, 2009.

[3] Zhang Hua, Yang Ruijiang. *Film and TV performing arts*. Zhejiang University Press, 2007.
[4] Zhao Qian. *Analysis of fine animation films*. Chongqing University Press, 2007.
[5] Liu Zhizi. *Methodology of film performance control*. China Film Press, 1999.
[6] Li Xinshe. 'Digital Entertainment Technology'. *Digital Entertainment Technology Magazine*, 2011-8-5.

Inheriting History, Retaining Nostalgia – The Protection of Cultural Heritage and Digital Development

Yao Huarong
(Beijing Xicheng District Institute of Cultural Relic Protection, Beijing 100035, China)

Abstract: Nostalgia consists of memories of people, events, and things from one's hometown. It is a mixture of memories, feelings, and sadness that are impossible to forget. Heritage buildings in historical and cultural cities are manifestations of history, the stories that surround them, the people involved, and the culture they carry. Like characters in the same book, they become a memory that cannot be erased. For the protection of such cultural artefacts, we need inheritance. Cultural resources must be excavated, and for this we need innovation. In terms of conservation, utilization, and publicity, we must integrate modern digital means, keeping pace with the times, adapting historical inheritance to modern development, and retaining nostalgia within inheritance and development.

Keywords: inheriting history; retaining nostalgia; cultural heritage; protection advocacy; digitization

Nostalgia consists of memories of people, events, and things from one's hometown. It is a mixture of memories, feelings, and sadness that are impossible to forget.Like characters in the same book, they become a memory that cannot be erased. For the protection of such cultural artefacts, we need inheritance. Cultural resources must be excavated, and for this we need innovation. In terms of conservation, utilization, and publicity, we must integrate modern digital means, keeping pace with the times, adapting historical inheritance to modern development, and retaining nostalgia within inheritance and development.

1. Cultural artefacts must be passed down

Carrying forward national spirit, historical and cultural inheritance are the main tasks of the museum as an inheritor of outstanding historical and cultural heritage, shouldering the heavy responsibility of cultural heritage valuation. Article 23 of the Law of the People's Republic of China on the Protection of Cultural Relics stipulates that a memorial or ancient building approved as a cultural artefact protection unit may act as a museum. There are 26 existing museums in Xicheng District, Beijing, involving social history, natural sciences, culture and art, as well as several comprehensive museums. These museums are carriers and inheritors of nostalgic memory, preserving memories in time and space, preserving history, and passing on the sentiments of nostalgia.

1.1 The Imperial Temple of Successive Emperors – A memorial to Chinese unification

The Imperial Temple of Successive Emperors involves royal temples of the Ming and Qing Dynasties, dating back more than 470 years. The Jingde Congsheng Hall is dedicated to 188 Chinese emperors – the legend of the Three Sovereigns and Five Emperors. Representing China's unification as a multi-ethnic country, it is a historical temple ritual system of the past, representing an endless line of Chinese forefathers. It contains the sacred hall where the Chinese people worship Emperor Yan and Emperor Huang, deepening the memory of national unification by hosting ancestral worship, including hundreds of family names.

1.2 The Former Beijing Residence of Li Dazhao – Commemorating the birth of the Chinese Communist Party

Li Dazhao made great contributions in the spread of Marxism, the creation of the Chinese Communist Party, and the establishment of the united front of the national revolution. He consolidated and developed cooperation between the KMT and the Communist Party, and led the revolutionary movement in the north. Li Dazhao's former home in Beijing was his longest place of residence outside his familial hometown, and this period was also the first golden age of his life. During this time, Li published more than 140 articles in various publications, and participated in 120 meetings, including three major Communist Party meetings, and one major Kuomintang meeting. He also received 30 visits from people from all walks of life, and delivered 30 lectures. No less, he engaged in teaching and revolutionary activities in Guangzhou, Shanghai, Wuhan, Luoyang, and Tianjin. During this period, many young people stayed at Li Dazhao's house, experiencing his care and guidance. Several important meetings of the Chinese Communist Party in the north were also held in his study. The home is witness to the most pertinent history of a series of revolutionary practices, such as Li Dazhao's dissemination of Marxism, the founding of the Communist Party of China, the leadership of the Northern Expedition, and the first instances of cooperation between the KMT and the CCP.

1.3 Xuannan Culture Museum – The origin of memories of intellectuals and the common people

Located in Changchun Temple, the Xuannan Culture Museum vividly displays the long history and rich cultural heritage of Xuannan. It is the starting point of Beijing's formation as the capital, and a fertile ground for the modern revolution in the pursuit of change. It is also the birthplace of Peking Opera, and a southern paradise for the common people. It is possible to trace the footprints and revolutionary stories of pioneers such as Kang Youwei, Tan Yitong, Xu Xilin, Qiu Jin, Sun Yatsen, Huang Xing, Cheng Duxiu, Li Dazhao, Mao Zedong, and Zhou Enlai. Visitors can also enjoy the performances and artistic careers of generations of Beijing opera masters like Cheng Changgeng, Tan Xinpei, Yang Xiaolou, Mei Lanfang, and Ma Lianliang. They can experience the bustle of Changdian and Tianqiao, browse local snacks, listen to folk songs, have fun, watch unique skilled artisans in action, and appreciate the folk culture and folk art of the capital. Many ethnic groups live there, co-creating the unique regional culture of Xuannan.

2. Cultural resources must be excavated

As the years go by, some historical and cultural artefacts and inheritances remain only in the depths of people's memories, and some gradually disappear into memory, but they must be excavated and preserved.

2.1 Xianliangci – The ancestral hall of noble statesmen

Located at 103 Di'anmen West Street, Xianliang Temple is one of Beijing's cultural artefact protection units. Established in 1730 during the Qing Dynasty, Xianliang was an ancestral hall for Qing Dynasty ministers who had made outstanding achievements – a total of 178 people, including representatives of the Han, Manchu, and Mongol ethnic groups. Amongst them were many well-known civil officials and a large number of famous generals. The existing Xianliang architecture and the main building have been mostly preserved in magnificent artistic style. This is an important site for the representation of national worship and sacrifice, and an significant historical artefact in the capital. Since the Qing Dynasty, Xianliang has been used for other purposes. Its original meaning and role have been forgotten over time, so they must be excavated for future generations. The Xicheng District Institute of Cultural Relics published *Into the Famous Ministers of Xianliang*, and the forthcoming *Compilation of Xianliang's Historical Materials* are preliminary steps in this excavation.

2.2 Shichahai – The preservation of a treasury of cultural resources

History is not written by emperors and famous ministers. It is rooted in the general public. Historical data is rarely recorded regarding the "activities" of the people. It lives on in the memories of the people. In order to consolidate these folk legends that are on the verge of disappearance, the Xicheng District Institute of Cultural Relics visited cultural celebrities and the populace living in the Shichahai area, and fully explored the historical human resources of the area to form the book, News in the Street Stalls — Interview with Shichahai. This is a preliminary exploration of oral history, and if one were to visit the 15 streets of Xicheng City one by one, it will be a thick humanistic history.

3. *Cultural artefacts require innovation*

In addition to a focus on the exploration of cultural resources, we cannot view cultural artefacts merely as objects. We must explore new ways of utilizing them, combining cultural resources with contemporary society, and creating a group of cultural brands with historical and cultural implications as well as temporal qualities, regional characteristics, and mass appeal. In this regard, the Chinese Anglican Church and the Pagoda of Monk Wansong are useful examples.

3.1 The Chinese Anglican Church – Giving new life to ancient architecture

The Chinese Anglican Church is located at 85 Tonglinge Road, Xicheng District (formerly known as Nangouyan). It represents one of six Protestant sects of the Anglican Church of England in China. After the sect's withdrawal from mainland China, the church building was used as a cluttered, crowded warehouse and office. In recent work to upgrade units, fuse old and new architectural culture, and bring together various cultures, a number of lectures, forums, salons, reading clubs, film clubs, and small concerts have been held, bringing new life to the ancient building.

3.2 The Pagoda of Monk Wansong – A public reading space

The Pagoda of Monk Wansong is located at the side gate of 43 West Sinan Street, and is the burial place of an old monk of the Jin and Yuan Dynasties. Zhuanta Hutong is named after it. In the course of history, it has been a private house, a mutton shop, and a photo studio. In order to innovate and utilize the existing space for cultural preservation, as well as to implement the task of the district committee and district government in building "literary Xicheng - a cultural homeland", the Xicheng District Cultural Committee used the existing Wansong Temple to construct a public reading space, exploring a new methodology for the full empowerment of spiritual, cultural, and practical values of historical and cultural heritage.

4. Updated means of promotion

Modern information technology is now deeply embedded in daily life, and the promotion, protection, and utilization of cultural artefact preservation must keep up with the times and adapt to contemporary developments.

4.1 The application of QR code technology in the promotion of cultural artefacts

As society pays increasing attention to the preservation of cultural artefacts, public awareness is not just about cultural artefact conservation units and their cultural implications. People are now interested in obtaining more information in a simple and efficient way. To this end, we intend to carry out pre-integration and post-processing of information regarding the historical evolution, architectural form, and protection status of all cultural artefact protection units in this district, installing QR code icons in obvious positions on identification plates of protected units. This will allow the public to retrieve information directly via WeChat on their mobile phones, or choose to download QR codes with a dedicated client according to their mobile phone traffic status and environment, making it possible to decipher comprehensive cultural information while serving the purpose of extensive publicity. Meanwhile, the QR code offers a voice explanation system, so even blind individuals can learn relevant information. Statistics are also installed in the background server, so that we can tap into the impetus to protect cultural units for further analysis and research.

4.2 The broadcasting of cultural artefact in a variety of places

We arranged for a team of experts to carry out thematic integration and digital recording of Xicheng District's cultural artefact protection units, combined with modern publicity techniques. On the basis of protection and research, we launched a promotional video series based on Xicheng culture entitled *Xicheng Story*. The series explores the historical and cultural resources of Xicheng, highlighting the regional cultural characteristics and key features, and displaying the cultural heritage in an all-round way. The film selects typical elements such as geography, royalty, folk culture, business, education, religion, architecture, academia, technology, literature, and art, striving to provide accurate and reliable information for the community to understand and experience Xicheng, while providing vivid, image-based information for local residents.

The promotional video is played in public places such as cultural centers, libraries, cinemas, theaters, shopping malls, and hotels. There is also a to a 3–5 minute app version that can be disseminated on WeChat.

4.3 The application of mobile app data in the maintenance and management of cultural artefacts

When a staff member sets or selects a building or exhibit to be inspected, the system uses the GPS module to guide the person via a mobile terminal. For example, at the Beihai Cloud Pavillion, an inspector found that the information in the database obtained in the previous year had changed. If the database is incorrect, corrective photos can be updated in real time. The back-end database immediately receives an update reminder, so background maintenance staff can review the new data. After successful review, photos or text given by the staff are updated to the database, and the mobile terminal directly updates results, achieving the dynamic management of cultural artefact information.

4.4 Cultural artefact information added to the Chinaso website

For over a year, we have been integrating various forms of information regarding regional cultural artefact protection units to form a relatively complete repository of cultural artefacts. In order to increase public understanding of cultural objects, at the end of last year, we cooperated with Chinaso to collect cultural artefact information within its web platform.

In recent years, in accordance with the 16-character policy to "protect with timely rescue; and appropriate use under strong management", Xicheng District has made significant progress in the protection and utilization of cultural heritage across the whole district via a three-solution initiative called "solve, liberate, interpret". We hope that heritage buildings will inherit the culture of Old Beijing and pass it on. Through these ancient buildings, people can trace the past, look to the future, and experience nostalgia. In future work, we must not simply inherit culture, but constantly explore its cultural implications and its accordance with modern society, to form a cultural consciousness that retains the nostalgia of society across various cultural platforms.

The Development of a Virtual Simulation Experimental Teaching Center for the Application of Cultural Heritage Inheritance

Zhu Kerong
(Beijing Union University, Beijing 100191, China)

Abstract: The virtual simulation experimental teaching center for the application of cultural heritage is a virtual simulation experimental teaching platform, founded in a national comprehensive liberal arts application and experimental demonstration center at Beijing Union University. Themed around cultural heritage protection, inheritance, and application, the digital infrastructure uses virtual reality, Cloud computing, new media, data mining, and other related technologies to form a multi-disciplinary and highly integrated virtual simulation experimental teaching platform. The center explores the organic integration of modern information technology and experimental teaching, continually improving the virtual simulation experiment environment, enriching virtual simulation teaching resources, improving openness and resource sharing, and striving to become the best in its field.

Keywords: cultural heritage; virtual simulation; experimental teaching center

Preface

In order to implement the *Opinions of the Ministry of Education on Comprehensively Improving the Quality of Higher Education* (Education [2012] No. 4), and the *Ten-Year Education Informatization Development Plan* (2011–2020) (Education Technology [2012] No. 5), in August 2013, the Ministry of Education issued the *Notice on the Construction of a National Virtual Simulation Experimental Teaching Center* (Higher Education [2013] No. 94). In 2013 and 2014, 100 such centers were established. The virtual simulation experimental teaching center for the application of cultural heritage at Beijing Union University (hereinafter referred to as the Center), was named as the National Virtual Simulation Experimental Teaching Center in 2014.

1. Background

China has countless objects of cultural heritage, and many are scarce resources that cannot be regenerated or replaced. The capital, Beijing, is a world famous historical and cultural city, with a history of more than 3,000 years. Having been the nation's capital for 800 years, it is rich in both material and intangible cultural heritage. As such, the protection, inheritance, and application of precious material and intangible cultural heritage in Beijing have been important tasks for a long time, and will continue to be in the future. Material cultural heritage is very precious, but due to natural weathering and human factors over thousands of years, many items are now on the verge of decay and destruction. The non-renewable nature of material heritage means these objects require as much protection as possible. The inheritance of intangible cultural heritage is also an urgent task at present, comprising the way in which a master teaches an apprentice. As national and municipal holders of intangible cultural heritage in Beijing grow older, their passing may lead to the loss of traditional craftsmanship.

To protect historical and cultural heritage well, it is necessary to cultivate personnel for protection, inheritance, and application. This requires mastery of relevant theoretical knowledge, and also a grasp of related technologies, which are largely learned through practical training. Unlike the large-scale cultivation of intangible cultural heritage protection and inheritance, the archaeological excavation and protection of material cultural heritage, and the identification, restoration, and display of precious cultural artefacts are difficult to repeat in real scenarios. Many of these operations are irreversible, posing high risks, high cost, and high consumption, and worse, the damage or destruction of national treasures.

In response to the challenges of special training for cultural heritage protection and application, the Center provides a virtual experimental teaching environment in laboratories and on computers, using virtual reality, multimedia, motion capture, human-computer interaction, databases, Cloud computing, and network communication technologies. It provides a highly-simulated experimental teaching environment to achieve heritage protection, inheritance, and application for high-cost, high-risk, inaccessible, irreversible or damaged cultural artefacts that would not be feasible to achieve in real environment. It also improves the cognitive, practical, and independent innovation capabilities of undergraduates with professional experimental training and innovative experimental design in real environments.

2. Content

2.1 Platform development

The Center takes disciplinary intersection, technical support, and the integration of virtual and real comprehensive application as its development concepts, based on a guiding ideology of coordinated planning, key focuses, resource sharing, and improved efficiency. It is a virtual simulation experimental teaching platform for the comprehensive protection and display of cultural artefacts, the inheritance of national "non-legacy" skills, the reproduction and application of cultural heritage scenes, and innovative application in the cultural heritage tourism industry for those with majors in history, cultural artefact and museum studies, archival studies, tourism management, exhibition economics and management, human geography, urban and rural planning, geographic information science, and related fields.

2.1.1 Comprehensive protection and display of cultural artefacts on the virtual simulation experimental teaching platform

This platform includes a virtual simulation laboratory for cultural artefact appreciation, a virtual simulation lab for the digital restoration of cultural artefacts, a virtual simulation lab for cultural artefact exhibition, and a "Three Mountains Five Gardens" multimedia Big Data management and cultural object mining lab, with certain functions complemented by practical teaching, partially functioning to expand experimental teaching practice. In the process of cultural artefact protection and experimental teaching, some essential links cannot be implemented as they are limited by difficulty, cost, and environmental risks. Instead, they are complemented on the virtual simulation platform. Equipped with information facilities, multimedia technology, and network technology, the platform allows the realization of remote experimentation in different areas, carrying out the process of cultural artefact protection and display at low cost and without risk. Through the combination of virtual and real elements, practices complement one another to achieve tasks that are difficult to carry out in traditional experimental teaching environments, increasing the impact of experimental learning.

2.1.2 A State-level virtual simulation experimental teaching platform for intangible cultural heritage

The platform is equipped with virtual simulation labs for the inheritance of ancient paintings and drawings, the inheritance of the "Eight Unique Skills of Yanjing", special Beijing art, and precious archive restoration techniques. It is an important carrier by which our university explores the protection, planning, and comprehensive utilization of cultural heritage areas. As a teaching practice platform for the Beijing Intangible Cultural Heritage Research Base, it is designed to serve the professional teaching and research of cultural heritage, providing necessary technical support for the protection, inheritance, and application of intangible cultural heritage.

2.1.3 Scene reproduction for cultural heritage, and its application on the virtual simulation experimental teaching platform

The platform features a cultural heritage scene reproduction and intelligent 3D model display lab, an urban factor analysis and innovation application lab, a space-time simulation and planning design lab for capital development, and a Cloud rendering and virtualization computing technology lab, using ARC GIS, ENVI, City Engine, AutoCAD, 3D Max, and other modeling software and tools alongside the ARC GIS Portal unified resource management system. Experimental equipment includes high-performance servers, visual workstations, electronic total stations, global positioning system digital acquisition devices, 3D data acquisition devices, and digitizers. The platform can collect, transmit, register and store multi-source data. It is capable of spatialization for multiple elements of nature, economy, society and culture. It has multi-angle, multi-level, and multi-person simulation, restoration, and reconstruction functions for each component and region of cultural heritage. It also has functions of spatial succession feature perception and protection planning process simulation for sites in different periods, as well as overall program evaluation.

2.1.4 Innovations in the cultural heritage tourism industry, and their application on the virtual simulation experimental teaching platform

The platform is not equipped with a rapid 3D model construction lab for digital scenic spots of cultural tourism, a virtual lab for the intelligent management of digital scenic spots, nor a virtual lab for creative product design. It uses a complete 3D high-definition active stereo projection system, a 360° panoramic seamless projection system, a seamless full cycle integration system, and an image recognition gesture recognition system. It also employs an infrared-sensing positioning system, digital restoration technology and a relevant platform, a 3D modeling system, a 3D printing system, a virtual broadcasting and rebroadcasting system, a face recognition system, an item monitoring system using the Internet of Things, and holographic display systems via iPad and Android mobile terminals.

3. Resources

According to the principle of combining the virtual and the real so that they complement one another in order to be true and not imaginary, the Center is able to carry out processes of cultural heritage protection, inheritance, and personnel training that cannot be completed in the real environment due to high costs, high consumption, inaccessibility, and irreversibility. It uses technologies such as virtual reality, multimedia, human-computer interaction, databases, Cloud computing, and network communications to construct a highly simulated experimental teaching environment, finally forming a virtual simulation teaching system with two levels, three links, and four modules.

Two levels refers to the professional training series and innovative expansion series of experimental projects that are at the base and embedded in different levels of experimental courses. The professional training series of experimental projects mainly cultivates students' professional application ability with cultural artefact simulation experiments, cultural artefact protection and restoration virtual experiments, cultural artefact display experiments, traditional painting and calligraphy simulation experiments, filament mosaic technique simulation experiments, cloisonné enamel simulation experiments, precious file restoration experiments, and historical scene reproduction experiments. The innovative series of experimental projects mainly cultivates students' comprehensive practical abilities and innovative spirit, involving the virtual simulation design of cultural artefact exhibitions, and the virtual design of cultural heritage protection planning.

Three links refers to cultural heritage protection, inheritance, and application, which together constitute the whole process and life cycle of cultural heritage alongside the sustainable development of society today. Protection is the basis of inheritance and application, while inheritance is the premise and reason for protection and application. Application is required for better protection and inheritance. The implementation of experimental teaching in these three links is designed to help students acquire relevant skills and abilities to participate in the processes of cultural heritage protection, inheritance, and application.

Four modules refers to the modules within the virtual simulation experimental teaching system: comprehensive protection and display of cultural artefacts, national level "intangible cultural heritage" technology inheritance, cultural heritage scene reconstruction and application, and innovative cultural heritage tourism industry applications. Each experimental module involves multiple projects. These projects and modules are independent yet interconnected, combined with traditional practical experiments to form a complete experimental system in cultural heritage protection, inheritance, and personnel development.

4. Main features

4.1 Serving the capital; strengthening application; constructing a virtual simulation experimental teaching platform that integrates a cultural heritage protection, inheritance, and an application platform; and creating a distinctive system with two levels, three links, and four modules

In order to meet the needs of the capital city and the country in the cultivation of personnel for material cultural heritage and intangible cultural heritage protection, and to aid Beijing in its role of a national cultural center, addressing challenges in the development of application-oriented professionals, the Center has established a virtual experimental teaching environment, using technologies such as virtual reality, multimedia, human-computer interaction, databases, research and network communications to construct a highly simulated, experimental teaching environment and a "two level, three link, four module" system. Embedded in the traditional experimental teaching system, it solves the problems of experimental teaching that cannot be carried out on traditional platforms, playing a crucial role in the cultivation of practical and innovative student abilities.

4.2 Merging integration, improving synthesis, and achieving interdisciplinary, multi-disciplinary, multi-technology, and multi-field fusion with cultural inheritance and application at the core

Focusing on the theme of cultural heritage inheritance and application, the Center has established a multi-disciplinary, multi-technology, multi-field cross-integration virtual simulation experimental teaching platform for a number of related majors in service history, cultural artefact and museum studies, archival science, tourism management, exhibition economics and management, human geography and urban and rural planning, geographic information science, and English language (including for international students and those from Hong Kong, Macao, and Taiwan). It has set up a variety of experimental projects, serving to train personnel in different disciplines, and also gathering technical strength across the whole school. It condenses the developmental direction of the discipline, carrying out simultaneous scientific research, personnel training, and social services in regards to cultural heritage, while realizing the integration of disciplines, technologies, and fields.

4.3 Combining the virtual and the real, while expanding benefits and creating a solid foundation and tradition for resource sharing at the Center, strengthening open resource sharing across the experimental teaching platform

At the Center, a good foundation and tradition of high-level resource sharing has been inherited, with complementary experimental teaching and interaction between teaching and research that expands its positive impact. The Center offers experimental teaching for undergraduates and postgraduates, as well as related research work for a variety of related majors, while making full use of the digital campus' broadband network to open the experimental platform and its resources to other students across the school, as well as the 21 students from the Beijing Xueyuan Road university community zone (Coll & Univ. T.C. Coll Rd. Zone), extending its radiation to other colleges and universities in the area. The Center is able to open up and share with users in China and abroad. In future, it plans to open a shared experimental teaching platform and resources for use by social enterprises, research institutes, enterprises, and other social users both inside and outside Beijing, in order to expand its benefits even further.

5. Conclusion

Virtual Reality (VR) is a comprehensive integrated technology that emerged in the 1980s. It is used in the fields of science, engineering, agriculture, and medicine, but rarely in the liberal arts. In 2013, the Ministry of Education began a review of the national virtual simulation experimental teaching center, encouraging liberal arts labs to begin thinking about how to use information technology to develop virtual simulation experiments. Therefore, at present, liberal arts majors are basically in the exploratory stage of virtual simulation experiments. Compared with virtual experiments in fields of science, engineering, agriculture and medicine, they are still immature. It is thus necessary to learn from the experience of other disciplines, building a unified virtual simulation platform with more abundant resources in order to cultivate personnel.

References

[1] Li Ping, Mao Changjie, Xu Jin. 'Carrying out the construction of national virtual simulation experimental teaching centers to improve information levels of experimental teaching in colleges and universities', [J] *Research and Exploration in Laboratory*, 2013, 11: 5–8.

[2] Wang Weiguo. 'Thoughts and suggestions on the construction of virtual simulation experiment teaching centers', [J] *Research and Exploration in Laboratory*, 2013, 12: 5–8.

[3] Wang Xiaodi. 'Understanding and discussing eight relationships in the construction of virtual simulation experiment teaching centers', [J] *Laboratory Technology and Management*, 2014, 08: 9–11.

[4] Xu Jin. 'Summary of the construction of the national virtual simulation experiment teaching center in 2013 and the application proposal in 2014', [J] *Research and Exploration in Laboratory*, 2014, 08: 1–5, 25.

[5] Qu Weina. *Research and implementation of virtual laboratories based on virtual reality technology* [D]. Northeast Petroleum University, 2003.

[6] Li Han. On *the relationship between "virtual" and reality* [D]. Capital Normal University, 2008.

[7] Wang Jianmei, Zhang Xu, Wang Yong, Zhao Yunhua. 'The status quo, policy and revelations of American Virtual Reality technology', [J] *Science and Technology Management and Research*, 2010, 14: 37–40, 56.

[8] Song Wei. *Virtual experimental research using virtual reality* [D]. Chongqing University, 2005.

[9] Li Deren. Application of virtual reality technology in cultural heritage conservation', [J] *Journal of Yunnan Normal University (Philosophy and Social Sciences)*, 2008, 04:1–7.

[10] Jiang Xuezhi, Li Zhonghua. 'Research status of virtual technology in China and abroad', [J] *Journal of Liaoning Technical University*, 2004, 02: 238–240.

Digital Management of Immovable Cultural Heritage in Districts and Counties

Wang Na
(Research Office, Institute of Cultural Relics, Xicheng District, Beijing 100035, China)

Abstract: With the rapid development of information technology, digital technology is being widely used in the Chinese cultural field. In the field of movable cultural artefacts, museums in various localities and industries use collections as their main resources. From collection management systems and collection display technologies to online museums and mobile guides, the development of digital museums is largely mature and showing great vitality. However, in the field of immovable cultural artefact management, the application of digital technology presents various gaps at different levels, and the integration of information technology in the management of immovable cultural heritage is urgently required in order to improve the level and efficiency of cultural artefact management. This paper explores applications and practices of digital technology in the management of immovable cultural artefacts as a carrier of dynamic cultural heritage management within districts and counties.

Keywords: cultural heritage; immovable cultural artefacts; digital; dynamic management

With the rapid development of information technology, digital technology is being widely used in the Chinese cultural field. In the field of movable cultural artefacts, museums in various localities and industries use collections as their main resources. The development of digital museums is largely mature, from collection management systems and collection display technologies to online museums and mobile guides, showing high-tech and diversified technological features. However, in the field of immovable cultural artefacts, the application of digital technology presents flaws at a variety of levels. Some State-level cultural protected units are open museums, and with the opportunities for digital development of museums, historical resources of immovable artefacts have begun digital processing for digital display, protection, and management. Large-scale, dynamic cultural heritage monitoring work has begun on immovable units of cultural heritage, in accordance with the requirements of World Heritage Cultural Site assessments. However, most municipal, district, and county-level cultural artefact protection units are still in a state of disorder. Digital management of cultural heritage has not yet been carried out, and the integration and management of immovable cultural heritage via information technology is urgently required for the improvement of cultural artefact management levels and efficiency. This paper explores applications and practices of digital technology in the management of immovable cultural artefacts as a carrier of dynamic cultural heritage management within districts and counties.

1. The current state of digital management of immovable cultural heritage in China

China's immovable cultural artefacts now rely on a hierarchical protection policy, that is, implementing a three-level management system of state-level, municipal-level, and district/county-level cultural artefact protection units, and corresponding administration agencies. Some of these units are open museums, and some are still used as offices or residences. Looking at the current status of immovable cultural heritage, there are three conditions.

1.1 Individual success stories – Open units making use of museum digitization

After years of development and conservation, many immovable cultural artefact units have been turned into public museums. With the help of digital museum development, digital technology has been applied to the display of immovable cultural artefacts and cultural resources. For example, at the Imperial Palace, the State-level cultural artefacts protection unit has digitally explored and researched several projects such as ancient building modelling and 3D display, cultural artefact restoration, and ancient construction and repair techniques, achieving remarkable results. In 2012, Beijing established the Yiheyuan World Cultural Heritage Monitoring Center, using scientific technologies such as advanced electronic mapping and 3D scanning to collect basic data from the Summer Palace for heritage monitoring, using professional computer software to create digital mechanics for ancient buildings. The modelling and heritage conservation monitoring of the remains of the Summer Palace involved data collection, analysis, research, and guidance of actual conservation work. By integrating digital technology with cultural heritage resources, disappearing cultural heritage is recreated, and precious relics can be conserved and managed more effectively.

1.2 Comprehensive top-level design – Dynamic monitoring of global cultural heritage

Among the three levels of cultural artefact protection units in China, many national units have been included in world cultural heritage. In accordance with requirements for the protection and management of world cultural heritage, cultural heritage protection needs to adhere to international standards. In 1994, UNESCO revised its statute of the Convention for the Protection of World Cultural and Natural Heritage, including monitoring and early warning among the duties of the World Heritage Committee, requiring countries and governments with World Heritage sites to establish corresponding monitoring systems and submit a six-yearly report. In February 2004, the General Office of the State Council issued the Notice of the General Office of the State Council on Forwarding the Opinions of the Ministry of Culture, the Ministry of Construction, and the Bureau of Cultural Relics on Strengthening the Protection and Management of China's World Cultural Heritage, proposing "the use of science and technology to strengthen world cultural heritage protection. It is necessary to strengthen the study of regular protection and management of world cultural heritage… to establish a dynamic information system and early warning system for the management of world cultural heritage in China as soon as possible, and to strengthen the monitoring of world cultural heritage protection." In 2006, the State Administration of Cultural Heritage promulgated the *Measurements for the Management of World Cultural Heritage Monitoring and Inspection*, and the *China World Cultural Heritage Expert Consultation Management Measures*, which clarified the monitoring procedures, duties, and contents of standardized world cultural heritage monitoring activities. In 2008, the first draft of the *World Cultural Heritage Monitoring Regulations* was completed, and the world cultural heritage in China was effectively protected and managed with the establishment of national, provincial, and district level monitoring and inspection mechanisms. In 2012, the State Administration of Cultural Heritage prepared for the establishment of a China World Cultural Heritage Monitoring Center at the China Cultural Heritage Research Institute. Based on specific planning, construction and development principles, the National World Cultural Heritage and Early Warning System was launched. In January 2015, the China World Cultural Heritage Monitoring Center was formally established as an internal organization of the China Cultural Heritage Research Institute, carrying out independent work. The China World Cultural Heritage Monitoring and Early Warning Platform consists of an overall platform function module, monitoring data and indication system, early warning information and disposal, and monitoring business processes and functions. It has been initially established, identifying 17 categories and 59 items of monitoring data and indicator systems that have an impact on the management of cultural artefact protection. Meanwhile, three local platforms –

the Grand Canal, the Classical Gardens of Suzhou, and the Longmen Grottoes – were launched to connect to the general platform, realizing the interconnection of national heritage across two platform levels. The opening of China's World Cultural Heritage Monitoring Platform marks the realization of dynamic management and monitoring of domestic immovable cultural heritage, greatly promoting deeper protection and research.

1.3 A neglected majority – The lag in digital work on domestic immovable cultural heritage

In terms of quantitative proportion, the above forms of immovable cultural heritage that have achieved initial digital developments are still a minority. National security units not included in world cultural heritage, and a large number of municipal and district-level cultural units are still basically untouched in terms of digitization. These protected units are still archiving and collating the "Four Requirements" in the traditional way. However, a large number of resources such as text, images, drawings, rubbings, and transcripts have been accumulated in the "Four Requirements" archives. Not yet integrated with digital information, they are simply stored as "files". Judging from consultations of such files, some of these resources are old and have never been updated. Some lack systematic analysis reports, making it difficult to make scientific assessments of historical and use value of the unit, while some cases are missing items and page numbers. The storage of paper files also requires office space, with the inconvenience of the review process also hindering the utilization and use value of resources.

2. Analysis of the necessity of digital management of immovable cultural heritage in districts and counties

According to Articles 11 and 12 in the *Guidelines for the Protection of Chinese Cultural Relics and Monuments*, "the main content of the assessment is the valuation of cultural artefacts, the state of preservation, and the conditions of management, including the analysis of historical records and the investigation of the status quo. Determining the cultural artefact protection unit and its level must be based on the assessment conclusions and announced by governments at all levels by law. The units identified as protected cultural artefacts must meet the 'Four Requirements', that is, a reason for protection, descriptive signage, a record on file, and a specialized agency or person responsible for management." The assessment of immovable cultural heritage must take into account historical conditions and current conditions, putting higher demands on dynamic monitoring and management of cultural heritage. The establishment of a cultural heritage monitoring and early warning platform with digital technology offers dynamic monitoring, and provides scientific and technological support for the formulation of protection planning and the dynamic management of cultural heritage.

It is not only world cultural heritage that must be monitored, but also local cultural heritage. First of all, the establishment of archives for cultural heritage and the systematic integration of resources and information are conducive to management departments in grasping the dynamics of cultural heritage at any given time, so as to develop scientific protection plans according to preservation status and the development of cultural artefacts. Secondly, the monitoring and early warning system can offer preventative protection based on analyses of monitoring data, applying high-tech means to cultural artefact protection and objectively formulating corresponding emergency plans for fire prevention, anti-theft, and anti-explosion measures. It plays the role of preemptive prevention and rapid response, enabling managers to find and solve problems in a timely manner. Furthermore, through the integration of local cultural heritage resources in the information technology system, the establishment of a cultural heritage database and unified platform for monitoring means that managers, users, and volunteers can all work towards common goals and implement them across municipalities, districts, and streets following a community linkage mechanism, in order to achieve multi-level, multi-dimensional protection of cultural heritage.

3. *Analysis of digital management feasibility for immovable cultural heritage in districts and counties*

3.1 A good resource base – Complete data in "Four Requirements" files and three surveys

Since the implementation of the grading policy for immovable cultural artefacts in China, it has been stipulated that cultural artefact protection units at all levels must establish "Four Requirements" files. Article 9 in the Law on the Protection of Cultural Relics states: "The cultural artefact protection units at all levels shall be demarcated by the people's governments in provinces, autonomous regions and municipalities directly under the Central Government, and the people's governments in counties, autonomous counties, and municipalities delineating the necessary scope of protection, making a detailed sign, creating a record, and setting up a special agency or special personnel to manage the unit independently. Article 13 of the *Standard of the National Cultural Artefacts Protection Unit for the Scope of Protection, Marking Instructions, Records, and Archives (Trial)* states, "the file must include records of the protected cultural artefact unit itself and related historical materials. The content is divided into scientific and technical data, and administrative management documents in the form of text, photography (photos, slides, film), video, drawings, rubbings, transcripts, hard disks, and other information carriers." Record files are divided into main, secondary, and test volumes. The main volume is based on protection management work records and scientific information. The secondary volume contains relevant administrative documents and daily work. The test volume contains information relevant to cultural artefact protection units, and is available for reference. All of these files are produced according to certain specifications, with the same classification and data standards, to be used as the basic resources of the cultural heritage database. In addition to the "Four Requirements" files, the third national cultural artefacts survey (completed in the previous two years) established more comprehensive cultural artefact data, reflecting the latest changes and developments of various cultural heritage sites in recent years, also providing resources for the establishment of a cultural heritage database.

Xicheng District is an example. As an institution for the management and research of immovable cultural artefacts, in recent years we have consistently completed compilation of "Four Requirements" files for district-level cultural artefact protected units. By organizing textual historical materials, photographs, drawings, developmental changes, and other information, files have been created for each unit. The information is stored and organized with electronic documents and paper files managed in a standardized way so as to facilitate protection and research. After several years of work, our institute has basically completed the recording and archival work for district-level cultural protection units. Meanwhile, with the support and assistance of industry units, we have also collected the "Four Requirements" files for national and municipal cultural artefact protection units within our jurisdiction. These resources complement the results of the third national cultural artefacts survey, providing a solid foundation on which to deepen the historical and cultural implications of historical artefacts. However, these resources are based on individual cultural protection units. Integration in a database will facilitate our understanding of the horizontal connection between units. It will be conducive to an analysis of the similarities and differences between the cultural protection units in our jurisdiction, providing analysis and judgment across the jurisdiction, and making it possible to adjust or develop more scientific protection plans and directions for development.

3.2 Mature examples to draw on – Successful cases of individual units and the opening of the World Cultural Heritage Early Warning and Monitoring Platform

In November 2014, the China World Cultural Heritage Monitoring Training Course was held in Beijing, directed by the State Administration of Cultural Heritage and hosted by the China Cultural Heritage Institute. More than 200 people from more than 140 heritage sites and research institutes in 24 provinces, municipalities, and autonomous regions participated in the training course, marking the development of the China World Cultural Heritage Early Warning and Monitoring System by the World Cultural Heritage Monitoring Center of the China Cultural Heritage Research Institute. Its launch followed two years of research, development, and commissioning. As the national research center and general platform for the monitoring and protection of China's world cultural heritage, the China World Cultural Heritage Monitoring Center officially entered the growth stage, as an important symbol of the nation's maturing protection management capabilities. The Heritage Monitoring Center released the World Cultural Heritage Monitoring Platform Data Index, including important parameters such as metadata tables and monitoring indicators, which will play an important role in the promotion of scientific management of China's world heritage. The training course also discussed many successful case studies, such as the Imperial Palace Cultural Heritage Monitoring Report, the Suzhou Garden Cultural Heritage Monitoring Platform, and the China World Cultural Heritage Early Warning Monitoring Platform, as well as the achievements of the Potala Palace Cultural Relic Monitoring Technology. These successful examples provide useful reference points for us to carry out cultural heritage digitization.

3.3 Preliminary practices – Xicheng District's practices in the monitoring of cultural heritage of the Grand Canal

Xicheng District is located at the northern end of the Beijing section of the Grand Canal, and the main portion of the Tonghui River Beijing Old City is located in the Xicheng area. The Shishahai and the upper Chengqing Gate (Wanning Bridge) are two heritage sites in the area selected for the Grand Canal's list of applications as world heritage. Each is what we call "one point and one river section". When the Grand Canal is declared a world heritage site, under unified Beijing deployment, Xicheng District will monitor the water level in the western section in real time, and submit a monthly report on Shishahai Heritage Monitoring to track data on the Grand Canal Cultural Heritage Management Platform. Due to the linear characteristics of Grand Canal's cultural heritage, data submitted at each heritage point is independent, and the Cultural Heritage Management Platform comprehensively analyses and evaluates data based on individual information from various heritage sites.

4. Key points and challenges in the management of immovable cultural heritage digitization in districts and counties

Immovable cultural heritage carries the traditional culture of the nation, precipitating historical memories, and passing on cultural memory. For protected immovable cultural heritage resources, counties must strictly follow principles of not changing the original state of historical buildings and the maintenance of historical authenticity, as formulated by the World Heritage Convention, so as to achieve comprehensive monitoring, real-time monitoring, data analysis, emergency command, and continuous protection at various levels of cultural heritage within the jurisdiction. Today, with the rapid development of informatization, we must fully empower new high technology for the protection and management of cultural heritage, improving the scientific and technological content of cultural heritage protection and management, and promoting the scientific and digital development of immovable cultural heritage in districts and counties. It is necessary to take note of one special point regarding immovable cultural artefacts, that is, each protected unit generally comes under the jurisdiction or use of different administrations, which poses certain difficulties for unified management. Therefore, it is possible that local cultural artefact departments may take the lead in setting up special agencies with sufficient personnel to carry out the development of a cultural heritage database, early warning monitoring, and cultural artefact inspection, integrating cultural heritage resources in the region, building a unified platform, and opening up new developmental prospects for the protection of immovable cultural heritage.

CHAPTER 3

Sharing and Utilization as the New Status Quo

– Digital Development and the Sharing of Museum Resources

The Digital Collection and Data Management of Museum Objects

Ji Hong, Wang Jing, Li Jianwen, and Zeng Zhaohui
(Department of Virtual Reality, Beijing Computing Center, Beijing 100124, China)

Abstract: In order to efficiently and scientifically manage increasing museum collections, and to extract data quickly and accurately when required, the digital collection of museum artefacts is of great significance for the systematic management and permanent preservation of collection-related information. Taking the digital collection process of natural museum specimens as an example, this paper explains the workflow, methods, tools, and data management specifications in digital information collection, focusing on the processes of 2D and 3D data acquisition in massive collections, and analyzing the work completed thus far. Finally, key issues in the collection process will be summarized and countermeasures provided, proposing an improved solution for future digital acquisition.

Keywords: museum, data collection, 3D scan, data management

Preface

The digital museum is a concentrated expression of the research and application of digital technology in the museum field, undertaking tasks of collection, protection, management, transmission, display, and use of historical, cultural, and natural heritage information resources [1]. It is a product of the organic combination of multimedia technology and the artefacts and museology industry in the information age – an excellent contribution to the promotion of sharing and passing on human history and culture. Photography and 3D scanning are key technologies for the digitization of museum collections, and are an important means of obtaining 2D images and 3D point Cloud data. Taking the digital collection project of the Beijing Museum of Natural History as an example, this paper explains the data collection and processing workflow, technology, and method in regards to large-scale museum specimens. Conducting process management on data acquisition and processing guarantees the progress of project implementation and the quality of the data collected.

1. 2D image acquisition

The digital collection project at the Beijing Museum of Natural History involves complete high-definition photographs of over 15,000 specimens over the course of 10 months, which was a time-sensitive and a critical task. The collected specimens include plants, fossils, insects, shellfish, pottery, and mammals of various types and scales. Plant specimens were collected on a 2D plane, and others on a stereoscopic 3D plane. 2D photography of mammalian specimens is the most complicated, as they need to be shot from multiple angles in order to obtain complete image information.

1.1 Personnel and equipment

Due to the large number of specimens to be documented for data collection, the members of the photography group were divided into four collection groups for simultaneous shooting, with each group assigned one photographer, one library administrator, and one recorder. The duties of each member were as follows: 1) Photographers — taking

Table 1 Equipment Specifications

Device	*Model*	*Configuration parameters*
Sinar	HY6	Full digital autofocus system, 6 x 6 and 6 x 4.5cm formats, film/digital compatible conversion, electronic control shutter speed of 32s ~ 1/ 1000s B release, shutter priority exposure mode and aperture priority exposure mode, automatic flash function
Canon	IDSMark III	Professional 135 digital SLR camera body (135 full frame, 21.1 million pixels) with 21.1 million effective pixels, ISO range 50~3200, 45 point screen focus, 100% field of view viewfinder, magnification up to 0.76 times in viewfinder, LCD size of 3in, resolution of 230,000, supports continuous shooting function, 5fps.
Canon	5DMark II	Full-frame digital SLR, effective pixels 21.1 million, display 3in, high-definition camera with full HD mode (1080)
Elinchrom	Digital remote light and shadow creation system	Reflector, light barrier, beam hood, softbox, flash system for outdoor shooting, etc.
Schneider	AFD Tele Xenar 2.8/180PQ	Autofocus, all digital macro lens, maximum lens resolution over 60 pairs

photographs, checking photo quality, management of extracted specimens; 2) Assistants — photo renaming, data sorting and backup, assisting in the transportation of large specimens; 3) Library administrators— printing specimen lists, handling specimen entry and exit procedures, specimen management, assisting with photography.

Each shooting team was assigned one set of SLR camera equipment and one digital remote control lighting and shadow system. Specific equipment parameters in use are shown in Table 1.

1.2 Shooting process

According to the shooting specifications of cultural artefact collection images, the planar specimen must be photographed in front view, recording the image as accurately as possible with one photo. The stereoscopic specimen is photographed from the front side as the main representative, also taking a forward view photo, choosing appropriate angles to collect top view, side view, and partial view, according to the shape and grade of the specimen.

Team members work together to ensure that the filming is carried out in an orderly manner. The library administrator generally picks out the collected specimens the day before filming and makes a specimen photography list. The team "records while shooting", so each time a photographer shoots a specimen, the recorder renames all collected photos according to the naming conventions, storing single specimen photo information independently, naming folders based on the specimen registration number, and storing all photos of the specimen in this folder, including front, top, rear, side view images. Each time a specimen's details are collected, the recorder must register its number, name, photographer, shooting time, and number of photographs on the shooting list so as to ensure that collected information can be searched efficiently. The workflow of 2D data acquisition is shown in Figure 1.

A digital camera is used to capture specimen photos in order to accurately and completely reflect the artefact' shape, color, form, and texture. Composition must be based on the principles of prominent subject, simple composition, and cohesive color. The background should be a solid, tasteful, and neutral color. Photographic lights and reflectors are used to create strong 3D effects on photographed specimens [2]. A specimen administrator must work with the photographer to conduct a phase review of the photo quality. Only when the quality meets the established requirements will it be organized into the database. Otherwise, it has to be reshot.

The original photos cannot be used directly, but must be processed with Photoshop editing software, adjusting brightness, resolution, noise, and background modification. Retouching the photo is not only for aesthetic appeal, but also ensures that it reflects the actual appearance of the specimen to the maximum.

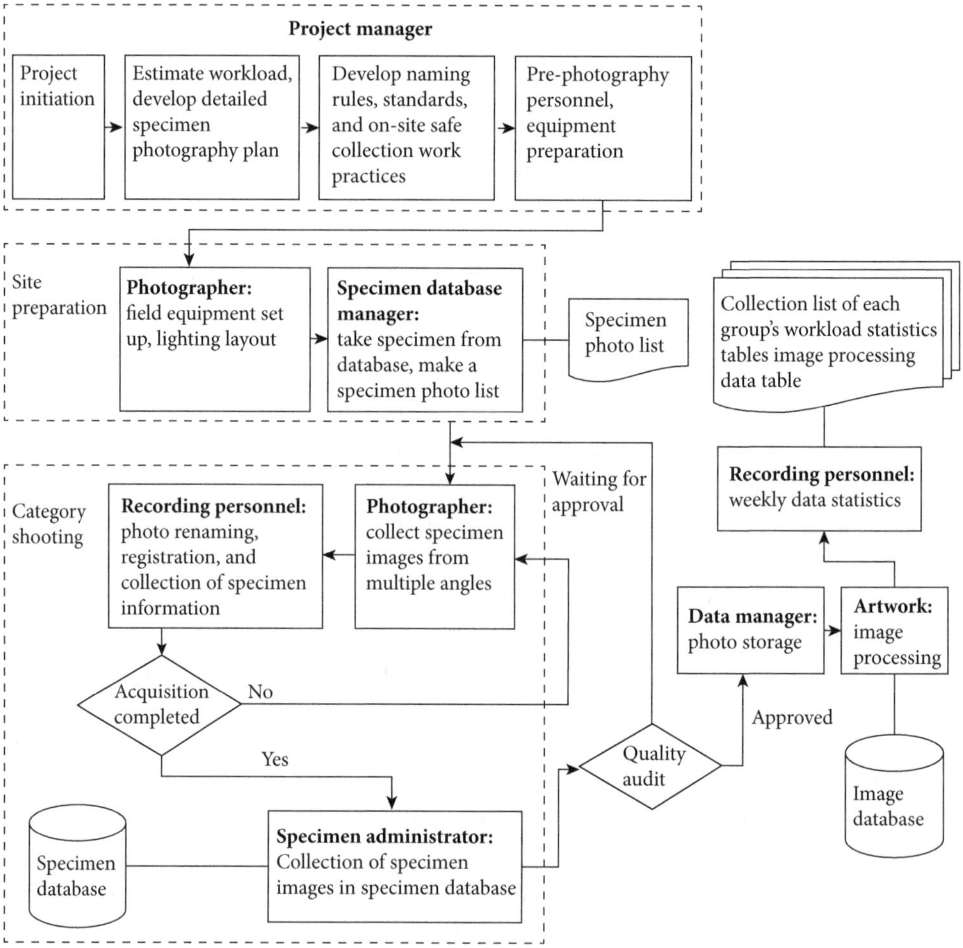

Figure 1 2D data collection flow chart

1.3 Massive data management

Specimen photos are stored in three formats for different purposes:

1) RAW format: original file.
2) TIFF format: saved in TIF file format for maximum quality after retouching.
3) JPG format: 80–100 KB, to be uploaded to the collection management system; 2–4 MB, used in the National Cultural Relics Bureau cultural artefacts survey.

1.3.1 Naming standards

The naming of photo files is: general registration number — position number — sequence number.

1) General registration number: the registration number of the collection under a current collection unit in the "Museum Collection General Register". If there is a special symbol such as a colon, forward or backslash, or asterisk in the general registration number, it must be input in full.
2) Position number: the angle of the photo, expressed in English capital letters, as shown in Figure 2:
   A: front view, B: top view, C: side view, E: partial view, F: bottom view, H: rear view
3) Sequence number: the picture sequence number from the same framing angle, represented by a half-width numeral.

Example:

The registration number of a collection in a certain museum in the Xicheng District in Beijing is 01701254. If the first image is a front view image file in .jpg format, it should be named: 01701254-A-1.jpg

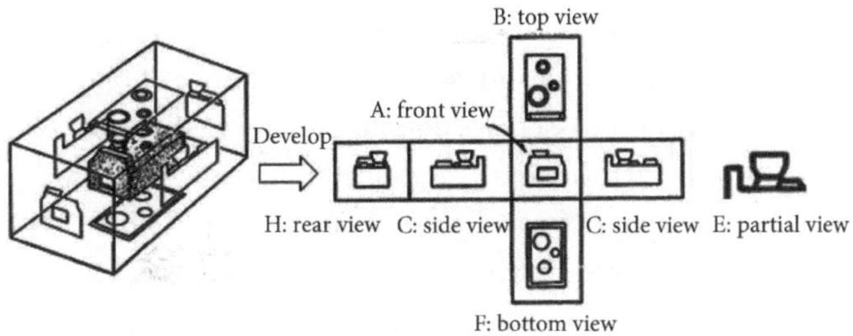

Figure 2 A naming orientation reference image from the Natural History Museum's photo naming convention

1.3.2 Document storage structure

The collected photos are stored according to the photo storage format in classified groups. The first-level directory contains four folders, separately storing the photos of all corresponding specimens based on type. The photos of each specimen are then stored in separate folders, forming a secondary directory, and the folder is named after the general registration number. The document storage structure is shown in Figure 3.

1.3.3 Data recovery and management

The recorder is responsible for daily data backup work, with data saved on the local computer and mobile hard disk. Local data is retained for one week, and mobile hard disk data is summarized once a week and uploaded to the server. At the end of collection, specimens extracted that day are sorted and counted and returned to the warehouse, and the number of photos collected is counted.

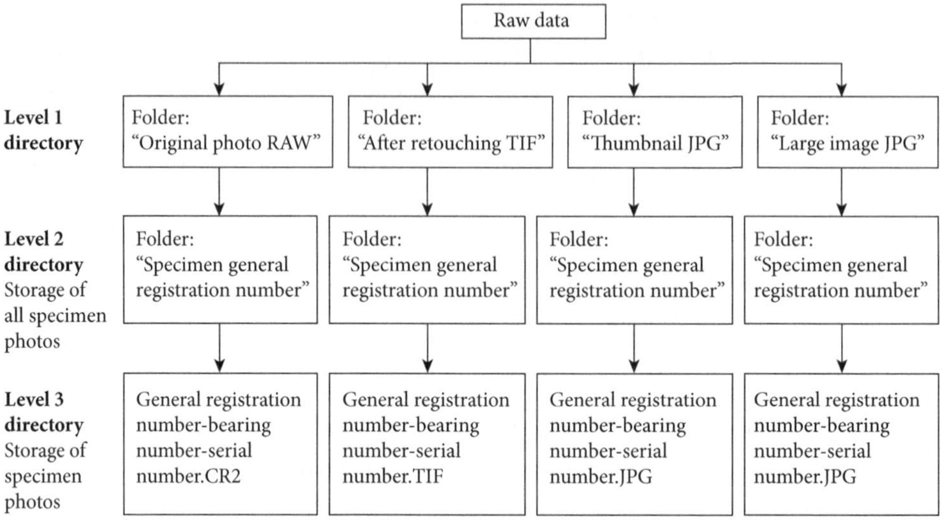

Figure 3 Document storage structure

The number of specimens sorted and counted, the number of photographs taken, and the number of photographs processed are counted on a weekly basis. As photo shooting and image processing are performed at the same time, processed photos are stored in separate places in the database. In order to avoid data loss and irreparable damage, the data and its copies are stored in different areas of the disk array – a double backup provision. Weekly work statistics are so refined that the name, number, angle, and time of the collected photos of single specimens photographed by every small group are collected as intelligent statistics in real time in the form of Excel spreadsheets, ensuring correct data association by setting various constraints. The aggregated weekly workload and the project's overall progress are reported to management staff, enabling the project manager to monitor project progress in real time.

Table 2 2D Data Collection Statistics

Group	*Collection specimens (pieces)*	*Collection photos (amount)*	*Input time (weeks)*	*Average shooting efficiency (pieces per day)*
Fossils	2,748	7,113	18	22
Insects	4,264	11,173	23	26
Precious items	2,076	9,482	30	10
Plants	4,007	4,414	11	52
Ancient humans	2,103	5,172	11	27
Total	15,198	37,354	–	–

1.4 Achievements

2D data acquisition has encompassed a total of 15,198 specimens, collecting 37,354 original photographs. See Table 2 for details on fossils, insects, rare and fine birds, herbarium plants, and ancient humans. See Figure 4 for photograph examples.

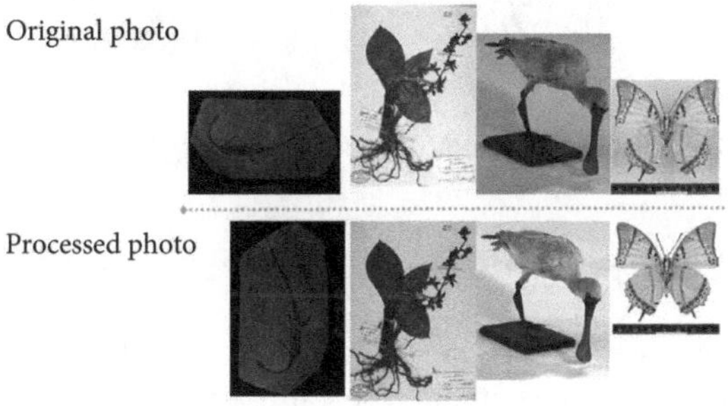

Figure 4 Photographs of specimens: original photos (top); processed photos (bottom)

2. 3D Data Acquisition

3D scanning has universal applications in the process of museum digitization, characterized by fast acquisition of 3D point Cloud data of stereoscopic specimens, meaning that 3D specimen models can be established conveniently using scanner data processing software. The created specimen model not only permanently saves specimen appearance information [3], but also facilitates specimen restoration and reproduction. In addition, the virtual display and interaction of the specimen can be realized through the 3D engine, offering a more vivid and novel means of displaying information about popular and natural science.

2.1 Data acquisition tools

As a variety of forms are to be collected, different sizes require a choice between two scanners, as shown in Figure 5. Due to its small size, the hand-held 3D laser scanner is light and easy to use, so is commonly used for 3D scanning of small 3D specimens in the digitization of museum collections, such as small animal bones and animal specimens.

The large-area laser scanner can quickly capture a wide range of high-definition and inductive 3D images. It is mainly used to obtain 3D data from large specimens in the digitization of natural history museum collections, such as Asian elephant and Mamenchisaurus fossils.

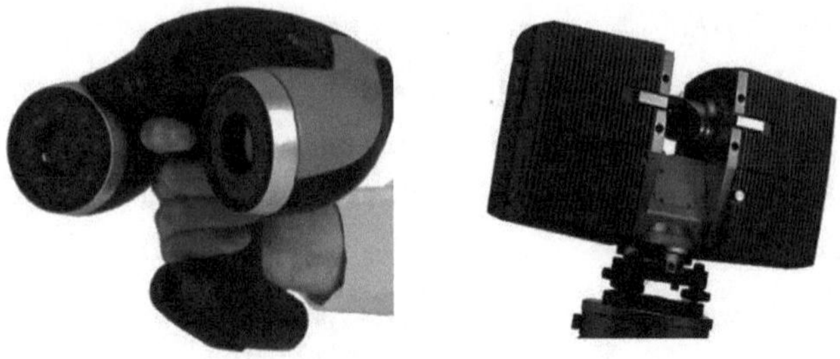

Figure 5 Handheld laser scanner and FARO large scene laser scanner

2.2 3D data acquisition and processing flow

First, the scanning device must be selected. The scan direction and order are based on the shape and size of the specimen. The calibration point should be pasted on the specimen while scanning, marking the boundary of the scan data at each point. After 360° omnidirectional scanning of the specimen, the multi-site Cloud data must be spliced and synthesized by data processing software provided by the scanner, constructing a 3D model of the specimen. The data acquisition process is shown in Figure 6.

Texture and animations can be added to the specimen model, which can then be imported into a virtual 3D scene, displayed in multiple perspectives, zooming in to see details or rotating the model at different angles, virtual 3D rendering offering a more realistic visual experience.

2.3 Principles of 3D scanning

The use of structured light for the 3D measurement of an object's surface contours is the most widely used visual measurement technique, mainly based on the principle of laser

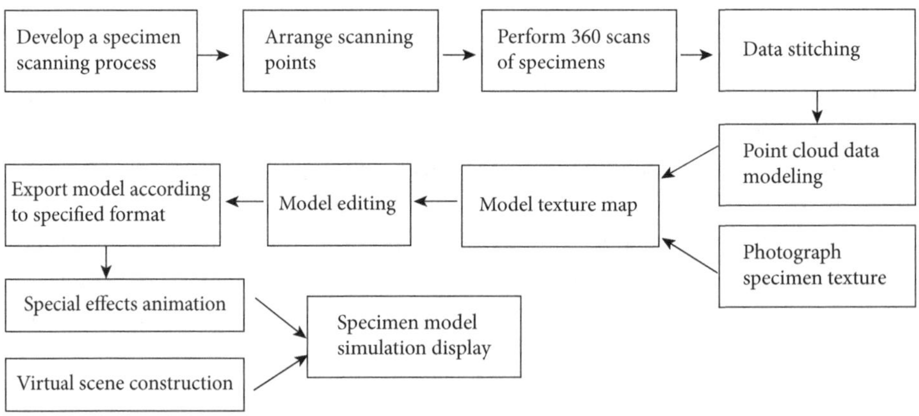

Figure 6 3D specimen scanning process

triangulation measurement, in which the light emitted by a laser and the normal line of the measured surface of the specimen at a certain angle are in incident on the measured surface of the specimen. Meanwhile, a camera is used to trap scattered or reflected light from the surface of the measured object, obtaining three-dimensional coordinates of the measured specimen according to the moving distance of the scattered or reflected light on the sensitive surface of the photodetector. The principle of 3D scanning is shown in Figure 7.

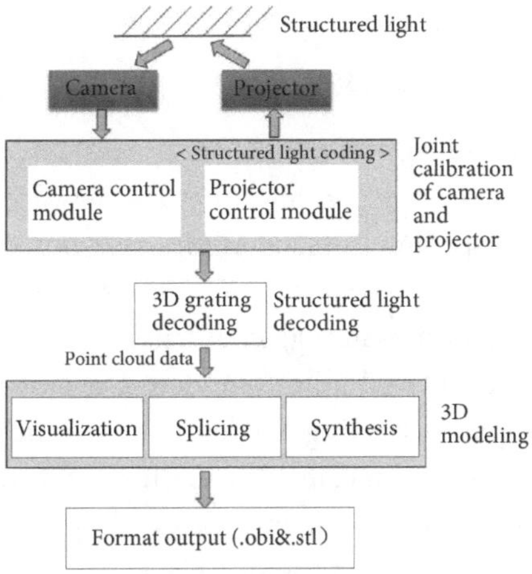

Figure 7 Principle of 3D specimen scanning

2.4 Results (see Figure 8)

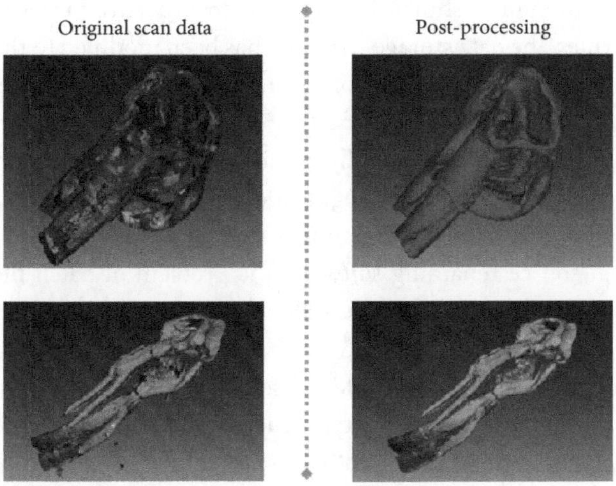

Figure 8 3D scanning point cloud and model for a bone specimen

3. Digital problems and solutions

The digitization of museum collections generally involves large-volume collections, meaning that the amount of data is huge, and there is a wide variety of data types and formats. In addition to typical 2D image data and 3D point Cloud data, there is also multimedia data such as animation, audio, and video. Scientific and appropriate data naming standards and storage management specifications are prerequisites to ensuring the smooth progress of the digitization process. The following are a few points regarding typical problems encountered in project implementation.

3.1 Image and folder naming issues

During the specimen shooting process, a difficulty was discovered in distinguishing between specimen photos strictly according to "front view, top view, side view, partial view, bottom view", as some specimens (such as dinosaur egg fossils, animal specimens) have irregular shapes, and there are no specific definitions to distinguish these viewpoints. For example, a front view of a dinosaur egg could also be a back view or side view, and the top view may also be considered a bottom view. Deviations in understanding can easily cause confusion in photo naming. In response to this problem, the person in charge of the project must communicate with the library during the process, making timely adjustments as required. Examples include dividing the specimen into front and back based on specific details, choosing the side view label when the angle is difficult to distinguish, or keeping the sample angles for the same type of specimen consistent. Such measures ensure the unification of naming to a certain extent.

Once data collection is complete, folder renaming has also resulted in a lot of duplication, because the requirements are perhaps not clear enough, and the user of the photos has poor communication with the recorder. The document storage structure originally proposed by the museum is shown in Figure 9. In order to conform to the library's database upload rules, the data storage structure has been modified to the form described above in section 1.3.2. The change of storage standards involves the "reshuffling" of the entire storage structure: Firstly, reclassifying 35,000 specimen photos by image format, a step that is easy to implement; Secondly, placing all photos of the same specimen in a folder named according to the general registration number; however, manually selecting photos of 15,000 specimens is overwhelming for staff.

With the open source renaming software, the problem of batch file extraction and automatic folder naming can be solved as follows:

1) Download and install "Files 2 Folders" software;
2) Select all photo files of the same type, such as .JPG photos. Right-click and select "Files 2 Folders", putting each photo into a folder with the same name, and rename all folders with renaming software Flash Renamer, the folder named after the

general registration number;

3) If there are multiple photos of the same specimen, the renaming software will only rename one folder, filtering out folders that have not been renamed, putting them in a new folder, and going through general registration number naming again using the renaming software, copying all renamed folders to the original folder. Folders with the same name will merge, and all photos of the same specimen can be placed in corresponding folders.

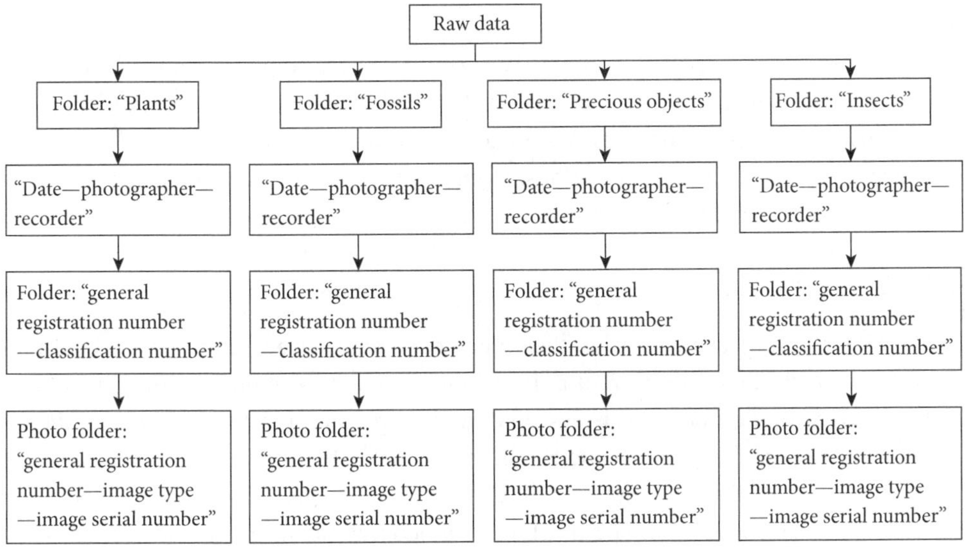

Figure 9 Raw document storage structure

3.2 Monitoring photo quality

Unlike artistic photography, which focuses on light and shadow, specimen images for archival purposes require clear photography that reflects the original appearance of the specimen to the greatest extent. The first few attempts at specimen shooting had an excess of rendering effects, lacking realism. Following communication with the museum, the principles of picture collection were determined, so that specimens can be displayed as accurately as possible.

Due to the lag in photo quality reviewing, the phenomenon of photo reworking caused by photo quality issues occurred later on, in a case involving hundreds of pottery specimens, which seriously affected the shooting process. During the process of data collection, it is necessary to arrange for special quality management personnel to perform quality monitoring of size, pixels, viewing angle, and shading effects in photos, to find and correct photo quality issues in a timely manner.

3.3 Data backup management

In order to ensure that collected data is not lost, a dual backup strategy is adopted, i.e. a recording personnel terminal (computer and mobile hard disk) backup and a management server (disk array) backup. In order to save collected data, the project team prepared 10 2TB mobile hard disks, opening up 20TB of storage space on the central server. The data collected by each group was uploaded to the central server once a week. Since the specifications of data management of recorders are not uniform across groups, an increase in collected data causes issues with repeated backup and some incomplete data backup.

Therefore, it is important to develop scientific, appropriate, and unified data management specifications so as to ensure data collection management. Furthermore, there must be effective data supervision to ensure that data recorders and managers operate according to these specifications. Lightweight automated data management tools will play an increasingly important role in future museum digitization.

3.4 Model mapping

Today's 3D scanning technology has the capacity to obtain surface texture data of cultural artefacts through the acquisition of point Cloud data, but for models with high surface texture clarity, model mapping is essential, and maps captured by professional digital cameras can ensure the accuracy and brightness of model texture. For large-scale cultural artefact specimens, there is a specific difficulty in one-to-one correspondence between separately collected images and point Cloud scanned fragments. In our experience, when taking photos and scanning simultaneously, numbering pictures and point Cloud data sequentially will provide twice the results with half the effort in post-layout.

4. Summary

The smooth implementation of data collection at the Beijing Museum of Natural History demonstrates the application of a divide-and-conquer strategy in terms of large data collection projects. Formulating scientific and appropriate data collection and management practices, and ensuring that these specifications are enforced, are prerequisites for the smooth progress of museum digitization. To do a good job, workers must first sharpen their tools. In this way, advanced shooting and scanning equipment and automated management tools can make collection work much more effective. In addition, the development and upgrading of databases will become an indispensable part of future museum digitization.

Data collection and management are the core issues in museum digitization. Therefore, the development of new methods, equipment, and technologies to solve these two core problems are the future development directions of museum digitization.

References

[1] Shen Xiaofeng, Wang Jian, Wang Qing. 'Data collection and information management of digital museums', [J] *China Information Technology* 23 (2009): 109, 122.

[2] Zeng Zhide. 'Image materials and museum digitization', [J] Chinese Museum 4 (2005): 50–51.

[3] Zhou Mingquan, Shui Wuyang, Wang Xuesong et al. Key technologies in cultural artefact digitization and its application in digital museums [C]. *Beijing Digital Museum Seminar* (2009): 250.

[4] Wang Peng. *Research on key technologies using structured light in 3D automatic scanning systems* [D]. Tianjin University, Feb 2008.

The Scientific Management of Digital Museum Resources

Lu Min
(Shandong Museum Department of Information, Jinan 250014, China)

Yuan Fangbing
(Juxian People's Hospital Department of Information, Rizhao, Shandong 276500, China)

Abstract: The rich resources of a museum cover many different fields, the distribution, autonomy, dynamics, and diversity of which increase the difficulty of digital resource management. Given these characteristics, this article reviews the classification, collection, storage, and management of museum resources, discussing related technologies and concepts.

Keywords: museum; digital resources; digital collection; database

Digital resources are an important aspect of the digital museum, and their effective management is the core issue in future development [1]. At present, digital museum development has altered its focus from the application of digital technology to thinking about the nature of the museum, paying more attention to resources and the fundamental attributes of collections, research, and education in the context of information technology developments.

With the progress of China's artefact and museology industry, the digital development of a number of museums provides a new arena for the protection and display of their collections. However, there is a wide variety in the resources of digital museums, such as geographical distribution, storage heterogeneity, morphological diversity, and management autonomy. Resources cover a variety of different fields, making management extremely complex. Therefore, it is important to make good use of the digital information resource platform of the museum, maximizing the museum's social value, and scientifically managing its resources.

1. Classification of digital museum resources

Digital museum resources are the building blocks of digital museums. The resources in different museums have a variety of characteristics in terms of distribution, autonomy, dynamics, and diversity. In order to integrate the resources of various museums, there has been a proposal to classify them from the perspective of digital form and multimedia display characteristics.

1.1 Text-based digital resources

Text resources of museum institutions are the most basic resource type, covering explanations (names, locations, circulation, age, texture, uses, ownership, preservation, collectors and collection time, and evaluation references), character information (such as memorialization), historical materials, professional dictionaries, news, reviews, books, periodicals, manuscripts, music scores, and other text-based data resources, as well as the museum's administrative and business history archives.

1.2 Image-based digital resources

Due to the characteristics of the museum itself, a large quantity of resources are reflected in the form of pictures, including of cultural artefacts, historical figures, manuscripts, literature, rubbings, calligraphy, paintings, murals, and statues, as well as photographs of visits, events, and business engagements. These photos are used directly in business operations such as collection storage, exhibition, social education promotion, and the publishing of scientific research, with extensive value of use.

1.3 Audio-based digital resources

In the development of digital museums, audio is essential in elements such as captions, descriptions, traditional artefacts related to opera styles, and academic conference recordings. With the continuing improvement of exhibition forms, interesting information about sources, textures, implications, excavation processes, and simulated sound effects have become indispensable aids. The same is true for online digital museums, with factors such as background music on the website, narration about collection images, virtual exhibition guidance and introductions, explanations of virtual 3D models, and the announcement of academic achievements of professional experts.

1.4 Video-based digital resources

In the domain of information, as compared with traditional museums, multimedia video display in contemporary museums has great advantages. The combination of real video footage and physical displays allows visitors to be immersed in the cultural implications of exhibits. Video resources recorded at the time of an event provide useful supplements to exhibits. For example, videos of archaeological excavations, operation processes and various emergencies at the scene, and recordings of lectures and academic reports from museology forums can be spread to the public via online video. Furthermore, videos can act as exhibition aids, showing how a traditional chime is played, the breeding process of a bacteria in a genetics exhibition, or the ceremonial rituals in an exhibition on Confucius.

1.5 Virtual 3D models

Virtual 3D models used in museums mainly involve 3D cultural artefacts, virtual 3D exhibitions, and virtual 3D buildings. A 3D model of an artefact is created to be as realistic as possible, with the help of 3D modeling. Information is acquired using 3D scanning technology, and a model is directly recreated, with the shape, color, pattern, and texture of the artefact recreated in full. The 3D models generated by this technology are generally saved in VRML format, and visitors can browse them using a dedicated VRML player. With this player, 3D artefacts can be displayed from 360°, and visitors can zoom in and out within a certain pixel range. Most virtual 3D buildings are also saved and displayed in this format, while virtual 3D exhibitions usually implement 3DMAX modeling technology, which is used by many digital museums now. The purpose of the virtual 3D museum hall is to achieve an immersive, interactive display, emphasizing the fact that the audience is at the core of the exhibition, highlighting the human aspect to provide an immersive experience. The changing elements act on the psychological changes of the audience, resulting in information transmission and feedback between the two, and establishing the interactive relationship, which strengthens the psychological role of the audience.

1.6 Digital animation resources

Digital museums strive to present collections in colorful ways, and animated displays are being used more frequently. Animations display information about exhibits in a dynamic way, solving the shortcomings of lack of variation and information in static information techniques such as text and pictures, making information more readable and accessible. For example, animation has been used to interpret the excavation process, with redundant information from the real process discarded, demonstrating excavation in a more concise and understandable way. For some cultural artefact collections, Flash animations may also be used to display the intrinsic quality of functional applications and production techniques, exhibiting ancient science and technology and recreating the production process. In this way, animation not only provides knowledge, but does so vividly, assuming a profound role in the promotion of historical and cultural essence while realizing the social education function of the museum in a rich and intuitive way. Some of the exhibition hall panoramas also use Flash animations to display 360° views of the exhibition, allowing visitors to obtain information and experience the hall in three dimensions.

2. *The collection and storage of digital resources*

2.1 The collection of digital resources in museums

The collection of information about cultural artefacts is a key component of digital museum resources. Currently, the digital collection of cultural artefact information involves recording and storing text according to unified format guidelines. Meanwhile, it utilizes advanced 2D and 3D scanning, digital photography, 3D modeling, and various image processing technologies to realize the high-precision acquisition and preservation of cultural artefact information, such as structure and texture. The goal is to secure text resources and high-fidelity digital models on the computer to provide accurate digital materials for the purposes of sharing, conservation, restoration, exhibition, publishing, scientific research, viewing, and development. In order to meet increasing demands for image quality, collection and processing must use high-definition images. With high requirements for digital fidelity, special targeted research must also be carried out.

In the digital collection of cultural artefacts, they must be classified in the first phase, and collection must be carried out following appropriate technical means. According to their spatial dimensions, cultural artefacts can be roughly divided into 2D planar artefacts and stereoscopic 3D artefacts. For 2D artefacts (such as ancient books, calligraphy and paintings, murals, embroidery, and paper cutting), high-resolution digital cameras are generally used for shooting, with digital scanners used to collect and archive the artefacts. For 3D artefacts (such as jade, pottery, porcelain, sculpture, and wood), digitization must be carried out in three-dimensional space, and a non-contact 3D laser scanner and multi-angle photos can be used to obtain data. The 3D laser scanning technology can instantaneously measure the spatial coordinate values via laser principles (including pulse lasers and phase lasers), obtaining complete spatial 3D information about the scanned object in point Cloud data acquired by the 3D laser scanning technology. It can then use it to build a 3D model of the artefact, with the potential to map the surface of the solid artefact onto the model in post-processing, making it more realistic. Such capabilities are unmatched by traditional 3D modeling software, and the use of this 3D laser scanning technology is both time and labor efficient, meaning that it is widely used in the digital collection of cultural artefact information. This technology offers permanent preservation of cultural artefacts, copy and repair, information sharing, and archaeological research.

It can also be applied efficiently in displays for digital museums, so as to realize cultural and historical value [2, 3, 4].

The digital collection of other resources in the museum is generally carried out in the following five ways:

1) For the collection of text-based resources, tablet input, keyboard input, network download, scan recognition, and voice input can be utilized to obtain data. Commonly used devices include computers and scanners.

2) The collection of image-based resources involves a process of sampling, quantifying, and conversion to digital images to be input and stored into the process of system memory. Commonly used devices include major categories such as photo scanners, digital video cameras, industrial cameras, digital cameras, and USB interface cameras. Image resources generally assume GIF and JPG formats, which are commonly used on the Internet.

3) Audiovisual multimedia content can be collectively called streaming media. A complete streaming media system includes:

a) an encoding tool for creating, capturing, and editing multimedia data to form streaming media;
b) streaming media data;
c) servers for the storage and control of streaming media data;
d) a network suitable for multimedia transmission protocols or even real-time transmission protocols;
e) players by which clients can browse streaming media files.

Some of these five elements are required by websites, and some are required by clients. Streaming standards and solutions will differ to some extent from company to company. At present, the mainstream online streaming transmission systems include the Real Networks Real system, Microsoft Windows Media Technology, and Apple QuickTime. Audio resources generally involve WAV, MP3, MIDI, and streaming formats, while video is mostly stored in a streaming media format such as .ASF.

4) Digital resources for virtual 3D models can be generated manually, or via a certain algorithm. 3D models refer to three-dimensional representations of an object, data formed by gathering data points, and other sets of information. At present, there are three methods for modeling objects: 3D modeling software, well-known forms of which include 3DMAX, Softimage, Maya, and AutoCAD; instrumental and measurement modeling, the main tools of which are being coordinate measuring machines for 3D measurements (CMM), and digital 3D scanners for actual 3D object modeling; image or video based modeling, mainly using image-based modeling and rendering techniques (IBMR technology).

5) Animations are moving pictures that operate by playing a continuous series of images. In digital animation resources, animations can be produced with software such

as AutoCAD, Painter, Photoshop, Flash, and 3dsMax. Animation formats include GIF, Flash, AVI, FLI/FLC and QuickTime. The GIF and Flash formats are widely applied in digital museums.

2.2 Storage of digital museum resources

In order to preserve digital resources more effectively, museums must establish databases for different categories, such as a cultural artefact archive database, an ancient books database, a cultural artefact protection database, a literature database, professional database, and an electronic publication and OA database [5].

The cultural artefact archive database consists of a document section and multimedia section, including both 2D planar images and stereoscopic 3D images. In terms of the document content of the database, based on the completeness of existing sources and the design goals of different databases, content from general cataloguing can be expanded to include identification reports and records of cultural artefacts, circulation status, collection records, repair records, exhibition and publication records, as well as related descriptions. Meanwhile, cultural artefacts are carriers of ancient culture, containing not only information about shape and texture, but also written information, information about manufacturing processes, and information regarding chemical and physical composition. The cultural artefact archive database must pay special attention to data security issues such as data formatting and access rights.

The ancient books database is largely composed of text and written data, but ancient books are also products of history. Whether it be an ancient cooking vessel, a rock from the Qin Dynasty, a bamboo slip from the Han, a carving from the Song, or journal from the Yuan, besides textual implications, the material carrier is itself part of the history of human civilization, with important value equal to that of textual implications. Therefore, ancient books have the dual characteristics of words and cultural objects. The ancient books database must contain multiple levels of information, including both text and image formats.

Scientific and technological conservation is aimed at investigation, identification, and research regarding the ontology of cultural heritage, so as to carry out physical and chemical scientific intervention on it, and to control and rectify environments for the purpose of proper preservation and restoration. This database consists of the compositional structure of cultural heritage, the qualitative mechanisms of change, the buried underground environment, the preservation environment of the museum, maintenance methods, restoration technology, dating processes, reproduction, and natural specimen production – a true record and accumulation of experience in cultural artefact protection. Therefore, the establishment of a cultural artefact protection database, as well as the improvement and development of related science and technology, are important indicators in the measurement of modernization levels for museums.

Academic research is an indispensable function of the museum, and a large collection of literature is an indispensable reference for this field. The management of professional books and periodicals, supplemented by a vast amount of electronic literature, provides sufficient research resources for researchers. Knowledge resources are easy to retrieve and borrow, and a corresponding literature database must be established. The contents of the literature database generally include the library's holdings and the material purchase of document resources for the social knowledge resource library. Secondary classification is carried out according to the business and research direction of the museum, so as to form a knowledge resource database system with specific characteristics.

The professional database contains multimedia resources such as audio, video, virtual 3D models, and animations. It should be built with consideration of the characteristics and needs of individual museums. Sections of the professional database may include an archaeological excavation database, a 3D ceramic appreciation database, an animation database, and video recording database of academic events.

Electronic publications are portable carriers of digital information, with a large capacity for information, vivid multimedia means, and interactive reading capabilities, and they have undergone major developed in recent years. With regard to museums, electronic publications cover a wide range of topics. They can be combined with current exhibits to produce a variety of topics, and may also be published in conjunction with catalogues of temporary exhibitions. In addition, explanations of various collections can be published in certain categories in the process of research into collections. When an academic conference is held, electronic versions of paper publications can be published simultaneously. The various digital resources of museums can be converted into electronic forms for publication and distribution.

OA (Office Automatic) is widely used in enterprises and institutions today thanks to its capabilities in integration, intelligence, multimedia, and electronic data exchange. The museum OA is based on the individual characteristics of a certain setting. In accordance with the procedures of various traditional office processes, the museum's publication of office information, drafting, modification, review, issuance, circulation, and filing are carried out by a computer connected to the museum database, with office information stored in the OA database.

3. *The management of digital resources*

Digital resource management [6], also known as electronic resource management outside of China, refers to practical activities and software systems used by information agencies (such as libraries) to track important information about digital information resources, especially Internet-based information resources such as e-journals, databases, and e-books. In theory, digital resource management is the organization and management of digital resource collections across two aspects: organization and management. Simply put, the organization of digital resource collections is the process of appropriately arranging, protecting, and accounting for obtained digital resources according to certain requirements. Digital resource management is the control and adjustment of digital collection resources, with the goals of feeding back information regarding resource utilization, statistics, and evaluation, effectively controlling the direction, speed, range, and density of resource operations so that the back-end system and user system work in an orderly manner, at an optimum state of equilibrium saturation.

The implementation of digital resource management in museums must first attain the digital conversion of all museum resources, ensuring that each digital collection has corresponding metadata. The museum then implements a library resource management system that stores and manages its own digital resources. In certain museums, digital resources are stored in the database and file system via the library resource management system. The management system associates digital resources with their metadata, structurally integrating resources and automatically generating management metadata (including storage information, origin information, and authorization policies). This ensures the integrity of the entire digital collection, and guarantees that every digital collection is suitable for use in the digital process. The resource management system manages all museum resources in a unified manner, providing an interface by which the upper user service module can access digital resources. The structure of the museum's resource management system is shown in Figure 1.

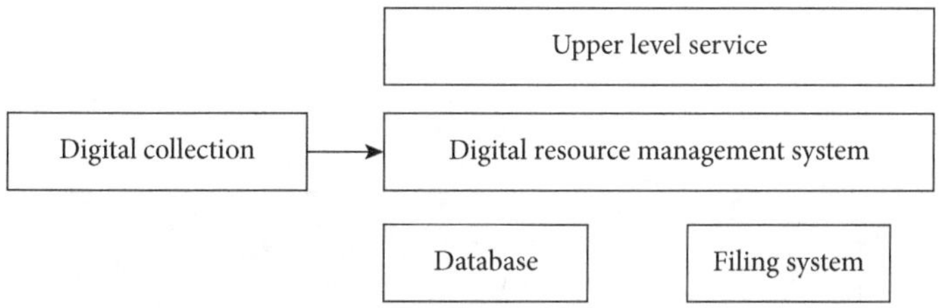

Figure 1 The museum's own resource management system

The museum's digital resources do not belong to itself only, but also to the nation, country, and humanity. So, digital resource management has to consider the question of data sharing. Building a distributive digital resource database and carrying out integrated management of resources will help to answer this question. The framework of digital management is shown in Figure 2.

The main modules in the data center system framework are described below:

1) Storage module: metadata and related information about all digital collections are stored in the database, with various types of digital resource files stored in the file system;

2) Resource management module: manages digital resources to provide support for the addition, preview, editing, and acquisition of digital collections;
3) Collection classification management module: achieves correspondence between digital collections in various places and centers, carrying out uniform identification analysis;
4) Search module: supports full-text search in Chinese and English;
5) Security management module: ensuring secure system access and controlling access rights to the collection;
6) Communication management module: responsible for the transmission of digital collections and related data across the local digital museums through the network;
7) Display module: based on XML (Extensible Markup Language) and XSL (Extensible Style Sheet Language) technology, different styles of digital museum display pages are customized and launched;

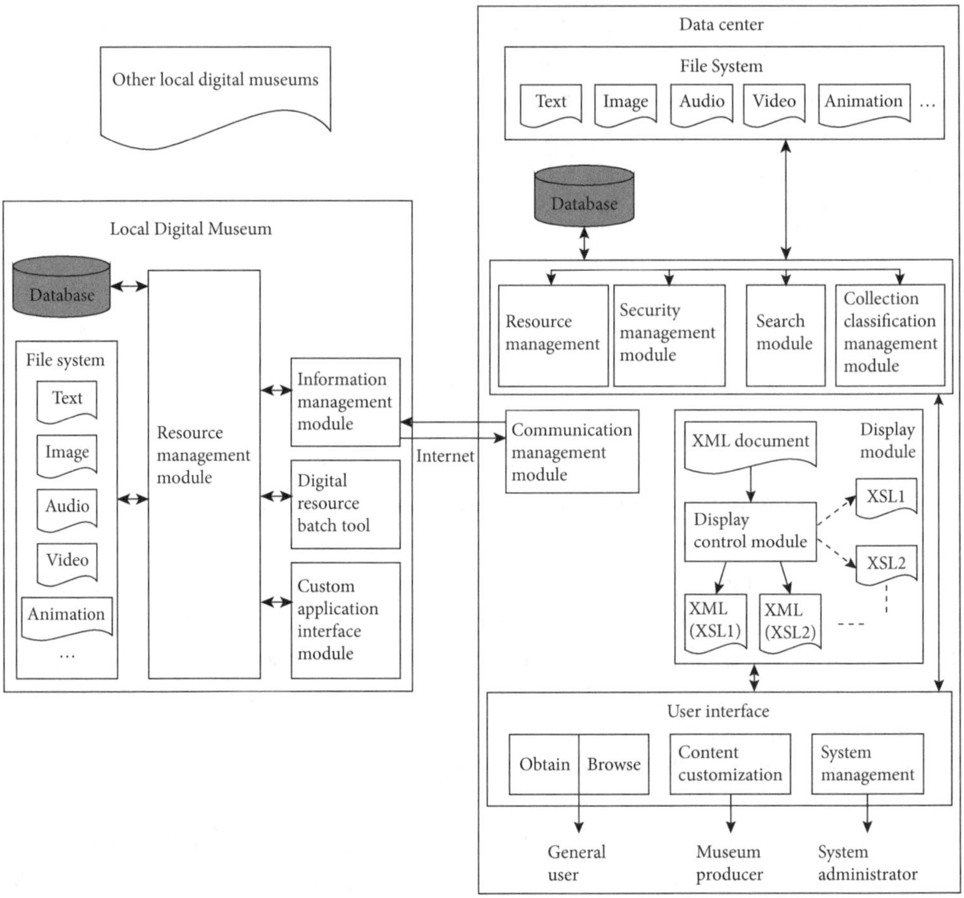

Figure 2 Architecture of the distribution management system for museum resources

8) User interface: providing different user services, such as browsing and retrieval of various exhibit information for ordinary users, digital museum display content and style customization services for administrators, as well as overall system and digital resource management services.

The main modules in the local digital museum system framework are described below:

1) Storage module: metadata and related information of all digital collections are stored in the database, with various types of digital resource files stored in the file system;
2) Resource management module: manages digital resources to provide support for the addition, preview, editing, and acquisition of digital collections;
3) Information management module: responsible for the transmission of digital collections and related data with the data center through the network;
4) Digital resource batch processing tools: allowing local museums to conveniently edit, store, and identify digital collections;
5) Application custom interface module: provides interfaces for local museums to use the various resources maintained by the system for the purposes of application development, such as the development of personalized local museum displays.

At present, a typical example of digital resource management in museums in China is the Digital University Museum project [7]. This project used a large-scale distribution system, including two centers in the north and south, connecting local sites to form a huge virtual museum group covering four disciplines. Users will not detect the difference in the physical storage location of collection. In the Digital University Museum, the museum's resources are managed with a hierarchical information model, dividing the museum into several top-level structures and resource communities, then dividing the collection structure into a series of collections and resource communities. This model solves issues with differing collection structures across branches. Outside of China, in 2002 the US Hewlett-Packard Company and the Massachusetts Institute of Technology Library jointly launched the open source system DSpace, which is a digital resource storage platform within the digital library field, and a warehouse to accommodate digital objects. The DSpace system is used for the long-term preservation and management of various digital resources of educational research institutions, using Dublin Core metadata standards including structural, management, and descriptive metadata. With the DSpace system, digital museum resources can be integrated to achieve structured and systematic management. The Fedora (Flexible Extensible Digital Object and Repository Architecture) system, launched in 1997, was jointly funded by Cornell University under the US Defense Advanced Research Projects Agency and the National Science

Foundation, undertaking research projects to address the content management, asset management, and preservation of digital resources. In 1999, Cornell University and the University of Virginia used Fedora to realize the application of a relational database to manage background data in the University of Virginia's digital library system, using Java to improve system functions, and making Fedora more applicable to the Web environment. Fedora has since been updated, with its 2.2 version released in 2007.

As two popular open source digital management systems, DSpace and Fedora provide a wealth of features that meet the general needs of submission, preservation, management, and distribution of digital collection resources. In China, the Virtual Reality Lab at Beijing University of Aeronautics and Astronautics and HP Labs have jointly developed DM-DSpace – a general digital object warehouse system for digital museums, its main purpose being to manage the mass of digital resources of the Chinese University Digital Museum. The distributed DSpace alliance breaks through the constraints of DSpace's independent operation, realizing communication between each DSpace. There is also TPI – a digital resource management system developed by Tsinghua Tongfang Optical Disc Co., Ltd., and an open source mass digital resource management system developed by the Department of Mechanical and Electrical Engineering and Applied Electronic Technology at Tsinghua University, as well as Hangzhou Maida Electronics Co., Ltd.'s digital resource management system, and Chaoxing's Digital Library Resource Management System. Internet-based digital museums display massive amounts of digital resources to visitors via the Internet, meaning that copyright protection and security management are also an important issue. At present, the main technologies for digital resource protection are digital watermarking technology and remote rendering technology. Security management includes digital resource backup and disaster recovery, server security control, prevention of malware attacks, and virus prevention measures.

Chinese museums today are inconsistent in terms of the development and management of digital resources, and most are still in the process of digital development. Only by perfecting digital development, and realizing effective management of digital museum resources, can the social effects of the museum be maximized, becoming an organized continuum that better serves the public.

4. Conclusion

The museum is an institution for the collection of cultural artefacts. It is also a scientific research institution, and a place for dissemination and education. It has an important role as a cultural body for public welfare, and must maximize its advantages in collection, scientific research, exhibitions, and publicity. Collection information, archaeological materials, exhibition information, scientific research information, and books and literature owned by the museum should be fully displayed to the public on the digital museum platform. The museums of the future will be more open, energetic, and capable of meeting the growing needs of cultural life. Only through the scientific management of digital resources can museums shoulder the role of social education and the promotion of Chinese culture.

References

[1] Dong Shaochun, Chen Hongjing, Xu Shijin. *Introduction to digital museums* [M]. Shanghai Science and Technology Press, 2007.

[2] Zhang Rui, Luo Yanlin, Zhou Minquan. 'Key technologies in cultural artefact digitization', [J] *Journal of Beijing Normal University*, 2007, 43(2).

[3] Information Center of the Beijing Science and Technology Association, Beijing Digital Science Association. *Digital museum research and practice in 2009* [C]. Communication University of China Press, 2009.

[4] Information Center of the Beijing Science and Technology Association, Beijing Digital Science Association. *Creative technology assists the digital museum* [C]. Communication University of China Press, 2011.

[5] Wang Hongjun. *The foundation of Chinese museums* [M]. Shanghai Ancient Books Publishing House, 2008.

[6] Bi Qiang, Chen Xiaomei. *Digital resource construction and management* [M]. Science Press, 2010.

[7] Qi Yue, Shen Xukun. *Management and display of digital resources in museums* [M]. Shanghai Science and Technology Press, 2008.

Research into the Application of CIDOC CRM in the Digital Museum

Di Qiong
(School of Applied Arts and Sciences, Beijing Union University, Beijing 100101, China)

Liu Hongzhe
(School of Information, Beijing Union University, Beijing 100101, China)

Abstract: Digital museum development aims to effect system and resource sharing in a distributed network environment, and to provide effective data for the decision-making and scientific management of the museum. CIDOC CRM is a relatively complete, unified global data model that is widely used in the field of cultural heritage. The application of CIDOC CRM to the development of digital museums proposes to realize the integration and sharing of a museum's information resources.

Keywords: CIDOC CRM; digital museum; integration; sharing

Preface

The development of the Internet and information technology has effectively promoted the digitization of the cultural heritage field, namely the development of digital museums. Simply put, the digital museum is a museum in digital space. It involves the digitization of collections, research, exhibitions, education, and other physical museum resources. Through the appropriate integration of these digital resources, museum services become more extensive, service content becomes more comprehensive, and museum resources are used more efficiently [1].

With changes in public access to information and the growth in the requirements of spiritual and cultural life, providing more comprehensive, professional, and systematic information services has become a major issue for museums. Major domestic museums have already begun the digitization process, establishing their own collection management databases, and websites. However, due to the lack of overall standards, the exchange and sharing of museum resources has yet to be achieved. This is an extremely unfavorable circumstance of digital museum development.

1. Analysis of the status quo of museum information

The development of digital museums relies on digital methods. Digitization is mainly achieved through the establishment of collection databases, collection management information systems, office management information systems, and museum websites. Amongst these, the development of databases is the most basic and important for museums in the digital age.

Collection databases contain digital text, images, audio, videos, animations, and 3D models pertaining to all collections. This is the basis of the digital museum. The database contains a large amount of descriptive information resources about cultural artefacts, recording their details. In the era of the physical museum, this information was manually recorded on cataloging cards and in archives. The catalog card was used to record the most basic cultural artefact information, containing only small amounts of information. The amount of information in archives, however, is relatively large, with standards promulgated by the State Administration of Cultural Heritage. This makes archives much more formal than catalogue cards. In that time, there was little exchange and sharing of cultural artefact information resources between museums, and there was no obstacle to communication and information exchange caused by inconsistent recording standards [2]. Our entry into the age of the digital museum has proved particularly important for educational functions in society. People wish to obtain authoritative, comprehensive, in-depth and systematic information via the museum, and this demand has forced museums to change management philosophies and seek cooperation and exchange with other museums. To this end, museums must declare digital resources so that resources across museums can be fully utilized. However, the lack of standards for records related to cultural artefact information seriously hinders the integration and sharing of museum resources.

Many international research institutions and organizations in the field of cultural heritage have developed standards to address this issue. Among the more influential museum standards are DC, CDWA, VRA Core, CIDOC CRM, ABC Ontology, Iconclass, AAT, TGN, and ULAN.

2. CIDOC CRM

Before discussing CIDOC CRM, we will explain the specifications of other relevant metadata and the scope of its applications.

Dublin Metadata Core Element Set, commonly known as DC metadata, is a metadata standard based on records. DC metadata involves a network information resource description format consisting of 15 basic elements that can be reused or used selectively, with possible subtypes and submodes. However, in the real world, there can be various relationships between data, and the current 15 data elements of DC metadata are fixed formats, with no connection between data. In addition, because DC metadata uses only 15 elements, and there are numerous types of cultural artefacts, DC metadata is greatly limited in terms of descriptive ability and range in relation to cultural artefacts [3].

The CDWA (Categories for the Description of Works of Art) [4] metadata standard is designed for art history research, art management personnel, and information technology experts. It is used in the museum community to describe art and images, to provide retrieval objects and images, to determine vocabulary choice, and to provide correspondence between art information systems. The elements of CDWA metadata include 512 categories and subcategories, and core elements consisting of very few categories, with 381 categories and 27 core elements representing the basic item information required. VRA Core (Core Categories for Visual Resources) was designed and developed by the Visual Resources Association in 1996, establishing metadata standards regarding art, architecture, prehistoric artefacts, folk culture, and other art-like visual resources. It is a standardized categorization designed with reference to CDWA, and is a template for the creation of system development or recording structures in local sharing environments. It has been updated to its fourth version and involves 19 elements [5].

AAT (Art & Architecture Thesaurus, Art & Architecture Index), TGN (Getty Thesaurus of Geographic Names), ULAN (The Union List of Artist Manes), and Iconclass have some effect on the sharing and integration of digital resources in specific fields. However, these metadata description specifications generally lack connections between concepts, having only concepts without descriptions for the relationship between concepts [6]. Museums are rich in collections and diverse in variety, so if the above metadata description specifications are used as the descriptive format, issues may arise in descriptive ability and the limited description range. Most importantly, connections between information items cannot be established.

CIDOC (International Committee for Documentation) is one of 20 committees that make up the ICOM (International Council for Museums) organization. CRM (Conceptual Reference Model) is a formal entity of CIDOC, which has spent over 10 years completing it to promote the integration, transfer, and exchange of diverse cultural heritage information sources. CRM conforms to international standards in form and specifications, having passed certification of international standards. The CIDOC CRM model is widely used in the cultural heritage field. It covers historical facts, humanities and art, archaeological remains, time and place, people, and copyright notice specifications of cultural artefacts. The main goal of CIDOC CRM is to achieve the exchange and integration of various heterogeneous cultural heritage information resources, ensuring that the details and accuracy of information provided meet the expectations and requirements of museum professionals and researchers [7].

CIDOC CRM is a formalized language that helps us to identify common information content contained in different formats. Through the implicit and dominant concepts and their relationships involved in the information resources of cultural artefacts, the CIDOC CRM model provides definition and formal description, and a common, extensible semantic framework to achieve a common understanding of cultural heritage information. It offers a common language for the exchange of information and the establishment of information systems in the field of cultural heritage [8].

The first version of the CIDOC CRM model was established in 1998. In the years of development that followed, the descriptions of cultural artefact information resources have been expanded numerous times. Currently, it contains 90 entities and 148 attributes to basically cover the information of historical facts, humanities and art, archaeological remains, time and place, human information, and copyright notices for cultural artefacts.

In the CIDOC CRM model, entities are connected with attributes. Entities at both ends of the attribute are the domain of definition and range of the attribute, and the attribute represents the connective relationship between entities. There is a hierarchy between entities, and by connecting attributes, scattered entities can be connected. Scattered knowledge elements may also be connected to facilitate digital storage and query of cultural artefact information resources. The CIDOC CRM model outlines a basic system of knowledge in a certain field through standardized descriptions of concept, attributes, and their interrelationships. Based on CIDOC CRM, the domain knowledge structure is clarified, and is suitable for knowledge storage and reuse. Even when the knowledge structure of the domain changes, it can be adjusted by ontological evolution, maintaining the consistency of the system [9].

3. The status of research into CIDOC CRM application in China and abroad

Application research can be divided into theoretical and practical applications. Theoretical applications involve using the CIDOC CRM information reference model for systems that are still in the research and demonstration stages. Practical applications are used in projects that have already produced results. The applicability of the CIDOC CRM information reference model has been verified.

There is a great deal of research on theoretical application in China, for example, the CIDOC CRM information reference model combined with the MPEG-7 (Multimedia Content Description Interface) cultural artefact metadata model. MPEG-7 can be used to describe production information, storage methods and usage of multimedia data, underlying content features, and high-level semantic information. It can also support fast and efficient user search for information of interest, effectively complementing the shortcomings of the CIDOC CRM information reference model in the description of multimedia data – a solution to the issue of heterogeneous interoperability in digital museum systems and resources. With the addition of metadata based on MPEG-7 standards, also makes it convenient for users to share, reuse, and query digital information regarding cultural artefacts, making it possible for digital museums to realize simultaneous semantic-based and content-based retrieval under a unified metadata model.

The CIDOC CRM information reference model is used in the virtual museum semantic network structure, with entities and attributes selected according to system requirements. While selected entities and attributes can be different among individual museums, they are all based on a unified information reference model, and the overlaps between them can be shared and exchanged. Within the semantic network architecture of this system, CIDOC CRM is used as the basis for cultural artefact resource data descriptions to form a semantic representation of cultural artefact data [10]. In addition, a metadata integration scheme based on CIDOC CRM works as an intermediary mechanism, and DC is an example to introduce the mapping between DC and CIDOC CRM [11]. Research based on the CIDOC CRM semantic query system realizes a query method based on the characteristics of painting content and a query method based on association, solving the problem of accurate painting query by major non-art users [12].

In terms of practical application, there are many cases worth referring to both in China and abroad. In 2002, Europe launched a three-year SCULPTEUR (semantic and content-based multimedia exploitation for European benefit) project [13], proposing a novel framework for concept integration, retrieving museum information based on metadata and content browsing. This project involves the description of collection information resources of several famous art galleries around the world, with the goal of providing a universal and versatile method by which to obtain collection information, including 2D images, 3D models, and other multimedia representations. Using CIDOC CMR as a reference model for museum collection information description, multimedia museum information is imported and managed through the CIDOC CMR-based semantic layer, including data integration/derivation, cleaning, mapping and enhancement, with the system expanding certain categories based on the CIDOC CMR. Another innovative part of the system is the ability to display the Concept Browser in the Ontology parts at will, while targeting interesting parts, acting as a tool for browsing the ontology to understand the structure of data and assist the user in query completion. Some foreign research organizations and institutions are also using the CIDOC CMR information reference model, such as LODAC (link open data in academia) established by Japan's National Institute of Informatics. Using resources gathered across many domains in Japan, with the mapping mechanism of CIDOC CMR, users can search for resources of each resource library, and also perform association search between resources. The Polish Digital National Museum project was developed by PIONIER – a digital library consortium of cultural universities, with information from more than 70 universities, libraries, museums, archives, and research institutes, all under the CIDOC CMR structure.

In 2000, the National Institute of Information Engineering at the Taiwan National University of South China began using the Lanyu Rendao Digital Museum, the Alishan Mountains and Zou Culture Digital Museum, and the China (Taiwan) Executive Yuan Cultural Construction Committee "Old Photo" National Cultural Database. In conjunction with CIDOC CRM, the data collected by the three digital museums went through data description mapping, completing the query system for the automatic retrieval and filtering mechanisms of digital collection. CIDOC CMR is used as a resource description structure for museum artefacts, with file descriptions in RDF file format, and corresponding examples of cultural artefacts stored in the RDF database [13].

Beijing Jiaotong University used the CIDOC CRM information reference model as a description standard for Xu Beihong's painting collection, developing a semantic retrieval and display system for a Xu Beihong Digital Museum based on CIDOC CRM and ABC Ontology [14]. The system carries out information classification of the paintings in the Xu Beihong museum according to user query, analyzing and storing them combined with semantic image features. This will enable non-art professional users to understand the various creative techniques used by the painter, as well as characteristics of works at each stage and the underlying implications and themes [15].

4. Conclusion

CIDOC CRM is of great significance for the realization of digital museum development, and conducive to the integration and sharing of digital museum resources. CIDOC CRM is a unified, global data model, with the capacity to completely and systematically display the collection information resources of cultural heritage and museum field through entities and attributes. For heterogeneous data resources, the mapping mechanism can be used to integrate and share data resources of different databases, so as to effect reasonable and efficient application of resources.

References

[1] Chen Gang. 'Analysis of the concept, characteristics and development model of the digital museum', [J] *China Museum* 3, 2007: 88–93.

[2] Chen Hongjing. *Principles and methods of the digital museum* [M]. Fudan University Press, 2002.

[3] Chen Yan, Zhou Xin. 'Metadata integration of cultural heritage resources using CIDOC CRM — Mapping of DC metadata as an example', [J] *Modern Information*, 5, 2010: 60-63.

[4] CDWA standards. http://www.getty.edu/research/conducting_research/standards/cdwa/ http://www.vraweb. org/ projects/vracore4/index.html.

[5] Yu Jia. 'Development and current status of VRA core metadata', [J] *New Century Library*, 2008, 4: 75-78. DOI: 10.3969/j.issn. 1672-514X.2008.04.024.

[6] Huang Yongxin. 'Reference model in the field of cultural heritage information', [J] *Research on Library Science*, 2012, 11.

[7] Crofts, Nick, Martin Doerr, Tony Gill, Stephen Stead and Matthew Stiff. *Definition of the International Reference Committee's Conceptual Reference Model (CIDOC CRM)*. Translated by Wu Yihui, 2010.

[8] ICOM/ CIDOC documentation standards group, definition of CIDOC conceptual reference model. Version 5.0.1[EB/OL].ICOM/CIDOC CRM Special Interest Group (23 Nov 2009).

[9] Hao Tinglei. *Research on the construction and representation of traditional festival information ontology* [D]. Huazhong Normal University, 2011.

[10] Liu Lina, Liu Hongzhe. *Cultural artefact metadata model based on CIDOC CRM and MPEG-7* [C]. Chinese Database Conference, 2005.

[11] Liu Hongzhe, Bao Hong, Yu Jiehua. \Virtual museum semantic network architecture based on CIDOC CRM', [J] *Journal of Computer Applications* 4, 2006.

[12] Chen Yan, Zhou Xin. 'Metadata integration of cultural heritage resources based on CIDOC CRM — DC metadata mapping as an example', [J] *Modern Information*, 2010, 5: 60–63.

[13] Huang Yongxin. *Research and implementation of semantic query systems based on CIDOC CRM* [D]. Capital Normal University, 2006.

[14] Tan Guoxin, Zhong Zheng. *Digitalization and industrialization of national cultural resources* [M]. Huazhong Normal University Press, 2012

[15] Wu Jing. *Design and implementation of an ontology-based painting retrieval system* [D]. Beijing Jiaotong University, 2006. DOI:10.7666/d.y1082872.

[16] Liu Xuerong. 'Research on cultural artefact metadata in the digital museum', [J] *Software Guide*,2009, 4: 155–56.

[17] AAT.AAT-TAIWAN Introduction. http://aattaiwan.teldap.tw/p/about.html.

[18] TGN.TGNstandard.http://www.getty.edu/research/tools/vocabulary/tgn/about.html.

[19] ULAN.ULANstandard.http://www.getty.edu/research/conducting_research/vocabularies/ulan/about.html.

[20] Iconclass. Iconclass standard. http://www.iconclass.nl/index.html.

[21] Dong Yan. 'A preliminary study on the metadata standards of digital museums', [J] *Journal of Beijing Union University: Natural Science Edition*, 2005, 2: 61–65.

The "Moon Walk" – an Example of the Application of Somatosensory Technology in Science and Technology Museums

Han Qingqing
(School of Humanities and Social Sciences, Beihang University, Beijing 100191, China)

Abstract: Methods of public learning in science and technology venues are moving past the form of pure text panel display, using digital technology with a focus on participation, experience, and interactive learning. In order to cater to visitors' needs, science and technology museums have changed traditional display forms, introducing somatosensory technology and the fully utilizing display functions of this interactive technology, to great public acclaim. This paper addresses the Beijing Planetarium's simulated moon walk experience using somatosensory technology to explore and assess the application model, as well as the effects and future development prospects of somatosensory technology in science and technology exhibitions.

Keywords: science and technology museum; somatosensory technology; popular science; interactive form

Preface

The 21st century is an age of high informatization. With the rapid development of science and technology, the way the public acquires scientific knowledge within science and technology venues is gradually moving away from pure text panel display forms, with more emphasis on participation, experience, and interaction. The main purpose of science and technology museums is to popularize scientific knowledge amongst the public, which in turn proposes to improve overall national standards. Science and technology venues focus on allowing visitors to engage practically and get involved, thinking and learning about science through entertainment and play, and paying more attention to the value of experience within exhibits and exhibition design.

With the rapid updating of Internet technology, somatosensory technology has also developed quickly as one of the most popular human-computer interaction methods. The Beijing Planetarium displays new ideas, new technologies, and interesting knowledge, making full use of somatosensory control technology and exploiting the display functions of its interactive form offers to innovate and optimize the expression of science and technology museums, and to enhance the image of science and technology exhibitions, thus expanding the scope of popular science knowledge, broadening channels of knowledge transfer, and increasing its influence. Evidence also shows that the application of this technology has been unanimously welcomed and recognized by visitors.

This article assesses the simulated Moon Walk exhibit, which uses somatosensory technology in the Experience of the Moon hall of the Beijing Planetarium, in order to explore the application of somatosensory technology in science and technology venues, and its future prospects for development.

1. An overview of somatosensory technology

1.1 The concept

Somatosensory technology, also known as motion sensing control technology, is based on dynamic capture [1]. First, the audience gives instructions to the machine with natural and intuitive movements [2]. Then, the machine recognizes and analyses a user's sensory actions in a special way, and produces machine feedback according to predetermined sensing models [3]. Successful implementation relies on a combination of two aspects of somatosensory equipment and systems. Somatosensory equipment is also hardware, with the chief goal being to recognize human motion gestures, and then input into an interactive game design based on somatosensory technology by means of information data [3]. The somatosensory system (the software) can stably and quickly perform hardware information processing, so that the machine side can feed back continually, with the key factors being rapid processing of motion information and motion recognition calculations [3].

1.2 Model categories of somatosensory technology in exhibitions

The application of current somatosensory technology in popular science venues can be summarized in three modes: operational interactive display mode, experiential interactive display mode, and immersive theme theater display mode [4], classified according to technical bases and presentation effects. See Table 1.

Table 1 Modes of Somatosensory Technology Applications in Exhibits

Model type	*Technical basis*	*Rendered effect*
Operational interactive display mode	Relies on peripheral somatosensory systems to identify and match gestures of the audience and determine user motion intention, then combines with computer virtual technology to achieve required operations	Multi-person, emotional communication realized through special dynamic interaction, enabling audiences to gain scientific knowledge more intuitively using both body and mind
Experiential interactive display mode	Simulates and restores scenes within the exhibition space, using new forms of the somatosensory interactive system and holographic audio-visual effects to comprehensively display and interact with human body movements and exhibit content	Realistic and ambient audio-visual experience using a variety of holographic effects, providing audiences with feelings of immersion
Immersive thematic display mode	Exhibition uses scenes to simulate the somatosensory interaction with peripheral devices, allowing the audience to participate and immerse themselves in different scenes via coordinated body movements	Through setting model and lighting effect, the exhibit's expression is improved and the interactive plot with the audience is developed

2. The embodiment of somatosensory in the Moon Walk exhibition

On February 7, 2015, the Exploring the Universe exhibition in the Lunar and Meteorite Exhibition Aréa of the Beijing Planetarium was officially opened the public after a year if preparation. Since that day, it has attracted many visitors. A new exhibition hall, it abandoned the old single-form exhibition format that relied on display panels to provide knowledge about astronomy, instead introducing advanced technology and equipment. Finding ways to spread esoteric facts about astronomy to audience in proper form, and to stimulate visitors' exploration of the mysteries of the universe have always been an issue for science museum, and are also the core ideas behind new exhibitions at the Beijing Planetarium. The new layout of the lunar walk display fully reflects these design ideals, combining visual presentation with a participative experience to change the way information is presented, including entertainment and storytelling to create the atmosphere of a Moon landing and simulate a walk on the Moon, using multi-angle, panoramic display of basic astronomical knowledge with cutting-edge dynamics.

2.1 About the Moon Walk exhibition

The Moon Walk exhibition explains the characteristics of the Moon and major Moon events such as the first successful landing of the Apollo 11 manned spacecraft on July 21, 1969, with the first human footprint left by the US astronaut Neil Armstrong at the landing site [5]. This feat was the first time that humans had visited a celestial body, giving it epoch-making significance. Gravity in the atmospheres of different planets is different, and the surface of the Moon has only one sixth of the gravity of the Earth's surface, which means that the experience of walking is different. First of all, walkers would experience instability and floating, and due to wearing spacesuits, they would be unable to walk normally, having to jump like a kangaroo.

The educational purpose of this exhibit is to enable visitors to participate and interact via somatosensory technology. It is also designed to spread and popularize astronomical knowledge (such as the experiences of landing and walking on the Moon) in a novel form that provides scientific knowledge, enhancing visitors' interest and response, and ultimately achieving educational entertainment that optimizes the effects of popular science education.

2.2 The overall process of environmental simulation in the Moon Walk exhibition

Firstly, the exhibit provides detailed steps for the experience of lunar walking. There are five main aspects:

1) Standing in the "footprints";
2) Raising ones arms to begin the simulation;
3) Jumping in place to control walking movements;
4) Turning shoulders to control direction;
5) Only one person at a time.

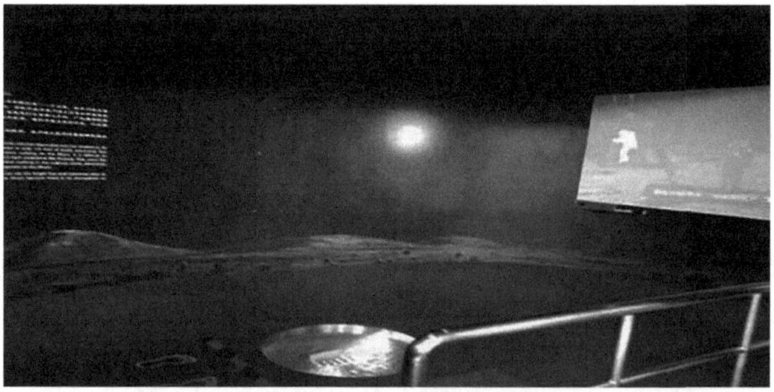

Figure 1

Secondly, in front of the footprints that operate the game is a concrete reproduction of the first footprint left by the American astronaut Neil Armstrong. Right in front is a big screen, below which is a somatosensory control device – an Xbox360 model, which is used to locate the human body and for convenience.

Finally, the walls conform to the style of the exhibition area. They are largely dark blue, with only a few stars, resembling the night sky.

A panorama of the Moon Walk exhibit is shown in Figure 1.

2.3 Application of the somatosensory machine model

The somatosensory machine model used in this exhibit is an Xbox360, and the development of Microsoft Xbox products for somatosensory games overseas is booming [6]. As the second generation of home video game consoles, its powerful features make it the only game console with a timing function, which makes it easy to set game times, making necessary interactions with the audience [7]. These features and technical advantages are sufficient for a wide range of functions that are applicable to science and technology venues. Xbox360 hardware devices include game consoles as well as external devices such as joysticks [8].

2.4 Experience links based on somatosensory technology design

First of all, visitors must stand on a footprint provided. Then, when the somatosensory device is fixed in position, a prompt will appear on the screen:

Please lift your hands to start this incredible journey.
Begin the Moon walking experience
Do you want to walk on the Moon?
Try jumping with both feet.
Do you see? Hurry up! Bend and explore.
Just turn your shoulders and you can look around at the Moon.
Ok, now it's time to return to the lunar module.

The experience is then complete, with time limited to 60 seconds. The interactive mode of the somatosensory technology used in this exhibition belongs to a second type of experience described above — experiential interactive display, reproducing the lunar surface structure and shape via simulation, and using the somatosensory interaction system to for exploration. Physical movements enable visitors to interact with the contents of the exhibition, offering an immersive feel.

3. *Application analysis and suggestions for improvements in somatosensory technology in the Moon Walk experience*

3.1 Analysis of strengths

1) Popularity
Somatosensory technology integrates learning and entertainment via new displays, free expression, and easy interaction. It is not difficult to imagine this combination being easily recognized and loved by visitors. As such, the newly renovated Lunar and Meteorite Exhibition Hall has attracted a large number of spectators, especially children and teenagers, though older visitors have also taken part in the experience.

2) Participation
Taking into account the intellectual and psychological characteristics of children, this exhibition uses somatosensory technology to allow them to participate directly without the need for a controller such as mouse or keyboard, extending the scope of audience needs and age range, and allowing people to participate directly in the experience regardless of age.

3) Interactivity and fun
The effects of popular science can be optimized by using the human-computer interaction function of somatosensory technology, considering and combining the needs and feelings of visitors, and furthering the entertaining and interactive elements of the exhibition.

4) Operational logic and immersion
Somatosensory technology offers relatively free interactivity and diversified instruction form – a virtual reality experience environment operated in a natural user interface by simple gestures or actions, offering an immersive viewer experience. Many people who experience it would like to participate a second time.

5) Parent-child interaction
A child is inevitably accompanied fully by a parent, and this simple and easy-to-use experience creates an environment in which parents and children can have equal exchanges in a relaxed and enjoyable environment, achieving effective cooperation and interaction and furthering the parent-child bond.

6) Education through entertainment
In changing the monotonous and boring display form, the addition of multimedia technology has greatly improved visitors' sentiments and raised levels of interest. The somatosensory technology provides a more novel display method, combining the mind and spirit in the context of the game experience, with the variety of perspectives offering

a unique attraction that increases audience interactivity and offers education through entertainment.

3.2 Suggestions for improvement

A major problem still exists in the use process. Because the somatosensory technology equipment used in the simulated lunar walking exhibition is the most basic and simple, the card machine in the experience process is often slow to respond, meaning that visitors must wait in line for a long time to access the experience. Inconvenience of this kind will affect the success of popular science communication, weaken the powerful function of somatosensory technology and its interest, and ultimately hinder the success of education through entertainment. Therefore, according to the design concepts, education purposes, and application effects of this exhibition, it can be improved in the following ways:

1) Object participation
The lunar walking exhibition is mainly aimed at children. Parents visiting with children can let them participate in the experience, so it can be worth considering adding parent-child activities such as somatosensory games that meet venue requirements and have astronomical content, to be played by parents and children.

2) Experience content
From the basic explanation of this case, it can be seen that the content is singular and the form is simple, having only to jump and walk on the Moon, bend over and explore, and then return to the lunar module. Perhaps an activity could be added after landing on the Moon, for example, inspecting the landing of the lunar launcher, collecting samples of lunar soil and moonstones, setting up a monument on the Moon, installing scientific instruments, or flying the Chinese flag. [9]

3) Length
The length of the experience in this case is 60 seconds. Visitors are sometimes unable to complete the content within this time, so the experience time must be lengthened.

4) Technical application
Considering the shortcomings, rudimentary quality, and slow response of somatosensory equipment, we must keep up with technological developments and use advanced technology to offer the best experience to visitors. Most importantly, the human body's motion be mapped onto the game, and the actions in the game must also feed back to the human body, to achieve true "interaction". This aspect is worthy of innovation and exploration.

4. Conclusion

Somatosensory technology will mature with development and the progress of the times. Although its development is only just beginning, it is expected to take the lead by virtue of its significant advantages in natural interaction, strong vitality, and diversity. Science and technology museums take into account the principle people-orientation, and understand that visitors do not wish to acquire information about astronomy in the form of presentation boards. Therefore, we must grapple with the technology trends of the times, using the latest modern high-tech technology to popularize scientific knowledge and serve the public. The addition to functional sensory technology, the Moon Walk exhibit allows the theme to be clearly displayed, meeting visitors' needs, capturing and mobilizing audience initiative and participation in the understanding of astronomical knowledge, and optimizing popular science education. Somatosensory technology will be an unstoppable development direction in the future, with broad prospects in popular science education activities. IT is worthy of in-depth development and exploration as a major direction for the future in the field of popular science.

References

[1] Qi Guohong, Dong Mingsheng, Wang Suilou. 'Somatosensory — the general existence of gods', [J] *Journal of Agricultural Biotechnology*, 2007, 4: 694–697.

[2] Zhu Mingming, Jing Hong. 'Gesture tracking and recognition based on somatosensory technology', [J] *Journal of Computer Systems*, 2014, 8: 228–232.

[3] Ma Jianrong, Zhang Sujing, Li Feng. 'Design and implementation of parent-child interactive games based on somatosensory technology', [J] *China Educational Technology*, 2012, 9: 85–88.

[4] Rong Zhu, Luo Xiaoxi. 'Application of kinetic somatosensory control technology in the innovative design of popular science exhibitions', [J] *Modern Machinery*, 2013, 4: 93–96.

[5] Shi Lei. 'A smile that never fades — Pondering Armstrong, the first man on the Moon', [J] *Space Exploration*, 2012, 12: 22–23.

[6] Wu Qian. *Scene design based on somatosensory XBOX360 games* [D]. Guangzhou University, 2013.

[7] Pan Nannan. 'Adding material to life', [J] *24 Hour Science*, 2013, 9: 39–40.

[8] Zhang Shichao, Qian Dongming. 'Research on the status and development of somatosensory technology', [J] *Journal of East China Normal University (Natural Sciences)* 2014, 2: 40–49, 126.

[9] Song Jiayu. '"Footprint on the Moon" design teaching', [J] *Literature Education*, 2008, 2(5): 70–71.

The Application of Panoramic Video and Personalized Distribution in Digital Museum Exhibitions

Wang Guangsheng, Gao Xin, and Sun Shuwen
(Microelectronic MachineTechnology Museum, Beijing University of Technology, Beijing 100022, China)

Abstract: Panoramic video is a form of virtual reality technology that is more advanced than panoramic images alone, using several cameras to jointly capture surrounding 180° or 360° scenes, and putting together several videos from several cameras via panoramic stitching to synthesize a panoramic video that truly reproduces the live panorama around the camera cluster. This paper covers both in-depth theoretical algorithm research and innovations in computer technology architecture, and also includes developments in panoramic video application and discussion of multi-camera group deployment.

In terms of computer engineering, this paper proposes a high-performance computing architecture for panoramic video generation, using a graphics processor as a coprocessor by which to transfer most panoramic video generation algorithm work to the graphics processor, so that computing tasks that would otherwise require dozens of central processors can be performed by a graphic processor, greatly reducing the number of CPUs used. In terms of panoramic video applications, this paper establishes the panoramic live broadcast system, which combines panoramic video with IPTV technology, sending generated panoramic video to remote visitors through steps such as cropping, compressing, encoding, and transmission. With panoramic video, remote visitors are able to see areas of interest when watching live shows, freeing themselves from the restraints of the camera controller's vision, allowing visitors to look around independently. This technology stands to change the form of traditional live broadcast TV, creating a new method of panoramic broadcast, and providing a new form for digital museum visits.

Keywords: panoramic video; general GPU; multi-camera; live broadcast system

The development of digital museums in China began in the 1990s and has since developed rapidly. For physical exhibitions, digital exhibitions, panoramic videos, and the personalized distribution exhibition model proposed in this paper, achieving the most rapid and visual methods is the goal, using different modern means to focus on the audience and meet the varying needs of visitors of different ages and demographics.

1. The status and development of the digital museum

The digital museum refers to an innovative exhibition model that uses digital means to carry out all-round, multi-form digital collection, storage, and processing of exhibit information. It utilizes virtual reality technology, 3D graphic image technology, computer network technology, stereoscopic display systems, interactive entertainment technology, and special effects, so that through the sharing of information about cultural artefacts, the physical museum that exists in reality may exist on the network. It realizes scientific research and social education, building an exhibition hall for virtual, remote viewing via the network and forming a brand new model for innovative visits and exhibitions in the era of the Internet [1].

The fly in the ointment is a lack of immersion. When updating exhibits, a lot of effort is required to improve digital content, and the price is high.

2. The advantages and disadvantages of current remote video exhibitions

Countries outside of China have begun to adopt remote video exhibition modes. For example, an article entitled *Robots Walking the Art Gallery After Dark* published by the Guangming Daily on August 28, 2014, describes the launch of the five-day "After Dark" event at the famous Tate Britain gallery in the UK. From 10pm to 3am on August 10 to 17, four robots were put into operation. This mean that art lovers all over the world could remotely control the robots by logging into the official Tate Britain website. Without leaving home, they could see the "500 Years of British Art" exhibition, featuring work by Picasso and Monet. Digital technology allowed visitors to remotely control a self-controlled robot as a guide to the Tate Britain after dark.

The first person who was authorized to manipulate the robot was former International Space Station Commander Neil Hardfield. Sitting in his home in Toronto, he held an iPad to command the robot in a satisfying endeavor no less exciting than exploring space. He called the experience "very magical", saying "you will forget that you are controlling the robot by hand, it becomes purely an extension of your mind. This is what technology should be."

By operating the remote robot agent, users visit the exhibition as if they were on site, saving valuable time and travel expenses. However, there are certain shortcomings in the Tate's remote agent robot exhibition mode.

First, four robots as tall as penguins wander around in the dark hall with searchlights on their heads. The person handling the remote control is like a driver, and other web users are passengers. However, a robot can only be operated by a single person at a time, and the rest must passively accept the route whether or not they are interested in the content of the exhibits the controller explores. Therefore, for most remote visitors, the essence is still traditional passive acceptance of video broadcasting. However, instead of photos taken by a photojournalist with a shoulder camera or camcorder, a single remote viewer manipulates the robot to shoot, meaning that not every remote visitor can fulfil their wishes in a purely personalized visit.

Second, the Tate Britain's remote agent robot exhibition model is costly. The four robots are like small coat racks, with two slender stems supporting a tiltable "head". The "head" is equipped with two cameras and LEDs, and each robot utilizes a circular wheeled base that can move in any direction. Aside from not going up and down stairs, these small robots can come and go freely around the museum, like a pair of remote eyes, allowing visitors to feel that they are present in the Tate after dark. Each robot is also equipped with a powerful battery capable of supporting five hours of continuous operation, an anti-collision bumper, and sonar detection technology to ensure they do not touch items within the exhibition hall or collide with one another. This is very expensive.

Third, the robot cannot go up and down stairs, which makes it unsuitable for larger exhibitions. Further, the Tate chose to use only four robots after considering the possibility for congestion or collision. This is a very limited number. Therefore, it is impossible for thousands of people to remotely make personalized visits to the exhibition at the same time.

In summary, the mode in which the Tate Britain robots explore exhibits remotely is not the best model. If one were to use a cylindrical, spherical, or hemispherical panoramic "compound insect eye" camera in the exhibition hall of a Picasso or Monet show at the Tate, collecting panoramic video of strips of space in a certain area, and thousands of remote client visitors logged onto the video booth created by the panoramic camera, then everyone could pursue their own wishes, choosing a perspective of interest, choosing a specific work, or switching perspectives and looking around, realizing personalized visits. This would be possible with the panoramic video and personalized distribution system designed in this paper.

3. *Panoramic video and personalized distribution in digital museums*

Figure 1 shows the application of panoramic video and personalized distribution systems in the museum.

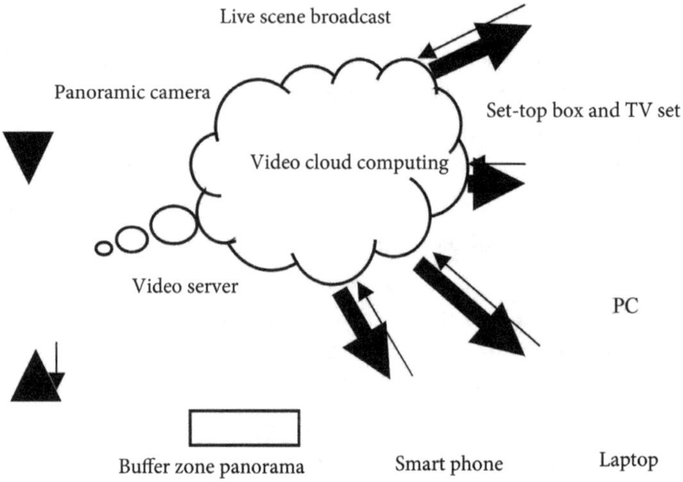

Figure 1 Schematic diagram of the 180° panorama and exhibition hall distribution at the Micro-Electrical Technology Museum at the Beijing University of Technology

The system (in trial operation) above is installed in the exhibition hall of the Microelectronics Technology Museum at the Beijing University of Technology. As shown in Figure 1, the system consists of a panoramic camera, a video server, and multiple clients. The panoramic camera is arranged in front of the scene to be shot, and is responsible for shooting various locations and areas within the scene, then transmitting multi-channel video information to the video server. The video server is a graphics workstation server, and is responsible for synthesizing multi-channel video into panoramic video, then processing the panoramic video further with clipping and encoding, finally sending it to the client over the network. The client can be a personal computer, a home TV with a set-top box, or any display device with interactive functions and Internet access. The client is responsible for receiving live video sent by the video server and decoding it in a display device. If the client wants to change the viewing area, client software can be used to send a location message to the server via a mouse, joystick, or remote control, directing the video server to the client's desired viewing area. Based on the position message, the video server crops the panoramic video to the area of user interest, coding it again and resending it to the client, completing the function of personalized client viewing selection.

3.1 Technical description

3.1.1 Camera cluster hardware

The panoramic camera is a multi-lens reciprocal omnidirectional camera that simulates the spherical, multi-section, compound structure of the insect eye through bionic principles, collecting 180° and 360° columnar, hemispherical, and spherical images of regions around the camera, using a special light path design to eliminate defects caused by intersecting optical paths.

3.1.2 Video server software

The central video server is mainly involved in the collection, generation, editing, encoding, and transmission of the panoramic video, receiving and processing location information uploaded by the client, and tailoring the panoramic video to the corresponding area of interest according to client viewer requirements. The workflow of the server is as follows:

1) Panoramic video collection
The panoramic video is captured using a camera cluster to capture all areas of a scene from multiple angles in multiple video formats; using video capture mode, i.e. Source Filter, to interface with camera driver software, with video data captured by an external camera imported correspondingly to the filter chart via driver software; the generation of panoramic video requires simultaneous collection via multiple cameras, so in the filter chart, the multiple video capturers correspond to multiple cameras; data transfer in video collection involves DirectShow, that is, the "push mode" Microsoft development kit for streaming media processing on the Windows platform, meaning that the source filter "pushes" video data on a downstream filter flow.

2) Panoramic video generation
Multiple videos are involved in a panoramic video synthesizer via multiple video collection devices; the panoramic video synthesizer has multiple framebuffers, corresponding to each input, using a multiple video stream synchronization scheme model: each camera's video frame will enter the corresponding framebuffer area and wait until there is at least one video frame in each buffer area, only then being released for panoramic video synthesis [2]. After the panoramic video frames have been generated in the framebuffer area, the system reads video data in the display buffer back into memory, treating the data as a new video frame, which is stored in a new memory area that has the capacity for one panoramic video frame. Then, the panoramic video maker time-stamps the newly generated video frame for time synchronization, and while realizing multi-channel video, simultaneously performs color fusion and perspective transformation to synthesize into a panoramic video. Panoramic video synthesis uses a cooperative GPU (Graphics Processing Unit) operation [3, 4].

The panoramic video maker recursively passes the panoramic video to the video distributor via the output channel. The video distributor is a one-in-multiple-out filter that copies the panoramic video outputs from the maker into multiple channel videos, then transmits them separately through each output for subsequent processing. Whenever output of a video distributor is connected via downstream filter, an unconnected output is generated automatically, waiting for connection with other downstream filters.

3) Panoramic video cropping

The video clipper is a transformation filter that clips the panoramic video sent by the video splitter, and sends the cropped result to a transform filter in the video encoder. The area of interest to each visitor is different, so the video clipper tailors desired panoramic video for each viewer, as shown in Figure 2.

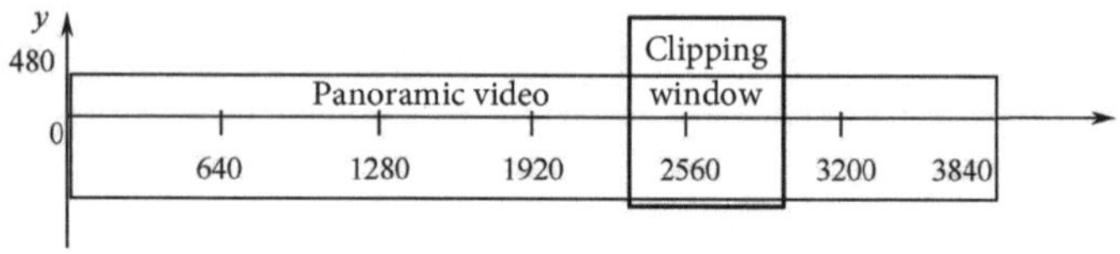

Figure 2 Schematic diagram of panoramic video position coordinates and the clipping window

4) Video coding

Video coding is a necessary component of any video transmission; it can greatly reduce the size of the video and make it suitable for transmission over the Internet. This system uses the Divx video encoder, which is based on an MPEG encoder, using a subset of the MPEG4 compression algorithm. Divx is not only a full-featured encoder, but most importantly, has been specifically optimized for video servers with multi-core processors, with significantly improved coding efficiency.

5) Video transmission

Before beginning video transmission, the server first starts and runs a "socket" streaming module, setting up the communication protocol. That is, the UDP protocol and address protocol of the socket, associating the socket with the network IP address of the local server, waiting for the server to inform the client IP address module, and when the client end has similarly set up a socket, communication will be established between server and client, receiving each other's IP addresses, at which point the server will inform the video sender the client IP address. Then, between the client and server, they can directly send and receive data by employing the "sendto" and "recvfrom" functions, as shown in Figure 3.

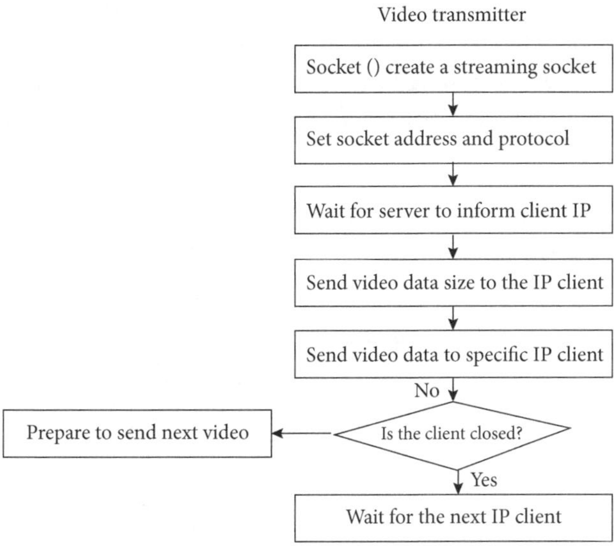

Figure 3 Flow chart showing the work of the video transmitter socket

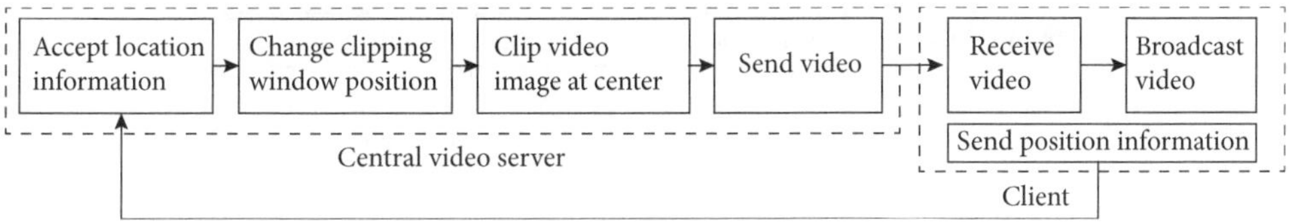

Figure 4 View control flow chart

6) Reception location

After receiving location messages sent by the client across the network, the video server analyzes the client ID sending the location message, such as the name and ID card, then tailoring the location module of the clipping window to a new position in the panoramic video according to the desired location. Then, the area of the new location is clipped, encoded, and sent to the corresponding client, as shown in Figure 4.

3.1.3 Client software

The client refers to a variety of display devices with interactive functions and Internet access, with basic functions of video reception, decoding, display, and the transmission of location information. The client's workflow is as follows:

1) Reception

When the video is received, the client first establishes a socket module, corresponding to a socket on the server-side of the video transmitter, that is, a socket flow, as shown in

Figure 5; setting the protocol module to run, after successful connection with the server, before receiving video data, data regarding video size and characterization is sent by the video transmitter. Then, according to the data volume, the client requests recvfrom() in order to receive video data sent by the transmitter, until the video data of the entire frame is received.

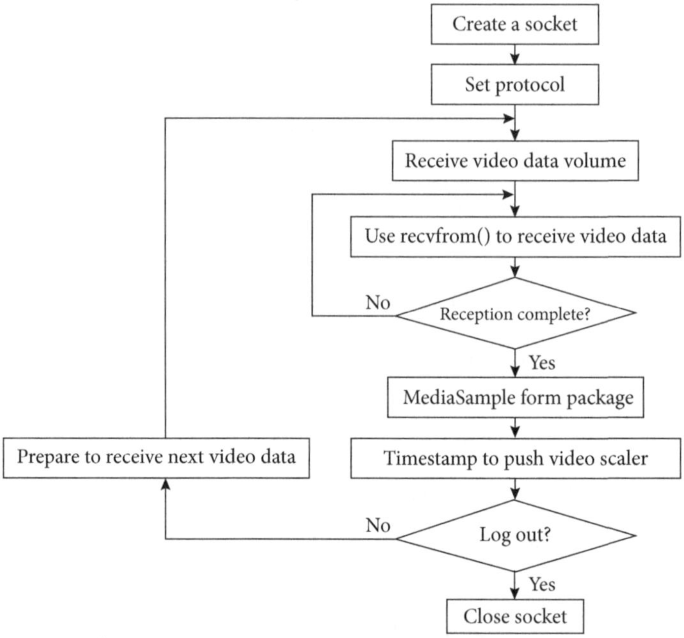

Figure 5 Client socket work

2) Decoding

The decoding link corresponds to the encoding link, restoring the video frame encoded on the server. The client video receiver corresponds to the server video transmitter, receiving video data from the transmitter and pushing to the downstream video FFDShow decoder through the output terminal. The FFDShow video decoder combines multiple decoding facilities into one, and is able to decode videos in multiple compression formats, including Xvix, Divx, and X264. In this filter chart, as the upstream filter output is in Divx-compressed format, and the media video decoder used during media negotiation is a Divx decoder. The client receives the video sent from the central video server and sets a video framebuffer in order to ensure the integrity of the video.

3) Video display

Video display is the process of displaying video, playing it on a screen through the graphics card. In Windows, video display uses Overlay, a digital video display technology used to optimize the mode of video playback.

4) Sending location message

When the client viewer requests a change in view angle or to view other areas in the scene, a location message module is sent with a corresponding location message to the server by manipulating a mouse, joystick, or remote control. The location message records the location of the client's requested viewing area from the central video server, where "location" refers to the location of a portion of pixels within the panoramic video generated on the video server. Each pixel in the panoramic video has a corresponding coordinate point within the coordinate system. As the location message is more important than the video data, and the data volume of the location message is relatively small, a network with a larger overhead but a high reliability TCP protocol is selected [5].

The real-time panoramic video broadcast system is based on a system of video splicing technology and Internet Cloud distribution technology with no dead zone or crossover. Viewers of various remote clients may choose their outlook at the same time without affecting each other. Terminals include PCs, smart TVs, and smart phones, and the system software for selecting viewing angle allows for moving and switching in real time via a keyboard, remote control, mouse, or joystick, just as with a live audience, looking around or glancing right or left [6], achieving personalized on-site visits.

4. Complementary advantages of the multiple mode real-time panoramic video broadcast system

Digital museums cannot completely replace physical museums. In the real-time panoramic video broadcast system proposed in this paper, content and function involve the combination of digital and physical museums, encouraging and promoting one another. Focusing on the audience and meeting the varying needs of visitors of different ages and levels, the system features immersive on-site, personalized, remote interaction and remote control functions, and can be combined with video surveillance systems to reduce costs. In this way, it offers a third way between the virtual and real methods, for visiting a museum via real-time video.

References

[1] Thoughts related to digital museums on Bobao Art Network.

[2] Li Han, Niu Jizhen, Guo He. 'Fully automatic seamless image stitching methods based on feature points', [J] *Computer Engineering and Design*, 2007, 9.

[3] In-depth analysis of a new chapter in GPU general-purpose computing. http://www.shbear.com/2/lib/200701/04/20070104335_2.htm.

[4] Zhang Jianlang. 'CPU is no longer the only core; GPU moves towards general computing'. *New Computer Science*, 2007, 4.

[5] Tang Ying, Zhang Hongxin, Zhang Meiyu. 'Texture image coding and real-time rendering algorithms using graphics hardware'. *Journal of Computers*, 2007, 2.

[6] Hjelsvold, Rune and Andrei Ouglov. 'Panoramic video in video mediated education, storage and retrieval methods and multimedia applications.' *Proceedings of the SPIE* 5682, 2004: 326–336.

[7] Jiang Dazhi. *Research and application of 3D reconstruction in Computer Vision*. Nanjing University of Aeronautics and Astronautics.

Exploring Micro-Control Technology in Museum Network Nodes

Jing Hailiang
(Ministry of Information, National Palace Museum, Beijing, 100009, China)

Abstract: All museums have a common goal, which is the preservation and interpretation of physical forms of social and cultural awareness. Under this premise, museums generally form a situation that is relatively closed corresponding to the physical form, with an absolute openness corresponding to the social and cultural function.

The network environment created in the course of museum informatization thus demonstrates relative characteristics. Cultural artefacts are the core, the openness of the collection is the background, and establishing a network structure that combines information research and application is what allows multiple environments to exist within the network architecture. This entails a certain openness in the process of cultural artefact informatization, and there are certain problems with closure.

How can we reconcile the contradiction between the openness and closure of information? This article will explore how to use micro-control technology in the network node to coordinate and solve certain issues with museum informatization.

Keywords: the museum, cultural artefacts information, micro-control technology

1. An Overview of the Informatization System of the Palace Museum

After more than 10 years of development, the informatization system at the Palace Museum has become the world's largest cultural network for public welfare. Due to the extremely rich content and large amount of information involved, this informatization system is relevant to people from all walks of life, all over the world. The number of daily visitors to the Forbidden City network exceeds 5,000, and the average daily hit rate tops 880,000, carrying out external text and image information publishing of exhibits, the publication of tourism related information and exhibition information, as well as providing related information retrieval services regarding the Palace Museum's exhibits and ancient buildings.

Furthermore, the informatization of the Palace Museum also involves an administrative management system, which consists of two subsystems: office automation, and budget and assets. The office automation subsystem carries out automated office business, such as official document circulation, project contract review and approval, supervision, work contact, information services, resource integration, file management, and work notes. The budget and asset subsystem carries out financial business such as annual budget filing, financial budget allocation, financial review, and asset management.

In addition, the key business management of the Palace Museum has been divided in the business management system of the Palace Museum, which includes seven subsystems. The cultural artefact management subsystem carries out cultural artefact registration, exhibition information, inbound and outbound records, and approval management; the document resource subsystem involves the Palace Museum's document records, query index, review, and approval; the ancient architecture management information subsystem holds records of ancient architecture information, maintenance work, and repair plan development and approval; the drawing resource subsystem carries drawing records, query index, and review and external use approval of drawing documents; the image resource subsystem involves the processing, query index, and use approval of all image data for cultural artefacts and buildings; the AV resource processing subsystem holds all video production processing, query indices, usage approval, and other services for video and audio data; the cultural display management subsystem involves the release and management of all final exhibitions related to the Palace Museum, the release and management of video and audio materials, the broadcast of exhibition route images, audio and video, and the final approval and testing of information published on the external website.

The above business system is inseparable from the creation and development of the Palace Museum network, the network structure is shown in Figure 1.

After years of development, a network system has been formed with a certain scale. The Palace Museum network has adopted physical isolation between the internal and external networks, effectively protecting the intranet business system and administrative system from Internet security threats.

In the external network, the Palace Museum website entails a website system, which provides a service network and user access network. The server providing the service network to the website system is connected to a H3C5510 switch, and the load balancing server provides reliable external services. The user access network allows end users of the Palace Museum to access the internet through the access switch and firewall NAT address translation. A firewall device is deployed on the network for access control and address

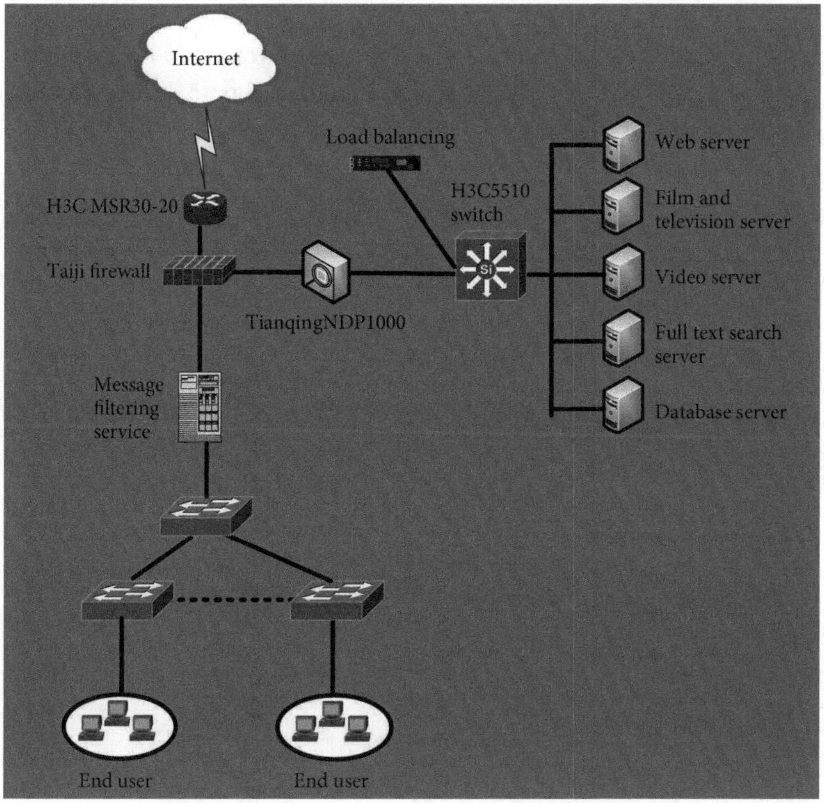

Figure 1 System network of the Forbidden City website

translation, an IPS device is deployed for intrusion detection protection, and a network version of the antivirus system is deployed to implement overall virus protection on the network.

In the intranet, the Palace Museum administrative management system includes two main subsystems: office automation and budget and assets. The business management system includes the cultural artefact management subsystem, the document resource subsystem, the ancient architecture management subsystem, the drawing resource subsystem, image resource subsystem, audio and video resource processing subsystem, and the cultural display management subsystem, as shown in Figure 2. The network structure is divided into two parts: the core layer and access layer. The core layer includes Huawei S8512 and Huawei S6506 switches, while the access network layer includes switches at each main level. The system servers are connected to the Huawei S8512 switch, and end users access these servers through access switches. A network version of the antivirus system is deployed to implement virus protection for the entire network.

In such a complex network environment, the network must be isolated and controlled. This necessitates the abandonment of the hierarchical method used in traditional network development, and requires technicians to introduce new control logic for network segmentation, dividing the complex network through nodes to form new and simple subnetworks. This is conducive to the smooth progress of museum work, and also facilitates precise troubleshooting. More importantly, under this subnet mode, we are able to introduce network micro-control technology to analyze, filter, and control network messages.

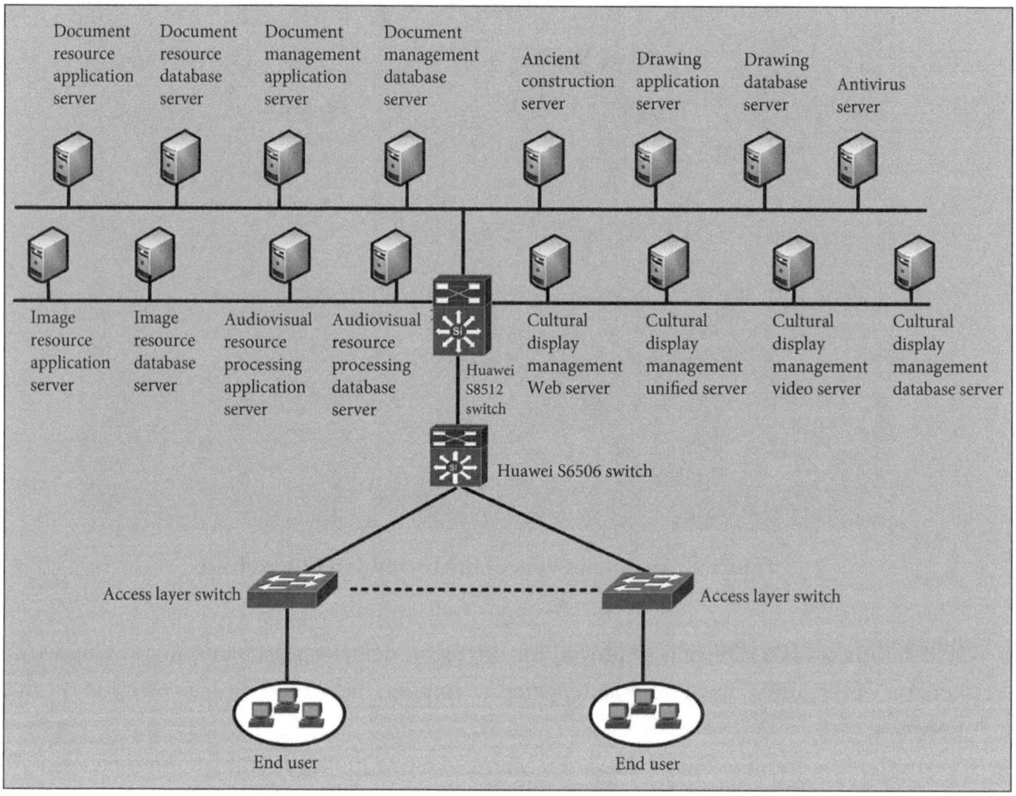

Figure 2 Business management system network

2. Network message filtering and precise control – achieving a huge pay-off for a small effort

In the context of the intricate and complex network required for the informatization of the Palace Museum, the network must be isolated and controlled, requiring technicians to abandon traditional network development methods, and to introduce new control logic and network segmentation to form new network nodes.

Within the new network nodes, we can introduce micro-control technology to analyze, filter, and accurately control network messages. Here, we provide a practical example, subtly improving the network node, and adding a message node host to the exit of the external Palace Museum Network, using the node host to capture and analyze messages, as shown in Figure 3.

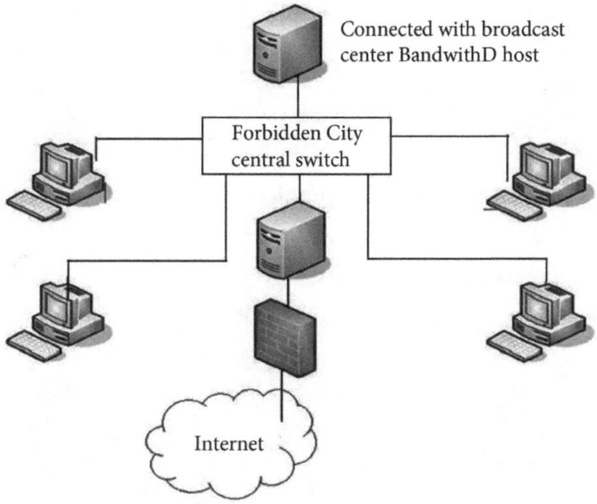

Figure 3 Improved BandwithD monitoring structure

In general, we use Wireshark and BandwithD to capture messages, define the DDoS source address, then combine switching equipment for BAND. Most DDoS attacks appear to be very substantial, and have been considered major "violent crimes" in the hacking world. However, as the size of the network expands, the size and bandwidth of available botnets continue to increase. Therefore, the launch of DDoS attacks has become relatively easy, and the scale of attacks is expanding, so it is easy to paralyze the target network at a lower cost. This has become one of the most widespread and harmful security threats on the Internet.

Using Wireshark filter tcp.flags==0x02 to check the data packet distribution, as shown in Figure 4, the data packet distribution changes significantly when a SYN Flood attack occurs, and the proportion of the SYN packet increases significantly.

Wireshark Static → Ipv4 → Endpoint can be used to analyze the address distribution of the data packet source. When a DDoS attack with a forged IP address occurs, the correspondence between the number of packets and the source address in the packet capture file changes significantly. As in the example in Figure 5, it can be found that in addition to the destination IP being attacked, Wireshark counts a smaller number of packets corresponding to the source address, and the byte size of the packet (bytes) is almost identical.

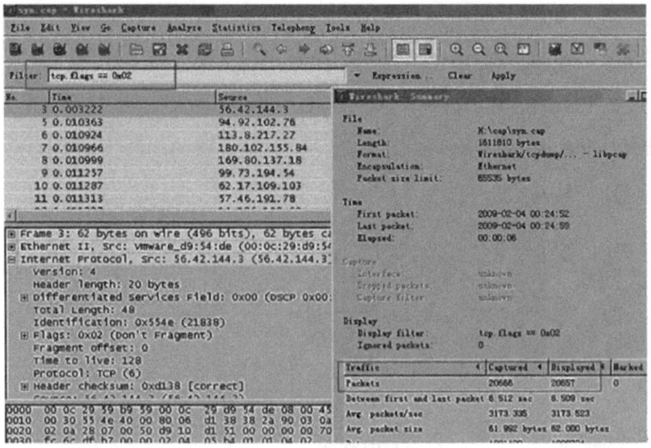

Figure 4 Packet distribution changes significantly when a SYN Flood attack occurs

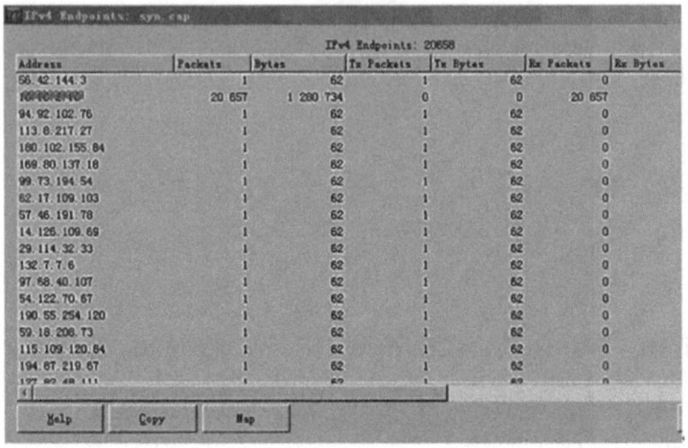

IPv4 Endpoints: syn.cap

IPv4 Endpoints: 20658

Address	Packets	Bytes	Tx Packets	Tx Bytes	Rx Packets	Rx Bytes
56.42.144.3	1	62	1	62	0	
[illegible]	20 657	1 280 734	0	0	20 657	
94.92.102.76	1	62	1	62	0	
113.8.217.27	1	62	1	62	0	
180.102.155.84	1	62	1	62	0	
169.80.137.18	1	62	1	62	0	
99.73.194.54	1	62	1	62	0	
62.17.109.103	1	62	1	62	0	
57.46.191.78	1	62	1	62	0	
14.126.109.69	1	62	1	62	0	
29.114.32.33	1	62	1	62	0	
132.7.7.6	1	62	1	62	0	
97.68.40.107	1	62	1	62	0	
54.122.70.67	1	62	1	62	0	
190.55.254.120	1	62	1	62	0	
59.18.206.73	1	62	1	62	0	
115.109.120.84	1	62	1	62	0	
194.87.219.67	1	62	1	62	0	

Help Copy Map

Figure 5 Packet number statistics corresponding to each source address

Although a fake source address is used when using a random source for DDoS attacks, the attacker cannot forge the positional relationship between the attack host and the target host. Sometimes, by observing the change in the TTL value of the data packet, it is also possible to obtain clues about the attacker and provide auxiliary support for attack defense. The attack procedure shown in Figure 6 does not modify the TTL value of the attack packet. All attack packets use the same TTL value.

The main purpose of the UDP FLOOD attack is to block server bandwidth by sending a large number of UDP packets. For Internet services such as DNS, voice, and streaming media, there will also be application layer attacks hosted through UDP. Figure 7 shows the attack packet generated by the well-known Snail Attack. Clearly, the payload of this packet is fixed. The attack is undertaken using large UDP packets and real source addresses.

Like UPD FLOOD, ICMP FLOOD attacks mainly block server bandwidth. However, unlike UDP FLOOD, ICMP usually does not carry data services, and is easier to defend with switch ACLs or server IP tables. Meanwhile, in order to use the zombie host more

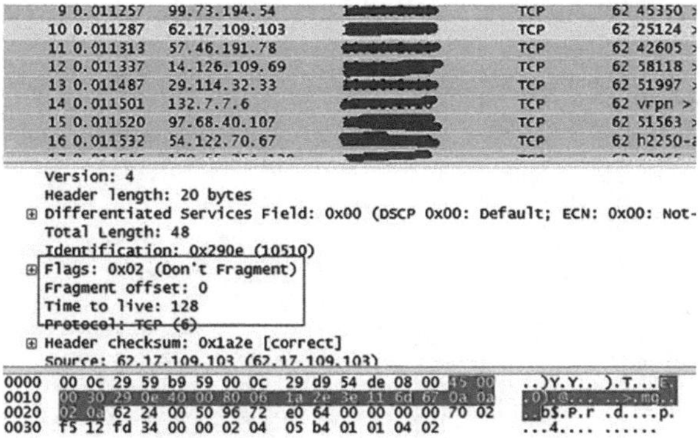

Figure 6 All attacker packets use the same TTL value

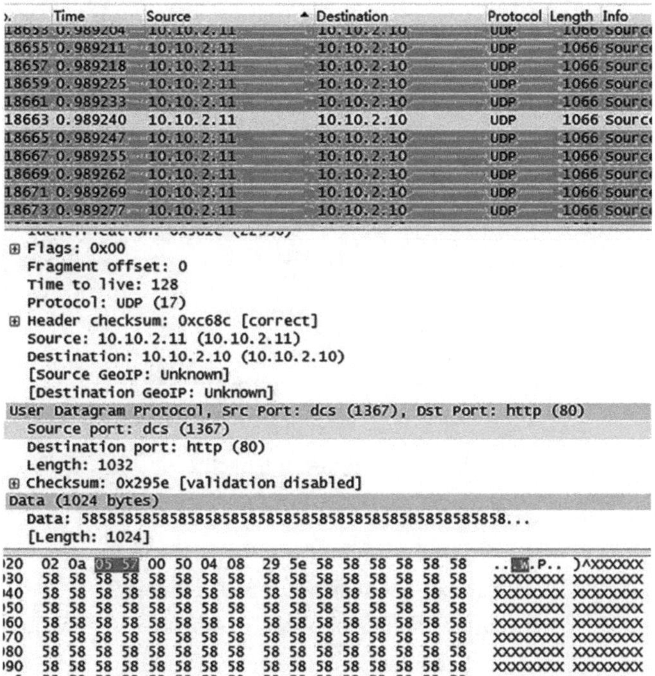

Figure 7 Attack packets generated by snail attacker

effectively, the attacker will use the larger packet, attacking with a fragmented IP packet. An attack like this bypasses the ACL without IP layer allocation, and also increases the burden of server allocation reorganization. It is worth mentioning that the ping packets sent by the Windows client have specific formats and sizes, and the use of general blank defense logics is effective in most cases.

The packet in Figure 8 is a typical UPD FLOOD attack packet that is 1,514 bytes in size, with a fixed payload and a fragmentation sign.

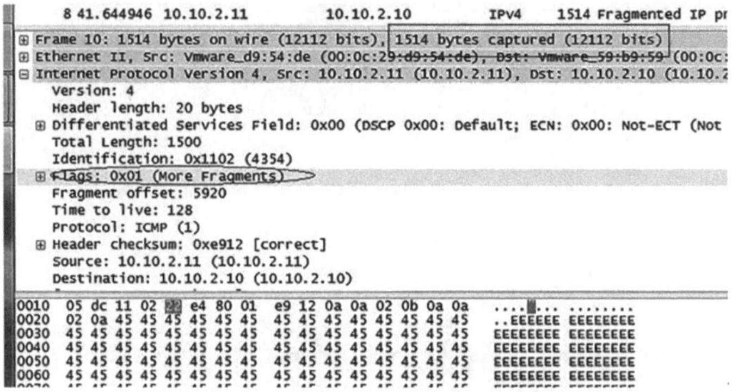

Figure 8 A typical UDP FLOOD attack packet

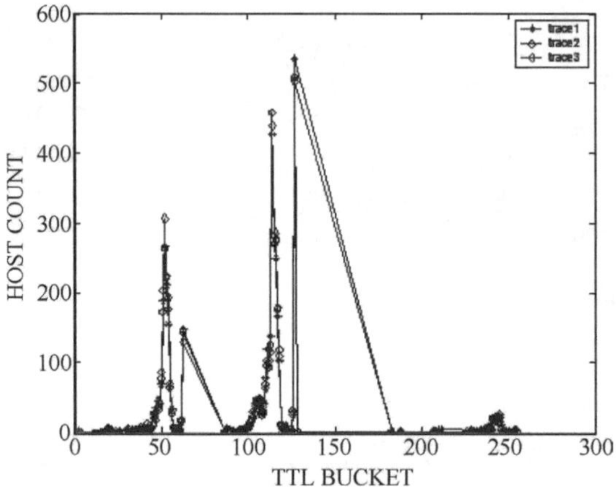

Figure 9 Micro-control chart

By measuring and analyzing the traffic within the existing micro-node network, the characteristics of DDOS packet can been summarized as follows:

The distribution of inbound packet source IP addresses along a certain link is maintained with different TTL values, but the measurement of the total number of inbound packet source IP addresses over the entire observation interval shows that in the current network environment, the number of inbound packet source IP addresses does not increase to saturation in a relatively short period of time, but maintains an approximate constant growth rate over a relatively long period of time.

Based on the micro-control chart (see Figure 9), it can be concluded that the packet distribution based on the final TTL value is bimodal when the link is not under attack. Final TTL values for most inbound packets are concentrated between 64 and 128, as most foreign hosts participating in communication use default values of 64 and 128 as initial TTL values. Meanwhile, the coincidence of distribution curves proves that the distribution of the number of inbound IP addresses based on final TTL values remains stable over the same measurement interval, with little change over measurement time.

In analyzing the monitored packets, we can uncover potential faults in the network, analyze attacks and attack directions within the network, and then perform troubleshooting and policy customization to achieve precise micro-control of the node network. In other words, a careful analysis of the CAP file at the time of attack, finding out what is ignored by the attacker, and finding out where the attack packet diverges from normal traffic flow can all play a part in exerting minimal effort for a huge pay-off.

3. *Establishing a transparent high-performance bridge firewall to achieve full control of passing traffic*

In 2008, our museum's Internet export underwent major bandwidth allocation. A bandwidth flow control system was designed to manage this problem. Since 2008, 100MB of Internet export has been a rare bandwidth in China. Corresponding flow control technology has been under research and development, with all security vendors having yet to introduce transparent bridge flow control products. In order to solve the problem of export bandwidth allocation under the technical means of the moment, it was necessary to completely alter the Internet export structure of the Forbidden City, which required changing the overall website structure and network structure. This solution involved the transformation of the whole Forbidden City network, which was a difficult task to implement in a short period of time. It affected normal access to the Forbidden City, and was not an effective solution to the problem.

Without changing the structure of the Forbidden City's Internet export, the development of a flow control system must first clarify the most effective methods required to solve such difficulties in the current environment, which can be implemented in a relatively short period of time. The Internet export of the Forbidden City is responsible for Internet applications for work purposes and, most importantly, the official website of the Palace Museum. The official Palace Museum website is the world's largest public welfare and culture website. Thanks to its rich content and large amount of information, it is utilized by people from all walks of life, all around the world. The average number of daily visitors exceeds 5,000, and the average daily click rate exceeds 880,000. As the official website of a major establishment at the national level, it represents the image of traditional Chinese culture, and has great social influence. Obviously, it is not feasible to carry out lengthy, large-scale network transformation in such an environment.

Using the production network micro-node to establish a self-made firewall in a transparent bridge mode, using token-bucket traffic supervision, and "penalizing" excess traffic through the supervision of traffic rates entering the network port (this penalty can be removal or delayed transmission), traffic entering the port is limited to a reasonable range, fully mastering all external network export traffic and thus solving the problem of export congestion.

The specific principles are as follows: the TCP message cannot use more than 50% of the network bandwidth; or, the token-bucket traffic supervision can choose to discard the message, or reconfigure the priority of the message. The token-bucket traffic supervision is performed using the token bucket mechanism. Here, the "bucket" refers to an internal storage pool of the network device, and the "token" refers to the virtual information packet that fills the "bucket" at a given rate. The bucket can be understood easily as a physical bucket, and the token as water flowing through a pipe into it.

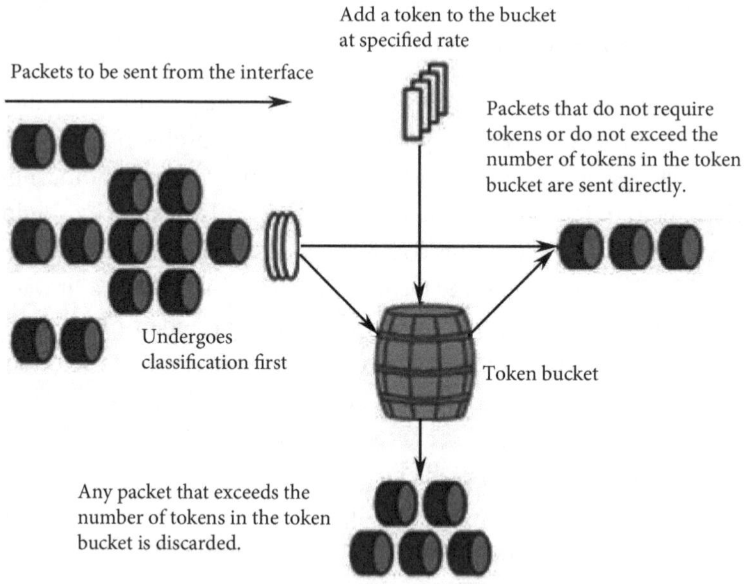

Figure 10 Controlling network bandwidth usage with token buckets

The switch adds a token to the bucket when it receives each frame, but there is a hole in the bottom of the bucket, which constantly gives out tokens according to the rate held as the average communication rate (in b/s). Each time a new token packet is added to the bucket, the switch checks if there is enough space in it. If there is not, the packet is marked as non-conforming, and will be policed (discarded or marked) under specific supervision.

The token bucket control mechanism is premised on whether any tokens exists in the bucket, to indicate when traffic can be sent. Each token in the token bucket represents a byte: if a token exists in the bucket, traffic can be sent; if a token does not exist in the bucket, traffic cannot be sent. According to the token bucket algorithm, batch processing has been performed in the Linux host. Due to the complexity of its deployment, specific production processes have been omitted here, only demonstrating the core strategy and algorithm overview, as shown in Figure 11.

1. Rate limit

2. The command structure for traffic-police is as follows

police <bps> <burst-normal> <burst-max> conform-action <action> exceed-action <action> [violate-action <action>]<bps> CIR

(Defines the average rate at which tokens are injected into token buckets)

<burst-normal>:BC (Defines the size of the BC token buckets)

<burst-max>:BE (Defines the size of the BE token buckets)

<action>: Defines the behavior, which is optional as follows:

drop—Drops the packet.

set-prec-transmit new-prec—Sets the IP precedence and sends the packet.

set-dscp-transmit new-dscp—Sets the DSCP value and transmits the packet.

transmit—Sends the packet.

The initial packet reaches a 450byte, and after 0.25s, reaches to a 900byte packet.

police is associated in the output direction of fa0/0. Router (config) # class-map access-match

Router (config-cmap) # match access-group 1

Router (config-cmap) # exit

Router (config) # policy-map police-setting

Router (config-pmap) # class access-match

Router (config-pmap-c) # police 8000 1000 conform-action transmit exceed-action drop

Router (config-pmap-c) # exit

Router (config-pmap) # exit

Router (config) # interface fastethernet 0/0

Router (config-if) # service-policy output police-setting

The initial token buckets are filled with 1000 byte tokens, and if the packet reaches 450 byte in the initial stage, then the tokens of 450byte need to be issued from the token bucket. At this point, the remaining (1000 – 450) byte=550 byte of tokens in the token buckets is ample, This will trigger comform-action and transmit a 450 byte packet. After 0.25 s, a 900 byte packet arrives. Within the interval of 0.25 s, (8000bps/8) × 0.25 s=250 byte tokens will be injected into the token buckets. At this time, (550 + 250) byte=800 byte tokens will be left in the token buckets. This time,there are not enough tokens in the token buckets, and exceed-action will be triggered and the 900byte packet will be dropped.

Figure 11 Overview of core strategies and algorithms

Following further research, the Linux operating system installed on the PC has been streamlined. After half a year of streamlining the embedded operating system, it was able to be installed in a 256MB DCOM card, to realize the transformation of a flow control host from a hard disk PC storage, to a high-efficiency workstation with an embedded x86-based hard disk-free chip. This embedded operating system has been effective for more than six years in the Forbidden City's network export, and has continued to work without issue for 183 consecutive days.

The equipment we developed is extremely stable. The second transparent bridging host has been built in the Jianfugong network transformation project, which has also been stable for several years now. A web management interface has been developed to get rid of command line control. In 2012, after further improvement it was miniaturized and made into an x86 form of switch. This transparent bridge flow control device is a result of many years of work, and it was a very proud moment when we solved a major issue in the Forbidden City's network with flow control equipment developed by the author of this paper. The equipment quietly and continually performs traffic control and message filtering at the exit of the Forbidden City's network, and has been in use for more than six years. Finally, in July 2014, other ACQ equipment was installed in place of the weather-beaten Linux host.

4. Establishing a micro-control node

Through the examples given, we are aware that the use of node control can provide effective solutions for practical problems, and also understand the necessity and feasibility of network node intervention control. This leads some technical engineers to ask if it is possible to form network nodes for micro-control intervention in the wireless network environment as well. Or is the micro-control technology only suitable for applications in the network environment? In fact, in the process of informatization, our information flow does not rely on the network alone. There is a substantial amount of interaction within the operating system. Node micro-control technology is not limited to network or wireless networks, but exists in every interactive corner of information transfer. A few application examples will be offered here.

We know that there are venue restrictions in museums, making it difficult to deploy a wired network within the exhibition hall, and making it hard to meet ever-changing network needs. A flexible and convenient wireless network is the optimal solution to this problem.

We also know that the use of wireless networks can extend museum functionality, but also raises issues of access security. Due to the openness of wireless network access processes, some visitors may access the network and conduct malicious scanning and attacks.

Here, we must perform a structural analysis of the wireless network. First, at the wireless access level, the wireless AP plays a role as a link (see Figure 12). Before carrying out partitioning of nodes for wireless, the AP exists as the first node between the visitor access and the application networks. Then, we must establish ICMP packet shielding on the wireless AP, and use AP with ICMP message and SYN Flood control. The connection of wireless devices in the exhibition hall is complicated, and if the basic SYN and ICMP message control and response cannot be satisfied, the wireless gateway will be unable to be accessed. This is because the device encounters a large ICMP packet attack or improper handling after a SYN half-handshake, resulting in a system crash. This is like a situation encountered in the development of a smart city, in which a wireless AP in a city bus had a signal but was unable to respond to connection requests. This problem is widespread. Most of the wireless networks in smart cities in China may have signals, but no connection.

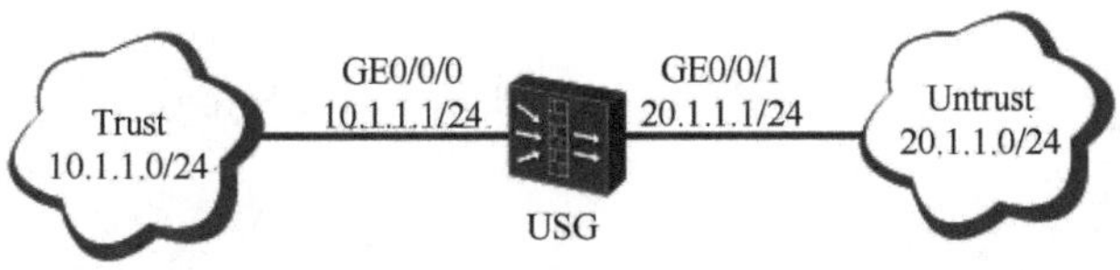

Figure 12 Analysis of wireless network structure

Thus, the first step is to prevent malicious ICMP and SYN half-handshakes, meaning that in certain large-scale wireless access terminals a layer of IDC nodes must be built between the wireless router and core network to implement access control and prevent malicious attacks.

Beyond this, when a wireless or wired network passes information to the application server, certain attacks can be allowed to occur at more levels.

For example, there is a risk of behavioral attacks in the system, meaning that the operating system level uses the network to interact maliciously and launch attacks. This requires us to further refine the relationship between the network and the application server, along with the application protocol, controlling the node between file upload and download, and data stream type, as well as direction.

First, we establish a network topology map, and then a data topology map (see Figure 13). After comprehensive analysis, ideas can be organized in order to build a node strategy map.

In other words, in the museum content application and network overall, we must first refine and differentiate the direction and category of each piece of information, conducting micro-control intervention for all categories, considering the data flow direction, making judgements about malicious access blocking and effective access release, and simplifying complex data and network environments by forming control nodes.

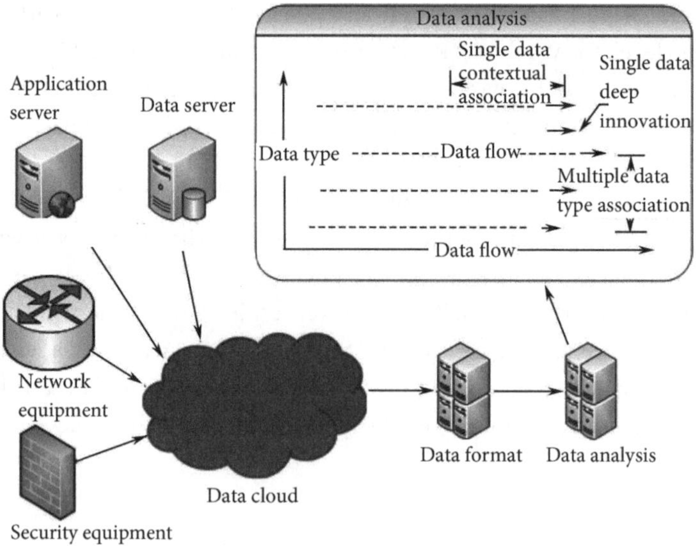

Figure 13 Network and data flow topology

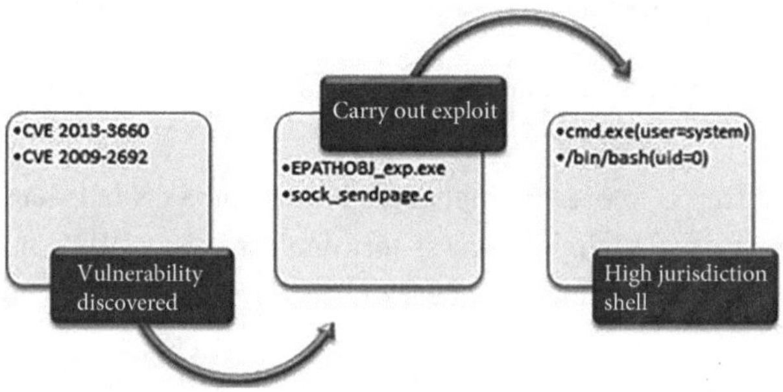

Figure 14 Vulnerability overflow attacks data flow direction

Figure 14 demonstrates a very specific vulnerability overflow attack on the data flow direction. Through three nodes, the attacker exploits the vulnerability of the operating system under the network environment, scanning in the first node, finding exploitable vulnerabilities and remote exploit overflow in the second node. It uses the sock_sendpace.c problem to attack, and the third node performs privilege escalation to complete the attack.

From this behavioral description, we see that it is easy to find three nodes of attack. Even if attack techniques are ever-changing, micro-control of these three nodes cannot escape key point control of micro-control nodes. We can use the three steps of scan blocking, content protection, and permissions to achieve complete control, changing the resulting attack, and achieving combined control of the operating system network and local policies. Of course, this requires technical engineers with a lot of practical experience, who are familiar with attack processes and key points of new vulnerabilities. This is a huge technical challenge for museum information personnel. In the face of this challenge there is no rest, as the development of information technology introduces changes with each passing day. We must continue to learn and accumulate experience in order to achieve secure application and research around museum informatization.

5. Prospects and conclusions of network micro-control technology

The rapid development of information technology today is bringing a variety of new technologies and applications. Tri-network integration is a major change in the Internet, and the field of museum information application will present refined applications as well as broad, diverse development. The integration of smart devices into everyday life has brought about diverse and complex devices. Given its current form, the next stage of the Internet will involve an interconnected era of networked objects.

These circumstances require that our information personnel have the ability to simplify and establish key nodes in information, following clear ideas and key micro-control nodes, making all connected things controllable, and not falling into an embarrassing situation in which objects are connected yet too difficult to control. As such, the micro-control node technology explained in this paper will certainly be applied widely and be developed in future.

Following this, as museum information personnel we must acquire knowledge about informatization regarding cultural artefacts, master the implications of these artefacts, and grasp key points in the direction of information flow. We are not limited to technology, as our working environment and premises involve using materials to spread wisdom and culture, so that cultural artefacts can bear the weight of civilization. This is not limited to the traditional, but also includes the modern, as combining culture with modernity is becoming a common practice.

A Brief Analysis of a Typical Open Content Project in the Museum Field – The Google Art Project

Meng Li
(Beijing University of Aeronautics and Astronautics, Beijing, 100191, China)

Abstract: With the development of informatization, the application fields for open content are becoming wider and wider. The Google Art Project adopts an open content model to create an online museum – a typical open content project in the museum field. This paper analyzes the project's concept and technological applications, and discusses the application of open content in the digitization of museum resources.

Keywords: museum digitization; open content; Google Art Project, online museum

Preface

Open content is a product of the Internet age. Given the development and popularization of the Internet, increasing numbers of resources are beginning to show digital, networked, and open characteristics. In this context, the kinds of open content and areas of involvement are becoming more diverse. To cope with the challenges of the Internet era, there is constant exploration and breakthroughs being made in terms of the digitization of museum resources, and applications of the open content model present clear examples. This article discusses the Google Art Project as a typical open content project within the museum field, analyzing its design and technical applications, as well as exploring the application characteristics of open content in the digitization of museum resources.

1. An overview of open content

Open content, also known as content open, was proposed by David Power in 1998 to describe original works that can be copied and/or modified by the public. Open content allows others to access information without special permissions or payments, allowing reproduction or even recreation to form derivative works [1]. Open content is thriving firstly because it can cope with the high cost of traditional publishing resources, and secondly because it can satisfy the actual needs arising from a lack of educational resources in certain areas [2].

Open content mainly offers users the right to freely and permanently engage in the five Rs: retain, reuse, revise, remix, and redistribute [3]. These rights allow people to use, modify, adjust, and mix content. However, the rights allowed for specific content depends on the actual openness of the content.

At the end of the 20th century, the Internet achieved rapid development, posing both opportunities and challenges for education. In 2001, in response to the challenges brought about by the Internet era, MIT altered the traditional education model, and proposed an innovative solution — the open courseware movement (MIT Open Courseware). In this campaign, MIT provided thousands of pieces of high-level courseware for free use by the public around the world. Other internationally influential open content projects include Wikipedia, Project Gutenberg, Linux Open Plan (for writing content documents), Khan Academy, BioMed Central (BMC), the Public Library of Science (PLoS), Librivox, Flickr, and Open Clip Art Library [4].

2. About the Google Art Project

The Google Art Project is a successful example of the use of open content in the museum sector, launched by Google on February 1, 2011. Using Google Street View technology to capture real museum scenes, and using gigapixel imaging technology to capture museum exhibits, Google built an online museum providing all actual and exhibit photos online for free. With just a computer and an Internet connection, visitors can make a museum visit without leaving home. In addition, Google Art Project collaborated with street art in June 2014, beginning online exhibitions of street artists' work collected from around the world. As of May 14, 2015, the Google Art Project had collected 91,495 exhibits from 551 museums including the American Museum of Modern Art, the Metropolitan Museum of Art, the National Gallery of Britain, the National Palace Museum in Taipei, and the Hong Kong Museum of Art.

The Google Art Project has integrated the concept of open content extremely well. In the past, museum visits were field-based, restricted by time and place, and sometimes entailing high ticket prices. As a result, many people have been discouraged from visiting museums. Some have never attended one, and this runs counter to the educational function of the museum. It is in this context that the Google Art Project was born. The Project allows visitors to visit online exhibits, permanently free of charge. Visitors are no longer limited by factors such as time and place. They can attend exhibitions relevant to their own interests in their own time, anywhere, whenever they like. Google Street View and ultra-high resolution pixels provide visitors with a nearly-real experience. Meanwhile, the Google Art Project offers them the opportunity to engage in online discussions with friends according to their areas of interests.

3. Analysis of the Google Art Project's concept

The concept of the Google Art Project is reflected largely in promoting cultural democracy and revolution, shortening the distance between art and people, providing a better visitor experience, enhancing the educational function of the museum, and attracting more visitors.

1) Promoting cultural democracy and revolution
With more information available online, we exist in a moment of intellectual democracy. Museums and art galleries are no longer exclusively for the elite and the nobility. Anyone who has access to the Internet can visit the online art program led by Google Art [5]. The Google Art Project provides equal rights and status in the context of museums.

2) Shortening the distance between art and the people
Unlike traditional museum visits, Google Art allows visitors access without leaving home. The online visitor mode saves visitors time and money, and also allows them to bypass certain factors such as distance, weather, and transportation. This has greatly reduced the barriers to democratic museum visitation, bringing art to more people.

3) Providing a better visit experience
As Amit Sood, the founder of the Google Art Project, has stated, the purpose of the program is not to replicate the museum visit experience, but to complement it. Although visitors are invited to observe the exhibits at relatively close range during a visit to the museum, there are certain nuances that are invisible to the naked eye, meaning that they are still liable to miss a variety of details. Google Art plans to show details that are impossible to see with the naked eye via gigapixel imaging technology. In addition, due to the collection limitations of individual museums, works by the same artist can be scattered across different museums. When visiting a museum, it is difficult to see multiple works by the same artist together for comparison and research. However, via the online "comparative learning" feature provided by the Google Art Project, users can do so online. In addition, the Project provides a "user collection" function, so that users

can save favorite works of art to their user collections according to their interests, thus establishing a personal online museum.

4) Enhancing the educational function of the museum

An important function of the museum is education. However, due to factors such as geography, financial costs, and time costs, many people have yet to visit museums. With the online museum provided by the Google Art Project, people can remove the influence of the above constraints and view exhibitions via the Internet, thus realizing the educational function of the museum.

The Google Art Project is a great example of the immersive learning philosophy. Immersive learning refers to the creation of a fully immersive learning environment for students via virtual technology [6]. The Google Art Project offers a virtual museum experience on the Internet via Google Street View, in which visitors are exposed to the environment and can interact naturally with it, creating a sense of immersion in which to devote themselves to learning [6]. In addition, when visitors use the Google Art Project to browse museums, they can attain visual and auditory impressions. Sharing favorite works of art and interacting with friends proposes to enhance the educational effect.

5) Attracting more visitors to the museum

As previously mentioned, Google's intention is a complementary experience to museum visits, rather than a copy or substitute. Online visits have decreased the distance between people and art. When members of the public visit online, interest in the museum increases, encouraging people to visit the physical museum.

4. A technical analysis of the Google Art Project

The main applications used in the Google Art Project include Google Street View technology and billion-pixel imaging technology.

Google Street View is a feature developed by Google Inc. that applies to Google Maps and Google Earth, providing 360° and 180° horizontal panoramas, and allowing users to view selected cities on ground level, showing different street locations and surrounding scenery [8]. Thanks to the 360° panorama, when users view the street landscape online, it is like walking down the real street. Google Street View obtains a panoramic image by using eight cameras on one plane, splicing and deleting photos to create a panoramic 360° photo [9].

A gigapixel image refers to a high-definition image with up to 109 pixels. The Google Art Project uses this gigapixel imaging technology to capture ultra-high definition exhibits when shooting museum exhibits. Gigapixel photos are generally made up of a large number of photos, or by the scanning of multiple negatives. For example, a 3.2

billion pixel photo is a combination of 612 photos [10]. In addition to its application in the Google Art Project, gigapixel imaging technology is widely used in the fields of astronomy and geology.

5. Summary and outlook

The Google Art Project subverts the traditional museum visit model, using Google Street View technology and gigapixel imaging technology as well as the Internet platform to display a new form of visitation in the information age. Via online visits, users have free and convenient access to museums around the world, and can enjoy art from different countries. Through the collection options in the Google Art Project, visitors are able to pick their favorite exhibits, collecting them in personal virtual museums. With comparative study options, visitors can put together several exhibits of interest, and perform comparative studies. The Google Art Project has helped museums create online exhibitions, and has also relieved the monotony of traditional visits by means of unique Internet characteristics, attracting more people, providing an improved visitor experience, and enhancing the educational function of the museum.

The Google Art Project is the result of a combination of open content and the physical museum. It is an innovation in the museum field as an adaptation to the Internet age. Crowds of people are no longer an issue, and distance has been eliminated with the high-quality recreation of exhibition halls and 7 billion pixels of infinite detail and texture coming to the forefront [11]. The Google Art Project offers a better museum visit experience, and also solves the limitations of physical venues. At present, the number of museums that have reached cooperative agreements with Google is limited, and methods for cooperation with more museums is the next challenge for the Google Art Project. In addition, the creation of both online and offline tours presents an important future issue for cooperation.

References

[1] Chen Chuanfu. 'Types of open content and intellectual property management', [J] *Journal of Library Science in China*, 2003, 6: 9–13.

[2] 'Part III of the New Media Alliance Horizon Report', [J] *Journal of Beijing Broadcasting University*, 2013, 8: 56–78.

[3] http://www.opencontent.org/definition/

[4] Fu Rong. 'Open content: sharing and cooperation', [J] *Library*, 2013, 2: 70–72.

[5] http://en.wikipedia.org/wiki/Google_Art_Project

[6] Hao Liying, Qi Xin. 'Transforming Chinese-English intercultural communication in immersive learning', [J] *Times Literature*, 2008, 6: 37.

[7] Cheng Yongjun. 'Using Virtual Reality technology to create an "immersive" network learning environment', [J] *China Educational Technology*, 2002, 4: 62–63.

[8] https://en.wikipedia.org/wiki/Google%E8%A1%97%E6%99%AF

[9] Wan Jintao, Wu You. 'Google Maps: Magic street scene shooting', [J] *Focus* 2010, 6: 50–53.

[10] Sun Bing. 'Looking at Van Gogh in 7 billion pixels – how cool,' [J] *China Economic Weekly*, 2011, 9: 74–75.

[11] Sun Linlin. 'What museums are busy doing on the Internet', [J] *Reader's Taste*, 2013, 6: 14–17.

The "24 Solar Terms" Digital Display at the China Agricultural Museum

Zhu Lin
(China Agricultural Museum Information Center, Beijing 100125, China)

Abstract: While the "24 Solar Terms" exhibition moves towards the application of world cultural heritage, its profound cultural significance and close connection to agricultural production and life have attracted increased attention. The China Agricultural Museum is undertaking this application work, while also introducing new ideas for daily exhibition design. It uses digital means to display the lunar calendar's "24 Solar Terms" in an advanced form of three-screen theater, presenting the cultural heritage created by the working people of China to the world in a novel and flexible way. The Museum uses digital means to protect cultural heritage, with far-reaching significance for digital museum development.

Keywords: 24 solar terms; digital; three-screen theater

Preface

The "24 Solar Terms" originated in the Yellow River Basin during the Spring and Autumn Period when ancient Chinese sages set the four seasons of Zhongcun (mid-spring), Zhongxia (mid-summer), Zhongqiu (mid-autumn) and Zhongdong (mid-winter). In the Qin and Han Dynasties, the "24 Solar Terms" were established. As fundamental cultural heritage created by the working people of China, they reflected changes in the seasons, guiding farming activities, and affecting the basic necessities of thousands of households. Li Xiusong, deputy of the National People's Congress, has stated that the "24 Solar Terms" represents wisdom of the Chinese nation, and a cultural heritage created by the working people of the country, and claimed that they ought to apply for world heritage status.

China is a country of world-leading cultural heritage, dating back 5,000 years. In recent years, awareness of cultural heritage conservation has been strengthened. With Xi Jinping as the general secretary, the Central Party Committee has stressed the importance of cultural heritage protection, which is of far-reaching significance. Culture is the core element of the nation's soft power, and an important indicator of national prosperity, national rejuvenation, and public happiness. Cultural heritage is the foundation of culture, containing spiritual values, ways of thinking, and creative imagination that are unique to the Chinese nation. China much strengthen the protection of cultural heritage, and improve the inheritance system of its traditional culture for future generations.

1. *About the "24 Solar Terms"*

Ancient Chinese people divided the trajectory of the solar year into 24 equal parts. Each aliquot was one "solar term", and they were collectively referred to as the "24 Solar Terms". This system is a cognition of the knowledge and social practice formed by the changing laws of seasons, climate, and phenology within the year, guiding traditional agricultural production and daily life. It is an important part of the traditional Chinese calendar system and its related practical activities. This cognitive time system is known as the "fifth greatest invention in China" in the international meteorological community.

The "24 Solar Terms" passed from generation to generation are a unique Chinese knowledge system, profoundly affecting people's ways of thinking and codes of conduct. Based on each seasonal node, cultural communities organize farming production, arrange the order of daily life, and hold vibrant seasonal festivals and folk activities. They also organize basic necessities for themselves and their families. This heritage is an important carrier of cultural identity in the Chinese nation.

The "24 Solar Terms" are generally accepted by the Chinese public, and their influence can be seen everywhere in daily life, with certain combinations with folk culture becoming fixed public holidays. The most famous are Qingming (tomb-sweeping), Lichun (beginning of spring), Lixia (beginning of summer) and the Winter Solstice, which are integrated into the calendar of festivals. Meanwhile, the "24 Solar Terms" are closely related to daily life, with winter solstice dumplings and summer solstice noodles, first term dumplings, second term noodles, and third term egg pancakes, and sayings like "training during the hottest days in summer and coldest days in winter". During these festivals, vibrant folk activities often take place.

Formed in the Yellow River Basin, the "24 Solar Terms" were based on observations of changes in celestial phenomena, temperature, precipitation, and phenology in the region, with a time guide for agricultural production and daily farming life gradually adopted throughout the country, across ethnic groups. This heritage project is rich in content, including proverbs, songs, and legends, as well as traditional tools, domestic utensils, handicrafts, and paintings, plus a festival culture closely connected to seasons, production ceremonies, and folk customs. The "24 Solar Terms" are a concrete manifestation of ancient Chinese agriculture, with high research value in terms of agricultural history and culture.

2. *Cultural heritage protection*

2.1 Background research

1) Status of foreign research on the protection of intangible cultural heritage
After assessing intangible cultural heritage in countries around the world, it has been discovered that with the trends of globalization constantly strengthening, the cultural aggression of superpowers, the promotion of international standardization, ongoing armed conflicts, and the surging development of the global tourism industry, the worsening environments are seriously threatening the protection of traditional culture. As a legacy of human civilization, intangible cultural heritage has considerable cultural value, and is a precious element in the study of ancient human civilization. With increased demand, research into specialized information technology for the development and utilization of intangible cultural heritage protection has attracted widespread attention. It is important to consider how to strengthen the protection of art related to the cultural heritage mentioned above, as well as its cultural forms, with the latest digital information network technology and virtual reality technology.

2) The status of Chinese research into the protection of intangible cultural heritage
At present, the digital conservation project for intangible cultural heritage in China has achieved some remarkable global achievements. For example, since 1997, the National Key Laboratory of CAD & CG at Zhejiang University has conducted several years of research into the digital protection technology for Dunhuang Art, achieving significant results. Various projects in China provide useful methods and experience for the protection, inheritance, and re-creation of endangered cultural heritage through information technology. They include the Virtual Palace Tours at Zhejiang University, the digitization of the Forbidden City at Peking University, the Microsoft Research Institute for the Terracotta Warriors, and various domestic digital museums such as the Digital Exhibition Hall of the China Agricultural Museum, the Shandong University Digital Archaeological museum, and the China International Friendship Museum project.

3) Comparing Chinese and foreign cultural heritage protection
In Western society, there is a general issue in the imbalance of cultural heritage protection. The protection of intangible cultural heritage in these countries has been attended to only under the protection of attached material heritage, and large cultural resource databases

in Western countries are based on material cultural heritage databases. China is a unified, multi-ethic country with a rich and vibrant cultural heritage formed by 56 ethic groups. With the strengthening of globalization and the acceleration of the modernization process, China's cultural ecology has undergone tremendous change. Intangible cultural heritage has been affected, and the strengthening of its protection is imminent. Unlike Western society with its lack of a systematic, comprehensive protection system, China has always seen the importance of the protection and development of resources for intangible cultural heritage. Its intangible cultural heritage protection work has entered a stage of comprehensive development, and a protection system has been initially established in line with national conditions.

2.2 The digitization of cultural heritage

Alongside strengthening awareness of cultural heritage protection, the museum is also responsible for the protection and inheritance of cultural heritage while collecting and preserving cultural artefacts. With the continuous penetration of information technology, museums are using digital technology more broadly, which also offers diversity in the display of cultural heritage.

1) Digital storage methods
A digital museum's collection is an information resource. Information resources (objects, pictures, periodicals, documents, sounds, images, and other multimedia materials that the museum accumulates over long periods of time) are stored and processed by computers, and the digitization of the system is implemented using a variety of digital devices. This digital storage method reduces the storage space of information resources, facilitates remote user search and query, expands the scope of information utilization, and rationalizes and maximizes the use of museum resources.

2) The organization of information
The digital museum is largely realized through the network, and the organization of information is transformed from the linearity of physical museums into a direct, networked structure. Every page of the digital museum provides a variety of links, between which visitors can freely jump according to individual needs, increasing their capacity for independent choice.

3) Organic integration of multimedia information
In the physical museum, most collections have become very fragile due to age. Therefore, in order to protect collections, it is necessary to maintain a certain audience distance, often making it difficult for visitors to observe collections at close range. In the digital museum, multimedia collection information, including images, graphics, sounds, and

words, is organically integrated through multimedia technology, allowing visitors to understand the historical and cultural information.

In the agricultural science exhibition area at the China Agricultural Museum, digital audio technology is used to produce the sounds of rain, insects, and frogs, providing visitors with a feeling of immersion in nature. The audience can also use realistic and digital 3D models to observe multi-angle and close-up views. Space breeding and animal cloning videos can also offer more vivid and realistic agricultural science education.

3. Digital display of the "24 Solar Terms" at the China Agricultural Museum

According to the definition of the museum, it must function as a "public education institution". So, while collecting and preserving cultural artefacts, the museum also shoulders the functions of communication and exhibition. However, it is also limited by display, visit time, and the cultural preservation environment. This means that traditional museum exhibitions are mostly dull and conservative, not allowing for a positive visitor experience in terms of promoting in-depth understanding and appreciation of traditional culture. As the only national-level museum with an agricultural theme in China, the China Agricultural Museum continues to develop and innovate in the field of digital museums. Following reorganization, the museum consists of five permanent exhibitions and an outdoor exhibition garden. The permanent exhibitions include Chinese agricultural civilization, Chinese soil specimens, traditional Chinese agricultural tools, ancient agriculture in painted pottery, and popular agricultural science for adolescents, with an exhibition area of 6,500m2. The outdoor exhibition garden features a traditional agricultural garden, a modern agricultural garden, and the "24 Solar Terms". Across the entire museum, rich exhibits, vivid scenes, novel artistic techniques (especially the use of skilled digital, technological means) demonstrate the achievements of Chinese agriculture.

The three-screen film "24 Solar Terms", jointly produced by the China Agricultural Museum and Crystal Stone Company, uses digital 3D technology to display China's intangible cultural heritage. The images vividly describe the formation process of the "24 Solar Terms", explaining how ancient scholars used the "24 Solar Terms" to carry out agricultural production, the impact of the "24 Solar Terms" on people's lives, and their important role in agricultural production and life. It uses modern technology to demonstrate the important contribution China made to agricultural science and technology, and the hard work and wisdom of its people, offering an unprecedented visual experience. A three-screen theater was set up independently, using technology to integrate domed, circular, and flat screens, creating a three-dimensional virtual space using multi-screen projection system technology. A panoramic, 330° vertical and 360° horizontal cinema is formed with 25 ultra-brightness projectors, 14 simulation workstations, and a super surround-sound system with up to 21 tracks. The linking of the image across three screens surrounds the audience's entire field of vision with high-definition digital images, creating a unique visual impact. The surround-sound stereo effect has the ability to reproduce sounds from the real scene for a highly accurate and immersive experience. The film describes the roots of the "24 Solar Terms", their relationship to celestial phenomena, and their important role in guiding agricultural production.

4. Conclusion

When the author was a teenager, their father taught them "24 Solar Terms Songs", which opened up a new perspective in their perception of the world. The practical appeal of cultural heritage protection evokes the cultural consciousness of younger generations, and establishes cultural confidence in a move towards cultural self-improvement.

The "24 Solar Terms" represent a relationship between mankind and nature. Modern life keeps us away from nature. According to incomplete statistics, alongside social and economic development, the proportion of agriculture has gradually declined, with China's agriculture accounting for only about 15% of GDP. In the next 10 to 20 years, China's economic growth will accelerate, and industrial society will replace agricultural society. It is possible that the "24 Solar Terms" will be forgotten. This should not be allowed to happen, as China is an ancient civilization with a rich history and culture. The traditions left behind by older generations are worth recollecting and studying in modern generations.

The "24 Solar Terms" concept, which is currently applying for World Heritage status, combines astronomy, climate, farming, and folk customs, having been used for thousands of years. Especially in the pursuit of a three-way win-win situation in modern China for "production, life, and ecology", the ancient wisdom contained in the "24 Solar Terms" is of great value. Compared with the recent hotly discussed Korean application for its warm brick bed, the "24 Solar Terms" is richer in culture. Its contemporary nature and reference to the development of other countries around the world are also more profound.

The World Heritage application is being made by the China Agricultural Museum. To this end, the "three-screen theater" is one of the museum's renewal projects. It is a film and television platform unique both in China and abroad, as well as a core broadcast engine platform with independent intellectual property rights. The museum display uses digital technology as an aid in order to highlight the content and cultural significance of the "24 Solar Terms". The choice of three-screen cinema technology is the icing on the cake. Developments in digital technology certainly provide improved applications for museum display, and the China Agricultural Museum will do its bit to protect the "24 Solar Terms".

References

[1] Striving to improve the digital and intelligent level of the museum — An interview with Lin Chengbin, Director of the Information Center of the China Agricultural Museum. *IB Intelligent Building and Urban Information*, 2004, 4.

[2] Zhang Aiwu. 'Modernization of traditional Chinese culture', *Journal of Henan University of Technology (Social Science Edition)*, 2010, 2.

[3] Li Yifan. 'The application and design of new media in museum exhibitions'. *Communication University of China Press*, 2009, 3.

[4] Ma Shuangshuang. 'Research on the development of Chinese museums'. *Northwestern University*, 2014: 2–3.

*Exploring the Development of a Taxonomy System for the Standardized Digitization Protection of Cultural Relics**

Luo Wei
(Beijing University of Chemical Technology, Beijing 100029, China)

Yi Junkai
(Beijing University of Chemical Technology, Beijing 100029, China)

He Kunpeng
(Beijing Gehua Group, Beijing 100007, China)

Abstract: By investigating the status of the R&D of various taxonomies in China and abroad, this paper analyzes the development processes of multiple taxonomies, tabulation rules, inter-word structures, and related information. We reach certain conclusions about the necessary inter-word structure in the preparation of a taxonomy. Then, following Chinese and foreign tabulating standards, and the classification standards for Chinese museum collections, we combine the characteristics of the current status of the domestic museum collection classification and China's cultural artefact system itself, referring to research into certain classification methods for cultural artefacts. Finally, we propose a development process for a taxonomy system for the standardized digitization protection of cultural artefacts, including the methods and techniques involved in each part of the process.

Keywords: taxonomy; cultural artefacts; standards; classification

* Funded project: "Research on the Framework of a Cultural Relic Digital Protection Standard System" (2014BAK07B01) funded by the National Science and Technology Support Program.

Preface

The digital conservation of cultural artefacts is an important method for regenerating artefact conservation. The huge amounts of data generated in response to this are posing a challenge to traditional methods of cultural artefact document information retrieval. At present, there are no taxonomies in the cultural artefact field for use in China. It is difficult to formulate standards for the theme words in the description, labelling, and retrieval of a great deal of documentation and other information resources in the cultural artefact domain, making it hard to share information resources regarding domestic cultural artefacts. This is not conducive to the long-term, sustainable development of digital conservation work in China.

Nowadays, the information for the registration and classification of museum collections in China is independent for each museum. It is therefore difficult to form normative standards, with inconsistent descriptions of similar collections across museums of different levels. This poses an obstacle for information exchange between museums. The development of a unified taxonomy for collection information will achieve the unification of the information resources in the cultural artefact field, including museum collection information. Therefore, the development of a taxonomy for the digital standardization of cultural artefacts is an urgent task to provide standardized retrieval tools for information resources for all cultural artefact protection and research units. It is urgently needed for the realization of digital sharing of cultural artefact resources, and also a necessity for the unification of standards and norms for national cultural artefact resources.

Developing a taxonomy for the cultural artefact digital standard system, and providing standardized retrieval and indexing tools for related cultural artefact fieldwork, helps to solve the issues of inconsistent conservation methods, repeated information, and information islands that occur in the process of digital cultural artefact conservation. It is also conducive to the establishment of a system for the protection of digital cultural artefact information and information resource sharing, offering basic support for the comprehensive elevation of the level of digital protection, artefact conservation, management, and utilization.

1. Related research

1.1 Status of the compilation of major taxonomies in China and abroad

1.1.1 Status of taxonomy compilation in China

The Chinese Thesaurus is the first large-scale comprehensive language word list for information retrieval in China, covering all disciplines, with a large collection of words and style specifications. This taxonomy plays an extremely important role in the promotion of the development of subject indexing work in China, bolstering the establishment of computer literature databases and the organization, development, and improvement of a professional taxonomy [1].

The Chinese Classification Thesaurus is an integrated search vocabulary compiled on the basis of *The Chinese Library Book Classification* and *The Chinese Thesaurus*, combining a classification index and a subject index. This taxonomy is a new, exploratory, and creative search vocabulary, which is of great significance to the library business and intelligence industry [2].

The Aeronautical Science and Technology Information Thesaurus is China's first topical taxonomy. It is the result of long-term exploration and research into search vocabulary for China's aviation science and technology industries, allowing descriptor language to be promoted and widely applied within the information system [3, 4].

The Social Science Search Vocabulary is a taxonomy that regulates the concepts of social science terminology, providing a basic reference book with which social science research institutions can conduct document indexing and retrieval work, as well as establishing a computer information management system [5].

The Chinese Archive Thesaurus is a dictionary consisting of standard words that reflect archive contents, and a word control tool for the indexing and search of subject words in archives, which can be used by archive, record, and document processing departments for the indexing and retrieval of files, documents, and materials [6].

The vocabularies of the *National Defense Science and Technology Thesaurus* [7], *The Military Thesaurus* [8], *The Agricultural Thesaurus* [9], and *The Chinese Chemical Thesaurus* [10] are aimed at their corresponding fields, providing search indexing tools for the specification and application of vocabulary.

Through research and analysis of the aforementioned vocabulary editing processes and methods, the vocabulary structure is concluded through the combination of professional needs and use in related fields. Specific vocabulary details are shown in Table 1.

1.1.2 Status of foreign taxonomy compilation

The Library of Congress Subject Heading (LCSH) is a vast, comprehensive collection of subject headings. It is the largest and most widely used in the world, occupying an important position in the history of search language development and current library subject cataloging [11].

The UNESCO Thesaurus was designed for UNESCO's Computer Documentation System (CDS). It is used for the indexing and retrieval of various documents and publications, with UNESCO documents, publications, and collections, and is also used by libraries and intelligence agencies with collection scopes similar to UNESCO [12].

The NASA Thesaurus is a taxonomy compiled by the National Aeronautics and Space Administration. The vocabulary index has been used in *The Aeronautical and Aerospace Science and Technology Report* (STAR), *International Aerospace Abstracts* (IAA) and the NASA Science and Technology Information Database (NASA STI) [13].

Table 1 Specific Methods and Rules of Various Thematic Vocabularies

	Vocabulary categories						*Indexing*					
	Theme	*Facet classification*	*Semi-faceted classification*	*Integrated theme classification*	*Primary table*	*Appendix*	*Category index*	*Word family index*	*Wheel index*	*Comparative index of other languages and Chinese*	*Other index*	*Inter-word relationship*
Domestic thesaurus												
Chinese thesaurus	√				√	√	√	√		English-Chinese		YDFSZC
Chinese classification thesaurus				√								
Aviation science and technology thesaurus	√				√		√	√			Subject word index/ model theme index	YDFSZC
Social science search vocabulary	√				√		√		√	English-Chinese	Index of proper nouns	
Chinese archival thesaurus	√				√		√				First stroke check index	YDS
National defense science and technology thesaurus	√				√		√	√		English-Chinese		YDFS
Military thesaurus	√				√	√	√			Chinese-English/ English-Chinese		
Agricultural thesaurus	√				√		√			English-Chinese		YDFSC
Chinese chemical thesaurus	√				√		√			English-Chinese		YDFSC

	Vocabulary categories						*Indexing*					
	Theme	***Facet classification***	***Semi-faceted classification***	***Integrated theme classification***	***Primary table***	***Appendix***	***Category index***	***Word family index***	***Wheel index***	***Comparative index of other languages and Chinese***	***Other index***	***Inter-word relationship***
Foreign thesaurus												
LCSH	√				√							YDFSC
UNESCO thesaurus				√	√			√	√		Classification descriptor word list	DFSZC
IRTE	√				√		√		√			YDFSZC
TSTT	√				√		√	√	√			
TJCST	√				√		√					YDFSZC
NASA Thesaurus	√				√				√			
Root Thesaurus	√				√		√					
Subject Heading For Engineering	√				√		√				Heteronym comparison list	YDFSC
TEST	√				√		√	√	√			YDFSC
MESH	√				√		√					
Thesaurus of Facets		√			√						Facet classification	

The Thesaurus of Engineering and Scientific Terms (TEST) is a taxonomy for engineering and science intelligence systems. The NTIS (National Technical Information Service) bibliographic database (BDB) uses its vocabulary indices [14], and the National Technical Intelligence Service of the US Department of Commerce was first to use the TEST keywords, which are currently applied by many government agencies and civil society.

Other important topic vocabularies in the international realm also include the *Information Retrieval Thesaurus for Education* (IRTE), the *Thesaurus of Scientific and Technical Terms, the Thesaurus of the Japan Information Center of Science and Technology, Root Thesaurus, Subject Headings for Engineering, Medical Subject Headings* (MESH), and the *Thesaurus of Facets* [15].

Analyzing and using the rules and methods of foreign subject vocabulary editing will contribute to more comprehensive and systematic development of Chinese taxonomies, as well as understanding detailed rules for the compilation process of aforementioned foreign topic vocabularies, also shown in Table 1.

1.1.3 Comparison of taxonomy compilation in China and abroad

In comparing the nature, main tables, appendices, index layouts, and inter-word relationships of vocabulary rules in taxonomies in China and abroad, specific methods and rules for various taxonomies are shown in Table 1.

TSTT in Table 1 is an abbreviation of the *Thesaurus of Scientific and Technical Terms*, while TJCST is an abbreviation of the *Thesaurus of the Japan Center of Science and Technology*. These are not internationally approved abbreviations, but are convenient for comparison purposes.

Given the above comparison points, the following conclusions can be drawn: most taxonomies use topic types, while a few use faceted, semi-faceted, or classification subject integration methods; each vocabulary has a main table (word table), most use a category index, some use a word family index, a few use a rotated terms index, and some use a targeted index based on characteristics pertaining to a specific professional field; almost every vocabulary shows a relationship between terms (Y), generation (D), components (F), genus (S), family (Z), and reference (C), in the foreign taxonomies the relationship between words uses (USE), natural language processing (UF), narrower term (NT), broader term (BT), related term (RT), top term (TT), also using (Y), generation (D), components (F), genus (S), family (Z) and reference (C) in comparison.

1.2 Correlating taxonomy standards in China and abroad

1.2.1 ISO 25964 [16]

ISO 25964 is an international standard for taxonomies, consisting of two parts: information retrieval of the topic vocabulary, and interaction with other vocabularies. The first international taxonomy standard was ISO 2788, published in 1974, and in 1985 the supplementary standard ISO 5964 was added. In 2011, ISO 2788 and ISO 5964 were cancelled and replaced by ISO 25964.

1.2.2 US standard ANSI/NISO Z39.19 [17]

ANSI/NISO Z39.19 provides rules for content, presentation, construction, testing, maintenance, and management of multilingual vocabularies. The standard focuses on controlled vocabularies presenting KOS (knowledge organization systems) content, which contains lists, synonym rings, and taxonomies. The main purpose of vocabulary control is to achieve coherence in content object descriptions and facilitate information retrieval. It is implemented using three methods: defining scope, annotations, and terminology; using equivalence relationships to link synonyms and synonym groups; and distinguishing among homographs.

1.2.3 British standard BS 8723 [18]

British standards BS 5723 and BS 6723 are identical to ISO 2788 and ISO 5964. Following years of development, the new standard BS 8723 (which combines BS 5723 and BS 6723) emerged in 2008 consisting of five parts: definitions, symbols and abbreviations, taxonomies, vocabularies outside the taxonomies, interoperability between vocabularies, and exchange of formats and protocols.

1.2.4 GB 13190-91 [19]

The Chinese rules for the preparation of taxonomies give the definitions of related terms and explanations of inter-word relational symbols, proposing word classification control, and principles designated by descriptors such as term selection range and related methods, and inter-word relationship processing. The standard specifies the structure of the taxonomy and the rules for the main table, appendices, and index, and the further management and revision of the vocabulary.

1.3 Classification status of Chinese museum collections

In the research into classification methods for museum collections, many scholars, experts, and organizations have conducted in-depth discussions and studies, proposing a variety of practical methods. In 1914, Zhang Jian divided all collections into 31 categories under four main groups; three categories of natural products, 14 categories of history, 11 categories of art, and three categories of education. This laid the foundation for collection classification within the Chinese museum community [20]. In the 1950s and 1960s, Fu Zhenlun conducted a related study on the methods of collection classification according to texture in his book *The Storage of Museum Collections*. In the 1980s, Song Boyin and He Zhigang proposed the *Four-Four Decimal Categorization System* [20] and the *Three-Three*

Categorization System. In 1985, the State Administration of Cultural Heritage presented a discussion paper titled *The Classification of Museum Collections*. At the second academic seminar of the China Museum Society's Professional Custodial Committee in 1988, a discussion paper was proposed on the classification of museum collections. Today, most museums adopt the concept of categorization by object, and classification by quantity, using types and quantities of cultural artefacts within the collection as a classification standard [21]. Precisely because each museum has its own classification method, it is difficult to form unified standards, which in turn leads to difficulty in the sharing of national cultural artefact information, resulting in industry issues such as information islands.

Cultural artefacts are the legacy of humanity in the course of historical development, including all historical and cultural artefacts both movable and immovable. They are not limited to ancient times, but are inclusive of the modern and contemporary [22]. Cultural artefacts are grouped together according to certain standards, and different categories are formed, with cultural artefacts in different categories clustered further in a hierarchy. Classification methods for cultural artefacts include time classification, form classification, texture classification, utility classification, attribute (property) classification, source classification, and value classification [23]. Due to the complexity of cultural artefacts, research into the methods of cultural artefact classification can be used as reference for the classification of categories in taxonomies, combined with research into museum collection classification systems.

1.4 Classification standards for Chinese museum collection

In 2001, the National Museum of Cultural Relics issued the *Museum Information Index System Specification (Trial)*, including three indicator groups, 33 indicator sets, and 139 total indicators, covering aspects related to collection ontology, management, and research. The *Standards for Describing Information Indexes of Precious Cultural Relics in Library Collections*, implemented during the comprehensive promotion phase of the cultural artefact survey project, set 28 core indicators for each collection login, including the item name, age, category, level, completeness, size, quality, characterization, and conservation record.

The above standards and norms are based on current laws, regulations, and regulatory documents, laying the foundation for the standardization and normalization of collection registration. However, there are still some indicators in these specifications that are not detailed enough to meet digital requirements. Based on the reference to the norms mentioned above, the category of taxonomy tables created in this paper hopes to further revise and improve indicators, and complete comprehensive and accurate implementation.

2. *The Development of a Taxonomy System for the Standardized Digitization Protection of Cultural Relics*

2.1 Overview

The taxonomy created in this paper is based on the study of taxonomy development in China and abroad, investigating and analyzing the classification status of Chinese museum collections and the classification methods of cultural artefacts. With strict adherence to Chinese and foreign taxonomical standards, laws, and regulations, as well as regulatory documents and domestic museum collection classification standards, this paper puts offers suggestions for tabulation specifications, vocabulary structure, and vocabulary use in the taxonomy system for the standardized digitization protection of cultural relics. The taxonomy established by this paper combines the characteristics of the cultural artefact system and domain, covering all of the content in the cultural artefacts system, and defining the relation and concept relationship in the taxonomy according to the relevance relationships produced by all areas involved in every segment of the cultural artefact domain, to achieve semantic relevance among inter-word relations, thus raising the recall rate and accuracy during retrieval.

2.2 Overview of the process

We first determine the scope of the vocabulary and the principles of term selection, selecting official and informal keywords on this basis, and arranging all formal and informal keywords in order of Chinese Pinyin to form a term list. According to related research into cultural artefact classification, theme words from categories of the same nature under the same category are clustered into a hierarchy forming a category index, and all subject terms are translated into English according to translation standards, constituting the English-Chinese index. Since some proper nouns such as artefact locations, ages, and characters are quoted in large amounts, the formation of an attached list will be required.

A flow chart demonstrating the development of the taxonomy compiled in this paper is shown in Figure 1.

2.3 Implementation rules for the taxonomy

2.3.1 The taxonomy system for cultural relics

Vocabulary sources of the taxonomy established in this paper are selected from various cultural artefact vocabularies used by museums and cultural artefact research units such as the National Cultural Relics Bureau, the National Digital Museum, the National Palace Museum, and the Dunhuang Institute. It also uses the National Bibliographic Database, the Chongqing Science and Technology Periodical Literature Database, and the Shanghai

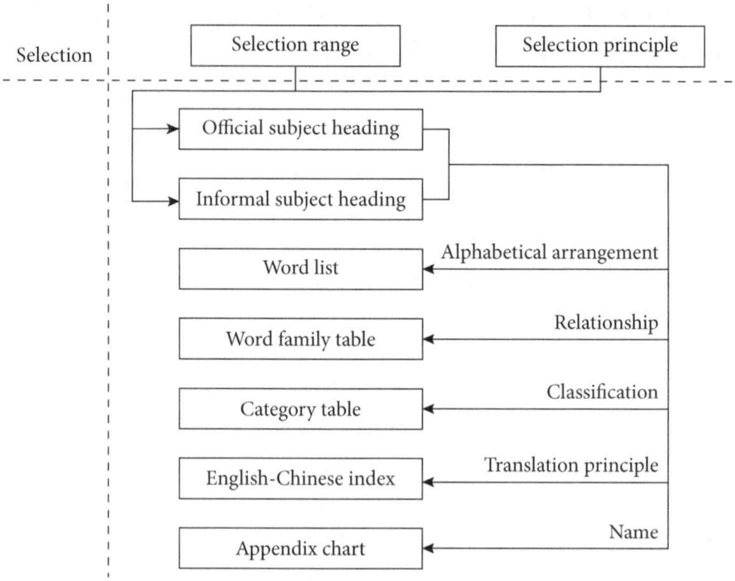

Figure 1 Flow chart explaining the construction of thematic vocabulary for the cultural relic digitization protection standard system

National Newspaper Index Database, as well as traditional paper documents and other resource libraries as data sources for the indexing of statistical subject terms. It is used to extract keywords or key fields of documents in the field of digitized cultural artefacts and all resource banks, using the frequency of indexed terms to determine selected interpretation values and term selection standards (the thresholds of books and journal libraries are determined separately). When selecting words from a range of terms, it is necessary to follow the principles of word selection, standardizing Chinese semantics such as word meaning, word class control, and phrase descriptor, to avoid complicated redundancy in the semantic environment and realize the accurate unification of selected subject words.

2.3.2 Rules for the taxonomy's main table (term list)

All subject terms obtained via selected terms include formal and informal keywords, each subject term corresponds to a taxonomy item, with the sequence of items arranged in the same form according to homophones in the Chinese phonetic alphabet to form the main table, which is the alphabetic list. The main table forms the body of the subject table. The taxonomy item is a basic unit of the main table, consisting of Chinese Pinyin, item terms, foreign language translation, reference items, and comment items. Informal descriptors cannot be used for indexing and retrieval, but are guide words for formal descriptors.

The format of the formal taxonomy is as follows:

			shanshuihua		← Chinese pinyin
	Entry term	→	山水画 (Landscape painting) [A06]		← Item category number
Reference items	Substitute symbol	→	D	Landscape	← Informal subject heading
	Item symbol	→	F	Green landscape Blue-green landscape Ink-wash landscape Light colour landscape 'Mogu' (no-outline) landscape	Key item words
	Attribute symbol	→	S	Traditional Chinese painting *	Key attribute words
	Family symbol	→	Z	Traditional Chinese painting *	← Family words
	Group symbol	→	C	Figure painting Flower-and- bird painting	← Related words
	Item comment	→		(Taking natural landscape of mountains and rivers as the main descriptive object)	

The format of the informal taxonomy is as follows:

		Zhongguohua		Informal word in Chinese pinyin
Item	→	中国画 (Chinese painting) [A06]	←	Informal item and category number
Item symbol	→	Y 国画 (guohua)	←	Official keywords

2.3.3 Family index

The word family index is a word family system displaying formal subject terms with affiliation relationships and partial and overall relationships in the main table classified according to regulations. The first term is a conceptual word that summarizes the head of a family of subject words. It is listed at the head of a family, followed by a * to indicate differentiation, sorted by Chinese Pinyin. Subject terms at the lower end of the family are arranged in a hierarchical manner according to concepts, and the hierarchical relationship is represented by the number of "•" present. The word family index and term list are based on Chinese Pinyin, and are linked by the "*" sign on the head word.

An example of a word family in the taxonomy is as follows:

guohua
Chinese painting*
- Figure painting
•• Taoist and Buddhist painting
•• Maiden painting
•• Portrait
•• Paintings of customs and traditions
- Landscape painting
•• Green landscape
•• Blue-green landscape
•• Ink-wash landscape
•• Light colour landscape
•• 'Mogu' (no-outline) landscape
- Ink wash painting
- Imperial court decorative painting
- Traditional Chinese fine brushwork painting
- Flower and bird painting
•• Fine brushwork flower and bird painting
•• Freehand flower and bird painting
••• Large freehand flower and bird painting
••• Small freehand flower and bird painting

2.3.4 Rules for category tables

The structure of the taxonomy's category table is based on the classification status of domestic museum collections and cultural artefact classification methods, referring to the classification index items of the *Museum Information Index System Specification (Trial)* for all keywords in the main table. According to research into cultural artefact classification criteria and requirements in the field mentioned above, terms are divided into three categories, and terms belonging to each category are ranked directly within the category in accordance with the words or the system. First-level categories can be divided into several major categories, each of which is divided into a number of secondary categories. The secondary categories are divided into three more categories, and the third-level categories display subject headings. Informal subject headings in the index refer to corresponding formal keywords with "Y", and the first term of a family is suffixed with "*".

2.3.5 The English-Chinese index

The English-Chinese index is used to index and retrieve English literary materials. The search sorts the English translations of all keywords in the main table in alphabetical order, with Chinese subject headings underneath. Informal keywords use "Y" to indicate formal keywords.

An example of the English-Chinese index in the taxonomy is as follows:

Bronze
铜器
China
瓷器
Chinese painting
中国画
Y 国画
Jade
玉器

2.3.6 Rules of the attached table

In the attached table, proprietary vocabulary such as age, region, Chinese cultural heritage, national key cultural artefact protection units, and historical and cultural sites (city/town/village/street) can be compiled into a separate table. The rules of the table are consistent with and complement the main table.

3. Summary

This paper proposes processes for the development of a taxonomy system for the digital protection of cultural artefacts, and the method of its vocabulary compilation. A lot research and work is still needed for the selection of specific subject terms in the taxonomy, vocabulary combination, and the detailed compilation of each vocabulary term. In the process of taxonomy compilation, we must also attend to its development towards ontology, producing forward-thinking work for the future development of taxonomies, in order to adapt to trends of the Semantic Web, while preparing for vocabulary maintenance and updates.

References

[1] China Institute of Science and Technology Information, editor of the Beijing Library. *Chinese Thesaurus* [M]. Beijing: Science and Technology Literature Publishing House, 1980.

[2] "Chinese Library Book Classification" Editorial Board. *Chinese classification thesaurus* [M]. Beijing: Huayi Publishing House, 1994.

[3] Compilation of the 628th Institute of the Ministry of Machinery Industry. *Aviation Science and Technology Information Subject Headings* [M]. The 628th Institute of the Ministry of the Machinery Industry, 1964.

[4] Zhang Kuiqing, Zeng Yaping, Qiu Zubin. 'Exploration and practice of aviation science and technology information retrieval language — the Aerospace Science and Technology Information Symposium 50th Anniversary', [J] *Intelligence Theory and Practice*, 2014, 10: 1–5.

[5] "*Chinese Archives Thesaurus*" editorial board. Chinese Archives thesaurus [M]. Beijing: Archival Press, 1988.

[7] National Defense Science and Technology Commission Intelligence Research Institute. *National Defense Science and Technology Thesaurus* [M]. 1978.

[8] *Military thesaurus* editorial board. Military Thesaurus [M]. Beijing: Military Science Press, 1990.

[9] Chinese Academy of Agricultural Sciences. *Agricultural Thesaurus* [M]. 1983.

[10] Editorial committee of Chinese keyword forms for the petrochemical industry archives. *Chinese Chemical Thesaurus* [M]. 1984.

[11] Library of Congress Subject Headings, 2001, 24.

[12] Hou Hanqing. 'Review of the UNESCO Thesaurus', [J] *Intelligence Science*, 1984, 4: 8–14.

[13] *NASA Thesaurus* [M] United States: Scientific and Technical Information Office, National Aeronautics and Space Administration, available through the National Technical Information Service, 1988.

[14] Wang Shaoping, Chen Zhaoshan, Chen Zhongming. *Dictionary of Library and Information Science* [M]. Shanghai: Chinese Dictionary Press, 1990, p. 20.

[15] Li Runyuan, Li Hongen. '20th century thesauri with great influence in China and abroad', [J] *Dunhuang Research Institute*, 2009.

[16] ISO 25964. Information and documentation. Thesauri and interoperability with other vocabularies. Thesauri for information retrieval [S] 2005.

[17] ANSI/NISO Z39.19. Guidelines for the Construction, Format, and Management of Monolingual Controlled Vocabularies [S] 2005.

[18] BS 8723. Structured vocabularies for information retrieval[S] 2005.

[19] GB13190-91. Chinese thesauri preparation rules [S] 1991.

[20] Song Boyin. 'On the classification of museum collections — Also on the "Four by Four Category Classification"', [J] *Southeastern Culture*, 1991, 6:326–337.

[21] Zhu Jingguo. 'Research on the classification of museum collections', [J] *Chinese Museum*, 1991, 29: 30–35.

[22] *China Encyclopedia — Cultural Relics and Museum* [M]. Chinese Encyclopedia Publishing House, 1993.

[23] Peng Tao. 'The construction of cultural artefact information databases and the preliminary study of cultural artefact classification', [J] *Shanghai Archaeology and Museology Review*, 2007, 2: 42–46.

Digital Development and Resource Integration of Temporary Museum Exhibitions

Li Shuangjiang and Chen Ying
(Information Center of the China Agricultural Museum, Beijing 100026, China)

Abstract: The temporary exhibition is an important exhibition mode for museums. Digital developments can make the "temporary" exhibition hall "permanent" on the network, and can also indefinitely expand and extend audience reception and viewing time. This article will explain the practices of the China Agricultural Museum in the digital development of temporary exhibitions, summing up the lessons learned and demonstrating how to use and share resources in the digital development of temporary exhibitions.

Keywords: temporary exhibition; digitization; sharing

1. The significance of the digitization of temporary exhibitions

A museum has a variety of resources, including collection resources and exhibition resources. Some large museums hold several temporary exhibitions each year, so the temporary exhibition is an important resource and exhibition mode. Compared with long-term exhibitions, the distinctive feature of temporary exhibitions is a short duration, usually a matter of days or months. Visitors are only able to visit in a relatively short and concentrated time period, after which the exhibition no longer exists. As a result, various adverse effects are possible. Short exhibition times may result in too many visitors per day, affecting each persons' visit. Moreover, a short exhibition time may mean that some people are unable to visit the exhibition before the time expires, resulting in regret and no alternative solution. The dismantling of the exhibition after a short period of time is not conducive to maximizing its function, and also causes a great waste of resources. Many exhibitions require specific booth constructions and often special effects, which, once removed, create a pile of discarded materials of little value.

How to uncover the maximum value of temporary exhibitions is an issue that museum staff must consider and solve. This way, people who do not have the opportunity to visit the site may also experience these exhibitions, and those wishing to see the exhibition can do so at any time. Modern, high-speed developments in computer network technology provide an answer and a solution, that is, the digitization of the exhibition. In this way, an exhibition can be preserved digitally when the physical exhibition has been taken down. Not only that, but digital development of temporary exhibitions and resource integration are important methods for resource sharing. Once the temporary exhibition is digitized, resources can be fully utilized to realize the maximum function of the museum.

2. The process of digitizing a temporary exhibition

The creation of temporary exhibitions includes the digitization of basic elements and the integrated development of resources. Digitizing basic elements focuses on the preservation of exhibitions in the form of images, videos, and sounds, and retaining original materials. Integrated resource development refers to the reorganizing and re-processing of various materials.

The specific processes can be divided into three stages, namely preparatory work in the early exhibition stage, digitization during the exhibition period, and information collection and resource integration in the middle and latter stages of the exhibition. The first two stages focus on the collection of data and basic information, while the later stage focuses on development, using professional software, adding design elements, making complete works, and realizing wider dissemination and sharing of resources through the network and CDs.

In the early stage of the exhibition, the main tasks include actively communicating and coordinating with organizers, and obtaining a variety of exhibition-related information including text, pictures, and videos, collecting and organizing materials as the foundation for later digital development. Some scripts and design content can be obtained before the launch, and when there is sufficient time, multimedia production for the exhibition can be completed in advance, introducing the layout of the exhibition and related content. This can be placed in the exhibition hall to facilitate search and inquiry, as well as enriching the form of the exhibition.

The exhibition period involves an important stage of digitization, with most exhibition information obtained during this stage. The main task is to complete information collection related to the physical exhibition, that is, the digital processing of the exhibition itself, including photos, videos, and even media reports. These resources are prerequisites for the digital development in the third phase, and can only be obtained at the time of exhibition. Once the exhibition is finished, information collection is no longer possible, so staff must be sure to do these things. Professional camera personnel must be sent to the scene to shoot the exhibition, record videos, collect relevant information, and save the exhibition information as comprehensively as possible, laying the foundation for the development and integration of resources.

The third stage is to focus on the development of exhibition information in the shortest time, striving to complete the virtual exhibition before the end of the physical exhibition, providing the audience with a new experience while relying on the promotion of the physical exhibition.

The China Agricultural Museum has been attending to the digital development of temporary exhibitions for more than a decade now, accumulating rich experience and exploring a variety of development forms.

3. *Specific development methods and forms*

The practice of digital development of temporary exhibitions at the China Agricultural Museum is mainly represented in the two forms of panoramic roaming and virtual exhibitions. From the perspective of means of communication, both forms can be displayed online without the restrictions of time and space, with visitors able to browse anytime, anywhere. In addition, in order to meet a variety of needs, the China Agricultural Museum has also made multimedia CDs for some exhibitions, the advantage of which is that they can be played locally. They are affected less by file size, and the amount of loaded information is larger.

3.1 Roaming

Panoramic exhibition hall roaming is accomplished through professional 3D panorama software, realized on the basis of virtual reality technology. 3D panorama simulates an interactive, imaginary 3D space scene with technology and equipment, simulating a 2D planar picture into real 3D space, presenting it to the viewer, and providing various functions for scene manipulation, such as zooming in and out in the exhibition hall scene and moving the scene in all directions to reproduce the effect of the real scene environment.

The production of panoramic exhibition hall roaming requires the use of professional equipment and software. First, a fisheye lens camera is used to shoot the exhibition hall to obtain panoramic photographic information. Secondly, panorama processing software is used to carry out piecing treatment and quickly generate a spherical or cuboid panorama. Finally, the spherical or cuboid panorama can be manipulated to create a full-scale interactive real scene, designing a variety of different styles of 3D panoramic roaming, to provide an immersive, 360° view experience.

The stand-out feature of the 3D panorama is the synthesis of various multimedia technologies such as text, image, video, and audio to simulate the actual environment so that visitors feel as if they are exploring the real exhibition hall. In addition, simple operation, low cost, short production cycle, and low resource consumption are also major advantages of the 3D panorama. It is a technology suited to the autonomous digital development of temporary exhibitions for museums.

3.2 Virtual exhibitions

The production of virtual exhibitions is more complicated and more personalized than the panoramic roaming of exhibition halls, necessitating the reproduction of exhibitions not simply as they are, but with added design elements on the basis of physical exhibitions, emphasizing comprehensive and detailed display of exhibition content, with a specific development process as follows:

1) Photographing exhibits and obtaining basic information
This is completed during the course of the physical exhibition by professional camera operators. The exhibits are photographed separately, obtaining original materials including textual introductions to the exhibits and other detailed information, with the purpose of completing digitization and saving all the information from the displays.

2) Obtaining other multimedia resources for exhibition applications related to displays and exhibits
Audio, video, animation, and other resources related to the exhibition theme are collected and organized, including various multimedia resources used in the exhibition hall. These resources enrich the display form of the virtual exhibition, and also expand the exhibition content to a certain extent.

3) Design of the virtual exhibition display form
Good design can provide enjoyment while highlighting the exhibition's theme. Virtual exhibition design, including interface design and exhibit display, is the blueprint for virtual exhibition production.

4) Processing original material and information, and completing the prototype of the virtual exhibition
Original information is processed to generate the various elements required to create a virtual exhibition, including the use of professional software to create a 3D exhibition hall panorama, processing exhibit images, and multimedia resources such as video and animation.

5) Completing virtual exhibition production according to the design to integrate various materials, including the addition of links, inlaid text, pictures, and videos.

6) Making modifications, perfections, and complete debugging to achieve online exhibitions

Virtual exhibitions and panoramic roaming are two different forms of temporary exhibition digital display. Panoramic roaming focuses on the reproduction of the real exhibition hall. It is simple and fast, and can be launched in the short term. Virtual exhibitions involve a variety of links, focusing on content expansion and diverse display means with added design elements. Both have their own advantages, and can be used to complement one another in the digitization process of temporary exhibitions.

4. Case study

The China Agricultural Museum began digitization early, and has done a lot of work in the last 10 years. At present, five basic exhibitions are available for Internet browsing, meaning that people can visit the museum without leaving their homes. Many distinctive temporary exhibitions have also been retained in digital form, to be reviewed at any time. For example, large-scale temporary exhibitions such as the Sannong Photography Exhibition, the China Agricultural Culture Exhibition, and the Rural Policy Since 1949 Retrospective Exhibition have employed professional software to create panoramic exhibition hall roaming, completely preserving the original exhibition appearance. With software upgrades and continuous technological improvements, the performance is impressive, not only showing the scene at the time of exhibition, but also adding a variety of digital means of display.

Taking the China Agricultural Culture Exhibition as an example, panoramic roaming of the exhibition hall has been realized, and various parts of the exhibition hall can be viewed at 360°, with the display form enriched on past foundations. Previously, panoramic roaming of the exhibition hall was completely static, playing only automatically or by manually selecting the view of a specific exhibition hall, showing a static image and no rich multimedia resources. It was fully panoramic and lacking in detail, not providing a clear view of every item in the exhibition hall.

Since 2009, breakthroughs have been made in panoramic exhibition hall roaming, adding various details and multimedia display elements, and embedding sound to make images more dynamic and realistic. The addition of hotspot links enhances visitors' appreciation of specific exhibits (see Figure 1), as they can click on corresponding hotspot links to zoom in and view corresponding content. This is not simply a panorama of the "plane" but also the "point", fully displaying details within the panorama (see Figure 2). Video and other enhanced display effects are also added to show the real exhibition hall, which can be embedded in the panorama (see Figure 3) and controlled by switches, to be opened and viewed at will, completely recreating the real exhibition hall. The hall can also be developed even further on this basis, embedding certain content not found in the real exhibition hall.

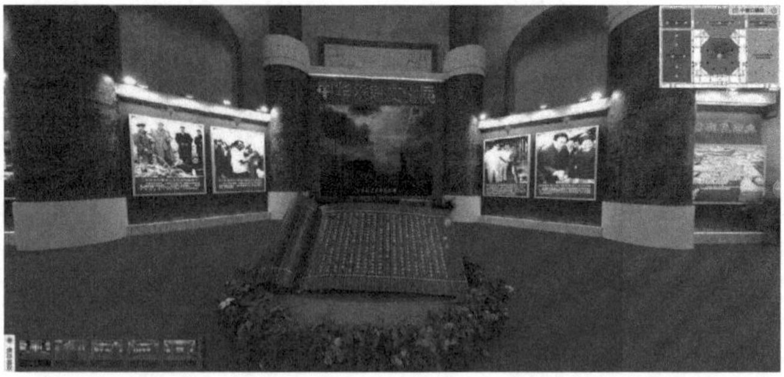

Figure 1 Panoramic roaming – clickable pictures

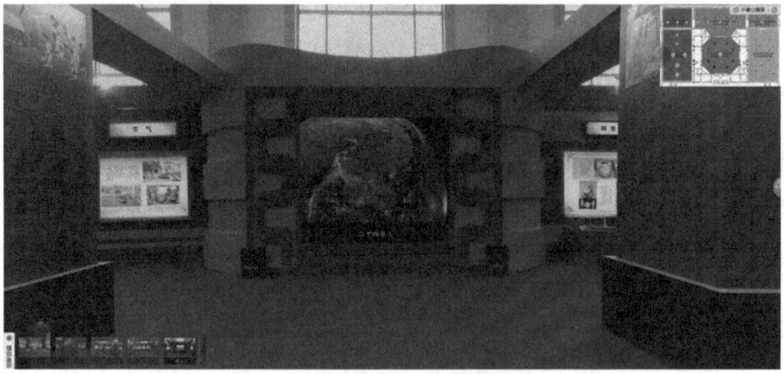

Figure 2 Panoramic roaming – video playback

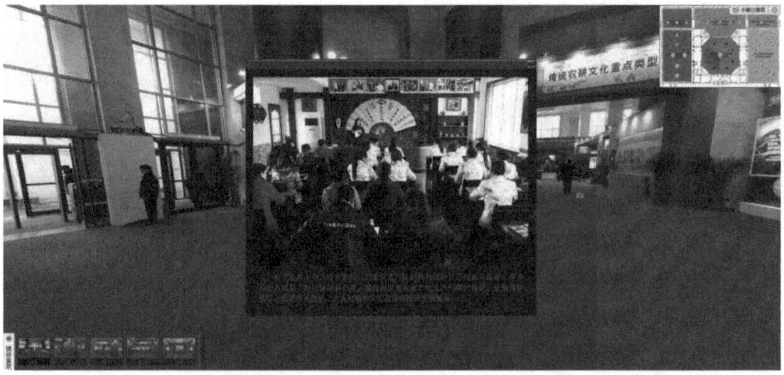

Figure 3 Panoramic roaming – click to enlarge the picture

In addition to the panoramic roaming of the exhibition hall, the virtual exhibition highlights digital development at the China Agricultural Museum. The development of the virtual exhibition requires the cooperation of various museum departments, relying on physical exhibitions (design scripts, design concepts, and design renderings), with further design and processing on this basis, expanding related content. Images of the Sannong Photo Exhibition in 2009 are shown in Figures 4 and 5, with pictures displayed in the form of a virtual wall – a new form designed according to the physical exhibition, with the exhibition's content divided into seven parts as per the form of the original exhibit. Visitors can click on a picture on the wall to zoom in, giving an impression of the real exhibition hall, and a useful exploration of the virtual display form.

More recently, the China Agricultural Museum has been experimenting with new virtual exhibition forms, integrating resources into virtual exhibitions and disseminating information via networks and CDs, realizing digital development and resource sharing.

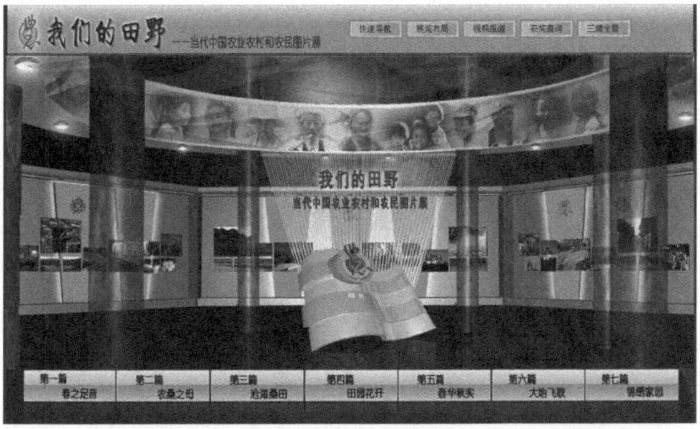

Figure 4 Sannong Photo Exhibition homepage

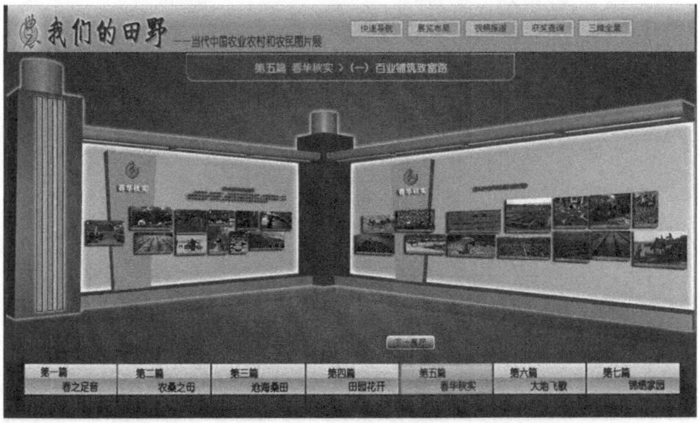

Figure 5 Sannong Photo Exhibition contents page

5. Summary

Temporary museum exhibitions often require a large amount of manpower, material resources, and financial backing. Once the exhibition is complete, it disappears, resulting in a great waste of resources. Therefore, it is often difficult to achieve the perfect unity of funding and social benefit for temporary exhibitions. Due to factors such as venue restrictions, these exhibits cannot be displayed for a long time The best way to resolve this contradiction is long-term preservation and networked display via digital forms. This kind of preservation is realized in two ways: first, the exhibition itself and related resources are further developed and integrated to form a virtual exhibition, so as not to be subject to time and space constraints; second, digitized material content has become a new source of wealth and resources for museums, not only preserved, but also conducive to continued dissemination and sharing.

References

[1] Qi Yue, Shen Xukun. *Management and display of digital museum resources* [M]. Shanghai: Science and Technology Press, 2008.

[2] Wang Ying. *Practice and application of 360° panoramic technology in virtual network display* [C]. The 16th China Association for Science and Technology's Annual Meeting: with a new vision of scientific development, striving to innovate science and technology education content forum, 2014.

[3] Xue Yingying. 'Application of Virtual Reality technology in the digital museums', [J] *Henan Science and Technology*, 2014, 6.

[4] Zhang Qiulian. 'Application of 360° panoramic display in virtual museums', [J] *Museology*, 2011, 19.

A New Digital Application for Checking the Condition of Museum Collections

Zhang Zhao
(Exhibition and Exchange Office of China Cultural Heritage Exchange Center, Beijing 100009, China)

Qiu Ke
(Shanghai Shangte Culture Communication Co., Ltd., Shanghai 200235, China)

Abstract: Using highly developed Internet technology and intelligent mobile terminals, we have jointly developed the first piece of professional software for reviewing the condition of museum artefacts within China, according to museum exhibition requirements, attempting to bring a new, working means to this purpose. Just one individual with a smart terminal device can complete the condition review procedure, which in the past has required the work of many people, with relevant exhibit details uploaded to an electronic database. In future, this software will be helpful for the integration of information about incremental exhibitions, as well as the secondary exploitation and utilization of exhibit data.

Keywords: checking the condition of artefacts; digitization; Internet technology; intelligent terminal equipment; database

Relying on their artefacts, exhibitions that are carefully planned and arranged can bring knowledge and visual enjoyment to their visitors. Artefacts are at the core of an exhibition. Today, inter-venue exhibitions are booming, and frequent movement increase the risk of damage. Clarifying safety responsibilities has become an important task in the relocation of exhibit objects.

The condition of artefacts is recorded in detail through text and photographs, as a basis for judging whether changes have occurred. However, inaccurate transcripts can be a problem. Currently, clear pictures are the best way of recording. In practice, most museums use paper forms alongside photos (see Figure 1).

However, current methods have the following key problems:

1) Text records and pictures cannot be integrated
Paper forms are generally A4 pages, containing a limited number of images. Meanwhile, paper forms are prepared in advance, and are only used with old images.

2) Paper text information cannot be electronized, which is not conducive to information accumulation
Information about an artefact's status, including each point delivery, is worth recording. However, traditional paper forms are not conducive to the timely recording of information, meaning that large amounts of information are ignored.

In order to solve the problems above, we have developed an "artefact point delivery system" (see Figure 2) by combining intelligent terminal equipment with Internet database technology. The features of this system are as follows:

1) High-tech equipment integration and intelligence
Current smart terminal devices integrate devices such as high-pixel cameras, large-capacity storage, multi-touch large-screen, and multi-standard network technology,

Name		Collection unit and number	
Measurements		Published books	
Unearthed information		Material	
Exhibition experience			
Status record:			
Image:			
Lending unit		Receiving unit	

Figure 1 List of Exhibits (example)

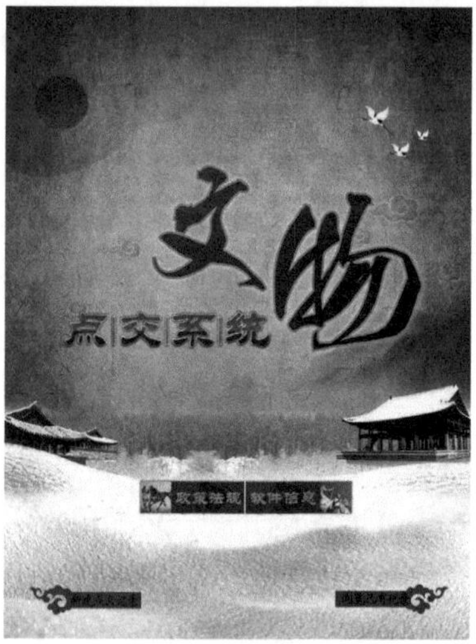

Figure 2 "Cultural relic point of exchange system" interface combined with customized software to provide a more convenient and comprehensive method than traditional recording methods. Our software can record text and photos taken by one person with one device, greatly improving efficiency. Meanwhile, mass storage and networking technologies allow more photos and text to be combined to form a complete record. The cross-checking system can greatly reduce labor duplication, running the individual processing of each department and each link in tandem, while also handling multi-step tasks in the process. Checking the input, review, approval and signature of all links is very convenient, as well as implementing queries on various links. It solves the problems of inefficiency and paper waste in traditional point delivery, and also solves issues with information investment that can be idle due to an inability to use a cumbersome modern office automation system.

2) Real-time information database
We designed a back-end database for this software, into which all information is uploaded, meaning that data is stored, utilized, and exported electronically. The database can also interface with the original exhibit data repository, and securely integrate incremental information with stock information.

3) Scalability
Excellent hardware and software platforms offer tremendous development potential and the ability to develop more work-ready features at a lower cost. At present, our software can realize functions such as real-time information input and upload, photo shooting and editing, and database information reading. As our work continues, more functions will be developed and used online.

4) Security
In order to ensure information security, our server uses the Central Party School server group. Exhibit information is owned by the museum, and the use of the software and database is limited to the user, not allowing organization access without authorization. Point delivery software can be closely integrated with each link, and dynamic electronic certification guarantees the unchangeability of the content of personnel signatures and the authenticity of their source. Users may also use dynamic authentication to achieve reliable identity verification, which is safer and more efficient than traditional keyboard passwords. The use of dynamic authentication rather than traditional password verification greatly increases information security in the system, and is also easier and more convenient to use.

5) Easy to learn and use
Annotations and signature information on the paper submission form can be introduced into the office automation network system. By defining the document circulation and

approval process, each user is able to manually edit, annotate, and sign documents on the network, so that various users can express opinions, and exchange, finalize, confirm, and sign documents in circulation. A user's hand-written notes and signature information can be saved and printed out. As long as the user can write, the office automation system's dynamic signature technology can be used easily, without having to learn a cumbersome keyboard input method, just like the daily review and signature of paper documents. This greatly reduces the amount of training required.

The digital exhibit check point system can free museum staff from tedious, low-end work, allowing them to focus on core affairs, and improving work efficiency and the controllability of information, as well as reducing office costs, improving the force of execution, and perfecting the management of point delivery work. Through the information development of point delivery work, the core competitiveness of the museum can quickly be improved. This involves:

1) Integrated resources
With network technology, resources related to a museum's external exhibitions (such as secondment units, outreach units, point delivery experience and cultural resources) are integrated and managed on a platform to realize the mutual promotion and appreciation of various resources, forming a set of formal, scientific, and open point delivery specifications to ensure healthy and sustainable development.

2) Speeding up information flow
The site work of the point delivery personnel can be communicated to relevant personnel in a few seconds without intermediate links. All staff member opinions and suggestions can be directly fed back to top leadership without obstructions, making it easy to identify issues quickly, improve processes, and discover talent. Those involved can communicate in a relaxed, direct, and fair manner over the Internet, building a cohesive team relationship and museum culture.

3) Standardized check point workflow
Establishing a close, coordinated, reliable, and simple check point work management mechanism, encouraging work, promoting sustainable and healthy development, clarifying staff responsibilities and authority, and delegating specific tasks to individuals to ensure accountability puts an end to haggling and shirking of responsibility, completely eliminating blockages, delays and distortions in information transfer, and guaranteeing correct and timely responses.

4) Improved work efficiency
Coordinating collaborative work between multiple departments can achieve efficient collaboration. Leadership can easily view specific work and progress of personnel, tracking

and monitoring to improve the efficiency and intensity of execution.

5) Saving energy and reducing consumption
The amount of paper used globally in office work is huge, and energy consumption is a problem that every member of museum staff must address in the pursuit of sustainable development. Electronic point delivery reduces the consumption of paper in terms of material energy, and also reduces the performance of trivial work and the valuable time each staff member spends on it.

We believe that the digitization of exhibit items will be the direction of future development. Having solved issues of information storage, reading, and security, digital technology and Internet technology provide great convenience and space for creativity.

We look forward to the next step in the development of this field.

First of all, exhibit checking will be paperless. The powerful hardware of intelligent terminal devices combined with highly customized professional software will replace the current paper form. Advantages such as convenient operation, highly integrated functions, an intelligent operating system, and a powerful database will meet the actual needs of exhibit checking.

Secondly, more equipment provides multi-faceted information about the status of an exhibit. With the development of industrial technology, more sensors will be integrated into smart terminal devices. Based on current models, more pieces of equipment and functions meet safety requirements, such as 3D scanning, surface temperature and humidity measurement, and high-precision microscopic detection. These devices will provide more information about an exhibit's status, providing more comprehensive, professional, and detailed data to ensure safety.

Once again, information within the "exhibition" node will be seamlessly connected. Internet technology will completely alter the existing pre-work preparation, on-site operation records, and post-mortem work patterns, opening up various isolated information sources and databases. In future, staff will collect exhibit information in real time on the spot, and will collect, confirm, and upload exhibit status in real time. All data will be shared with authorized personnel in real time, and the database will continually increase with incremental exhibit data. Data analysis and output according to demand will become a reality. Implementing the safety of exhibits and the real-time confirmation of exhibit status will become part of daily work.

Finally, a wealth of information about exhibit status will change the packaging and transport industry. The packing and transporting of exhibits is based on an item's conditions, and focuses on ensuring its safety. As more information on exhibit status is accumulated and utilized, more accurate, comprehensive, and timely data will be provided to packaging and transportation organizations, so that appropriate solutions can be formulated according to an exhibit's conditions. This will alter the current exhibit packaging and transportation industry, reflecting high-value, high-cost, and high-tech industrial characteristics.

Our work is just beginning. We hope that more people will pay attention to the developments in exhibit condition checking. With our work as a fulcrum, we intend to provide this industry with greater convenience.

Analysis of the Gameplay Elements in the Museum App "Emperor for a Day"

Zhang Wenjuan
(School of Humanities and Social Sciences, Beihang University, Beijing 100091, China)

Abstract: In the context of the widespread application of intelligent terminals, museums have begun to borrow the app format to promote popular science, enhancing the reach of popular science and expanding its social influence. This essay discusses the Palace Museum's *Emperor for a Day* app, analyzing its gameplay, including immersive elements, acting, tasks, action, and puzzles. The app was designed and developed by the development team at the Palace Museum, and is a combination of popular science and interest software, promoting an understanding of traditional Chinese culture for teens. In addition, this paper proposes suggestions for improvements to further enhance the reach of popular science.

Keywords: app; museum popular science; game elements

Preface

With the popularization of smart mobile terminals, the number of mobile Internet users has increased rapidly, and apps have become an important tool for both work and daily life, satisfying individual user needs in areas such as clothing, food, housing, travel, social interaction, gaming, and reading.

Smart terminals offer a new and unique experience to the public, offering new forms of museum education. Mobile phones and tablets have become high-quality media forms for one-to-one information transmission in all areas of life, and are suitable for the promotion of museum-related information. In addition, the colorful forms of presentation that are available can assist the public's understanding of museum exhibitions.

In the museum sector, apps have broad application prospects and significance. Exhibitions are a museum's primary cultural products, and are the main mode in which they serve society. Apps can also help us to can seize the opportunities offered by the development of the mobile Internet, using scientific and technological strength to innovate exhibition forms and content, expand promotion and communication channels, and enhance the penetration and radiation surface for the transmission of cultural products by museums [1]. Museums can also use apps to display collections at multiple levels and angles, comprehensively enhancing the reach of popular science and expanding the museum's social influence.

The Palace Museum's *Emperor for a Day* app uses game elements in an informative and interesting way, promoting public understanding of traditional Chinese culture, and assisting young people in understanding the various Forbidden City collections and China's royal history and culture, providing a new resource for scientific communication and cultural education in the era of the mobile Internet.

1. The app and its application in the museum field

1.1 The app

The term "app", an abbreviation of application, refers specifically to third-party applications for smart phones and tablets. They are a virtual product used by the majority of Internet users, and have become an indispensable part life, with many users are even willing to pay for the experience an app provides. At present, apps are divided into six categories: communication, media broadcasting, life support, leisure and entertainment, tool support, and industrial applications [2].

Leisure and entertainment apps provide users with enjoyment and entertainment, including games, travel, graphic entertainment, mobile audio and video. *Emperor for a Day* falls into this category. Leisure and entertainment apps are often combined with location capabilities, LBS services, and AR (augmented reality) technology to provide users with better service.

The most intuitive feature of an app's carrier mobile device is its mobility, which allows people to use their device at any time, anywhere. Mobility ensures that the software can be used to carry out learning at any time, offering great convenience.

1.2 Apps in the museum field

At present, museum apps can be divided into two categories. Some are made by museums for museum lovers, with maps and other tools, while others are commissioned to be produced and published by professional companies [3].

Guide maps, visit routes, and cultural artefact guides can be presented through an app, providing visitors with videos, pictures, and text. By downloading the app, visitors can learn more about exhibitions and artefacts in a comprehensive, clear, and fast way. For example, the National Museum has developed the *China National Museum Exhibition Newsletter* app, providing practical information on national museum profiles, offering visitor instructions, and providing monthly information and pictures of exhibitions.

Overall, a museum app involves the following aspects:

1) Special exhibit promotions, further information, and retrieval services;
2) Information such as profiles of curators, locations, and visit guides;
3) Display of artefacts, providing cultural information and audio and video;
4) Allowing users to forward and share content to WeChat and Weibo;
5) Collecting visitor experience information via questionnaires, prize-winning quizzes, and other activities, as a basis for improvement.

2. *Emperor for a Day*

Emperor for a Day is a children's iPad game app, using a timeline as a route, and leading children into the Qing court to learn about the emperor's day-to-day food and clothing, studies, office work, and leisure practices. It does so using an interactive map (see Figure 1), allowing users to play amongst important buildings of the Forbidden City such as the Qing Palace, the Yangxin Temple, the Royal Garden, and the Changyin Pavillion.

2.1 Background story

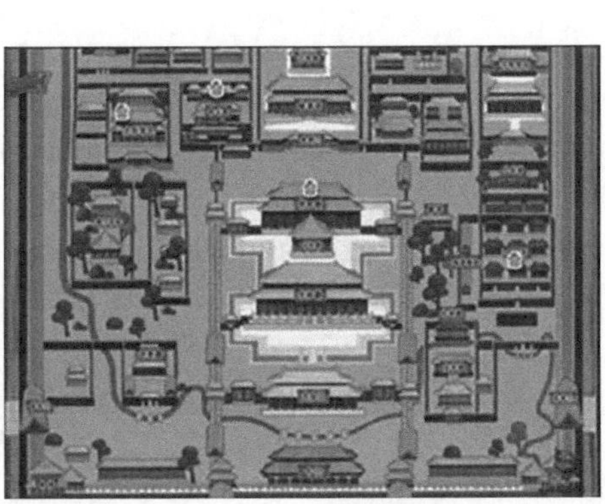

Figure 1 Panorama of the Forbidden City in the game

The background storyline runs through the entire game app, allowing a player to enter the game environment and gain knowledge through an immersive scenario. In the background story of *Emperor for a Day*, the little emperor wants to go and play outside the palace. The lion at the door finds a civilian who looks like the emperor to enter the palace to replace him, and teaches him how to act as the emperor. As the timeline unfolds and the story develops, the daily activities of the little emperor are revealed to players. The story is lively and compelling, offering a unique experience.

2.2 Users

The *Emperor for a Day* student, that is, the role player of the little emperor in the game, plays the game through interaction with the system. The target demographic is children aged between nine and 11 years, at the lower grades of elementary school when an interest in historical knowledge is beginning. In the game, the descriptive language and words are relatively simple, with a small, regular language vocabulary in line with the cognitive level of lower primary school students. In the e-book version, each chapter can be read in about 10 minutes, during which a primary school student can concentrate on experiencing and learning, aligned with cognitive characteristics.

2.3 Purpose

Unlike an ordinary game app, *Emperor for a Day* involves many historical and cultural artefacts as well as explanations of royal culture, aiming to enable young people to learn historical culture and gain knowledge about the Forbidden city, to exercise skills of observation and hand movement, and to cultivate a love for traditional culture.

The game meets a player's sense of accomplishment in the design of a challenge with achievable goals [4]. To achieve the goal of a game, it must be attainable. This game is simple to operate, rich in form, interesting, and easy to follow.

2.4 Rules

In *Emperor for a Day*, more than 200 large and small interactive points have been set up to trigger the plot, pushing forward the life trajectory of the emperor and restoring scenes of life from the former Forbidden City. In addition, the app combines popular game design features such as puzzles, collecting, sorting and hand-painting, to avoid boredom. The scenes include shooting an arrow through a willow leaf a hundred strides away, making

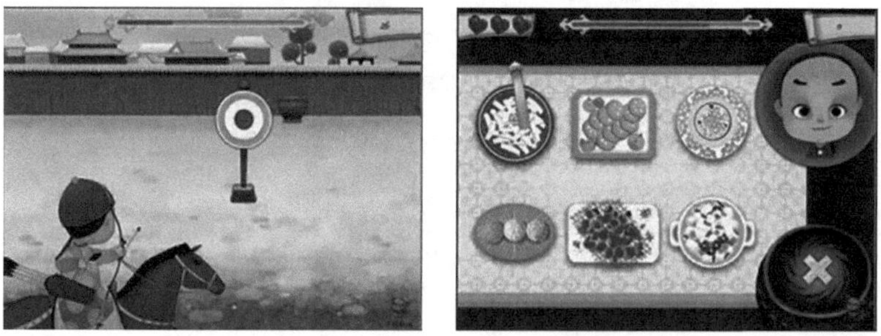

Figure 2 High-precision shooting, and a taste test

oneself up to go on stage, and silver medal poison testing (see Figure 2). The rules of the game are presented in simple descriptions that are easy to understand.

The game design involves a lot of detail. For example, users can click on the bird that sometimes flies across the screen to read allusions or minor pieces of information; they can look for hidden paw prints on the screen and click through to enter a "branch quest", such as the "treasure map" link; they can also click on the upper righthand corner on the Emperor's Handbook to review their achievements.

2.5 Help

In the *Emperor for a Day* app, help is provided on the spot so that users can solve problems at any time, without becoming disappointed or tired due to long-term issues. In addition, the design prompts users to proceed to the next step with a flashing on-screen icon.

2.6 Levels

Emperor for a Day divides information into a number of scenarios, combining historical knowledge, cultural artefact information, and place descriptions within the emperor's day, popularizing science through hands-on practice. In the game, levels are set by the plot, and the emperor's day is divided into several segments. The tandem level layout means passing the first level before moving to the next, a mode that can promote a child's continued exploration and participation in the game, step by step.

2.7 Rewards

The honors or material rewards given to learners who complete tasks is a key factor in attracting young people. As the game develops, participants have access to a variety of treasures, the prototypes of which come from artefacts collected in the Forbidden City, such as antler chairs and sandalwood. Cultural artefact information and knowledge cards obtained in the game can be read in the e-book of the *Emperor's Handbook* (see Figure 3).

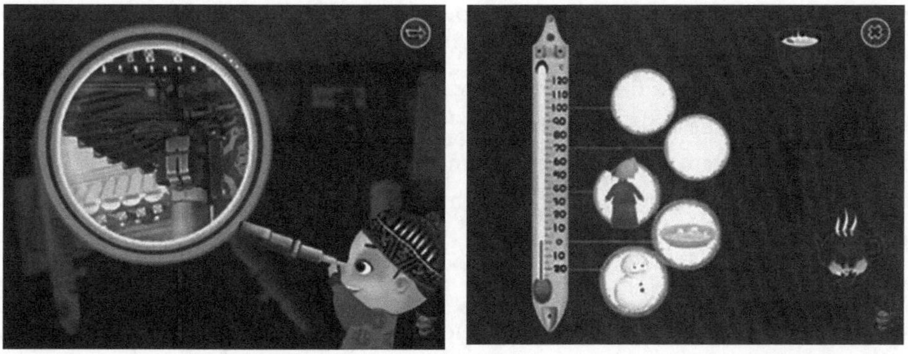

Figure 3 Section for information about cultural relics

In addition, after completing each task, participants receive corresponding achievements, such as titles of "a person of great talent", and "having the ability to recognize people's worth".

3. Game elements within the *Emperor for a Day* app

Game elements are the main embodiments of gameplay, including a variety of paradigms of interest to the player, and game factors that are exploratory, immersive, and interesting. The design and selection of game elements must align with the characteristics of users, information, and the goals of promoting popular science [5].

3.1 Immersive elements

Immersive elements are a commonality across art forms, in traditional media such as novels, movies, and dramas. *Emperor for a Day* creates realistic situations through the lively images of the little emperor, HD pictures of the Forbidden City, vivid scene recreation, and an upbeat musical rhythm, to create a realistic situation that immerses participants in the story, making them think that the Imperial Court events are real and not just entertainment. The app immerses children in the atmosphere of the little emperor. They forget that they are operating and learning as they become integrated into the situation, acquiring and recognizing artefacts, and understanding information that the designer hopes to convey.

Although the cartoon image is not a true image of the emperor, it is in line with the cognitive expectations of children aged nine to 11, as processed artistic images are more attractive to children; the smoothness of interaction allows for an easy experience, and imaginary resonance promotes immersion and learning through entertainment.

3.2 Play elements

The deliberate playing of roles, thinking, expressing, and moving about based on the situation of a character makes the experience more interesting. In this game, the participants play the role of a civilian acting as the little emperor, having the unusual experience of 24 hours in the emperor's life. Participants must consider problems from the perspective of the little emperor, which is an extraordinary experience. They are immersed in the drama as they operate and learn. Players regard themselves as little emperors, and from this perspective, they think about problems with their own minds – the most direct experience of role playing.

3.3 Mission

The mission provides players with a clear and challenging game goal to engage them and reward them with prizes and accomplishments. In *Emperor for a Day*, users earn rewards by completing tasks such as shooting an arrow through a willow leaf from a hundred strides away, and silver medal poison testing, to improve their hands-on ability, from which they gain a sense of accomplishment and victory. In addition, after completing a task, players can win honorary titles to motivate their progression to the next task.

3.4 Action

Action elements are the plainest form of gameplay, and the actions themselves are an experience worthy of repetition. Based on touch technology, page turning, and jumping, users can interact with the system through click operations. For example, in the silver medal poison test, participants must click on silver medals to find the ones that turn black, determining whether the food is poisonous. In the puzzle design, fragmentary, sorting, and supplementary puzzles are used, allowing users to appreciate the joy of action. These action elements exercise users' hands-on abilities and hand-eye coordination.

3.5 Educational elements

In the game, some of the problems have been designed to be solved with intellectual thought, to exercise users' thinking ability. Educational elements are especially important in popular science apps. For example, when adjusting the order in *Emperor for a Day*, players must mobilize their memories of ancient poetry. When choosing appropriate clothing for the emperor they must think about the characteristics of clothing in different environments. When shooting an arrow from a hundred strides away, they must consider the design of advancing speed. They must use their brains, and this is the embodiment of educational elements.

4. Conclusion

4.1 Popular science

Emperor for a Day uses the latest technology, modern game elements, a lively design, and learning through entertainment to attract children's participation in the experience and acquisition of traditional cultural knowledge. In addition, the software utilizes novel interactive methods and an attractive 3D displays, offering a very good introduction to the traditional culture of the Forbidden City and profound history and culture of China.

4.2 Suggestions

In order to realize the knowledge transmission function of *Emperor for a Day*, to play a greater role in popular science, and combine the characteristics of scientific projects at museums, this paper puts forward the following suggestions:

1) Regular updates
The times are changing with each passing day. Much of the software that was "hot" for a while has been replaced by new versions due to the speed of software updates. This requires developers to constantly inject new elements into their products and continue to enrich content and form. The exhibits at the Palace Museum provide inexhaustible resources for development and design. For example, the *Palace Museum Daily* is updated every day, so that learners can accumulate new knowledge on a daily basis. It is recommended that designers integrate knowledge about the Forbidden City and traditional culture into *Emperor for a Day*, continually enriching the game's software.

2) Enhanced interactive design
On starting elementary school, children's emotions landscapes alter, requiring affirmation and recognition from peers. App design must also provide affirmation, rewards, and encouragement in a timely manner [6]. Although *Emperor for a Day* supports sharing in Weibo and WeChat Moments, sharing only shows the app's poster. We recommend providing the options to add comments, communicate online, and share achievements so as to enhance interactivity. For example, the reading app *Baby Loves to Read* is highly interactive. Clicking the brush icon on the homepage allows young readers to create and upload their own drawings, and "like" the work of other children.

3) Integrating virtual reality
It is recommended that this software be based on existing technologies using LBS, GPS, and AR, combined with museum-specific conditions, such as locations and exhibitions, to connect the game more closely with the real Forbidden City. In addition, young people should be able to personalize works connected to the Forbidden City through photos and audio anytime and anywhere, and then upload and share them.

References

[1] Ciu Xianhui. *Research on the significance and future potential of interactive exhibits for public education in museums* [D]. Central Academy of Fine Arts, 2014.

[2] Chen Xijing. *Research on the applicability of the Science Guide App in the Chenshan Botanical Garden* [D]. Donghua University, 2014.

[3] Wu Changwen. 'Apps in the mobile Internet wave and their applications in museums', [J] *Fujian Fair* 2012, 2: 75–77.

[4] Chu Xuejuan. *Using game elements to construct simulated space in network experiments* [D]. Jilin University, 2006.

[5] Huang Shi, Fu Zhiyong. *Game planning and management* [M]. Beijing: Higher Education Press, 2012.

[6] Song Liqing. *Research on the interactive design of extracurricular learning apps for urban primary school students* [D]. Jiangnan University, 2014.

*Analysis of Metadata for Cultural Relics in Digital Museums**

Jia Yanxia and Zheng Xiaohui
(Tsinghua University Library, Beijing 100084, China)

Abstract: With the development of digital museums in China, cultural artefact organizations at all levels across the country have accumulated a large amount of data. Sharing and searching this data requires a standardized metadata system. On the basis of investigating the status quo of Chinese digital museum development, combined with the development of foreign metadata standards, this paper analyzes the ideas and needs related to creating Chinese metadata standards, proposing to build a metadata standard system related to the structures, values, and content of data resources.

Keywords: digital library; cultural works; metadata

The collection and management of cultural artefacts, along with their functions in publicity, education, and research, are based on the artefacts themselves. The use of cultural artefacts is essentially the use of cultural artefact information [1]. In the development of digital museums, the central task involves the management of collection-related information. The regulation of this information is the key to its management [2]. The State Administration of Cultural Heritage has carried out many surveys and reviews of cultural artefacts, and has also established a data sharing platform. However, there are certain difficulties in the actual task of identifying and recording cultural artefact collections. Problems related to non-uniformity are widespread. One of the major reasons for the

* This article is a result of the "Research on Digital Data Protection Standards for Cultural Relics" project under the National Science and Technology Support Program "Research on the Digital Data Protection Standard System and Key Standards for Cultural Relics". Project number: 2014BAK07B02.

difficulties in data sharing and information exchange across databases built by various kinds of cultural artefacts management units at various levels, is the lack of a unified national standard system for metadata. A complete set of standards describing artefacts involves descriptions of elemental structure, value, and selection. Standards will lead to better descriptive cataloging, helping with document consistency and record sharing, while enhancing end-user queries and resources. Based on an investigation into the status quo of digital development in China's museums, combined with the developmental state of foreign metadata standards for cultural artefacts, this paper analyzes the mentality and needs in the development of metadata standards for cultural artefacts in China.

1. Status and issues of standards for digital museum information

1.1 Informatization is widely implemented

The cultural artefacts management unit has carried out various levels of informatization. The State Administration of Cultural Heritage performed a census of movable and immovable cultural artefacts, establishing the "Cultural Relic Survey and Database Management System" as well as the "Third National Survey of Immovable Cultural Relics Information Management System". These tasks have clarified a "stock" of national cultural artefacts, establishing an electronic account for basic cultural artefact information, as well as issuing some operational guidance documents.

Some cultural artefact management units have also realized office automation and informatization of cultural artefact management in their actual work. For example, the Shanghai Museum's digital cultural artefact information management system consists of two subsystems: information management and a storage general ledger, comprehensively realizing the automatic management of basic elements such as collection cataloging, recording, research collection, media collection, protection and restoration, and circulation services [3]. Similar cases include the Zhengzhou Wenbo Information Network System [4] and the Heilongjiang Provincial Museum Collection Management System [5]. In addition to realizing automation, some cultural artefact construction projects such as the *Digital Palace* and *Digital Dunhuang* have also digitized cultural artefacts and used advanced information technology to complete digital image collection of 2D and 3D cultural artefacts, as well as achieving virtual display.

In the process of informatization at all levels, a large amount of cultural artefact data has been accumulated. At present, in the process of digital development at various museums, the information index of items recorded in the cultural artefact records, the bibliographic rules on which they are based, and the value criteria of elemental items do not follow unified standards or norms, and each has adopted different standards to meet their own needs for cultural artefact management and collection. The State Administration of Cultural Heritage has issued some specific guiding documents and requirements for data in the cultural artefacts census. However, due to differences in data or documents for various cultural artefacts at various levels, there are many issues with processing, batch harvesting, and submitting data, posing difficulties to data sharing.

1.2 Status of current standards

In the process of digital museum development, one key area is the standardization of cultural artefact information. The informatization of collections must solve the issues of standardization and normalization. This is the most crucial, fundamental task in the collection of information, and a necessary foundation and premise for the realization of computer-based management [6]. The National Cultural Relic Bureau and various museums and research institutes have carried out explorations in this regard. In terms of structural standards for cultural artefact information records, the most notable are the *Museum Information Index System Specification (Trial) [cultural artefact exposition {2001} No. 8]*, and *Museum Audio and Video Index Specification (Trial)*. The former established three groups of indicators, including collection information, management information, and related documents/audiovisual information, with three indicator sets and 139 specific indicator items under them; the latter intends to supplement and perfect the audiovisual indicator sets. In the specifications for recording cultural artefact information, there are the *Regulations on the Registration of Cultural Relics in the Database* (WW/T0017–2013), and *Specifications for the Recording of Museum Collection Information Indicators* (internal exchange documents). A team of experts from the China University Digital Museum Engineering department issued the *Humanities and Art Museum Information Index Specifications* [7]. There are also some guiding documents within the industry, such as the *Collection of Precious Cultural Relic Collection Data* and the *Indicators and Descriptive Rules for the Collection of Precious Cultural Relic Data*, published in the *Relic Investigation and Development of a Database Management System* project, as well as standards for identification, classification, naming, chronology, measurement, and information collection for immovable cultural artefacts based on the third national cultural artefact census [8], which can be used to guide personnel in the collection and recording of information about immovable cultural artefact.

1.3 Insufficient development of information standards

China has accumulated a large number of basic data and digital images of cultural artefacts, as well as other digital expressions and information specifications. Following investigation and analysis, it has been found that there are certain shortcomings in the development of standards for cultural artefact information in the process of building digital museums in China.

1.3.1 Collection indicators require improvement

The issues with collection of cultural artefact information are as follows: firstly, the existing information index specifications in the system are still incomplete; secondly, the structure of existing information index specifications must be improved; thirdly, in the practice of various cultural artefact units, standards are not implemented uniformly.

Although the existing *Museum Information Index System Specification (Trial)* and other guiding documents are attempting to solve issues of non-uniformity in information collection management standards, the above specifications are designed largely for movable cultural artefacts, without considering specifications in the area of collection information about immovable cultural artefacts. The specifications for cultural artefact information collection indicators are still incomplete in the system.

The existing information index specifications are largely for recording content attributes and the related management information of collection, the resource subject being the collection entity; indicators describe mostly appearances and physical characteristics, lacking overall thematic information and effective disclosure of individualized features such as creators, topic descriptions, national culture, and associated resources [9]. In addition, current museum collection metadata lacks consideration of the relationships between cultural artefacts when it comes to indicator settings and information recording. Certain relationships may exist between cultural artefacts, such as a part of a whole, a single piece and a whole set, or parts in the same series. The design of relationship-type or association-type elements helps to establish these relationships between artefacts. On the whole, the current structure of the index system for cultural artefact information must be improved.

In practice, museums and cultural artefact management units have different standards for the collection of cultural artefact information, and often set up metadata indicators and image collection indicators that are suited to their actual needs. For example, the collection information form established by Zhengzhou Museum involves collection, collection management, and collection document information, making a total of three indicator groups, more than 30 indicator sets, and 232 indicator items [10]. Each cultural artefact record of the Palace Museum contains six data types, including 40 pieces of data regarding basic cultural artefact information, cultural artefact descriptions, collection property information, collection location information, usage information, and research information [11], with different descriptive information collection indicators set for different collection types. The National Museum has developed a metadata specification system that is suitable for our digital resources, including seven element sets, 60 elements, and 342 qualified elements [12]. The differences in practice and implementation of different standards have brought about difficulties in the sharing of cultural artefacts.

1.3.2 Information recording specifications need improvement

In terms of the recording of cultural artefact information, currently there are *Guidelines for the Registration of Cultural Relics in Collections* (WW/T0017–2013) and *Guidelines for the Recording of Museum Collection Information*. These recording specifications are relatively simple, addressing some elemental content such as how to choose and construct ordered information content. However, they lack detailed explanations or guidance. Certain challenges and issues are difficult to address when actually collecting specific cultural artefact data. For example, in the investigation of data items issued by the National Cultural Relics Bureau and the investigation of existing cultural artefact management in the Forbidden City, the National Palace Museum found that several important information items were not clearly defined. In practice, it is easy to produce recording errors, such as names, classification, piece work, and condition [11]. On the whole, specifications for the recording of cultural artefact information must be improved.

1.3.3 A lack of element value criteria

In the recording of cultural artefact information, the values of certain elements involve professional terminology and vocabulary, such as geographical location, age, style, material, and techniques. In order to avoid inconsistencies in descriptive information among various information recording personnel, the Palace Museum has set up a drop-down menu option for many items. In the registering of various types of cultural artefact information, the system has established general categories for cultural artefacts. However, there is still no uniform or clear classification standard for cultural artefacts in China. Various museums and cultural artefact management units across the country have explored and established their own archival and classification systems. Some are divided based on cultural material, some by cultural artefacts function, and some by manufacturing processes, and there are many intersections of classifications and definitions. Issues found in actual operations reflect the lack of specifications regarding the values of elements within cultural artefact information specifications, such as terms or controlled vocabularies.

2. Development of foreign metadata standards for cultural artefacts [13–14]

Led by the United States, foreign countries have formed a series of metadata standards related to cultural artefacts, which have been widely used, and have reference value for the research and development of China's metadata standards for cultural artefacts.

Based on Murtha Baca's *Introduction to Metadata* (2nd edition, 2008), a standard should be a set of tools for guiding data structure, values, and content, its roles being to guide descriptive cataloging and to facilitate consistency in documents and record sharing, assisting end users in the query and acquisition of resources. Criteria for controlling data values and determining how to select, organize, and format data content are two different types of standards that must be combined under an agreed upon data structure.

Data structure standards are the logical first step in creating a standards system. In the field of art and cultural heritage, data structure standards have been fully developed, such as the *Categories for the Description of Works of Art* (CDWA) in the United States, and *Visual Resources Association Core Categories* (VRA Core), following developments in CDWA Lite and Objects ID. Launched in 1995, CDWA is a metadata scheme designed to describe works of art. It defines 31 categories and 501 subcategories that are currently widely used in museums, including 36 core elements.

Developing data value criteria is the second aspect in building a standard system. In the early 1980s, the Getty Research Institute began work on the *Art & Architecture Thesaurus* (AAT), in addition to the artist's joint directory (the Union List of Artist Names, ULAN), the *Getty Thesaurus of Geographic Names* (TGN), the *Thesaurus for Graphic Materials* (TGM). Along with the names and subject specification files of the Library of Congress, these vocabularies became the tools required to form a potential shared directory.

Cataloging Cultural Objects (CCO) is the third stage and standard for data content. The role of content standards is to instruct organizations and staff using information records how to perform tasks such as naming a work, and selecting age information. In the standard system for cultural artefact metadata, data structure standards such as CDWA and data value standards such as AAT have been full developed. Relatively speaking, content standards used to describe unique cultural resources in museums have appeared much later. The CCO was drafted in 2004, and in 2006, the *Cataloging Cultural Objects: A Guide to Describing Cultural Works and Their Images* was published, promoted, and applied by the American Library Association. CCO is a guide to the selection, organization, and standardization of data that guides catalogers in completing metadata elements in records. The roles of the CCO are to improve the quality of descriptive cataloging, and to facilitate the sharing of documents and the query and access of end users.

Metadata structure standards, elemental value standards, and content standards must be used in tandem to ensure a high degree of consistency in descriptions and satisfactory search results.

3. The requirements and ideas of metadata standards for cultural artefacts in China

After 2000, with the promotion of projects such as the Cultural Relic Survey, the Development of a Database Management System, and the Creation of Digital Museums in Universities, as well as the informatization practice within a group of museums represented by the Palace Museum, the development of digital museums in China has achieved gratifying results. However, in light of the introduction and analysis above, China has developed slowly in the development of metadata standards for cultural artefacts. At present, the consistency of national cultural artefact data and documents is poor, and sharing is difficult.

Referring to the development of the aforementioned foreign data standards, China's standard system for cultural artefact metadata has equivalent requirements in terms of structure, value, and content standards.

1) On the basis of the *Museum Information Index System Specification (Trial)* and the collection of cultural artefact information indicators based on actual needs, combined with domestic practices and cultural artefact census index items, reference should be made to the metadata description standard for international cultural artefacts, to build a complete metadata index system for describing cultural artefacts. Combined with the status of current domestic research and practice, it is necessary to determine the core elements of cultural artefact description to adapt to descriptive needs of core information for various cultural artefacts. It is also necessary to set up a complete metadata indicator system for describing artefacts, considering thematic cultural artefact content, history and culture, cultural artefact relations, and related images, to compensate for the shortcomings in existing research and practice.
2) It is necessary to study and formulate data value standards related to cultural artefacts. In terms of the descriptions of cultural artefact information, there are many elements that determine the normative terminology and control the standardization, accuracy, and uniformity of description. For example, in the cultural artefact information management system of the National Palace Museum, specified options are set for some indicators. The norms and consistency of recording are guaranteed by terminology, and the classification, unit group, age, ornamentation, texture, and size must follow terms or values listed in the drop-down menu. Some authors have also raised the issue of language specification in digital museum archives, believing that when cultural artefact information is recorded, there is a lack of standardized professional dictionaries, leading to inconsistencies in information records, and making it difficult for users to query and retrieve cultural artefact resources [15]. This reflects the fact that cultural artefact recording has a practical need for technical terms. In this regard, the *Art & Architecture Thesaurus* (AAT) compiled by the Getty Institute is worthy of recognition and reference.
3) In terms of metadata content standards, *Specifications for Cultural Relic Registration* (WW/T0017–2013) and *Guidelines for the Recording of Museum Collection Information* and other guiding documents and internal specifications for cultural artefact units already exist. What is still lacking are detailed bibliographic specifications and guidelines for collection cataloging practices. Referring to the CCO, combined with characteristics of different kinds of cultural artefacts in China, we can formulate

information content description rules suited to Chinese cultural artefacts, improve the normative nature of cultural artefact information, and facilitate the exchange and sharing of cultural artefact data.

With the development of digital museums in China, the issue of how to share cultural artefact information and meet user query demands for unified and high-quality cultural artefact information must be considered. The need for development of metadata standards seems more necessary and urgent. Relatively speaking, standards for data value and cultural artefact recording (data content) in China are primitive and insufficient, so efforts must be made to strengthen research in the future. We hope that the analysis provided in this paper will offer a reference for the future research and practice of metadata standards for cultural artefacts.

References

[1] Zhu Jingguo. *Conceptual thinking in museum digitization* [C] China Museum Society Professional Museum Digitization Committee. Proceedings of the Inaugural Meeting and First Academic Symposium of the China Museum Society and Professional Museum Digitization Committee. Beijing: Yanshan Publishing House, 2006, p. 100.

[2] Guo Canjiang. *Museum collection information standardization management and digital museum construction* [C] China Museum Society Professional Museum Digitization Committee. Proceedings of the Inaugural Meeting and First Academic Symposium of the China Museum Society and Professional Museum Digitization Committee. Beijing: Yanshan Publishing House, 2006, p. 108.

[3] Shanghai Museum. 'Digital cultural artefact information management system in the Shanghai Museum', [J] *Chinese Cultural Heritage*, 2004, 3: 82.

[4] Yan Shuguang. *Analysis of the Zhengzhou Wenbo Information Network System* [C] China Museum Society Professional Museum Digitization Committee. Proceedings of the Inaugural Meeting and First Academic Symposium of the China Museum Society and Professional Museum Digitization Committee. Beijing: Yanshan Publishing House, 2006, pp. 83–87.

[5] Chen Yu, Tong Qiang. *Discussing the needs analysis of museum collection computer management systems* [C] China Museum Society Professional Museum Digitization Committee. Proceedings of the Inaugural Meeting and First Academic Symposium of the China Museum Society and Professional Museum Digitization Committee. Beijing: Yanshan Publishing House, 2006, pp. 111–115.

[6] Lin Chengbin. *Reflections on the construction of the information system of the Agricultural Museum of China* [C] China Museum Society Professional Museum Digitization Committee. Proceedings of the Inaugural Meeting and First Academic Symposium of the China Museum Society and Professional Museum Digitization Committee. Beijing: Yanshan Publishing House, 2006, p. 63.

[7] Chen Hongjing, Wu Qinmin. *Standards and methods for resource construction in digital museums — Humanities Art* [M]. Shanghai: Shanghai Science and Technology Press, 2006, p. 2.

[8] The third national cultural artefact census implementation plan and related standards and norms [OL] (29 Jan 2015) <http://pucha.sach.gov.cn/Html/Uploaded Files/liuyihan/2008-11-24/disanci quanguowenwupuchashishifanganjixiangguanbiaozhunguifan.doc>

[9] Jia Junzhi, Shi Xuan. 'Research on the construction of digital museum metadata standards', [J] *Journal of Shanxi University (Philosophy and Social Sciences)*, 2015, 38(1): 114–119.

[10] Liu Zhongwei. *Experiences in the construction of a cultural artefact information database and the use of collection management software* [C] China Museum Society Professional Museum Digitization Committee. Proceedings of the Inaugural Meeting and First Academic Symposium of the China Museum Society and Professional Museum Digitization Committee. Beijing: Yanshan Publishing House, 2006, pp. 119–130.

[11] Hu Chui, Mei Xue. *Establishment of the basic data table in the cultural artefacts management system of the Forbidden City* [C] China Museum Society Professional Museum Digitization Committee. Proceedings of the Inaugural Meeting and First Academic Symposium of the China Museum Society and Professional Museum Digitization Committee. Beijing: Yanshan Publishing House, 2006, pp. 143–147.

[12] Feng Jiace. 'Construction and application of national museum metadata specifications', [J] *Museum Research*, 2013, 3: 33–38.

[13] *Cataloging Cultural Objects: A guide to describing cultural works and their images, online edition* [OL/M] (25 March 2015) <http://cco.vrafoundation.org /index.php/toolkit/cco_pdf_version/>

[14] Chen Heqin. 'Describing the standard content of cultural artefacts CCO', [J] *Education Materials and Library Science*, 2009, 46(4): 497–522.

[15] Cao Hong. *Language expression and normative issues in digital museum collection archives* [C] China Museum Society Professional Museum Digitization Committee. Proceedings of the Inaugural Meeting and First Academic Symposium of the China Museum Society and Professional Museum Digitization Committee. Beijing: Yanshan Publishing House, 2006, pp. 149–152.

Intellectualization and Materialization: Design Concepts and Technological Applications at the Ningbo Smart City Science and Technology Museum

Yuan Yuan
(School of Humanities and Social Sciences, Beihang University, Beijing 100191, China)

Abstract: In recent years, the word "smart" has appeared frequently, and the concepts of the "smart planet" and "smart city" have become popular. These emerging concepts are inseparable from the rapid development of various new technologies such as digitization, the Internet of Things, and Cloud computing. With innovations in internet technology, some smart museums and science and technology museums have emerged, with the core concepts of connecting people, objects, and venues. Using the example of the Ningbo Smart City Science and Technology Museum, which opened in September 2014, this paper analyzes its design concepts and applications of Internet of Things technology, demonstrating a successful case of the transformation of a popular science museum from traditional methods to informatization, intelligence, and materialization. Finally, based on existing design ideas and achievements, further suggestions and development prospects are proposed.

Keywords: intellectualization; Internet of Things; popular science venue; smart city science and technology museum

Museums, particularly those devoted to science and technology, are becoming important locations for public education services, scientific research, leisure, and entertainment. Nowadays, the exhibits in these venues are not static. Visitors are not met with cold display windows, but instead are shown beautiful low-carbon cityscapes invented by digitization. Experiences like this are now offered at the Ningbo Science and Technology Museum in Zhejiang Province, providing the intelligence and materialization of popular science that venues require to meet the social expectations of their visitors.

Although features of the Internet of Things are now reflected in many venues, the rapid development of IoT applications still tends towards concepts rather than reality. The use of IoT technology at the Ningbo Smart City Science and Technology museum presents breakthroughs beyond the use of the urban optical fiber network, wireless broadband, and mobile internet. Related to the Internet of Things, new terms such as "smart planet", "smart city", and "smart China" have emerged. The Ningbo Smart City Science and Technology Museum reflects the concept of being "people-oriented, at your fingertips", providing new ideas for the design and development of popular science venues.

1. About the Ningbo Smart City Science and Technology Museum

The Ningbo Smart City Science and Technology Museum is a typical example of the application of Internet of Things technology. Keeping pace with the times, it is the first large-scale, comprehensive, interactive popular science venue to be named after the "smart city" in China, with the theme of "Smart dreams, at your fingertips", and led by the

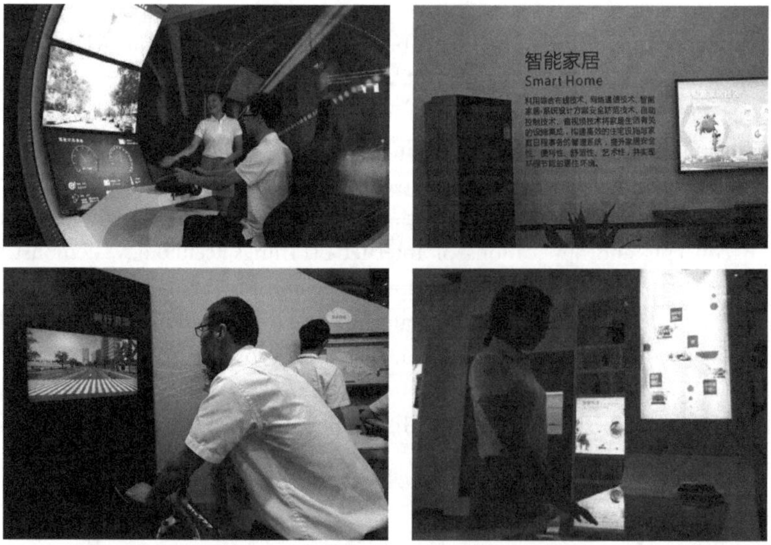

Figure 1 Examples of typical exhibits [9]

concepts of "low-carbon, green, and environmental protection". The museum is divided into five sections: "Panoramic Ningbo", "Harmonious Ningbo", "Free Ningbo, "Peaceful Ningbo", and "Synergetic Ningbo", covering the tangible experiences of smart cities such as smart living, intelligent transportation, and smart education [2].

The venue presents the Internet of Things and its key technologies, with strong information integration capabilities, for example, car windshields that can search for nearby gas stations, recipes for evening meals that can be arranged on a coffee table, text messages sent and received while dressing in the mirror. These kinds of efficient and convenient high-tech life experiences are just a few of the museum's exhibits (See Figure 1).

Many new exhibits are also included, for example, a mind-controlled remote control aircraft (see Figure 2) that collects brainwave signals through a helmet, which are transmitted to a mobile device that wirelessly controls the aircraft, thereby achieving its goal of increasing audience concentration whilst playing the game.

More interestingly, one can also operate the smart city management system as "the mayor" (see Figure 3). With an iPad connected to four large screens, in only a few seconds, information on the population, real-time traffic, infrastructure conditions, and major events appears on the big screen.

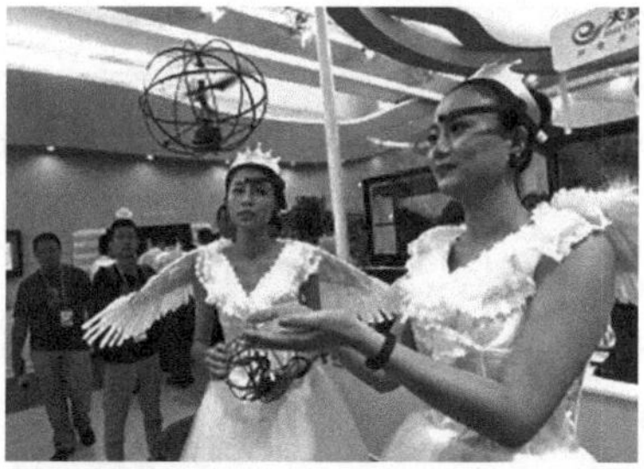

Figure 2 Mind controlled aircraft [9]

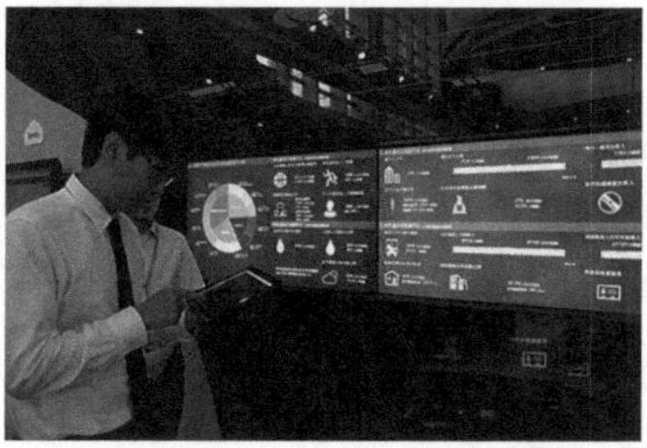

Figure 3 Staff in the Smart City operation management system [9]

2. *Design ideas and features of the Ningbo Smart City Science and Technology Museum*

2.1 Analysis of the design concept

2.1.1 From digitization to instrumentation

Prior to this, many science and technology venues preferred to adopt digital design concepts. Using a website as the carrier, venue information was made virtual and shared via network communication and resource digitization. However, the collection and processing of digital resources (especially for large popular science venues) is not a simple task. In fact, visitors prefer interactive experiences in real-world scenarios rather than static models and specimen displays.

In 2009, IBM in the USA proposed the idea of the "smart planet" to the world [5]. The concepts of "materialization and intellectualization" began to take shape, imbuing objects with "smart life" and the capability to "communicate" with human society.

2.1.2 Smart City, Smart Technology

In 2010, IBM officially proposed the "smart city" vision [5]. This was a step forward in the development of urban life and society.

The Ningbo Smart City Science and Technology Museum adopted the design concept of "smart city, smart life", integrating modern intelligence technology, and advocating "low-carbon, environmental protection" for a more convenient existence. Visitors can capture pictures with sound control, experience a head-mounted 3D theater, experience the smart home, smart transport, and smart travel. The whole city becomes a vivid organism thanks to the Internet of Things, realizing the intelligent materialization of life in the future.

2.1.3 The concept of the "people-oriented" venue

Visitors to the Smart City Science and Technology Museum are able to experience the thrill of human-machine communication in an efficient and convenient way, with just a single mobile phone. The application of Internet of Things technology enables visitors to feel "things" by watching, touching, listening, and thinking. This reflects the "people-oriented" concept of the venue, highlighting the museum's most essential purposes, namely education and inspiration. The materialization and intellectualization of the exhibits brings rich knowledge and high-tech pleasure to visitors, connecting the world of information with the real world.

2.2 Analysis of related characteristics

The Smart City Science and Technology Museum has achieved good results, enabling visitors to feel its "living power" across three characteristics.

2.2.1 Creating a new kind of situational, interactive learning

In educational psychology, "situational cognition theory" suggests that the essence of the cognitive process is determined by context, and learning is the process of active exploration in real situations [6], emphasizing the combination of the learning process and practice.

In order to imbue its displays with beneficial effects, this science and technology museums teaches via the creation of context based on situation cognition theory. One example is the exhibit that shows a car windshield searching for nearby gas stations. First of all, the chosen theme is closely connected to real life, and new technological knowledge is integrated into the interactive aspect of the exhibit to attract visitors to pause and experience it. Moreover, in terms of the generally entertaining experience of the game, visitors are transformed from passive visitors to active participants, resulting in meaningful learning. In addition, the context of the exhibit promotes communication, interaction, and cooperation, highlighting visual learning.

2.2.2 An interactive "human and material" experience

The design of many exhibits at the Smart City Science and Technology Museum is derived from ideas of "interactive design", referring to processes of interaction and influence between people and things. Almost every exhibit in the museum allows for touch or audiovisual audience interaction, with interactive forms including head-mounts, physical manipulation, and touch screens. Exhibits should be:

1) Learnable
Specific scenarios reflecting the knowledge structure and background information are displayed to visitors. In the interactive exhibit process, there are clear, complete, and convenient functional modules that facilitate the audience's understanding and use of interfaces while providing excitement and surprises.

2) Purposeful [6]
Exhibits have a clear purpose, allowing visitors to see the needs and goals of urban life, providing personalized needs service, and focusing on service effectiveness for future smart life.

3) Contactable
In the interaction between visitors and exhibits, visitors are told what they are doing and what to do next. The exhibits have a clear structure and theme to ensure smooth and fluid interpretation in their experience, putting smart life at their fingertips.

2.2.3 Forming the "ecological chain" structure of "human-object-venue"

Researchers develop the venue's management and conservation plan through feedback on audience behavior and demand analysis. It is then implemented on the display and delivery information of exhibits. Visitors are provided with interactive experience and knowledge sharing, which is designed to inspire and educate them. In this way, the "ecological chain" structure of "human-object-venue" is formed (see Figure 4). The three constitute a mutually influential and mutually promotive cycle [7]. This is also a special embodiment of the word "smart" used in popular science venues. The Smart City Science and Technology Museum uses the new and powerful Internet of Things technology and communications technology to realize an organic connection between the three elements. This "smart" quality fully immerses the audience – another progressive development for museums in this field. Furthermore, visitors are places the core of the ecological chain, providing them with an impressive understanding of the smart city.

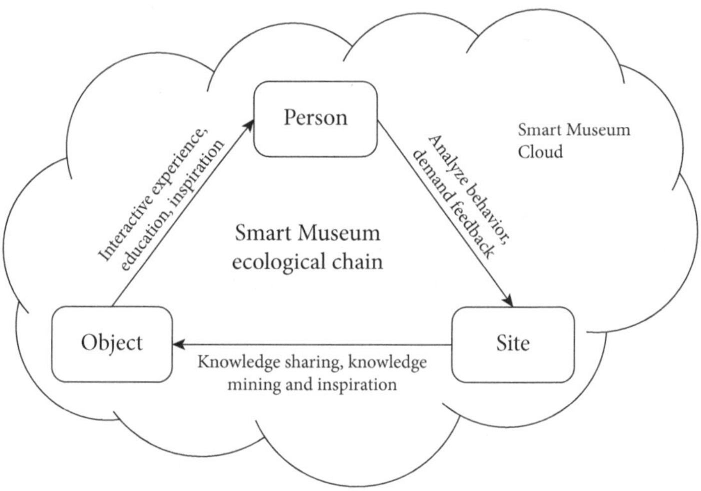

Figure 4 The "person-object-site" ecological chain model [7]

3. The application of Internet of Things technology at the Smart City Science and Technology Museum

The Internet of Things is a network that realizes the intelligent identification, location, tracking, and management of objects through the low-cost, high-performance network services of Internet-based sensors. IoT technology is used at the Ningbo Smart City Science and Technology Museum, meaning that visitors are able to easily and efficiently perceive information. They can then process it intelligently and transmit it reliably through a tablet or mobile phone, so that social life and physical systems are integrated. This foreshadows the smart cities of the future, in which production, social life, technology, and manufacturing will be more refined, more convenient, and more dynamic. The IoT can offer strongly interactive exhibits in science and technology venues. It can protect and manage exhibits while enabling venues to achieve their social and educational goals.

3.1 Applications of RFID technology

Radio Frequency Identification (RFID) is a communication technology. Its most prominent advantage is non-contact identification, penetrating a variety of extreme materials and environments to read embedded tags efficiently [8]. In the Smart City Science and Technology Museum, short-range wireless communication (near-field communication, or NFC) is a good example of its application. Through a mobile phone with wireless near-field communication functions, visitors can easily transfer information, pay for goods or tickets, and browse interfaces and other applications without direct contact.

3.2 Sensor technology

Examples of sensor technology at the Smart City Science and Technology include allowing visitors to arrange dinner recipes on a coffee table, and image sensing devices that monitor home security at all times, made possible through low-cost sensors connecting to wireless networks. Sensors perceive measured information and convert it into an electrical signal or another form of information, allowing stationary objects to slowly "come to life".

3.3 Cloud computing and Cloud storage

One example of this is the "Comprehensive Urban Management Office" at the Science and Technology Museum. In addition to monitoring various areas of urban life in the office of the future mayor, data can be stored and read at any time with tools such as mobile phones, tablets, and computers, allowing the mayor to perform city operations and command decisions efficiently and accurately, ensuring smooth, streamlined access to data anytime, anywhere. This technology unifies and systematizes the venue's smart identification and management of Big Data.

4. Evaluation and improvements

4.1 An evaluation of the effects of displays at the Smart City Science and Technology Museum

On the day that the Smart City Science and Technology Museum opened, visitor feedback was very positive, attracting more people to visit. This achievement is inseparable from the understanding and reflection of development designers in the new "smart city" field, and the experience they have accumulated over the years. The first priority is to promote the intelligent management of venues and smart exhibitions. The second is to achieve the integration of visitors and exhibition objects, using the Internet of Things to achieve informational teaching, smart education management, and human-object communication. In the future, there will be more venues like the Ningbo Smart City Science and Technology Museum.

4.2 Improvement plans

The Smart City Science and Technology Museum must be strengthened in terms of guidance and overall display. Two improvements and suggestions are presented below.

4.2.1 The "Life Science System of the Future"

The exhibits at the Smart City Science and Technology Museum are classified by theme, but all must achieve the purpose of stimulating visitors' interest and thinking. To this end, the "Life Science System of the Future" [3] has been proposed, focusing on interesting exhibits and the establishment of a systematic and Internet of Things-based experiential, interactive teaching environment, including:

1) An "education through fun" tour route;
2) Simulation of a real life experience for visitors, from the entrance of the venue to the exit;
3) In addition to experience and operation, it is also necessary to set up a Q&A session for the audience to promote further thought.

Examples include connecting "smart transport" and "smart home"; in terms of leisure, offering head-mounted Google glasses and 3D smart cinemas; in terms of service, setting up wireless charging devices with connections set up between each area, providing scene transitions and reducing fatigue.

The idea of "experience-thought-experience" makes the abstract concept of science more intuitive, vivid, and interesting. Improving scientific literacy with experience and interactive teaching methods, and inspiring the exploratory spirit of the audience are the meaning of a science education system geared towards learning through entertainment.

4.2.2 Building a networked virtual science museum

The structure of the physical venue of the Ningbo Smart Science and Technology Museum is novel and unique, and the collection of high-tech exhibits with the support of modern technology is refreshing. However, it is difficult to find further information on the Internet, and there is no official website providing specific information and services. In addition to the physical visit experience, 3D panoramic technology could be used to create a more complete "virtual network venue" that is open to the audience free of charge, providing a more reasonable and complete design plan. To this end, the Smart City Science and Technology museum has several possible improvements to make:

1) A complete online networked version could be established to provide more detailed venue information services;
2) In terms of exhibit display, preliminary parts of the virtual exhibit experience could be set up online, attracting visitors to the physical venue for subsequent exploration;
3) In terms of audience services, preliminary services could be provided such as ticket purchase, tour booking, and program downloads, allowing for personalized visit plans;
4) A management system must be set up, and the developer should have a personal account through which to implement simulated teaching activities and perform venue operations.

5. Conclusion

The Ningbo Smart City Science and Technology Museum is a typical example of the intellectualization and materialization of popular science venues. Internet of Things technology stands to improve the efficiency of the venue, providing a powerful interactive experience platform for the auditorium. It will also enhance visitor satisfaction and provide a diverse learning, research, and leisure experience [4].

As a popular science education and service base that is open to the public, the science museum must adhere to "people-oriented" design concepts, providing new modern technologies such as the Internet of Things at visitors' fingertips. Although there are still many bottlenecks in terms of technological applications in domestic science and technology venues, the realization of more informative and intelligent science and technology museums and venues will be the only way forward. Alongside technological advancement, this will also indicate opportunities and challenges for the construction and development of venues.

References

[1] *From digitization to intelligence* [D] (Oct 2014) <http://www.hn.xinhuanet.com/2014-10/30/c_1113039477.htm>.

[2] National survey of information network design. Ningbo Smart City Science and Technology Museum and the Opening of the 2014 China Smart City Application Technology and Products Expo Telecom Exhibition Hall [D]. (13 Oct 2014) <http://www.cidn.net.cn/show.asp?id=29310.>

[3] Sun Xinhua, Liang Danyu. 'The popular science Internet classroom', [J] *Shanghai Economics*, 2014, 7: 60.

[4] Yang Xin. 'Analysis of the applications of Internet of Things technology in digital museums' [A]. *Proceedings of the 2013 Beijing Digital Museum Symposium*, 2013: 336.

[5] Chen Gang. 'Smart Museums — New trends in digital museum development', [J] *Chinese Museum*, 2013, 4: 3.

[6] Chen Zhong. *Application research and implementation of Internet of Things technology in exhibitions* [D]. Beijing University of Posts and Telecommunications, 2012, pp. 13–15.

[7] Pan Zhipeng. 'A preliminary study of the Smart Museum' [A]. *Proceedings of the 2013 Beijing Digital Museum Symposium*, 2013: 299–300.

[8] Chen Junfeng. 'Application assumptions museum information and RFID technology in collection management based on museum research' [A]. *Suzhou Cultural Theory*, 2012.

[9] Image sourced from the Internet.

Research on Strategies for the Protection and Utilization of Data Resources in the Big Data Era – The Example of the Palace Museum

Shi Xiumin
(Palace Museum Information Department, Beijing 100009, China)

Abstract: This paper puts forward the basic concepts, techniques, and characteristics of Big Data, summarizing the types, storage methods, data volume, and future growth trends of existing data resources of the Palace Museum. It analyzes the status of data storage, protection and utilization at this museum, as well as research on the strategy for data resource storage and protection, along with management and utilization in the context of Big Data.

Keywords: Big Data, the Palace Museum, data storage and utilization

Emerging services such as Cloud computing, the Internet of Things, and social networks are driving the types and scale of data in society to increase at an unprecedented speed. The Big Data era has officially arrived, and the issue of how to manage and use it has become a general concern. The scale effect of Big Data has brought great challenges to data storage, management, and analysis, and changes in data management methods are already happening. After more than a decade of work, the Palace Museum's data resources are slowly adapting to Big Data. How to preserve these precious data resources against the background of Big Data, and how to excavate and use them, are key issues to be considered in the future by the Palace Museum in its management of data resources.

1. Basic concepts, techniques, and characteristics of Big Data [1]

With the emergence of new information distribution methods represented by blogs, social networks, and location-based services (LBS), and the rise of technologies such as Cloud computing and the Internet of Things, data is growing and accumulating at an unprecedented rate. The era of Big Data has come. Academia, industry, and even government agencies have begun to pay close attention to the issues is provokes.

So, what exactly is Big Data? It is a relatively abstract concept, literally meaning data that is huge in size. However, it is not obvious to discern differences between the concept of Big Data and previous ideas of "mass data" and "ultra-large-scale data", and there is currently no single accepted definition of the Big Data concept. Different definitions arise based on the characteristics of Big Data, and the explanations and generalizations of these features have been used to create definitions. Amongst these definitions, the most illustrative is the 3V definition, meaning that the data must satisfy three characteristics: scale, diversity, and high speed. The International Data Corporation believes that Big Data must also be valuable. IBM believes that Big Data must be authentic, while Wikipedia's definition is straightforward: it is a collection of data that takes an intolerable amount of time to capture, manage, and process when using common software tools.

From databases to Big Data, not only does the amount of data increase, but the variety of data types, and the relationships between patterns and data, data processing objects, and tools are all undergoing fundamental changes. In the era of Big Data, data is transformed into a basic resource, and should be used to collaboratively solve problems across multiple fields. As Google's chief economist Hal Varian has said, data is widely available, and what is lacking is the ability to extract knowledge from it. The core purpose of data collection is to extract useful knowledge from it based on demand, then apply it to specific areas.

The complete embodiment of Big Data value requires the coordination of multiple technologies, of which Cloud computing is the basic platform and support technology. In the Big Data era, the diversification of data management and analysis makes relational databases inapplicable in many scenarios, instead using Big Data specialized processing platforms such as Hadoop.

In summary, information in the Big Data era features source heterogeneity, wide distribution, and dynamic growth. It is these characteristics that distinguish Big Data from traditional data management systems, posing new challenges for data management.

2. Data resource forms at the Palace Museum

After years of informatization, the Palace Museum has acquired a large variety of data resources, and is still in the process of moving towards Big Data. The data resources here can be divided into basic cultural artefact data, cultural artefact management data, cultural artefact conservation monitoring data, administrative management data, archive data, and environmental monitoring data. Basic cultural artefact data is related to the attributes and resources of cultural artefacts – an important digital resource for the Forbidden City. Business process data includes management and process data required for internal system operations. The various forms of data resources at the Palace Museum are shown in Table 1.

Table 1 Summary of the Existing Data Resources of the Palace Museum

Data category	*Specific categories*	*Storage technology*
Basic data on cultural relics	1.8 million cultural relic pieces and catalogue data	Database table
	Ancient construction data	Database table
	Document drawings	Disk file
	2D images	Disk array
	3D images	Disk file
	Audiovisual	Disk array
	Historical image files	Disk file
Cultural relic management data	Cultural relic logistics data	Database table
	Resource utilization data	Database table
Cultural relic conservation monitoring data	Monitoring data	Database table
	Attachment file data	Disk file
Administrative data	Automated office data	Database table
	Financial management data	Database table
	Attachment file data	Disk file
Archive data	PDF office files	Database table / Disk file
	Document data	Database table
		Disk array
Environment monitoring data	Environment monitoring data	Database table

3. Analysis of the current status of data resource storage, protection, and utilization at the Palace Museum

3.1 Analysis of the storage and protection status

At present, the Palace Museum uses methods such as a relational database, a file system, and a disk array to store data resources, using both full and incremental backup systems to back up most of the data in the business system. The data backup method is based mainly on disks, with applications and databases backed up to the disk of the backup server, and image resources, drawing resources, and document resource data fully backed up using tapes and CDs due to the large amount of data. In terms of backup technology, in addition to CD backup, which requires local server backup, LAN backup is used. The Palace Museum also protects data resources with both hardware and software, using a raid-5 file storage format for data storage and protection. It uses permission settings on Windows file sharing to control restrictions to user access of important data, and uses anti-virus software to protect the server system and files from overall virus intrusion, replacing failed hard disks with online Hot Spare disks in the array.

In the data resources of the Palace Museum, besides a small amount of structured data that is stored and managed in the form of database tables, most of the resource data related to cultural artefacts is stored in file form, and a large number of attached files accompanying the business system are included. Internal Forbidden Palace system files are characterized by large quantity, varied formats, multiple file size, and fragmented use, meaning that storage management is highly complex. In summary, the main issues of current data resource storage and protection include the following aspects:

1) Data resources are too scattered and require uniform management;
2) The data backup mechanism needs improvement. Due to the failure to achieve real-time active standby, the problem of zero data loss has not been solved, and local and remote backup centers have yet to be established;
3) The long-term preservation specification of data resources has not been established. Although data preservation backup has begun to take shape in the Forbidden City, the long-term availability of these resources has not been addressed;
4) There is a lack of effective digital copyright protection methods. At present, the digital resources of the Forbidden City lack effective digital copyright protection technology;
5) Potential threats exist from hacker attacks, viruses, illegal operations, intentional destruction or tampering by internal personnel, misuse, hard drive destruction, and natural disasters.

3.2 Analysis of management and utilization status

The processing and utilization of data resources in the Palace Museum began relatively early, and the scale of data occupies a leading position in the industry. Existing data resources are clearly classified, and the storage technology is appropriate. A mature backup data mechanism has been formed, and a platform for images, documents, and drawing resource retrieval and utilization has been established. The image resource management system is used for the processing, retrieval, and utilization of image resource data, providing functions such as image collection, upload, storage, burning, and verification, as well as image retrieval and utilization application approval processes, attending to the processing and publishing needs of about 50,000 images for internal and external exhibitions. The drawing resource management system is used for processing, retrieval, and utilization of engineering drawings of various departments in the Forbidden City, realizing functions such as collecting, uploading, storing, and disk burning of drawing images, as well as functions such as drawing retrieval, and the examination and approval of applications to utilize drawings. The document resource management system is used for the processing and retrieval of image data of two documents from the Qing Dynasty historical archives – "Qin Cases" and "Qin Financial Files" from the Qing House of Internal Affairs. The audiovisual resource processing system is used to manage video and audio resource data, realizing the retrieval, application, review, relocation, production, and authorization of video and audio resources.

With the continuous growth of various data forms, the existing data resource management issues in the Forbidden City have become increasingly prominent. The main problems are as follows:

1) Different data resources exist in different information systems, with different storage formats
There is structured data managed by databases, and also static images, text, videos, 3D models, and animation products directly managed by the file system. This makes it inconvenient to achieve unified query and use;

2) Lack of a unified resource retrieval and utilization platform
Although different forms of data resources can be integrated and utilized to an extent, the system is far from achieving system-wide resource integration management and utilization;

3) Large room for improvement in data access efficiency
In terms of resource file access, since resource files are stored centrally in the server. When resource files are large and the number of files is high, major operational delays are inevitable. In terms of database access, there are problems such as slow query speed when facing large volumes of data.

4. The strategy for data resource storage, protection, management, and utilization at the Palace Museum in the era of Big Data

4.1 Storage and protection strategy

1) For the management problem regarding multiple disk arrays, a storage virtualization layer must be built to run above the disk array, separating logical mapping of resources from physical storage. The layer adopts centralized management on system administrators and user shielding of specific disk or tape details and storage paths, dynamically allocating storage resources to individual applications according to specific needs. In particular, virtualization technology allows for disk arrays to be used to simulate tape libraries, providing application with storage resources that are as fast as disks and as large as tapes.

2) Centralized backup of data resources through virtualization solutions

The Palace Museum has built over 20 business systems, developed at different times. In general, a single physical machine runs a single application, providing low server load efficiency. As the number of servers increases, the space in the equipment room becomes saturated, and the electric power is overloaded, exacerbating problems. Using highly developed virtualization technology to integrate existing server resources allows the server to run more smoothly and efficiently. The existing backup strategy of the Palace Museum is tightly coupled with the business system, and cannot meet needs of virtual servers. In future, we must consider the full application of virtualization technology to build a virtual real-time disaster recovery backup platform to achieve the centralized backup of data resources.

3) Centralized storage becomes distributed storage, improving data access throughput by allowing concurrent access on multiple server disks

The traditional network storage system uses a concentrated storage server to store all data. This storage server becomes a bottleneck in system performance, and also the focus of reliability and security. It is unable to meet the needs of large-scale storage applications. The distributed storage system adopts a scalable system structure, utilizing multiple storage servers to share the storage load, and using location servers to locate storage information, which not only improves system reliability, availability, and access efficiency, but is also easily expanded.

4) The gradual establishment of a remote disaster recovery backup center, which will form a two location, three center data disaster recovery model

The central office of the Palace Museum serves as the main data center for the production and operation of business as a centralized monitoring facility. The local disaster recovery office in Beijing is used as a backup data center within the same city, and is used to switch part of the business or all application systems to the same city backup data center when there is a short-term unresolved fault or non-regional disaster within the primary data center. The remote disaster recovery equipment center serves as an off-site backup data center that is used in recovery of critical business application systems in the short-term or in the event of a regional disaster.

5) Exploring technical standards for the establishment of file formats for the long-term preservation of digital resources and their utilization

The long-term preservation of data resources has two implications, one being long-term storage, the other being long-term access. Digital resources accumulated by the Palace Museum are precious, and we must continue to attend to long-term preservation policies, regulations, technologies, and standards for international and domestic data resources, exploring the establishment of long-term preservation strategies for the museum's own data resources, and ultimately establishing a long-term preservation system.

6) Realizing content protection, identity authentication, and secure transmission of digital resources via DRM digital copyright protection technology so as to protect the copyrights of digital resources

At present, a series of relatively mature DRM software products and components have been developed both in China and abroad, which can be applied to various resource files such as 2D and 3D images, audio, and video.

7) Fully exploiting the use of equipment such as tape libraries, Blu-ray discs, SATA hard disks, and large capacity server hard disks to improve the security and efficiency of data resource storage and backup.

4.2 Management and utilization strategy

1) On the basis of the realization of data interconnection and intercommunication, we will continue to deeply analyze the consistency of data resources of various calibers, establishing a global unified storage and management system for various data formats stored in different locations. The mechanism will enable polymorphic data to support applications flexibly, avoid data redundancy, reduce data conflicts, and achieve effective organization of various data forms, as well as uniform management of various data forms and safe utilization of various data forms.
2) We will build a unified resource retrieval and utilization engine across all business resources, and an index based on the attributes of different resources, allowing business personnel with access rights to clearly understand and find various data content and formats required from a unified portal, obtaining the data resource they need to use from it. Furthermore, we will establish a data resource access log to trace the data usage process and ensure data security.

3) We will extend the search and utilization functions of digital resources to the Internet, displaying precious digital resources from the Palace Museum to partner museums, industry experts, and cultural enthusiasts in China and abroad, providing quality services for research, education, and learning.
4) Based on the deep exploitation and utilization of digital resources, we will create a museum knowledge discovery system. Through unique identification, in-depth data mining, and granular analysis of knowledge objects, we can establish different knowledge objects amongst various digital resources, with direct correlations that help users quickly form a structured understanding of the information required.
5) We will build a distributed private Cloud storage system to uniformly manage the attached files of all business systems and the working files of internal Palace Museum users. As a new application trend, private Cloud systems have the advantages of unified management, easy centralized backup and disaster recovery, easy expansion, easy upgrade, cost saving, and green energy saving. Users of private Cloud storage do not need to know specific details of Cloud composition. As long as the corresponding interface is known and the corresponding strategy is provided, the rest of the work can be completed by the Cloud.
6) Cache technology is considered to be one of the most effective ways to reduce server load, reduce network congestion, and enhance Web scalability. In future, we must begin with the two aspects of Web cache and database cache optimization, carefully sorting out data access characteristics of different business logics in the system, and carrying out full top-down design in system planning to improve the efficiency of data access.

5. Summary

The era of Big Data has arrived, and its correct use has brought great convenience to public life, at the same time bringing major challenges to traditional data management methods. Faced with the trends and challenges of this era, the Palace Museum regards the preservation and utilization of data resources as a top priority in informatization, and is constantly studying data resource storage and management technology within Big Data. It is continually imparting, inheriting, and exploring in depth, so that its precious cultural heritage resources can be presented to the people of the world more effectively.

References

[1] Meng Xiaofeng, Ci Xiang. 'Big Data management: Concepts, technologies and challenges', [J] *Journal of Computer Research and Development*, 2013, 50(1): 147–150.

Digital Museums and the Internet of Things

Liu Jia
(Palace Museum Ministry of Information, Beijing, 100009, China)

Abstract: The development of information technology centered around the Internet of Things is known as the third information technology revolution, and has received significant attention in the field of cultural heritage protection. This paper begins with the concept of IoT, its key technologies, and its development status in China, and goes on to discuss the application prospects for Internet of Things technology in museums.

Keywords: cultural heritage; museum; internet of things

Preface

The development of information technology centered on the Internet of Things has been hailed as the third revolution in information technology. In recent years, the Chinese government has attached great importance to the development of IoT, issuing a series of industrial policies and plans and raising it to the status of a national strategy. In the field of cultural heritage protection, IoT has also received attention, with the National Cultural Relic Bureau carrying out a series of deployments since 2009, organizing strategic planning research on cultural heritage protection based on IoT technology. Many Chinese cultural and historical institutions such as the Palace Museum, the Capital Museum, the Hunan Provincial Museum, and the Shanghai Museum have also focused attention on IoT technology, making useful efforts in regards to collection management, audience service, environmental monitoring, and security monitoring.

1. The concept and development status of the Internet of Things

1.1 What is the Internet of Things?

Our current Internet is a computer terminal interfacing another computer terminal, and in the Internet of Things, a "thing" refers to any object in the world. The Internet of Things is an "Internet connected by things", that is, all objects connected to the Internet via information sensing devices such as radio frequency identification (RFID), infrared sensing, global positioning systems (GPS), and laser scanners, for information exchange and communication, to realize smart identification, location, tracking, monitoring, and management. The Internet is the foundation of the Internet of Things, and IoT is an extension of the development of the Internet. EPOSS (the European Platform on Smart Systems Integration) performed an analysis in the *Internet of Things by 2020* report, and predicts that the future development of the Internet of Things will go through four stages. Before 2010, RFID was widely used in the logistics, retail, and pharmaceutical fields; between 2010 and 2015, objects were interconnected; between 2015 and 2020, objects will have entered semi-intelligence; and after 2020, objects will become fully intelligent.

At present, the Internet of Things has entered a stage of practice and development in China, with research on sensor components reaching maturity, and wireless communication networks covering vast urban and rural areas, forming a complete industrial chain from materials, technology, devices, and systems to networks. On August 7, 2009, then Premier Wen Jiabao put forth the concept of "perceptive China" in Wuxi, and the State Council listed the Internet of Things as one of China's strategic emerging industries. In November 2009, the State Council officially approved the development of the National Sensor Network Innovation Demonstration Zone (National Sensor Information Center) in Wuxi, which greatly increased policy support in the Internet of Things industry, and significantly accelerated the development of the IoT industry.

1.2 Analysis of the IoT's structure and key technologies

The Internet of Things is divided into a perception layer, a network layer, and an application layer (see Figure 1). The perception layer includes data acquisition devices such as sensors and sensor networks prior to data access via the gateway. This layer is the basis for the development and application of the Internet of Things. RFID technology, sensing, and control technology, as well as short-range wireless communication technology are the key technologies of the perception layer. The network layer is built on existing mobile communications and the Internet, with the Internet of Things connected to the mobile communication network and Internet via various access devices. Perceptual data management and processing technology in the network layer is the core technology in the data focused implementation of the Internet of Things. As a storage and analysis platform for mass sensing data, Cloud computing will be an important part of the IoT network

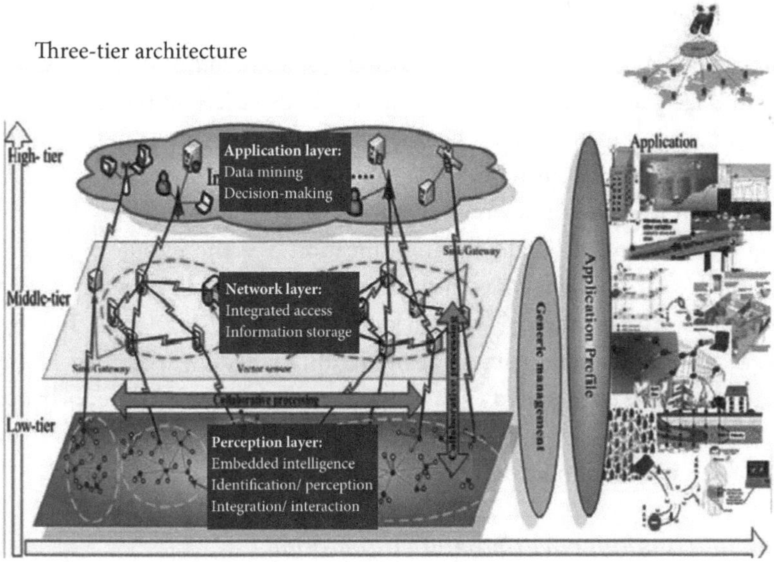

Figure 1 Introduction to the Internet of Things

layer, and the basis for many applications within the application layer. The application layer of the Internet of Things uses perceptual data that has been processed and analyzed to provide users with a variety of specific services, such as logistics monitoring, smart retrieval, smart transport, mobile wallets, and electronic tolls. The application layer is the purpose of IoT development, with software development and smart control technologies providing users with rich and colorful IoT applications.

Key technologies of the Internet of Things include IoT architecture technology, unified identification technology, communication technology, network technology, software services and algorithms, hardware technology, power and energy storage technology, security and privacy technologies, and standardization. The characteristics of the Internet of Things can be summarized as comprehensive perception, seamless interconnection, and high intelligence.

2. IoT application prospects in museums

In recent years, the concept of the Internet of Things has also been applied in the museum field, including features such as electronic ticketing, smart terminal navigation, audience interest behavior analysis, and indoor environmental monitoring. In summary, IoT technology stands to cover all areas of cultural heritage monitoring, management, and service in museums, including:

2.1 Wireless security and fire protection systems

The wireless deployment of security monitoring systems in a museum is ideal, as it does not alter building structure or interior decoration. It also accesses the local network with wireless terminal equipment, so that the status of various key areas can be attained in real time. Video information from monitoring points is collected by cameras and wireless video servers, and live video information from different areas is then transmitted to the monitoring center through wireless communication and the monitoring management center responsible for receiving video information at each monitoring point, performing real-time monitoring and storage, playback, and backup. Using IoT technology, information regarding the operation status of firefighting facilities in ancient buildings and hidden dangers can be transmitted to the command center in real time through wireless communication modules, so that staff can monitor the buildings around the clock, eliminating fire hazards immediately.

2.2 Indoor and outdoor environmental monitoring

The indoor and outdoor environmental monitoring system is used to meet environmental monitoring needs for the preservation of a museum's cultural artefacts, exhibition halls, and outdoor furnishings, maintaining protection parameters such as temperature, humidity, illuminance, radiation intensity, wind speed, air flow, and rainfall via IoT technology. It monitors these and other parameters in real time, and mines monitoring data to establish an environmental parameter database to provide technical support for the preventative protection of cultural artefacts, providing scientific bases for the development of protection measures.

2.3 Monitoring and management of movable and immovable cultural artefacts, including ancient buildings, animals, and plants

Due to the comprehensive perception, seamless interconnection, and high intelligence of the Internet of Things, the technology can be applied to the monitoring and management of ancient buildings, cultural artefacts, animals, and plants. For example, the use of IoT technology can achieve comprehensive monitoring and management of warehouse status, exhibition status, and utilization status, as well as the establishment of real-time monitoring systems for animals and plants via IoT, including remote real-time monitoring, warnings for the illegal movement, damage warnings, and animal and plant growth monitoring.

2.4 Visitor management and visitor services

In the smart museum, IoT technology can be used for electronic ticket scanning. Channels are unattended at the venue, and visitors simply enter and exit according to ticket allowance, with management personnel controlling, counting, monitoring, querying, and directing the channel from a management center elsewhere. It can also be used for minor electronic payments. Tickets can be recharged freely, to be used for dining and shopping within the venue. Another use is personnel monitoring and management, in which base stations are set up at the entrance and exit of regulatory areas, attractions, roads, and some necessary locations, providing coverage of the entire regulatory area to monitor personnel positioning, tour track management, search for lost visitors, and manage visitor density. Yet another use is to push visitor service information, including smart electronic tour guides, temporary delivery of service information to tourists, gaming interaction, and the automatic production of personalized cards.

2.5 Internal management of museum personnel, vehicles, and assets

The Internet of Things can also be applied to the management of museum employees, and the entry and exit of vehicles. Employees wear work permits containing RFID chips. These chips can also be installed on vehicles. Data is read by RFID readers positioned at entrances, then transmitted to the management center, performing functions such as personnel and vehicle identification. In regards to fixed asset management, IoT technology can be used to realize smart, dynamic, real-time tracking and centralized monitoring and management of the entire life cycle of an asset (including addition, allocation, inactivity, disposal, and maintenance).

3. Conclusion

In summary, the Internet of Things stands as the next major source of productivity in the promotion of rapid global development, and it has broad application prospects within the museum field. We must also consider that the IoT museum applications are still in their infancy. In addition to needing innovation in ideas and financial support, certain museum characteristics (such as conservation of cultural artefacts and ancient buildings) also pose significant difficulties and challenges in the applications of IoT technology in the museum field, and it is necessary to adopt development strategies with unified planning and step-by-step implementation. We have reason to believe that with technological developments, the introduction of standards in the sector, and strong support from national policies, the use of IoT technologies to build smart museums is no longer a distant prospect.

Online Museum Virtualization Solutions Based on Web 3D Display Technology

Wang Jia
(Beijing Hodojou Technology Co., Ltd., Beijing 100190, China)

Abstract: With the popularization of the "Internet +" concept, traditional museum resources can be spread more efficiently via the Internet, allowing museums to display and arrange their own exhibits, rediscovering the new value of archive-level exhibitions and breaking through the limitations of physical museums and time, allowing all interested parties to "experience" the attraction of a museum at close range. Online museum virtualization programs based on Web 3D display technology provide a partial solution, while the experimental website www.hodoso.com (also based on this program) tests the practicality and stability of the program in a practical way.

Keywords: Web 3D technology; museum virtualization; Internet +; digitization

Preface

In 2013, the concept of "Internet +" was proposed, using Internet technology, tools, and applications to upgrade traditional industries and create new formats, becoming a dominant development direction. Exhibition-related industries, especially the museum industry, are frontrunners in this respect, with various virtualization technologies having been applied to the displaying of collections over the last few years. With the advent of the Internet era, museum virtualization solutions are also undergoing comprehensive upgrades.

By the end of 2014, there were 4,165 museums across China, with a total of approximately 35,500 cultural artefacts, 20,000 exhibitions each year, and 600 million visitors [1]. Given the richness of offline exhibition resources, and the improvement of museum digitization, the effective use of resources has become a key problem that must be solved.

According to a statistical report released by CNNIC, as of December 2014, the number of Internet users in China reached 649 million, with an Internet penetration rate of 47.9%. In the face of such a large potential user community, sophisticated online virtual museums have become an urgent need.

Compared with traditional offline exhibitions, online virtual exhibitions break the limitations of time and space, expanding a museum's influence. With a variety of new display technologies, users can obtain a personalized browsing experience. For museums, virtual online display can increase display autonomy, and remove the restrictions of physical conditions and economic costs, offering users an improved experience with optimal results.

An online museum virtualization solution based on Web 3D display technology was officially proposed, designed, and implemented at the end of 2013. Ho Do So (www.hodoso.com) was officially launched online in 2015 as a proponent of this solution – a small-scale promotion operation aimed to push Web 3D-based browsing. Meanwhile, the program aims to return the control of exhibit and information management to the physical museum, with the government embassy no longer dependent on the system builder, making virtual museum development a truly autonomous long-term solution.

1. Web 3D display technology

Web 3D technology itself has no strict definition, but can be considered as 3D graphics technology that is reliant on the Internet. It is widely used due to browser support for 3D display, the pressure of heavy traffic on network bandwidth, and complexities in its early development. With the establishment of the HTML5 standard in 2015, WebGL technology made a big splash, allowing Web 3D technology to gain prominence. WebGL uses JavaScript to implement 3D graphics without the need for plug-in support, allowing web developers to write code to render 3D scenes and objects on a system graphics card (GPU). This technology has made development and application extremely simple, popularizing 3D display on the Internet.

However, because WebGL technology is still in its infancy, there is still a way to go at the level of widespread application, with Flash 3D, Unity 3D, and traditional Web 3D technologies based on the VRML standard [2] that is still used widely in the industry. This means that display solutions are not limited to one Web 3D display technology, and the display technology most suited for certain program implementation is selected based on specific situations and the scope of application.

2. Comparison of Web 3D and existing virtual display methods

Most existing online digital virtual museums in China use traditional, flat display methods (text, images, video). Some adopt 360° panoramic technology with certain 3D visual effects, and there are only a few cases of Web 3D display.

Unlike single item display, museum exhibition display emphasizes the fusion of exhibits and scenes. In this respect, common 360° panoramic technology display methods can achieve integrated display of exhibits and the environment, completely based on the shooting of physical scenes. This technology also stands out in terms of exhibition hall restoration.

However, since 360° panoramic technology scenes involve one-time imaging, the construction period is long and costs are huge. As a one-off project, if exhibits change, modifications have to be made by original contractors, making maintenance and management very inconvenient. From the production point of view, due to photo-based "splicing", requirements for the shooting environment and equipment are relatively demanding, with high production costs; in regards to user experience, users are limited to a fixed "visual" of the 360° panoramic technology browsing mode, and are only able to complete a viewing tour according to the creator's viewing angle and travel route.

Museum virtualization based on Web 3D display technology circumvents the above problems.

3D model scenes built based on real data are not inferior to the panoramic photo stitching in terms of displaying venue details, as real-time rendering technology allows Web 3D display to change light and shadow effects according to different lighting levels – an issue that the fixed images of panoramic technology cannot overcome.

Furthermore, since the construction of a 3D model is not an integrated structure, there is great freedom in terms of venue layout editing; it is possible to increase or decrease space and edit exhibits.

Interactivity is the most significant feature of Web 3D display technology, as only real-time rendering technology can provide a high level of interactivity. This so-called interactivity refers to the ability of 3D visitors to control and manipulate the virtual scene and its 3D objects, with visitors able to change their direction and speed of roaming in the virtual scene at any time, to open doors within the virtual scene, and even change the appearance of an object display if necessary. These features can be seen only as a black hole in terms of panorama technology.

Of course, the greatest advantage is still in the display of non-planar exhibits. If the display of flat exhibits is not enough to reflect advantages of 3D stereoscopic objects, then the ability to display 3D exhibits and non-static exhibits completely defeats traditional display. The seamless 720° experience of the "Thinker" and its object image at a glance are shown in Figures 1 and 2.

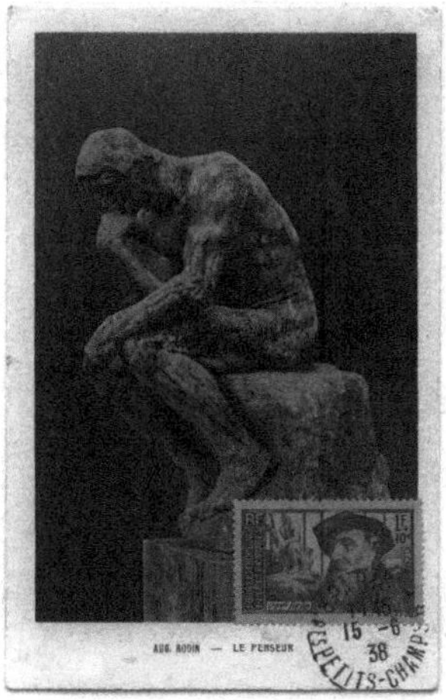

Figures 1 Postcard of The Thinker

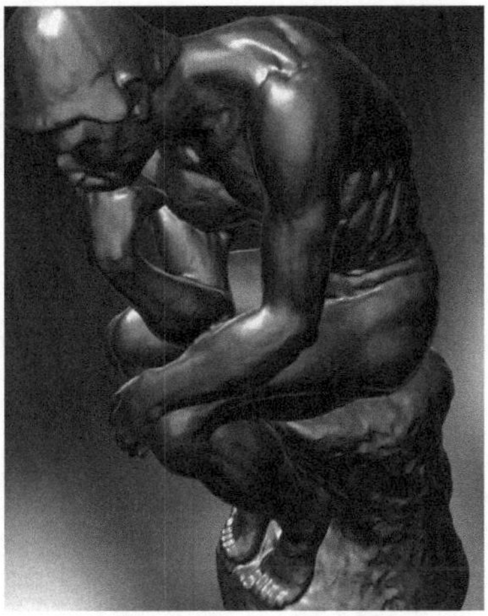

Figures 2 Model of The Thinker based on Web 3D technology

3. The overall architecture of online museum virtualization solutions based on Web 3D display technology

Web 3D technology has various advantages for vector display.

A brief framework for the Web 3D solution is shown in Figure 3.

The entire set of programs is divided into four parts according to functions: resource collection and management, resource storage, resource display, and data analysis (resource utilization).

The first three parts provide basic resources around data that can be used in Web 3D. In the final part (data analysis), valuable information is obtained from the feedback collected during the Web 3D presentation process to further enhance display effects and improve the experience of virtualization.

4. Solution implementation

The implementation of the solution will occur part by part according to the functional divisions in the framework design.

4.1 Resource collection and management

Resource collection and management will complete the most basic processes of the program, and everything is built on the precise completion of this step.

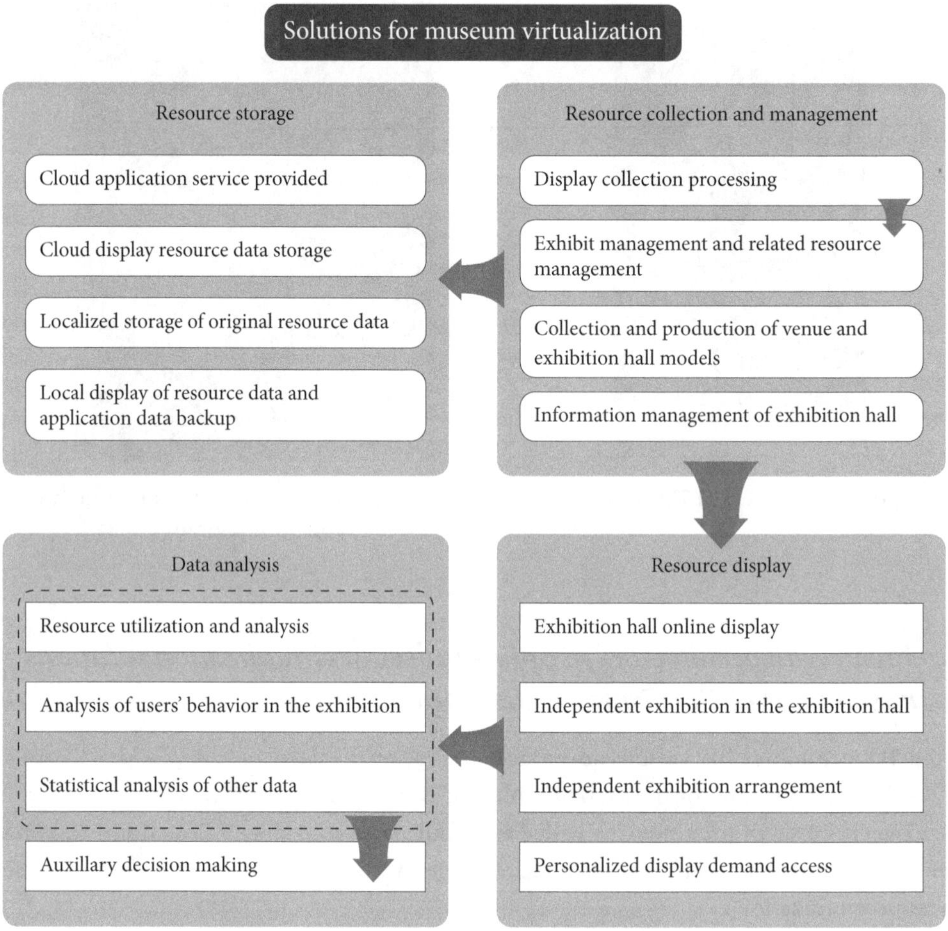

Figure 3 Brief diagram of the Web 3D solution framework

4.1.1 Exhibit collection and processing

Exhibits are the core value of every museum. According to statistics, in general, less than 50% of museum collections are exhibited, and in some places the proportion is lower than 10%. Consideration for the safety of artefacts is one of the reasons for this. Virtualization will solve this issue, as the display of exhibits after virtualization will not damage the physical objects, making them easier to share and spread.

From the perspective of existing exhibit resources, overall presentation is diverse, from common flat exhibits (photography, painting, calligraphy), to three-dimensional resources (sculpture, installation), and dynamic exhibits (graphics, sounds, behaviors). Obviously, unified acquisition and processing methods will not meet this need.

Planar exhibits are generally converted directly into the most conventional digital resources through specific scanning instruments or photographic equipment. However, due to large size, such resources are generally not directly usable in network display, meaning that these kinds of resources must undergo processing twice to achieve actual use. Figure 4 depicts the conversion process for conventional planar digital resources.

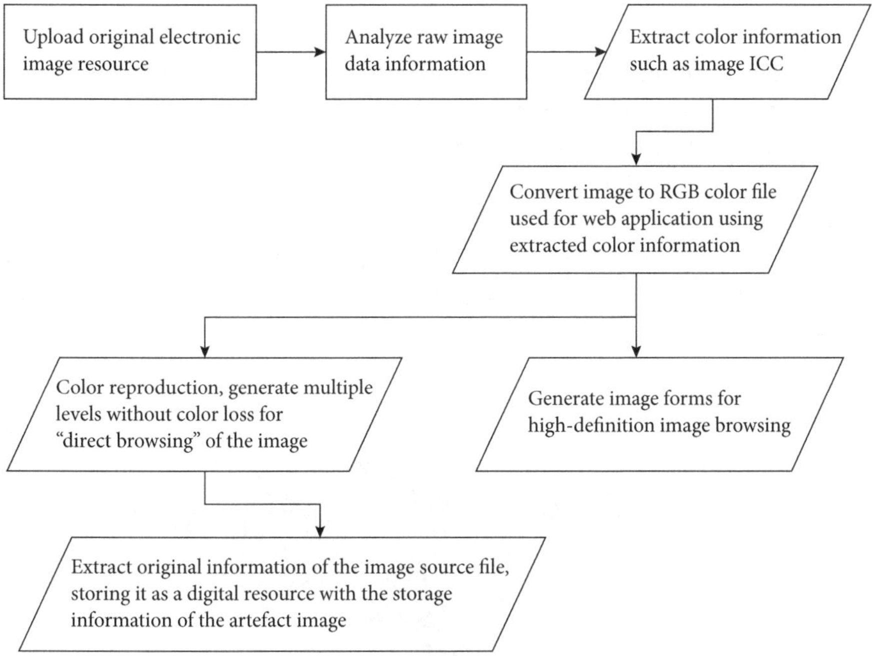

Figure 4 Simple flow chart for plane resource processing

The process shown here uses special image processing services to achieve efficient, low-loss processing, the core of which includes color reproduction technology and image preview slice preprocessing technology at the level of hundreds of billions of pixels.

Image processing services can also extract original data, such as the Exif information of a photo, which is included in the planar digital resource, achieving the most thorough acquisition of original data.

In terms of audio resources, automatic and manual processing methods are adopted. The automatic processing aspect is mainly applied to WebGL model display, serializing bitmap model scan data extracted by the device, the conversion of which results in a data storage format that is recognized by the browser program. For data processing demonstrated by use of other Web 3D methods, raw processing for models is completed using semi-automated model data import processing.

Currently, the processing of audio and video resources must be treated differently depending on the availability of information from original resources.

4.1.2 Exhibit management and related resource management

As the number of complete exhibits increases, confusion will inevitably occur in utilization, and therefore a relatively well-functioning resource management system offers essential assistance to virtualized museums.

A comprehensive exhibit management system will include functions shown in Figure 5, with identification of grouping and labeling effectively allowing the museum to quickly locate resources and select resources for future deployment and utilization. The editing of metadata is a necessary means for the scientific management of exhibit information. Authority allocation allows museums to safely control and share the management of exhibit digital information and data rights. Hooking up related resources is an important means for the enrichment of exhibit content. By expanding the resource chain of exhibits, it will be possible to link exhibits to one another and expand their depth (see Figure 6).

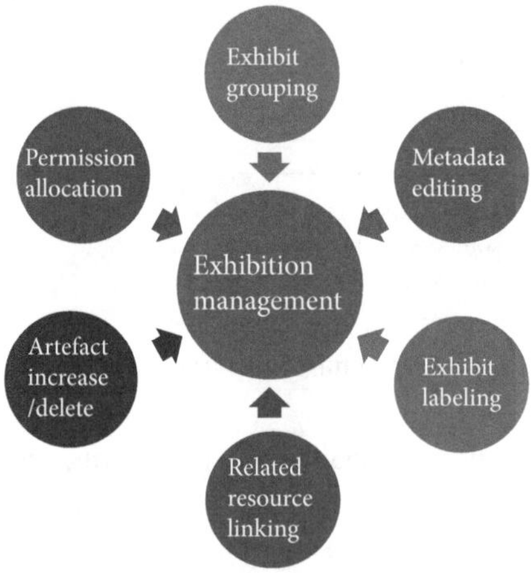

Figure 5 Functional display of the exhibition management system

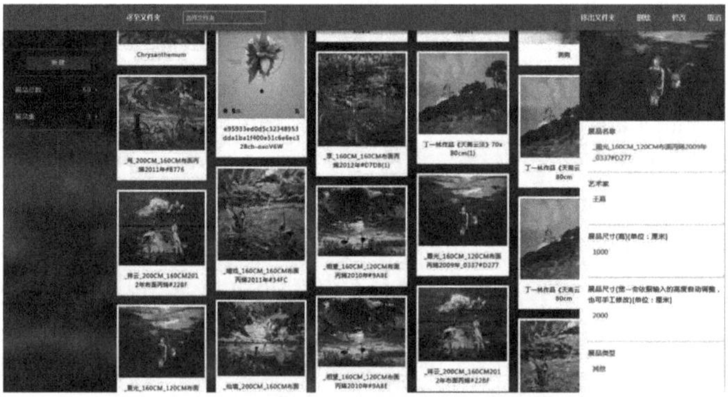

Figure 6 Exhibit management system in actual use

4.1.3 Collection and production in venues and exhibition halls

As a carrier, venues and exhibition halls are characteristic to each museum, and are also an important window to express a museum's cultural value. The collection and production of models are designed by professional personnel according to source data provided by the museum, with scenes collected on the spot to guarantee 100% restoration of the venue.

4.1.4 Management of exhibition information

Museum virtualization will provide a huge quantity of exhibition-based displays, with the museum able to fully and independently arrange interactive modes for virtual online exhibitions and physical offline exhibitions, to comprehensively enhance the its practical value. From a management point of view, systematic exhibition organization also represents professionalism, enhancing the management of a museum's activities through exhibition classification, precise information, and feedback statistics.

4.2 Resource display

Following the completion of resource collection and collation, the optimization of exhibit display has become the focus of the solution, giving rise to Web 3D-based resource display.

4.2.1 Online exhibition hall display

As one of the core methods for online display, museum exhibition halls must offer convenient access for users with lower-end and mobile terminals over the network anytime and anywhere.

The advantages of Web 3D display are as follows:

1) 3D exhibition hall browsing
As a basic function, visitors can freely browse virtual 3D scenes by manipulating the keyboard and mouse to obtain a realistic browsing experience (see Figure 7).

2) All-round display
Online display also offers an all-round display that is not available in an offline venue. In physical exhibitions, visitors see only what is available in the exhibition hall, while the map system of the virtual museum (see Figure 8) offers a full picture of the museum and the user's travel routes. Through the overall 3D appearance of museum browsing, users can obtain an intuitive understanding of the 3D space, at the same time appreciating the artistic sense of the museum's architecture.

3) Multi-version and cross-platform Web 3D display
Existing Web 3D technology has appeared in many applications, from the most avant-

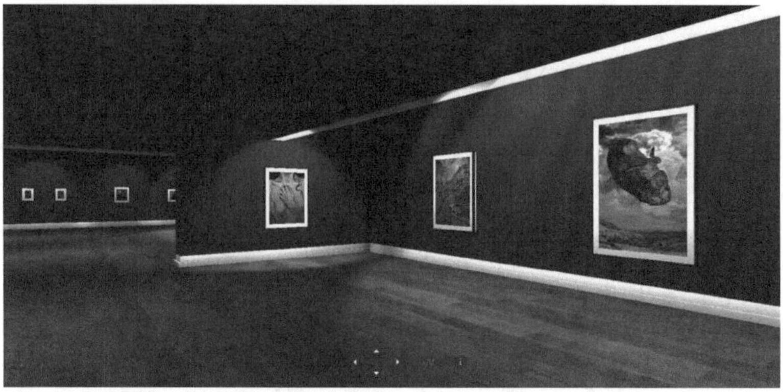

Figure 7 Immersive experience

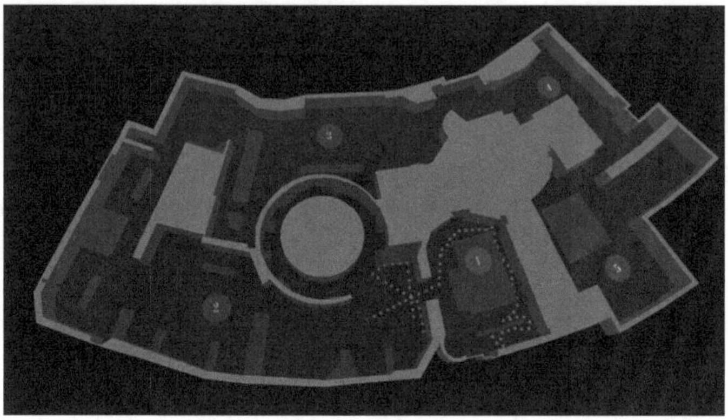

Figure 8 Exhibited map of the exhibition hall

garde HTML5 WebGL technology to Unity 3D, as well as early Flash 3D technology, with a variety of display coordinate systems being used for different presentations (with different display methods, spatial coordinates are different). The system provides a coordinate conversion method based on unique, relative coordinates, with the exhibition hall, its exhibits, lighting, and environment stored in unified standard, as shown below.

```
<?xml version="1.0" encoding="utf-8"?><root>
   <exid></exid>        <extitle></extitle>  <exdesc></exdesc>
   <roomid></roomid>
   ………..
<bgmusic></bgmusic>
   <walls>
              <wall name="xxxxxx">
                         <color>RGBA(241,241,241,255)</color>
```

```
            </wall>
    </walls>
    <floors>
            <floor name="dimian">
                    <color>RGBA(123,123,123,255)</color>
            </floor>
    </floors>
    <photoframe></photoframe>
    <ambient brightness="1.0"></ambient>
    <lights></lights>
    <works>
            <work>............
            </work>
</works>
</root>
```

Meanwhile, this data is also used for the self-service exhibition system of flat exhibits, which can display information about the exhibition hall and exhibition in a unified manner across a variety of terminals.

4) Automatic tour function
Visiting the exhibition according to preset routes frees visitor hands and optimizes the browsing experience.

4.2.2 Independent exhibit display

Another highlight of the digital virtualization of museums is the fact that "visitors" are presented with an experience in surprising ways that were previously impossible to realize, including the independent exhibit display function.

1) 720° exhibit display
Based on Web 3D display technology, visitors can see details of an exhibit from any angle, using the convenience of the network to discuss and share exhibits in real time with many followers.

2) HD detail display
Although 3D exhibits are generally displayed in detail, this feature has advantages for flat exhibits, with detailed display relying on the basis of powerful acquisition functions to display micro-details, as shown in Figure 9.

This kind of texture-level display relies on virtualization technology to allow ordinary visitors to admire and study exhibits at close range.

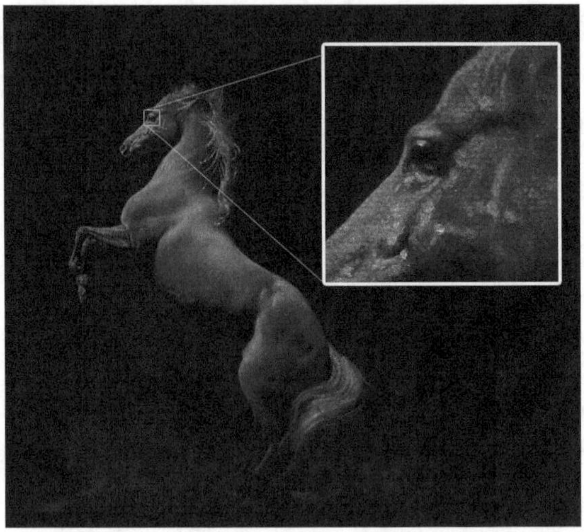

Figure 9 Planar exhibition in HD display

4.2.3 Self-service exhibition system

The self-service system and its supplementary functions improve upon the implementation efficiency of museum staff in virtualization. In view of the existing Hua Duo mass edition open self-service exhibition system, a good deployment system contains the following necessary functions:

1) Detailed exhibition information
Detailed and professional information is provided for the venue so that exhibitors can understand the parameters and overall structure of the exhibition hall in detail from the outset, facilitating more professional and appropriate exhibition planning.

2) Supplementary plane self-service system
The supplementary plane of the automatic self-service system quickly generates exhibitions according to the number of exhibits selected by the user, the spacing of exhibits, and the height of line of sight. The user can then make detailed adjustments according to individual needs in order to complete the display.

3) Professional parameter setting
For certain professional needs such as lighting adjustments, a professional parameter adjustment interface can be opened for museum personnel to set optimal special display effects. Figures 10 and 11 demonstrate a case of lighting adjustment in the implementation of an existing HoDoSo display.

By adjusting lighting, the look of the space can be improved, and the three-dimensional effect of the exhibit can be increased, as shown in Figure 10.

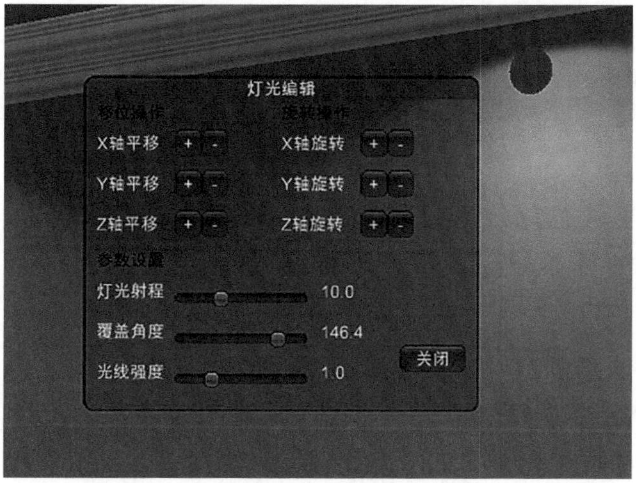

Figures 10 Lighting adjustment parameters

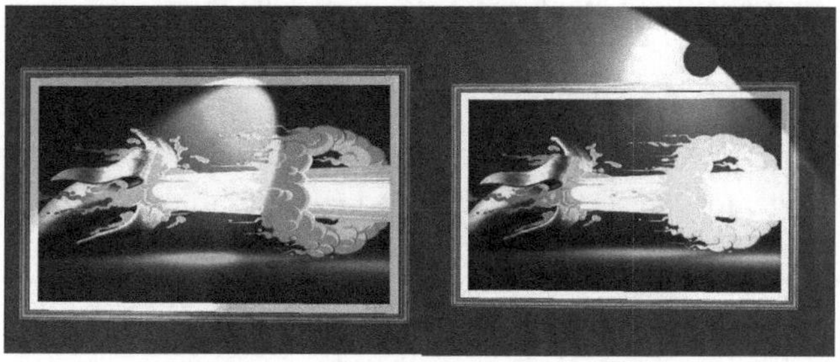

Figures 11 Lighting adjustment effect

The self-service exhibition system is a highly open access method to meet individualized customization requirements and different requirements in terms of the development of exhibitions and curation.

4.2.4 Personalized display demand access

In response to the individual needs of museums in regards to existing modernized displays, a data display interface is provided to help museums make appropriate and effective use of existing resources in the fields of mobile display, large-screen display, and VR interactive display .

4.3 Data analysis

Museum virtualization programs are not simply designed to help museums display collections in a better way. They also play an important role in the improvement of business and research.

4.3.1 Resource utilization and analysis

The virtualization and digitization of exhibits allows for convenient and quick search and statistical capabilities of artefacts. Museum can comprehend the quantity and other information of museum collection through simple statistics.

In exhibit and exhibition management, a standardized classification method has been adopted, also allowing for individualized labels to be set, realizing the combination of faceted and keyword searches, meeting the needs of various users, and even offering recommendations according to functional demands.

4.3.2 Analysis of user viewing behavior

As seen in the user behavior track record shown above in Figure 9, audience behavior undergoes analysis after it is collected and stored, analyzing data regarding each exhibition and the interest level, viewing time, and viewing frequency of each exhibit according to screening conditions. This provides a clearer picture of the values of education and popular science brought about by physical museum collections.

Meanwhile, the analysis of audience attention to long-term exhibitions, artefacts, historical exhibitions, and related categories can help museum personnel to decide on activities, exhibition layouts, and publicity, as well as adjust the display layouts of offline exhibits and arrange appropriate viewer diversions, guiding offline work via scientific data analysis.

4.3.3 Other types of analysis

Other analyses include the analysis of user review data, recommendations for appropriate museum exhibits and promotions, as well as commercial value-added push data.

4.3.4 Supplementary decision making

The combination of the above usage data and the management of existing data can appropriately and effectively improve service, research, and marketing, allowing the virtual museum to bring added value to the physical museum.

4.4 Storage of resources

Storage of resources is one of the most interesting aspects of museum virtualization, requiring both efficient application and security of original digital data. Therefore, the proposed solution entails a storage structure for data to be stored separately on the Cloud and local servers for different resources and applications. A simple structural diagram of the system architecture is shown in Figure 12.

With a core storage scheme similar to the structure above, high-definition digital resources are stored safely and locally, allowing the public to efficiently browse open resources with effective data backup. This maximizes the convenience of Cloud servers as well as the high scalability and security of local servers, at the same time providing the most reasonable optimization in terms of hardware cost and maintenance of the whole solution.

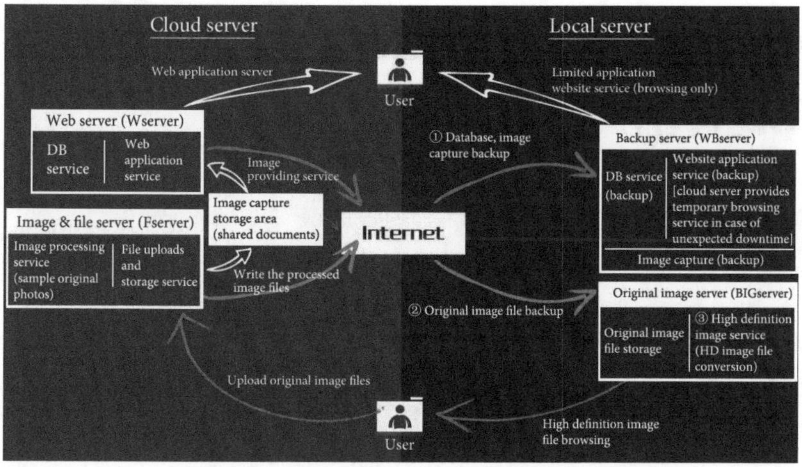

Figures 12 Storage structure distribution

5. Conclusion

Online museum virtualization solutions based on Web 3D technology provide new ideas for the museum and exhibition industries. With the continuous advancement of virtual reality technology, the online, virtual 3D exhibition can completely restore the reality of an exhibition, and also enable visitors to interact and experience unprecedented entertainment value. The introduction of new interactive technologies such as VR will allow visitors to experience museums across time and space. In the process of consistent improvements, this idea will bring about new visions for the foreseeable future of virtualization. The "Internet + museum" concept will bring a new vitality to the sector.

References

[1] Meng Xin. 'From digital to intelligent: the Smart Museum is ready to go'. Chinese Culture News (30 Oct 2014)

[2] Liu Yunzeng. 3D graphics technology on the Internet — Web 3D. *Computers and Networks*, 2013, 10.

Exploring App Construction at the China Film Museum

Liang Wenjuan
(China Film Museum Network Information Center, Beijing 100015, China)

Jin Wenxin
(China Film Museum Event Management Department, Beijing 100015, China)

Abstract: The number of mobile Internet users in China has exceeded 900 million [1], comprising the world's largest mobile Internet market. Applications (or apps) have been increasingly appreciated and applied by various organizations and institutions. The Internet is moving quickly from PC to mobile platforms, and with the further development of the museum's cultural communication, social education, and research functions, the question of how to use apps to promote services to the public is being considered and explored by museums both in China and abroad. This article focuses on how to use apps to serve the audience of the China Film Museum and national web users, exploring how we build our apps and enrich the publicity and services of our collection using new media. This article also provides guidance for app development projects at our museum, as well as offering references and inspiration for other venues.

Keywords: Apps for the museum industry, apps for the film industry, app development, app function positioning

Overview

Opportunities and challenges coexist in the app field. At present, all kinds of new media are developing rapidly, and new media public service platforms are receiving increasing attention from institutions and organizations. Meanwhile, with the fast development of smart terminals, China's mobile Internet user count has exceeded 900 million [1], forming the world's largest mobile Internet market. With the rapid development of 3G and 4G technology, as well as the promotion of WiFi, apps are becoming more popular amongst various organizations and institutions. The app makes full use of the mobile Internet and smart terminal processing to provide users with varied and practical services. According to data released by Apple WWDC (2015-6-9), the total number of apps in the App Store exceeds 1.5 million, with the total number of app downloads exceeding 100 billion. This data is still growing rapidly. The transfer of Internet use from PCs to mobile terminals is irreversible, and at present, the rapid development of the mobile Internet market poses both challenges and opportunities. All industries must seize opportunities for constant innovation, striving to be frontrunners in the mobile Internet field.

With further developments in cultural communication, social education, and research functions at museums, the use of apps to promote services is an issue that is being considered and explored by museums in China and abroad. However, at present there is no uniform methodology or guidelines to follow.

Therefore, this article examines how to use apps to serve visitors to the China Film Museum as well as national web users, exploring how to build better apps for our museum and enrich our publicity and services with new media. We provide suggestions, guidance, and inspiration for other app development projects in our museum, as well as references and inspiration for other institutions in the industry.

1. Analysis and comparison of app status in the same or similar industries

1.1 Status analysis

The author has conducted a large number of investigations into the status of apps in the same or similar industries. This article will focus on the analysis of apps in several related fields, as follows:

1.1.1 Research and analysis of museum apps

Domestic and foreign museum apps are generally divided into categories such as collections (or special exhibitions), multimedia guides, professional interactivity, and hybrids.

1) Domestic museums

Information display and education are the main purposes of an app. Due to spatial limitations, here is a brief explanation of the National Museum of China's special exhibition app.

Since September 2012, the National Museum of China has released 12 apps, as shown in Figure 1.

Except for the three items marked as free, the apps are subject to fees ranging from 12RMB to 18RMB. The vast majority of Internet apps are free today, so this price is high, especially in China. Figure 2 below shows a screenshot from the free "Ancient Chinese Porcelain Art at the National Museum of China" app, which is based on a series of cultural artefacts. In the app, the artefacts are approached in various forms, including graphic display, text caption, and voice description, offering multi-angle display.

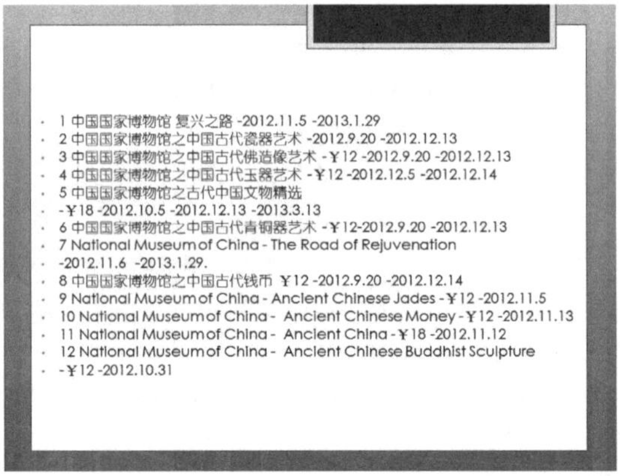

Figure 1 List of 12 National Museum of China apps

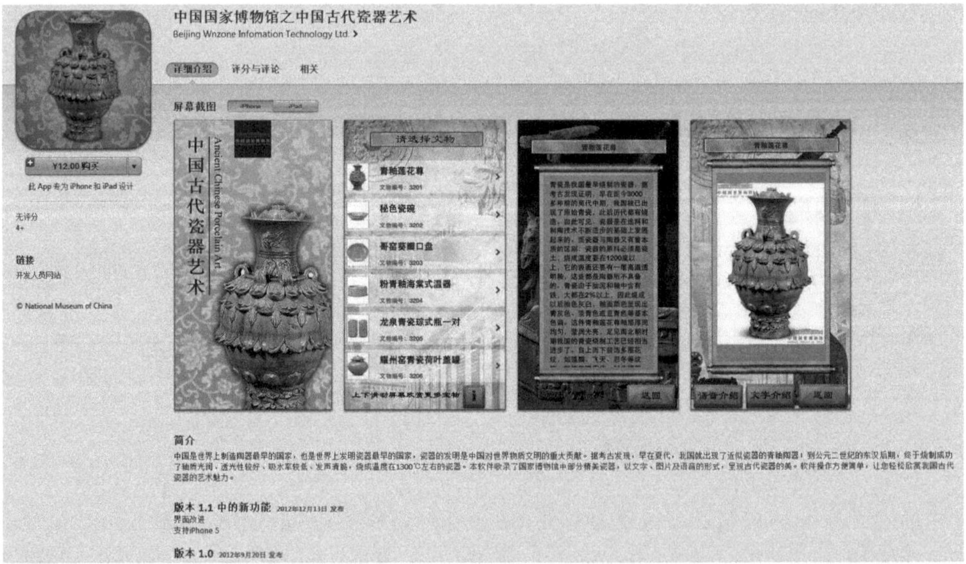

Figure 2 View of the "Ancient Chinese Porcelain Art at the National Museum of China" app

The disadvantages of this kind of app in comparison with mainstream apps are the relatively small download volume and unsatisfactory effective usage rate. Multimedia guide apps are also common among domestic museums, such as the Suzhou Museum, Wuhan Museum, and Zhongshan Ship Museum, but due to limitations of space, these will not be expanded upon here.

2) Many well-known museums abroad, especially those with extremely high-value collections such as the Louvre, the British Museum and the American Museum of Contemporary Art have built apps centered on their collections, displaying content, details, and stories for free or for a fee, often providing a free version with upgrades to the paid version.

The Louvre's app is divided into two versions, free and paid (12RMB). The app supports Russian, Turkish, German, and 12 other languages. It contains more than 400 paintings, and the collections of more than 200 painters, which can be sorted by artist or style for easy retrieval; resources can be collected or sent via email or shared on Facebook, enlarged, or added to a slideshow. The museum's most famous paintings, "Mona Lisa" (Leonardo Da Vinci), "The Coronation of Napoleon" (Jacques-Louis David), and "Marriage at Cana" (Paolo Veronese) can all be found in the app.

The British Museum's app also has free and paid versions. The free, streamlined version covers over 600 artefacts, including the world-famous Rosetta Stone and the collection of Ancient Egyptian mummies. Users can search, collect artefacts, choose tours, and view interactive maps. The paid version (25RMB) shows the entire museum, including rare artefacts from the Greek, Roman, Asian, Aztec, and Assyrian Empires.

In addition to the collection-centric apps mentioned above, the multimedia guide app is also popular overseas. For example, the Scotland National Mining Museum's app features multimedia audio navigation that covers an overview of the museum and its history, audio guides to certain selected collections, video introductions, and map tours.

Professional interactive game apps are also used by foreign museums. One example is the National Museum of Scotland's interactive museum treasure hunting app, "Museum Explorer", in which users look for the nine strongest, most ferocious, tallest, and most unpleasant smelling "top" collections in the museum during their visit, according to hints given within the app. Completing tasks allows users to collect up to nine medals within the app, to become the ultimate museum explorer.

This rich and varied content above makes these apps extremely popular. The artistic and historical value of collections also plays an important role. In comparison, Chinese museum apps are relatively unpopular.

1.1.2 Research and analysis of cinema apps

The cinematic app market contains many examples, such as Maoyan Movies, QQ Movie Tickets, Pudding Movie Tickets, Ha Tickets, Leying.com, Tuan 800 Tickets, Video News,

Movie Finder, Mobile Movie Tickets, YXT Film, and Mobile Movies. The number is huge, and all of the apps have very similar primary functions including information about movies, cinemas, as events, as well as personal modules. The majority are apps in the vertical field of films led in production by internet companies.

In particular, Douban Movies, Mtime, Maoyan Movies, and Wanda Movies are the most popular. Amongst them, the Wanda Movie app is the product of Wanda Cinemas, for movies only screened at Wanda cinemas and conglomerate venues. The first version of the app was released in April 2012, and was one of the earliest in this field.

KO Movie – China's first social movie ticketing app – offers real-time online selection of low cost tickets at 600 theaters in more than 100 cities across China. Regarding personal functions, in addition to providing standard services such as online seat selection, ticket scrambling activities, movie commentary, and professional film review, it also provides specialty content functions such as movie dating, voice review, and SOS, which can be used to issue an immediate ticket if an order cannot be completed due to network quality or insufficient balance, if the movie is about to start, or if a seat is about to be occupied.

1.1.3 Analysis of image-type apps

The author has also conducted a large number of investigations into the categories of image presentation and sharing apps.

1) Image presentation apps

According to download volume, the most popular mainstream image presentation apps are iQiyi, LeTV, Sohu Video, Baidu Video, Youku, and Cool6. Competition amongst these platforms is particularly fierce. Following a large amount of investigation, evaluation, and discussion, we believe that these apps run contrary to our development direction, so we will not explain them in too much detail here.

2) Image sharing apps

Mainstream image sharing apps include Weishi, Miaopai, Weipai, Paike, and Youku.

We believe that the development direction of short video apps in this field is worth exploring, and therefore our focus is on short-scale video footage sharing apps. In the current domestic field, there are two dominant apps – Weishi and Miaopai. The former was especially obvious, with extremely rapid development on the Tencent platform, and large amounts of advertising and promotion from famous artists, with frequent updates (see Figure 3). The main functions of these products are extremely similar, supporting the capture and sharing of short videos lasting eight seconds ("8 seconds of unlimited joy") or 10 seconds ("sharing 10 seconds of touch").

Of course, the cost of product development, operation, maintenance, and marketing is great, reaching at least tens of millions of RMB.

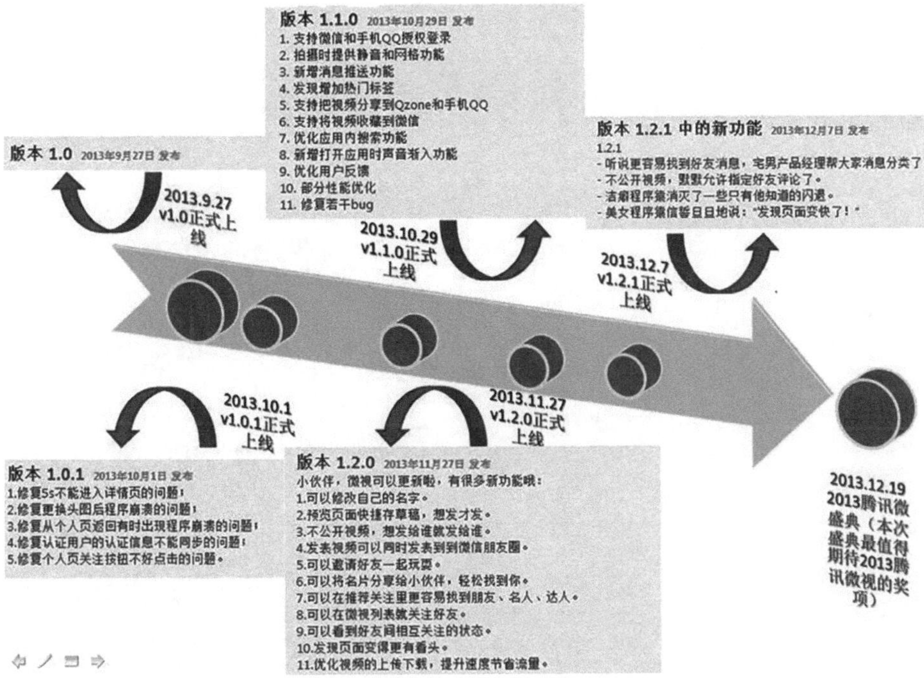

Figure 3 Timeline for Microvision product development

Furthermore, image sharing (or social video) app technology must overcome several basic problems. First, certain technical difficulties must be fixed, including (but not limited to) video capture, video upload control, breakpoint retransmission, video transcoding, video loading, and smart recommendations. Second, other issues must be addressed, such as the selection of video popularity indicators, the planning of video classification, the formulation of reporting mechanisms, and the selection of video playing devices (possibly incurring fees), with the need to perhaps implement basic functions such as private messaging, comment modules, favoriting, forwarding, and sharing. The costs for technical strengthening and time for R&D have become key considerations.

While examining the above three app forms, we have also found that the above apps generally present rapid iterations and short update development periods. Using Weishi and Miaopai as examples, Figure 4 shows the timing of upgrades to these two app products, proving that the apps have quick iterations for updating and upgrading. The aforementioned National Museum app series is also online in various iterations (see Figure 5 for details), and the release time is also relatively dense. This puts high demands on technical, human, and operational cooperation within R&D teams during app development.

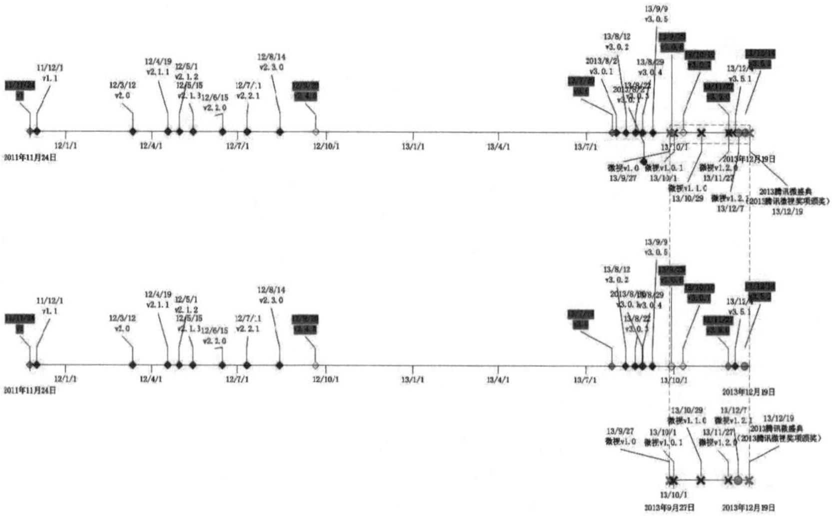

Figure 4 Timeline for WeSee and Miaopai online updates (as of December 2013)

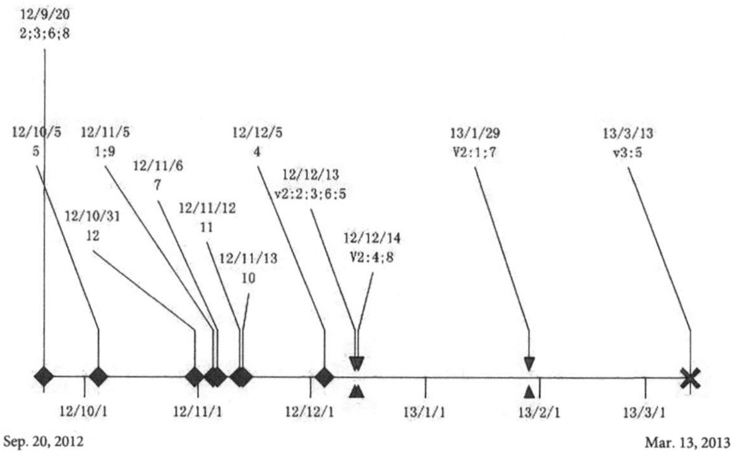

Figure 5 Product upgrade timeline for the China National Museum's app series (as of December 2013)

1.2 Comparison

The author has compared our institution with cinemas, Shiguang.com, Douban Movie, m1905, and other museums, but for the sake of space we will not go into detail here. Instead, we will provide a few overall thoughts regarding these comparisons.

1) Advantages of other museums
Well-known museums abroad, especially the British Museum and the Louvre, have collections that are extremely rich in resources, and high in artistic and historical value. Frankly speaking, we cannot compete. Apps that use their collections as their main content are very popular, such as the aforementioned apps by world famous museums such as the Louvre, while Chinese museum apps based on collections and introductions to institutions are not. Firstly, downloads are relatively minimal, and secondly, the actual usage rate is relatively low following download. Therefore, our primary museum apps will not be used for introductions to our collections or exhibition halls.

2) Advantages of our museum
As a national professional film museum, we have advantages in terms of rich and colorful brand activities, brand resources, and unique artist resources within the film industry, which should be utilized to the fullest.

2. Analysis of the online attention of China Film Museum visitors

According to the CNZZ analysis of traffic to the China Film Museum website in 2013 (see Table 1 for details), the following features stand out:

Except for the main page of the official museum website occupying first place for the most visited page at 31.87%, only two of the top 10 pages are non-movie related pages; fourth is the member club page (2.89%) and eighth is the museum map page (1.53%). Seven of the top 10 visited pages are related to movie and movie details, accounting for 32.99% of the total visits – even more than the home page. Amongst these, the top two video pages visited account for 17.42%. Generally speaking, our museum's visitors pays much more attention to information related to movies and screenings.

3. Development direction of the China Film Museum app

Combined with the aforementioned research and comparative analysis of apps in the museum and cinema fields, as well as the analysis of specific business and audience needs at the China Film Museum, plus an analysis of its own advantages and disadvantages, the following are the most suitable applications:

1) Basic film information platforms (film entertainment and information apps): connected to news industry categories such as industry news, allowing users to comment on and interact with information. At present, there is no single success story for these apps, despite high user volume. We believe that products such as Weibo's "Sina Entertainment" channel can be sufficient to meet the needs of target users.
2) Commentary and interactive platforms (social film apps): Displaying film directors, casts, plots, posters, and trailers in an interactive comment platform. At present, typical successful cases in the industry include Douban Movie and the Shiguang.com Movie app. In the early stages, we gained user resources from our loyal followers and members, as well as rich offline activities and industry influence (national museums). However, technical strength and abilities for continuous update iterations and rapid response are a huge challenge when up against Douban Movie and the Shiguang.com Movie app.
3) Interactive platforms for film topics (social topic film apps): Interactive platforms based on movie topics, allowing hot topics to be launched and web users to engage in

Table 1 CNZZ flow analysis of the China Film Museum website (2013)

Web address		*Views*		*Proportion of visits*	*Remarks*
http://www.cnfm.org.cn/	353,678	287,065	269,724	31.87%	First page
http://www.cnfm.org.cn/2012-07/23/cms13098article....	193,315	44,685	42,601	17.42%	
http://www.cnfm.org.cn/yppqs.shtml	68,110	19,871	20,530	6.14%	
http://www.cnfm.org.cn/club/club.shtml	32,080	8,572	8,860	2.89%	
http://www.cnfm.org.cn/2013-09/27/cms14800article....	30,727	22,862	23,888	2.77%	
http://www.cnfm.org.cn/yppq/Movielist.shtml	25,924	20,146	19,421	2.34%	
http://www.cnfm.org.cn/wlypsy.shtml	17,931	8,908	9,614	1.62%	
http://www.cnfm.org.cn/maps.shtml	17,011	4,877	5,216	1.53%	
http://www.cnfm.org.cn/node_993.shtml	16,965	5,283	5,474	1.53%	
http://www.cnfm.org.cn/2012-06/20/cms13830article...	12,954	7,773	8,207	1.17%	

heated discussions, forming online public opinions. A typical case is Sina Microblog, and we face the challenge of homogenized competition with Sina Weibo. Meanwhile, from the perspective of product planning itself, while the Douban Community app has been successful, it is far from Douban Movie from a user number perspective. Thus, how these apps ensure user volume is an ongoing issue.

4) "Quick Movie" micro-video interactive platforms (social video apps): This is a more innovative idea best suited to the focus of our museum, with Internet functions in all of the above four directions. Typical examples of successful comparative apps at present include Weishi and Miaopai, described above. Weishi developed rapidly in an inevitable direction after it was launched. With the effects of star involvement and grassroots comedy videos, as well as powerful television, subway, and Internet advertising, it is launching an offensive to quickly occupy the top spot in this field.

Although this direction is more innovative, it also faces great challenges. First, when compared with Tencent (one of the Internet's three giants), there is a huge gap in our technical strength, as well as our abilities for quick upgrading and updating. This will be the greatest challenge. Secondly, Tencent's huge promotion cost (estimated to be in the tens of millions) is something we cannot compete with. However, we have certain advantages; as a national museum in the film industry, we have unique resources such as senior national film artists who offer long-term support for and attention to the development of our institution, as well as offline activities for film industry brands. As such, we must make a choice for our direction.

4. Position analysis

The target visitors for each category are different, and product features are also positioned differently (Figure 6). For example, "Paike" records what happens around you, making everyone a reporter; the target group of a film information category is the cohort who often go or plan to go to the cinema to watch a film; topic discussion, such as with

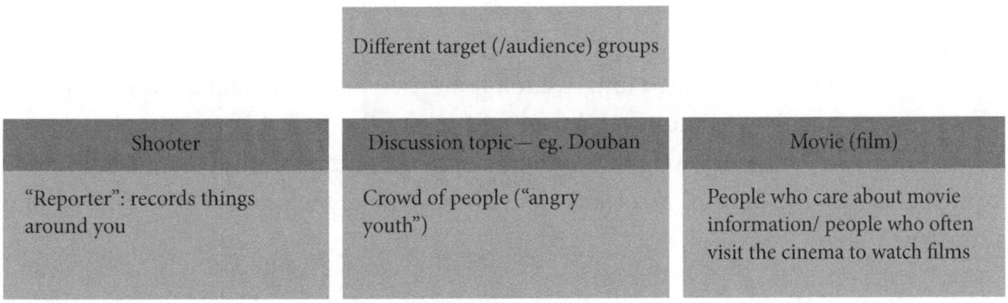

Figure 6

Douban, is concerned with web users. It is important to be clear about the positioning of the product.

The position analysis of the target group for the China National Film Museum (hereinafter referred to as CNFM) app is as follows:

1) Primary positioning

a) Provides services to the following "followers" of CNFM offline activities

- Loyal CNFM visitors: encouraging visitors to comment and communicate;
- Loyal CNFM members: encouraging members to comment and exchange after observing member specials;
- Loyal CNFM film critics: in conjunction with offline activities, encouraging critics to publicize film reviews, and release intense critical reviews and exchanges from Film Critic Club events to the app, allowing deeper discussion after events; exploring, spreading, and developing seed reviews;
- Students from CNFM's partner universities (especially film and television colleges): planned promotion of the app within film lecture halls, encouraging participants to interact, publish film reviews, and begin topic discussions;
- CNFM visitors, especially hardcore followers who often watch films at IMAX theaters;
- CNFM visitors (usually movie lovers).

b) Addressing online CNF "followers", such as Weibo and WeChat followers, providing targeted services and targeted spread, and utilizing excellent services to strengthen fan loyalty. (Note: the above two groups overlap with offline activity groups to some extent).

2) Further positioning

Services for film industry professionals: as a feedback platform, the app can obtain the latest audience feedback, thus establishing a bridge between the audience and the industry; providing services to major film industry media outlets: as a film industry communication platform, showing the latest trends in the film industry and promoting industrial exchange.

3) Final positioning

Services for the aforementioned long-term followers and experts; aggregating the power of web users, forming opinions, and feeding back to the industry. As a national-level professional film museum, we can use the new media function of the app to make due contributions to the improvement and flourishing development of exchange in the film industry.

5. Functional design

Every one of the four types of app development directions mentioned in Section 3 of this article can be used as a single, extended app development, eventually forming the China Film Museum App Group to serve a variety of visitor groups. For the first CNFM app, we must consider the "powerful" nature of expected competitors with function D amidst the project construction cycle and funding issues, considering the integration and development of functions B and C. Therefore, in the context of balancing funds and development cycles, the current app has the following five functions:

1) Interactive comments
The creation of a movie library containing film directors, casts, plots, posters, stills, trailers, and other film related information, as well as an interactive comment platform based on the movie library, allowing users to post comments about films anytime, anywhere. We hope to capture some classic reviews that will influence reputations within the film industry.

2) Film topic interaction
Allowing the launch of popular movie topics and heated discussions based on the topics, forming online public opinions and building a bridge between the audience and the industry.

3) Exhibition information interaction
Showcasing the latest exhibition information held by our museum, so that film followers can understand Chinese films. Allowing users to interact and comment on exchanges, creating collisions of ideas and forming beneficial interactions.

4) Real-time navigation
Supporting real-time navigation based on the user's current position in relation to the museum.

5) "My center"
Recording a user's purchases, comment interactions, and "movie of interest" collections. Allowing for the unbinding of association/membership cards with phone number/email, and cache clearing.

6. Other

The authors of this article also engaged in a preliminary exploration of the promotion of the first CNFM app. Due to the limitations of space, it will not be expanded on here.

Furthermore, in regards to the app design process, we read Zhang Xiaolong's *Behind WeChat's Products*, and found the proposed product design concepts very helpful to the app's design and ideas. We hereby express our gratitude to Zhang.

Finally, the development of the app field is extremely rapid, with constantly changing demands. Therefore, following the launch of our first museum app, it is necessary that we continue to attend to iterative improvements.

References

[1] China Ministry of Industry and Information Technology. 'Economic operations of the communications industry in August 2015' [Z]. (23 Sept 2015).

⇨ CHAPTER 4

Mobile Navigation for the Thousands

– How Emerging Media can Boost Communications and Services for Museums

Developing Museum Communications and Services around Smart Mobile Terminals

Yan Tao
(Department of Social Education and Information, Beijing Ancient Architecture Museum, Beijing 100050, China)

Abstract: In the face of traditional communication and service methods that fall short of the needs of their audience as well as social development, museums must develop new communication methods in line with the living habits of their visitors, providing services that are more suited to visitor needs. In recent years, the popularity of smart mobile terminals has provided renewed support for museum development. With this new lifestyle change, the museum has developed a series of applications to serve visitors and provide cultural communication. Based on the author's research and experience in museum informatization, this article discusses the progress of information technology's influence on museums, and how it is changing the track of museum development. The focus here is on discussion and research on smart mobile terminals and mobile phone applications, demonstrating the latest achievements in the development of modern museums and discussing the changes brought about by the advancement of information technology.

Keywords: museum; mobile terminal; communication; service

At the present stage of museum development, it is necessary to focus on highlighting promotion and improving the quality of services based on a traditional model of development, in order to meet social development requirements and public needs regarding cultural life. Museum publicity is not limited to the exhibitions and cultural artefacts themselves, but also involves the dissemination of culture, the creation of museum brands, and the dissemination of development concepts and cultural studies via branding. In order to expand the scope of museum communication and publicity, as well as to meet audience needs, the traditional publicity model that relies on print distribution is far from satisfactory, and museums must continue to innovate new means of communication. Meanwhile, they must gradually replace the existing traditional service model, improving the visitor experience, saving audience time, and reducing the cost of visits, while providing services with consideration of audience perspectives.

1. Traditional communication methods for museums

The longstanding traditional publicity method in museums utilizes print media such as brochures, exhibition catalogues, newspapers, and magazines. This model has a slow release period, poor effectiveness, and a limited audience, and is no longer aligned to modern people's reading habits. Contemporary society relies less on printed resources, instead turning to electronic reading, making it difficult for traditional museum promotion to adapt to the developing needs of today's society. With the advent of Internet technology and the proliferation of computers, websites have been on the rise for several years. As the most basic and earliest Internet-associated application, the website has grown into a major publicity tool for museums, and almost every museum has its own official site. As an authoritative publishing and promotion platform, the official website is still a standard for museum promotion, and continues to play an important role. Websites allow for the complete display of buildings, exhibitions, artefacts, and visitor guidance according to specific requirements. A website can accommodate a rich amount of content in a variety of forms, and according to its own characteristics and audience needs, a museum may create a unique web presence to serve its vast audience. However, the interactive nature of the website itself is poor, as the audience passively absorbs the contents of a museum-generated display, unable to interact or ask questions. In this way, visitors glean only what the museum hopes or guesses that they need. A website does not cater to the true needs of the audience, which makes website-based service increasingly limited, and unable to meet today's social needs. Meanwhile, the creation, operation, and maintenance of websites requires professionals, time, energy, and money. Updating and publishing content is cumbersome, and cannot fully complete a museum's functions of external publicity and service.

2. The popularity of smart terminals brings new opportunities

With the popularization of smart mobile terminals, the infrastructure of the Internet has been improved, and the development speed of mobile networks is accelerating. Applications that rely on the Internet have gradually shifted from traditional computer terminals to smart mobile terminals, such as mobile phones, which never leave people's hands. Public lifestyles are changing, and so is the model of social development. Therefore, museum publicity must keep up with social progress and changes in public living habits.

Daily public life is increasingly dependent on digital information equipment and networks. Life on all levels is supported by digital forms, and is becoming completely digitized and networked. We longer exist without a network, and cannot be separated from mobile terminals. Our mobile phones are constantly on. Therefore, every industry has to face up to this development trend, and must make full use of it to develop specialized services.

3. *Launching museum promotion on smart mobile terminals*

Faced with the rapid development of smart mobile terminals and the transformation of public lifestyles, the traditional cultural museum industry must constantly innovate, quickly adapt, and make full use of new technologies to create a fast-communicating, wide-reaching, and strongly interactive publicity platform.

3.1 Weibo promotion

Following the appearance of official museum websites, Weibo was the fastest to spring up in terms of methods of spreading information. The emergence of the influential Sina Weibo occurred more than six years ago, and the platform has since become a mature and widely used Internet communication tool. Before, the public did not have a wide-reaching and widespread social tool, and the desire for self-display and life sharing was difficult to achieve. Therefore, Weibo quickly garnered popularity as a social sharing tool. Its characteristics are abundant pictures and text, with short-form content, variable expression, and strong effectiveness. It is an effective tool by which the public can express themselves at any time in any place. An important advantage of Weibo is its interactive capability, which is a good method of satisfying public desire for attention and interaction when self-promoting. Weibo provides a successful platform on which users can show, discuss, and express themselves.

Similarly, the museum can use Weibo as an important innovation from traditional publicity, utilizing it as a platform to showcase itself and serve its audience. Compared with traditional websites, Weibo is time-sensitive, easy, and low in operating costs, with far lower requirements for staff posting than in the operation of a website. Most importantly, visitors now have a platform on which to interact with the museum. In the past, viewer queries and consultation services occurred on the telephone, which was much slower, with very few people providing services. Calling cost would be incurred as well. Now, with Weibo, visitors can interact with the museum at any time while learning about it and its exhibitions. Nowadays, Weibo is the main platform for daily museum publicity.

As content release on Weibo is convenient and fast, it makes up for the shortcomings of websites. Websites are difficult to update consistently, and new information only appears when there are important items, while Weibo offers content release any time without restrictions. Meanwhile, it features short-form content with concise language, which is far more likely to be noticed by visitors, as opposed to long-form articles that are difficult to read. Now, the Weibo client can use mobile terminals such as mobile phones and tablet computers, making it convenient to access anytime and anywhere, so as long as there is a network. Meanwhile, museums are now able to self-promote at any time. They are no longer limited to fixed spaces and venues, which is another advantage that cannot be matched by website updates. Nowadays, to see the latest museum developments, one does not look at the official website, but the museum's Weibo, as this is one of the most direct and efficient channels. Weibo has taken on greater responsibility than official websites, providing improved opportunities for publicity.

3.2 WeChat publicity and interaction

In recent years, WeChat has become the hottest social tool following the development of Weibo. It is the newest platform on which the public can connect with one another and self-promote, involving a wide audience. Compared to Weibo, WeChat has few publishing functions, but is also a chat tool, meaning that users can communicate in real time. Therefore, it is used more frequently and efficiently. Meanwhile, it is easier to share via WeChat, as it is a custom tool for mobile socializing that is more closely aligned with public habits and integrating more functions. Compared with Weibo, WeChat has great communication advantages. Weibo requires more attention, and is not a communication tool as such, meaning that many users only visit when they have the time to comment and interact on posts. Meanwhile, WeChat is a communication tool that the public cannot live without. It can be used anytime and anywhere, meaning that news published through WeChat is presented to the public more quickly than through Weibo. WeChat and Weibo are not the same. Although more content can be published on WeChat, there are more restrictions, and its biggest advantage is the potential for wider dissemination. As a quick sharing social tool, WeChat shares content instantly to every friend in each circle of friends. Therefore, even if certain people pay minimal attention to a museum's WeChat account, information is delivered firsthand, producing a radiating effect for very effective communication. This is a feature that Weibo lacks. The communication capacity of WeChat has been acknowledged by the public. Basically everyone with a smart phone has a WeChat account, and the public use of WeChat has become part of daily life.

By setting up a public WeChat account, museums can publish their latest developments and interact with their audience. A museum's public WeChat account has become an important platform by which it can publish exhibition news and announcements. It is the most convenient way for museums to communicate with visitors, and also the method that is best suited to audience usage habits. Through public WeChat accounts, museums can get closer to visitors, encouraging an affinity and providing a new platform through which they can understand and communicate with the museum. WeChat publicity is completely reliant on smart mobile terminals. Built around public user habits, it has greatly improved museum promotion.

4. Museum application services through smart mobile terminals

The application of smart mobile terminals is not limited to publicity. It can also provide greater convenience for the development of museums. Meanwhile, it provides further platforms for open museum work, and more ways by which visitors can understand and visit museums.

4.1 Smart terminal exhibition guides

The traditional forms of guided museum tours involve use of a map, tours led by guides, and assisted tours through audio guides. Each of these modes has its drawbacks. Map-led visits map little guidance, only showing general directions without offering further assistance. Tour guides are generally the best method, but there are certain issues, one being if there are not enough guides, an appointment cannot be made, and there is also no opportunity for a one-on-one tour. Usually, a group of people surround and listen to a guide, often without a balance between listening and seeing. The route and visit time cannot be chosen as per individual preferences, meaning that visitors do not always get the best experience. Audio guides are not intelligent, nor convenient to operate, nor intuitive. They lack visuals, and have limited content that cannot fully support an exhibition. Therefore, traditional guided tours are not well adapted to actual visitor needs.

Navigation through smart mobile terminals has been applied in certain museum exhibitions, with good results. Given the rise of network development, many museums have employed wireless networks to facilitate audience use, so that visitors can view exhibitions and connect to free wireless networks to assist in their visit. This includes interesting content at their fingertips, and using their mobile phones to search for information online, take photos, and share their visits with friends in real time. The provision of a network facilitates visitors and popularizes museums. This is the current status quo at most museums, but it is only one part of the network's role. There is still more value to be mined, such as the development of exhibition guide apps. On visiting an exhibition hall and connecting to the network, users can use their phones to install a navigation app to guide them without the use of other tools. Mobile phones are a convenient navigation tool. There is no need to make an appointment, no waiting time, and no cost. Guided mobile phone navigation has many advantages that are unavailable with traditional methods, making it the optimal navigation method. Guide apps are visual exhibition explanation systems – advantages that are unavailable with any traditional navigation methods. In the past, visitors were offered only audio-based methods to understand exhibitions, whether through tour guide interpretation or focused audio-guides. Both of these are incomplete forms of navigation, so visual navigation offers a huge improvement. The greatest benefit of visual navigation tools is their richness. Visitors are offered text and images that are consistent with the exhibition via their mobile phones. This is a very intuitive model. At the same time, navigation apps can store more content than is available in the actual exhibition, which is limited to displayed content, offering the potential for extended exhibitions. Ultimately, the content that is shown to the audience is very limited, as much of it cannot be presented completely. The navigation app makes up for this, providing more complete exhibition content on smart phones, and offering convenient understanding and educational knowledge related to the exhibition. Further great benefits of guide apps are that they are easy to use and manipulate. Whether visitors choose to view exhibitions in order, move to key exhibits of interest, or go back and forth repeatedly, they can navigate easily via the app, with visual navigation and convenient operation offering great convenience. It may thus be said that the installation of guide apps on mobile phones during visits will become an important method by which museums present guided exhibitions in future.

4.2 Smart mobile terminal services

4.2.1 Mobile ticketing and reservation

Entry to a museum involves ticket purchase for exhibitions that charge fees, and appointment bookings for free exhibitions, and this is the first window through which the museum faces its audience. The provision of convenient and time-saving admission services is the museum's first priority. Traditional ticket queues or long lines for registration create a large concentration of visitors at the entrance of the museum, which seriously affects their mood and the duration of their visit. In order to solve this admission issue, many museums have opened online ticketing and booking services, so that after reaching the exhibition hall, visitors can purchase tickets via smart phones or make reservations by phone without having to wait in line. The past practice of queuing to buy tickets and register has been gradually replaced by the simple operation of mobile phones, saving visitors a great deal of time and energy, and also creating a positive atmosphere.

4.2.2 Virtual museum tours

Museum visits are the most direct method of learning about exhibits, but they are not always possible. For museums to spread culture as widely as possible, we require new ways for visitors to understand museums – ways that are convenient and easy to access. As a replica of the physical museum, the virtual museum provides even richer content, as it opens a window for all visitors. A virtual museum mobile app goes beyond the computer screen, sending a virtual museum directly into the audience's hands. Viewers can browse exhibitions freely without space and time constraints. The virtual museum provides a platform for visitors who are unable to visit the physical museum. After installing the virtual tour mobile app, visitors can visit the museum at any time, in any place, free of charge.

5. Using smart mobile terminals for museum development

Smart mobile terminals have become an indispensable part of people's lives. Museums must seize this opportunity and make full use of these terminals to develop and promote their services in various ways. The museum has evolved from a traditional cultural venue to a cultural pioneer, and museums must learn to use the latest technological means for their own development. Smart mobile terminals are the prime public focus at this stage, but more new technologies are likely to emerge in the future. Museums must keep up with technological advancements, and continue to innovate and develop with changing audience demands and personal development needs.

Research into iBeacon Applications in Museums

Yu Hui
(Display Department at the Chinese Museum of Women and Children, Beijing 100005, China)

Zhang Yucui
(Office of the Chinese Museum of Women and Children, Beijing 100005, China)

Abstract: In recent years, with the development of the Internet of Things and related technologies, the demand for indoor positioning technology in various industries, especially the service industry, has become stronger and stronger, with many museum services evolving from "informatization" to "intelligence". iBeacon technology allows devices to use low-energy Bluetooth technology to transmit data across smart mobile terminals whilst obtaining all kinds of accurate location information, also providing message push functions to users through the app, meaning that iBeacon provides many desirable functions. This article explains the basic status of iBeacon, discusses its application prospects within the museum industry, and points out issues with promotion and application in the specific example of our museum, intending to provide a new perspective on the development of smart museums.

Keywords: iBeacon; BLE; museum intelligence; audience services

1. About iBeacon

With the increasing popularity of smart terminals, the practicality and necessity of location services has becoming increasingly prominent. Among existing positioning systems, GPS is the most widely used, with meter-level positioning accuracy that is able to meet the public needs of outdoor positioning. Indoors, for example in museums and libraries, occlusion and multipath effects within buildings mean that GPS signals are greatly attenuated, and cannot meet positioning needs. As a result, people are exploring a number of other technical methods for indoor positioning, but most have issues such as high cost and large engineering requirements. Therefore, research and exploration into indoor positioning technology continues to be a popular issue.

In 2013, Apple introduced iBeacon technology at the Apple Worldwide Developers Conference (WWDC). iBeacon is a new form of location-based technology, seen by many as a location-based alternative to GPS indoors. Bluetooth devices are the essence of iBeacon. With its underlying technology using low energy Bluetooth transmission technology (BLE), it can be seen as Apple's alternative name for BLE. Compared with traditional Bluetooth technology, iBeacon is lower-cost and lower-power, with an 800MAH micro-battery keeping the device alive for approximately two years. iBeacon functions like a beacon, setting itself as an information base station. When a smart terminal enters the specified area, the beacon device sends data packets at intervals of between 20ms and 10s. Once the smart terminal receives the broadcasted data, it initiates a scan request, to which the beacon device responds and simultaneously marks the location information of the terminal. Furthermore, because iBeacon technology uses Bluetooth broadcast channels to transmit signals, it eliminates the cumbersome pairing process.

Based on low-power consumption and high stability, iBeacon provides excellent functionality for retail and public service sectors. At the moment, United States industries using iBeacon technology include commercial promotion, tourism, real estate, property management and advertising media, with a variety of content. Target, the third-largest supermarket in the United States, uses iBeacon technology to implement location-based services within the supermarket. Users need only add their shopping lists in advance through the Target App (see Figure 1), then upon arrival to the supermarket, the app will

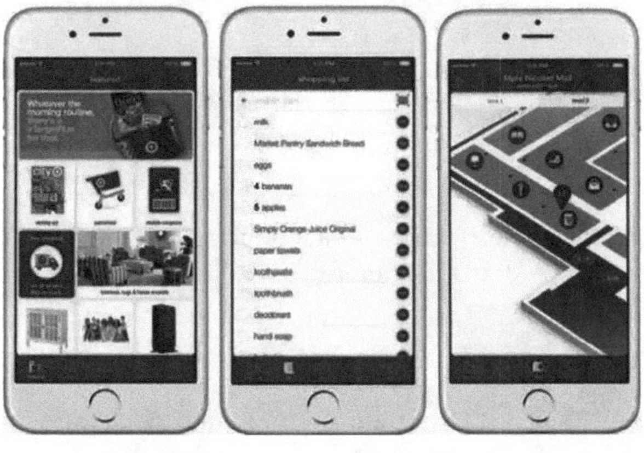

Figure 1 Target superstore's mobile client

display a shopping guide map, with the location of each item in the shopping list marked. The locations of alternative merchant-recommended products are displayed on the map. This function makes the shopping process more efficient, and is also a means of publicity for merchant offers and promotional information.

As an innovative technology in the development of the Internet of Things, iBeacon can be used in places like museums, as well as any venue featuring huge amounts of information, with infinite prospects.

2. iBeacon application prospects within the museum

According to the Fast Company report, many cultural institutions in the United States have used iBeacon technology to send announcements and related information to users. In China, some museums have begun using iBeacon technology, deploying base stations in the exhibition halls, combined with the WeChat shake function, to push more rich exhibition content to visitors. iBeacon technology was also used in the exhibition *Inheritance and New Trends — Chinese Traditions and Families* held at the China Museum of Women and Children at the beginning of the year. Visitors could use the WeChat shake function within the exhibition hall for more details about the content (see Figure 2). Of course, iBeacon applications within the museum industry go far beyond this, and can provide much greater assistance.

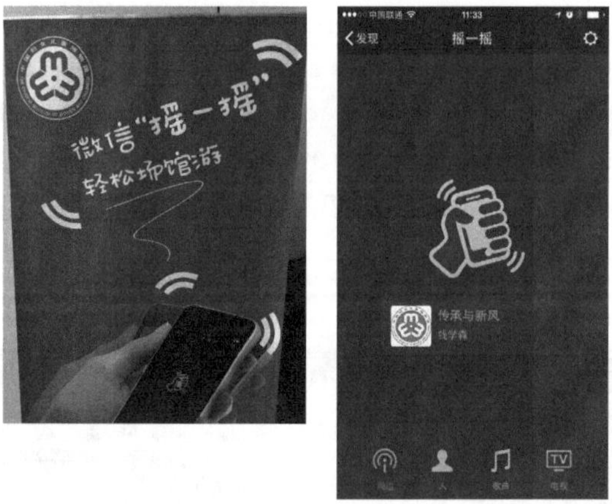

Figure 2 WeChat shake function for the "Heritage and new trends – China's positive family traits and home lives" exhibition

2.1 Indoor positioning

Location awareness is the most important and fundamental feature of iBeacon. iBeacon indoor positioning is similar to WiFi positioning, both calculating distance according to a model of RF field signal attenuation, the only difference between the two being that WiFi uses hotspot topology to broadcast WiFi signals; iBeacon uses Bluetooth base station topology to broadcast a Bluetooth 4.0 LE signal. iBeacon signal strength is represented by the RSSI value, and the distance between a smart terminal and the iBeacon device can be judged by the changing RSSI value. This value can be a rough message (close, far away, or out of range), and may also be accurate to a distance in meters. If an iBeacon base station is placed within every 20 square-meter space, positioning accuracy can be maintained at a decimeter level, which can fully meet the daily positioning needs of a museum. In general, the purpose of positioning in the museum is to make visits more comfortable and personalized.

The key needs of positioning within the museum are twofold: route planning and exhibit positioning. Route planning means that within the museum environment, the acquisition of real-time audience location allows the beacon equipment to push information regarding current human traffic in every exhibition hall and recommended visits to the audience through the app, so that visitors can arrange suitable tour routes accordingly. Exhibit positioning is largely used in the smaller environment of an exhibition hall. By placing small beacon equipment around exhibit objects, when visitors search for information about these exhibits, the iBeacon base station can accurately pinpoint their location within an exhibition and send information to the viewer's smart terminal. In this way, unnecessary struggles such as blind searching or asking around are avoided.

In addition, more complete and human-based positioning within the hall also allows for the selection of the most optimal visit process for visitors based on their actual situations. The planned length of a visit definitely affects the choice of content. Therefore, given planned visit time, making recommendations regarding the best route and content for the visit is the developmental goal of museum indoor positioning technology. Just like an experienced tour guide, based on a simple query and judgment of audience needs, the best tailor-made program will be provided. In perfecting the technology, the provision of a visitor's preferred visit length, and choice of information such as exhibits of interest, a tour route will be automatically generated to meet a visitor's needs. In just a short visit, all of the expected content can be seen without fail, while longer visits provide a fuller experience of the museum atmosphere and the exhibitions. Combined with Big Data and Cloud computing technology, iBeacon technology can optimize the choices for the visiting process according to actual audience situations.

2.2 Auxiliary function

iBeacon makes exhibitions more intelligent, presenting more comprehensive content to visitors. Explanation has always been an important means of museum education, with unquestionable importance. As of the end of 2013, the total number of visitors to Museums nationwide exceeded 600 million. Currently, most Chinese museums still rely on manual exhibit explanation. However, in the face of increasing visitor numbers, difficulties are often encountered in a shortage of personnel, especially in the case of special hours such as holidays, when the number of visitors far exceeds the capacity of museum staff. Although some museums provide electronic presentation equipment, such one-way models do not meet visitors' need for prompt replies to query. People expect museums to be more thorough, stylish, and interactive in their explanation. Nowadays, the iBeacon-based explanation service within the exhibition hall can alleviate personnel shortages to a certain extent. Via information push services, visitors can quickly access material such as voice, images, and even video content, to gain a more comprehensive understanding of an exhibition's background and content. This free form of intelligent explanation has gradually become popular within the museum industry, acting as a scholar who remains at the audience's side during their visits, helping them to interpret the exhibitions at a deeper level. This way, the "layman" can not only see artefacts, but also understand them.

Auxiliary functions provided for exhibitions by iBeacon are further reflected in the details. One example is multimedia videos within the exhibitions. These videos provide in-depth analysis and research on exhibitions of their artefacts, offering a direct understanding. However, the current status quo proves that most videos within exhibitions are a form without full functional value. Why is this the case? Most videos in the exhibition hall cannot adjust play progress as required, so visitors can only begin viewing from a random position; if one wishes to view the video from the beginning, they must wait for the end of the play loop. This wait feels extremely long, with few visitors willing to commit. iBeacon technology can be used to make use of visitors' mobile phone screens, meaning that they can instantly view multimedia video content within the exhibition, giving them the power of choice. In this way, technology has changed the status quo that leaves videos incomplete in function, while also providing more convenient conditions in which the audience can appreciate the exhibition.

2.3 Analysis of audience surveys

Audience surveys are an indispensable part of museum work. Using iBeacon technology to conduct them makes it possible to collect and analyze personal information and various forms of visit data, enhancing a museum's standards. In recent years, with changes in the exhibition concept, simple display is no longer the pursuit of curators, and audience service is gradually becoming the core value of the exhibition. Exhibitions no longer act on their own will. Emphasis is now placed on the satisfaction of a visitor's inner emotional and spiritual needs. Therefore, research on audience visit behavior has become an important basis upon which the museum improves exhibition services and formulates development strategies. The iBeacon device is used as a medium for user information collection, accurately recording the duration of a viewer's stay in various areas of the exhibition hall and their specific behavior during the visit. This data represents the actual feelings of the audience, and facilitates personnel in specific estimates and analysis of deep-seated details. This form of audience surveying is more convenient and efficient than traditional methods. More importantly, traditional observation of audience behavior only focuses on individuals or small groups of visitors, and has an approximate, random nature that is very error-prone and cannot ensure the stability or accuracy of data. Audience data collected by iBeacon is more like an overall data sample, and compared to sampled data, can more accurately evaluate audience behavior. It can also be integrated with a recommendation system according to viewer browsing, search, and preferences to recommend exhibitions or events that users may like. In future, a "big museum" can be created with integrated audience data from all museums, to realize the sharing of information and culture. This will have a profound impact on exhibitions, social education, services, and the whole museum industry.

3. *Limitations of iBeacon at the current stage*

Although iBeacon technology has extremely broad development prospects, there are still various problems and limitations in its application in museums.

First is the issue of construction cost, including of iBeacon base stations and the technical cost of additional functions. The creation of an iBeacon base station involves equipment costs, and must also account for manual deployment. All of the iBeacon functions mentioned above are based on a premise of precise positioning within the museum. This is an essential attribute of iBeacon, as well as its foundation. The iBeacon positioning solution is divided into single-point, two-point, and multi-point positioning, according to beacon equipment usage numbers. At present, the single-point positioning scheme is the most widely used, adjusting positioning accuracy by dot density. While it is the least difficult to implement, it has relatively low precision in the realization of positioning requirements within the area. Locating the precise location of a visitor in real time, or even navigating within the hall, is based on the deployment of a complete beacon device. Differing precisions correspond to differing construction costs, with more precise positioning requiring greater investment. The multi-point positioning algorithm is complex and difficult. At the same time, because Bluetooth itself uses the RSSI scheme, it is not easy to achieve high-continuity positioning. This is because the RF signal is easily interfered with or absorbed during propagation, so to use iBeacon to implement a stable positioning system requires a professional team to complete the base station deployment process. In addition, maintenance costs are a problem that cannot be bypassed by any project. Although some iBeacon devices are able to continually send signals for about two years, their batteries are generally non-removable, so when the device stops working, it must be replaced with an entirely new beacon, increasing the follow-up cost of this technology.

The technical cost of add-on functions includes app development, server set-up, as well as the cost of the algorithms that each function relies on. Many people mistakenly believe that push messages are sent by the beacon device itself, but they are not. iBeacon is simply a base station providing location services, and all additional features must be implemented via app software installed on mobile phones. Once app software receives location information broadcast by the iBeacon, a certain calculation is performed to obtain the current location. When the calculated location meets set conditions, the app requests corresponding content from the data server to be presented to the viewer. Only then is a push message completed. In other words, if the museum intends to use iBeacon to implement the most primary function of intelligent explanation, in addition to deploying the iBeacon base station, museums need software for at least one app, and one data server. The achievement of more functions relies on further technical support. For example, navigation functions within the museum depend on the support of a map algorithm, and recommendation functions require personalized coordinated algorithm. Taking a step back, iBeacon's original intention was to create more comfortable conditions for museum visits, to enhance the visitor experience. Regardless of app design or algorithms for various support functions, if visitors are not cared for and conveniently serviced, then location services lose their value. From this, it is also clear that such high dependence on related technologies has increased the difficulty of widespread use.

Second, regarding the audience, iBeacon offers visitors a new way to enhance their museum visit, which involves a convenient and efficient tour process. However, because iBeacon has an open agreement, outsiders can easily forge a base station with the same ID, and this security risk is an unavoidable problem for iBeacon. The collection and use of audience data requires a secure and confidential mechanism to prevent it from being stolen, so this security issue affects the large-scale deployment of iBeacon. In addition, the inherent public impression of Bluetooth technology is difficult to alter over a short time period. Although Bluetooth 4.0 incorporates deep sleep technology, meaning that the Bluetooth module will automatically enter sleep mode when in standby with negligible power consumption, most people still believe that Bluetooth is a power-hungry function, and turn off their phone's Bluetooth function to save power. As iBeacon relies on Bluetooth, encouraging visitors to keep their Bluetooth on is a basic necessity to ensure iBeacon's widespread use. Lastly, for iBeacon to achieve its maximum value depends on the overall attitudes of the museum industry. Information interoperability and resource sharing are themes in the current era, but the data collection and analysis of visitors to a single museum can only be said to take advantage of basic iBeacon functions. Only by achieving true information sharing across the museum industry as a whole can the value of iBeacon be maximized.

4. Conclusion

Any new technology will inevitably face challenges upon its appearance and promotion, and iBeacon is no exception. In any case, the idea of using indoor positioning technology within museums to serve visitors has gradually come to fruition. iBeacon is simply an opportunity in this new field of smart operations. There is still a long way to go where museums are concerned.

References

[1] Wang Zhu. 'Research and application of personalized navigation technology in smart museums', [J] *Computer Engineering*, 2009, 08.
[2] Li Lin. *Research on audience interpretation of museum exhibitions* [D]. Fudan University, 2009.
[3] Wang Ying. 'Discussion of museum navigation design', [J] *Ergonomics*, 2010, 3.
[4] Wang Hui. *Research and implementation of indoor location algorithms based on RSSI wireless sensor networks* [D]. Beijing University of Posts and Telecommunications, 2010.
[5] Wu Ningning. 'The revelation of Big Data in museums' [N]. *China Cultural Relics*, 2013, 9.
[6] Zhang Lan. *Opportunities and challenges for museums in the Big Data environment* [C]. Shanghai, 2013.
[7] Wu Donggan. 'Design and implementation of smart iBeacon navigation systems', [J] *Journal of Guiyang University (Natural Science Edition)*, 2014, 9.
[8] Chen Hongyu. *Application of indoor mobile navigation technology in science and technology venues* [C]. Beijing, 2014.
[9] Guo Limin. 'Research on iBeacon applications in museums', [J] *Research on Library & Information Work at Shanghai Colleges & Universities*, 2015, 1.
[10] Zhang Zhuosheng. 'Smart Museums using iBeacon positioning technology', [J] *Electronic Engineering & Product World*, 2015, 1.

WeChat Applications in Museum Guides

Zhao Yan
(Department of Science, Beijing Museum of Natural History, Beijing 100050, China)

Abstract: With the development of science and technology, we have entered a new era of digital information, and museums that collect ancient artefacts are applying more novel digital information means to display and services. Amongst them, WeChat has become a popular guide tool for museums due to its wide user base, convenient use, powerful functions, and free platform. This paper explains the navigation methods selected by some museums, discussing the feasibility and functional application of WeChat guidance, including the issues of using WeChat navigation within museums, in order to reveal how to best use WeChat to help in the dissemination of information and development of museums.

Keywords: WeChat; museum; guide

Remarkable innovations in communication in the information age have brought new media technologies to the forefront [1]. Smartphones, network coverage, and convenient communication permeates daily life, and these are exactly the elements required by the museum to fulfill their educational communication function. Many museums have kept up with technological advances, optimizing public services through emerging media, disseminating scientific knowledge, and presenting novel content to visitors in fresh and active ways, allowing them to "play".

1. The status of museum guide facilities

1.1 Electronic voice navigation devices

Traditional museum visits are generally divided into self-guided tours, lecturer-led guides, and rented audio guides [1]. In order to make tour routes smoother and at the same time obtain some information about the exhibitions, many visitors are willing to hire audio guides. Most museums and attractions have audio navigation equipment, and visitors can choose the appropriate language. Some navigation devices play associated voice introductions using exhibit numbers, while others sense viewer location and automatically discuss the exhibits in the area.

1.2 The Weibo era

With the rapid development of science and technology, the network occupies a large part of people's lives. Microblogging is short, succinct, time-sensitive, and interactive, becoming a platform for intimate exchange between museums and their visitors. Visitors are not only shown the dynamics of the museum, but can also comment and repost, opening up promotional opportunities. Exhibition information, event information, and popular science essays within official Weibo accounts enable visitors to gain more information and acquire knowledge. As such, official microblogging has swiftly emerged as a new force of the museum service system, closing the gap between the museum and the public, and becoming a highway by which the museum may communicate with the public [3].

1.3 Pocket apps

With the popularity of smart phones, handheld applications have entered daily life. As a result, some famous museums around the world have developed corresponding applications, such as the British Museum, the US Natural History Museum, the Louvre, the Tokyo National Museum, and the National Palace Museum in Taipei.

The pocket app guide has become a popular form of guided tour in many museums. The newly-built Shanghai Natural History Museum has set up seven augmented reality observation points, and via download of the mobile app, the perfect combination of virtual and real worlds can be realized, offering a novel experience. Augmented reality technology will be more widely used in the museums of the future. Although apps are strongly targeted, they require a certain amount of capital investment, and multiple versions are developed for various operating systems and terminal devices [4]. For free, open museums and exhibitions, costs must be calculated.

1.4 WeChat's public platform

The enthusiasm for microblogging has yet to subside, and WeChat has already spread rapidly across thousands of households. People use it to transmit voice messages and images, as well as scan QR codes, and WeChat operations have become an indispensable daily activity for many people. The museum has also seized the opportunity to follow in these fashionable and technological footsteps, using WeChat as a method of guiding. The WeChat tour function launched by museums is an attempt to use new media in the display and service of the museum, conveying audio, images, and text to bring visitors and exhibits closer together [1].

In 2012, the National Museum opened a WeChat public platform, and museums such as the Guangdong Provincial Museum and the Tianjin Museum also began using WeChat assisted tours [5]. Due to advantages like a flexible interface and no downloading, WeChat has won over more museums [6]. Visitors need only follow a museum's WeChat platform to access information.

2. The advantages of WeChat applications in museum guides

2.1 Extensive WeChat applications

According to a British media survey, at the end of 2014, WeChat users topped 500 million. In addition to development in China, WeChat also has an overseas version, which exceeded 100 million users as of 2013.

In addition to its large user base, WeChat software is also one of the most frequently used, enabling it to become a link between museums and visitors, assisting in publicity and communication. Meanwhile, it can also provide one-to-one services for self-guided tours. Its low application cost and easy operation have attracted large numbers of users, fully catering to the basic needs of museums [7].

2.2 Powerful WeChat features

With an open public platform, convenient QR code scanning, and strong targeted message push, WeChat stands to play a powerful role in communication [8]. Its text, voice, and image transfer, as well as interactive games are suited to the needs of museum guides. Visitors do not simply hear explanations; they can also see images and participate in interactive games, allowing the museum to bring collections to life. These fresh technological and fashionable elements are urgently needed.

2.3 A sense of audience participation

Participation is especially important in exhibitions, allowing visitors to absorb what they have learned. Within the exhibition hall, visitors can operate the WeChat interface on mobile phones, scanning QR codes, sending codes, and interacting through messages. In this way, WeChat has become a means for visitors to participate, with the museum at their fingertips in real-time interaction with exhibits and curators. Museums try to imbue the visitor experience with technology that centers on information about the collections [9]. Using WeChat navigation makes a visit more interactive and interesting, allowing visitors to gain more knowledge.

3. *WeChat applications for museum guides*

3.1 Visitor guides

The guide is an essential part of a museum. Many museums have free paper guides at the entrance, with information on opening hours, exhibitions, and highlights. Now, visitors can make appointments via WeChat, and can also gain access to the above information.

The Capital Museum's WeChat platform includes comprehensive information, with three categories clearly marked "Exhibition Appreciation", "Visit Guide", and "Service Information". Clicking on the "Visit Guide" section offers instructions and voice information. The interface also includes introductions to permanent and temporary exhibitions under the "Exhibition Appreciation" tab.

3.2 Exhibit explanations

Museums are based on exhibits, and the content of traditional museums is disseminated via graphics, explanatory plates, and limited text descriptions within the exhibition area [1]. As visit time is limited and museums are rich in content, visitors often do not spend much time reading text on display boards. WeChat tours are a good way to challenge this issue.

For example, at the National Museum and the Guangdong Provincial Museum, artefacts are given numbers. Visitors can access explanations by transmitting these numbers through WeChat Follow, including voices, pictures, and videos that offer interaction with the collection [2].

3.3 Interaction

3.3.1 Query messaging

Rather than passive reception of explanations, visitors may have questions about exhibits. Through the WeChat platform, they can take the initiative to ask questions, and staff and expert lecturers within the hall can answer them. For example, the Shanghai Natural History Museum has an "Expert Response" section on their WeChat platform, allowing visitors to upload images of unknown plants and animals to be identified by experts.

In addition, audience needs and suggestions regarding their visit may also be exchanged via WeChat. In this way, it acts as a sort of visitor book.

3.3.2 Interactive games

As a way to increase interest in exhibitions, multimedia equipment is usually set up in the exhibition hall, and visitors are invited to play interactive games via touch screens. However, when many young people are visiting an exhibition, it is inevitable that there will be competition, which will lead to facility failure or even damage. In these circumstances, the WeChat interactive game function can be maximized. In general, multimedia facilities in the exhibition hall involve short, simple games, which can be completed via WeChat. This allows young people to participate, and also increases interest from adults.

3.3.3 Feedback

Museums are built for visitors, and their feedback is taken seriously. Many museums use questionnaires to solicit feedback from the audience, and they can now set post-view feedback on WeChat, allowing visitors to rate the satisfaction of their visit and facilitate the collection of data and statistics. Meanwhile, the museum must also investigate what visitors learned during their visit, as well as their favorite artefacts, as this is beneficial to research and development.

In this vein, the Shanghai Natural History Museum has designed a "hands on" module in which visitors can take pictures of their favorite area or exhibit and share them to the platform.

3.4 Promotion

3.4.1 Self-promotion

Museums must improve their influence, not only drawing up plans behind closed doors. Beyond collecting artefacts, they must also carry out publicity and promotion. WeChat is a great platform for this, allowing large-scale events to be pushed to interested visitors via messages, and forwarded in order to expand influence. In this way, more interested individuals will gain access to event information, and museum activities are able to gather popularity. For example, the Beijing Museum of Natural History puts research progress and new exhibition information on their WeChat platform, which is widely forwarded and has attracted many visitors with an interest in animals.

3.4.2 Public interest promotion

In addition to pushing event information, the museum is also responsible for writing academic essays based on its research, so that even if visitors do not visit the physical museum, they are able to gain knowledge through reading pushed articles.

Public welfare publicity is a major aspect of the museum's social benefit. Compared with issuing posters, WeChat is a promotional method that is both resource- and cost-saving, also reaching higher audience numbers. Through public messaging of both images and text, it is possible to promote ideas such as the conservation of hydropower resources, the protection of cultural artefacts, and the prohibition of purchasing animal products, while also sending introductory messages for meaningful events such as World Environment Day and Earth Day. This way, while producing public welfare publicity, museums simultaneously enhance their influence in society.

4. Possible issues

4.1 Reaching groups without WeChat

Although WeChat's audience is extensive, many primary and secondary school students do not have it. To solve this issue, a museum can set up a teachers' area on WeChat, integrating the curriculum-style activities of primary and secondary schools as a reference for teachers who are leading groups.

The elderly also own fewer smart phones, which presents similar issues for access. Therefore, WeChat tours cannot completely replace manual and electronic navigation [5]. For a certain period of time, the museum's WeChat tour will coexist with other guided tours.

4.2 Network coverage

Another big issue with WeChat guidance is network coverage in the museum. If the signal in the hall is poor, it will affect WeChat's navigation function. Some visitors may not have enough mobile data to listen to audio guides or receive exhibit images. To answer this issue, museums should improve their WiFi, which requires financial support. When WiFi is available, it also stands to distract visitors from their visit, as a large amount of other content can be accessed over the network, including TV, videos, and online shopping.

4.3 Paying more attention to WeChat than the exhibits

Ultimately, WeChat's operability offers greater stimulation than viewing exhibits, so people are likely to put more time and energy into their mobile phones. In terms of exhibitions, technology is a means, not an end [10]. WeChat is, after all, a tool to assist people's understanding of exhibits, so this tool needs to be able to explain exhibits and also spark further interest that encourages people to return. Therefore, we can design specific interactions with exhibits in WeChat, for example, putting images of fake antiques in a game, making the audience compare these to the real artefacts in the exhibition hall to spot the differences, or using other interactive games that require them to find answers during their physical visit, encouraging them to refocus, and thereby deepening their memory and understanding.

5. Summary

The advancement of science and technology has promoted considerable lifestyle changes. The museum, which traditionally offers a serious, old-fashioned impression, has begun to pay attention to the application of new technologies in the information age in order to draw visitors closer [9]. WeChat guiding has achieved the improved integration of the social science and technology platform and museum education functions [5], offering pre-visit guidance, being a tour guide during one's visit, and delivering feedback after the visit. It is an intimate and interactive exchange platform for the museum and its visitors. Although there are still certain issues with the current status of WeChat guiding, it has great importance in the field of social education, public service, and publicity for museums, with scope for more careful research and development in the future [11].

References

[1] Wu Chunxia. *The use of new media technology in popular museum science education* [C]. The 16th Annual Meeting of the China Association for Science and Technology — 16 steps towards a new vision of scientific development, efforts to innovate science and technology education content forum proceedings, 2014-05.

[2] Xiao Peng. *A new platform for library information push and reader service in the micro-era — inspired by the Guangdong Provincial Museum WeChat guide* [C]. Passing the torch of a 50-year storm — Guangdong Academic Association Annual Academic Conference 2013 award-winning essays at the 50th anniversary of the society, Nov 2013, pp. 69–74.

[3] Li Xiuna. 'Weibo/WeChat: The museum experience in media applications', [J] *China Museum*, 2013, 4: 97–102.

[4] Wang Xueliang, Han Xiaoyan. The museum on WeChat [N]. Xinhua Daily Telegraph, 2013-04-12.

[5] Zheng Wenjun, Ren Xiaofei. 'Discussion on the application, advantages and disadvantages of WeChat navigation', [J] *Science and Technology*, 2014, 8: 55–56.

[6] Li Meng. *Can new technology "sweep" through the future of museums?* [N]. Chinese Culture News, 2012-12-31.

[7] Zhou Xinyi. 'Research on the development of WeChat applications in museums', [J] *Teachers*, 2013, 17: 118.

[8] Peng Siya. *WeChat: a new channel for the popularization of science education in science and technology museums* [C]. The Role and Responsibility of Popular Science — Exploration of Chinese Popular Science Theory and Practice — Proceedings of the 20th National Popular Science Theory Symposium, Oct 2013, pp. 389–395.

[9] Yun Fei. *'Museums at our fingertips' websites, WeChat, not one less* [N]. China Art News, 2013-05-27.

[10] Su Donghai. *Meditations on the museum (Vol. 3)* [M]. Beijing: Cultural Relics Publishing House, 2010.

[11] Xiang Kun, Liu Xinyang. 'The use of smartphones in museum navigation', [J] *Cultural Relics Annual*, 2014, 6: 40–45.

Analysis of Smart Museum Navigation Systems Based on WiFi Location Technology

Shen Hongjun
(National Palace Museum Department of Data Information, Beijing 100009, China)

Abstract: With the continuous development of wireless communication technology, indoor positioning technology has become diversified, including WiFi, ZigBee, Bluetooth, and Ultrawideband, which have been widely used across various scenarios. WiFi positioning technology features fast transmission, stable signal, and large data flow. Therefore, it can be integrated into smart terminals for display and navigation services in the museum field, forming an smart navigation system that uses technologies such as wireless positioning, wireless transmission, multimedia, and the Internet, making up for the shortcomings of traditional manual explanation services by satisfying visitors' needs and facilitating the digital promotion of the museum. This article explains WiFi positioning technology and the applications of smart navigation systems based on this technology within the museum.

Keywords: WiFi positioning; smart navigation; museum

Preface

As a cultural symbol of a city or country, the museum carries enough meaningful cultural content to satisfy the different needs of a variety of social groups, such as viewing art, the appreciation of historical artefacts, cultural education, leisure, and entertainment. Nowadays, with China's strong promotion of cultural development, museums are required to receive large numbers of visitors every day. In order to satisfy audience curiosity regarding history and culture, with the rapid development of information technology today, museum personnel must apply network, digital, and multimedia technology to improve the service quality of the museum, while creating a comfortable and pleasing environment for visitors, as well as satisfying their requirements for multifaceted understanding and learning.

Electronic navigation systems are now widely used in some museums, but the existing navigation system has many shortcomings. For example, automatic triggering and smart guidance cannot be fully implemented, and there is a lack of real-time statistical analysis of audience traffic. There is also a lack of an audience positioning system, an information push system, and an visitor information collection system. An smart navigation system based on WiFi positioning stands to solve these issues.

1. Definition of the WiFi-based smart museum navigation system

1.1 Principles of WiFi positioning technology

1) Each wireless AP has a globally unique MAC address, and the wireless AP is generally stationary.
2) When WiFi is switched on, the device scans and collect surrounding AP signals. The MAC address broadcasted by the AP signal can be obtained whether it is encrypted or connected, or even when the signal strength too low to display the AP signal in the wireless signal list.
3) The device sends data indicating the AP to the location server, which then retrieves the geographic location of each AP and merges the strength of each signal to calculate the geographic location of the device and send it to user equipment.
4) The location server constantly updates and supplements its own database to ensure data accuracy.

1.2 The smart navigation system

The smart navigation system is composed of a navigation content system and navigation equipment. With the current popularity of smart devices, a navigation device can be a handheld terminal provided by the museum, or a smart phone, tablet, and laptop belonging to the audience. The content server is designed and produced by a museum according to its own needs, and can be installed on navigation devices provided by the museum or a viewer's own smart terminal in the form of an app or web page. When visitors use their own smart devices, the smart navigation system can be installed via download from the official museum website or via on-site QR code scanning.

1.3 The smart museum navigation system based on WiFi location technology

The smart navigation system based on WiFi positioning technology is composed of a smart terminal with WiFi capabilities, a wireless access point (AP), a positioning server, and a content server, as shown in Figure 1.

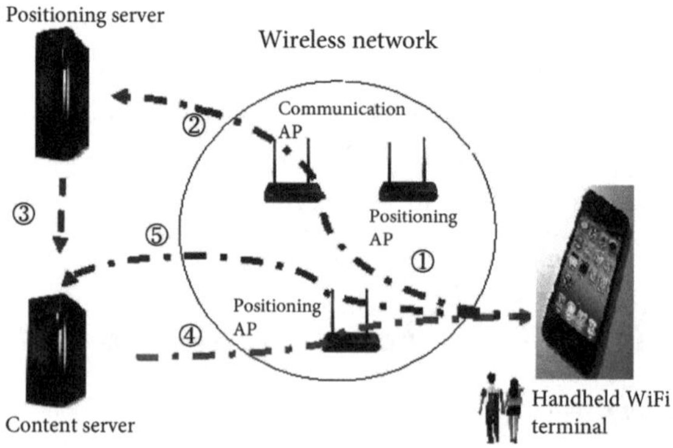

Figure 1 Smart WiFi positioning for a museum navigation system

2. Advantages of the museum navigation system based on WiFi positioning technology

1) At present, many museums have built or are in the process of building wireless networks, so navigation systems based on WiFi positioning technology can be used on the basis of original wireless network equipment, saving costs and avoiding redundant construction.
2) WiFi positioning technology has high precision – up to 3~5 meters depending on the specific environment. The positioning accuracy can be improved up to 1m in key areas, offering an accurate grasp of a visitor's active track and real-time positioning.
3) The wireless network has high bandwidth and strong scalability, and is able to realize various functions such as voice, video, positioning, and wireless sensing within the smart navigation system.
4) Big Data technology allows for the collection of visitor information and analysis of their needs, in order to conduct more targeted exhibitions.

3. Application of smart navigation systems based on WiFi positioning in museums

3.1 Automatic navigation service

3.1.1 Current navigation systems

Currently, the navigation systems of major museums in China still rely on explanation boards, electronic voice interpreters, and tour guides. These traditional navigation forms have a variety of disadvantages. For example, explanation boards can generally only mark out the name, place of excavation, age, and other basic content of the cultural artefacts, which cannot satisfy visitors' needs for an in-depth understanding. Electronic voice interpreters only offer audio explanations recorded beforehand, which is too rigid a form. Because of the low accuracy of the equipment, the explanation content may not match the real situations at times. There are also issues with the maintenance and cleanliness of equipment. Professional guides generally carry out group presentation services, and will not provide free presentations to individuals. When there is a big group of visitors, the surrounding environment may interfere with the guide, and the visitors may not be able to hear the explanation clearly, resulting in loss of interest in continuing to look. Some guides explanations remain at the level of reciting tour guide speeches, which is not sufficiently thorough or in-depth.

3.1.2 WiFi positioning in smart navigation systems

Through the smart navigation system based on WiFi positioning (with equipment provided by the museum or using visitors' own smart terminal equipment), visitors are able to see all museum exhibits, as well as detailed information and multimedia materials regarding the exhibits. When a viewer clicks on exhibits of interest, the system will use his or her real-time location with relation to target exhibit locations obtained by the WiFi-based positioning system, to direct the viewer towards exhibits of interest without having to follow a fixed route. During a visit, the system will automatically display surrounding information on the smart navigation terminal at the visitor's end, allowing them to learn about exhibits via text, voice, and multimedia. Using this system, visitors are able to walk freely around the exhibition using the smart navigation terminal, controlling the pace themselves without needing to follow a guide, making the learning experience both easy and independent.

3.2 Collecting visitor information

The visitor is the main object of museum service, also the embodiment of the value of the museum. In order to serve their visitors, museums must understand their needs. To do this, they must collect visitor data.

Through the smart navigation system using WiFi positioning, a large amount of visitor data can be obtained through interaction with the system, such as age, sex, origin, preferred exhibit type, time spent at each exhibit, and visit flow data of each exhibit, as well visitors' opinions and suggestions. Museums constantly collect and archive this data through the system, analyzing complex data using Big Data technology and thus studying audience preferences. This enables museums to provide more targeted services and change the state of exhibitions behind closed doors, attracting more visitors and promoting history and culture.

3.3 Real-time audience positioning

Currently, some large museums can become crowded due to public of visitors during peak periods. Sometimes, people get lost. In the past, museums generally used voice broadcasts to find lost people, but this method is inefficient. Now, visitors are able to mark their peers in the navigation system so that when scattered, they can use their own smart navigation terminals to locate people. Visitor groups can also utilize decentralized visits, setting up certain gathering points thanks to the efficiency and convenience of smart navigation.

The smart navigation system based on WiFi also offers security advantages. When a security issue arises, suspicious people can be located in the background of the system, attaining their active route in real time.

3.4 Push information

Whenever a museum has important events, independent exhibitions, or emergency situations, push messages can be sent to visitors via the smart navigation system. For example, when a museum is holding a VIP reception, messages can be sent to visitors' smart terminals through the smart navigation system, so that they can communicate with staff and plan their tour accordingly. When encountering critical situations such as fires or earthquakes, smart navigation systems can push safety information to visitors, indicating routes to emergency evacuation passages and locations within the exhibition hall, facilitating the transfer of crowds, and reducing risk factors. When visitors need help, such as in cases of trauma or illness, the communication function on the navigation system can be used to directly contact relevant departments and access timely help. In addition, the system also provides a range of service information regarding the museum, such as a maps of the surrounding area, public transport, hotels, and dining, providing improved services from an audience perspective.

4. Conclusion

The smart museum navigation system based on WiFi positioning technology offers an excellent visitor experience according to individual preferences. The guided navigation system using multimedia technology provides vivid and detailed exhibit explanations to visitors, allowing them to understand collections from many angles and acquire historical and cultural knowledge. Meanwhile, the museum is also enabled to analyze visitor preferences from collected data, in order to study audience needs and host more targeted exhibitions to enhance their appeal. At this moment of digital museum development, smart museum navigation systems powered by WiFi positioning will surely grow in popularity.

References

[1] Zhu Zhongyi. 'Application of WiFi-based indoor positioning technology in museums'. *Software Industry and Engineering*, 2013, 3(21).

[2] Wu Yuhang, Wu Caicong, Chen Xiuwan. 'Introduction to several indoor positioning technologies'. *China Surveyance and Mapping*, Jan 2008, 3.

Online Exhibitions and Public Services in the New Media Era

Xu Bingbin
(National Palace Museum Department of Data and Information, Beijing 100009, China)

Abstract: With the rapid development of economic and cultural undertakings in 21st century China, museums have entered the service era. From the perspective of online exhibitions in the new media era, this article explores the development and innovation of museum services in the Big Data environment, including guided tours, public education, and academic research.

Keywords: new media; museum; online exhibition; public service

Preface

With the rapid development of China's economic and cultural undertakings in the 21st century, the number of museums and supporting facilities has significantly improved, and the focus has turned to providing high-quality and diversified services to the public, indicating the museum's entry into the service era. As Professor Song Xiangguang of the Institute of Archaeology and Culture at Peking University has pointed out, "The main feature of the museum in the service era is that service now holds an important position and role in museum work, with museums building business processes based on the goals and course of service, organizing their resources according to service objectives. Museum service in the service era presents overarching, systematic, scientific, service goal-oriented, and service quality-oriented characteristics" [1]. *The China New Media Development Report (2014)* stated that since 2013, China's new media development has further demonstrated the trends of accelerated mobilization, integration, and socialization, with micro-distribution based on new media becoming a new impetus in the promotion of Chinese social development [2]. The visitor is at the core of the museum, so serving the public is a crucial responsibility and mission. In the new media boom, given the popularity of personal computers, smart phones, and smart handheld devices, and the rise of online applications like wireless networks, social media, and mobile clients, this has brought further inspiration for the innovative development and public services of online digital exhibitions.

1. Online display and guidance services

Museum service is based around collections, exhibitions, and hardware and software facilities, with the most basic being an overall guided tour. In the new media era, museum navigation services based on digital display have moved beyond traditional physical models, covering a global audience with quality service.

At a time when new smart terminal technologies are spreading rapidly, museums must keep up with the times and improve public service levels. In particular, the popularity of smart phones has changed daily human interactions and habits of information acquisition, more closely reflecting personal interests and needs, and forming a completely personalized communication platform. With the increasing maturity of mobile technology and handheld devices, digital navigation methods are more abundant and exciting, and the accelerated development of mobile self-guided navigation systems is inevitable in the development of digital museums. Members of the public are able to download a mobile museum app, subscribe to mobile newsletters, or join member forums, to acquire all kinds of information in real time. They can also make appointments for a visit or purchase tickets, design self-guided tour routes, scan QR codes to hear professional explanations, gain further background knowledge, participate in interactive games, buy souvenirs, and share their experience with friends. For example, the Handheld Palace App launched by the Palace Museum is a navigation-based application that can be loaded "in your pocket" to facilitate visitors' interpretation of the palace and its historical allusions within the open area. Using GPS for accurate positioning, the app facilitates self-guided route planning and free and easy tours, and offers the choice of a variety of recommended theme-based routes. Users can also take photos anytime, anywhere, to share their experiences with friends. The Palace Museum also plans to release 360° panoramic HD maps in open areas and even some closed-off areas, providing the public with better guide services and visitor experience. The diversified high-quality guidance services based on digital display can be said to further minimize the distance between the museum and the public.

2. *Digital display and public education services*

The director of the Palace Museum, Dan Jixiang, has said: "Education is the cornerstone of national rejuvenation. The museum embodies the essence of human cultural heritage, describing the progress of human history, showing the overall civilization and wisdom of mankind, and holding unique educational resources. Therefore, the incorporation of museums into the national education system is in line with the developmental trends of world museums and the need for museums to fulfill educational purposes, promoting the close integration of museum work, school education, and social education, as well as forming a more complete social education network. It also a requirement for the overall improvement of civil society in perfecting China's modern national education system, building a society of learning, and forming a lifelong education system" [3]. Digital display used in public education can play a greater role without the restriction of museum opening hours or physical space. Faced with visitors of varying age groups and educational backgrounds, museums must provide rich and diverse cultural and educational services accordingly.

Young people are curious, and have a strong desire for knowledge. They are also highly accepting of the Internet and new media applications, so education programs that use digital display are most popular amongst this group. For example, the Capital Museum has launched a children's edition website, with lively pages and interactive programs that attract young people's attention. A youth website for the Palace Museum titled a "Forbidden City Adventure" is also under development. In recent years, the content of youth education has become more abundant, with display means constantly being innovated, so museums must improve display and service quality in step. In particular, museums must build on their awareness and social responsibility to provide quality services to young people, and launch a variety of rich, colorful, and distinctive educational activities via various new media technologies. For example, the A Day of an Emperor iPad app produced by the Palace Museum in 2014 is a mobile app developed specifically for children (aged 9–11), leading children through the Qing Palace to learn about the Emperor's way of life, official studies, and leisure and entertainment activities. Fully hand-drawn in a cartoon style, it offers interactive mini-games throughout that provide children with historical knowledge, allowing them to experience the fun of an expedition within the virtual Forbidden City while learning about traditional culture. Youth education at museums should be integrated with school education, and in order to leverage the rich resources of the "second classroom" that is the museum, to provide quality school education services. A development team conducted detailed preliminary research before establishing the app, ensuring that it would be unanimously appreciated by students, teachers, and parents. After the app was released, it received rave reviews, winning the Best App award in the App Store in October of that year. Thus, it is clear this form of traditional cultural education is not only popular, but also a useful endeavor in the field of youth education within Chinese museums.

In addition to youth education, museums are also an ideal place in which to enhance the cultural quality of the whole population, and achieve lifelong social education. Dan Jixiang, the director of the Palace Museum, once pointed out that museum education has the features of nationality, sociality, eternality, intuition, richness, openness, autonomy, and pleasure. The interactivity, openness, and participation of the new media era have enabled more equivalent communication between museum personnel and visitors, and digital displays based on new media technology are increasingly attractive, leaving deeper impressions on the public. For example, the Palace Museum's Beauties of Prince Yong Zheng app, the "Forbidden City Superstitions" app, and the "Night Revels of Han Xizai" app for iPad and Android are all based around cultural artefacts and supported by academic research, naturally and ingeniously integrating rich media information such as voice, text, and video. They provide more interactive and entertaining experiences on the basis of appreciation and education, as well as in-depth interpretation of artefacts, spreading various kinds of cultural knowledge. These iPad apps are beautifully crafted and rich in content, and have garnered instant popularity, with downloads repeatedly hitting new highs and frequently topping the App Store rankings.

In recent years, due to the accelerated pace of work and life, people's leisure time has shown a tendency of fragmentation. Many visitors are no longer satisfied by spending a lot of time looking at museum exhibits or websites in order to appreciate exhibits. Instead, they want to use the gaps between daily work and life anytime and anywhere, to understand the cultural heritage behind one or two artefacts for just a few minutes at a time, and integrate cultural information into their own lives. Under such circumstances, the use of new media communication to provide the public with short, quick, and high-quality services while promoting the development of the industry has become a key concern for museums. To this end, following careful research and preliminary preparations, the Palace Museum has launched a lightweight "Daily Palace" mobile phone app for Android and iOS, introducing a single treasure each day, carefully selected from more than 1.8 million artefacts. It offers an appreciation of the Forbidden City's collection in mere square inches, exploring the amazing everyday details of ancient life, and showcasing the inexhaustible artistic value of the treasures that have been passed down. Opening the application, many of the museum's cultural artefacts appear within reach, some accompanied by detailed and engaging explanations of traditional crafts and their background stories. The "Daily Palace" app is in calendar form, with a fresh and elegant notes section in which users can record their thoughts at any time, and share them on mainstream social media. After registration and log-in, note content can be synchronized anytime, anywhere. This is an exploration of the use of new media to integrate cultural heritage into contemporary life, as well as a new approach by the museum to carry out public education and cultural dissemination. It has already proven to be very popular.

3. Online display and academic research services

The museum is the main location for the preservation, display, and interpretation of physical historical evidence, playing a hugely important role in academic research [4]. Museums have vast collections. The characteristics of scientific research chiefly combine social science and natural science, and the integration of objects, history, and theory. On the one hand, the museum must strengthen its own academic research and enhance the authority of its exhibition and spread of education. On the other hand, as a cultural institution providing a public service, museums must also publish important information about their resources online, as well as the latest in scientific research. This will provide a wealth of material for professional research across various disciplines, and will play a role in the promotion of academic research.

Collection resources are the basis for academic research in museums. Therefore, in the case of mature conditions, a museum collection must be announced in a timely manner in order to benefit public research and use, as well as to accept supervision from all walks of life to gain public trust. For example, the Palace Museum published the *General Collection of the Palace Museum* on their official website, including first, second, and third-grade cultural artefacts, as well as general artefacts and ceramic objects. Each cultural artefact is marked with a collection number, name, time, and related information. In today's increasingly mature Big Data environment, the retrieval, induction, and in-depth analysis of mass collection data and past research data will provide new discoveries for collection management and academic research, as well as other forms of professional research.

In addition, in the new media era, the Internet, mobile devices, and smart systems have made public participation and cross-disciplinary research possible. New media information resources are abundant, available anytime and anywhere for inquiry and reading, with ever-expandable space by means of links, with an almost endless capacity. New media is fast, and is not limited by factors such as printing, transportation, and distribution. Information can be sent and shared in an instant, heightening public enthusiasm for interaction. Wider public participation also offers new resources and inspiration for academic museum research, potentially enhancing research and promoting work in various museum businesses. Museums must follow the trend of the times and use new digital exhibition technology to publish academic achievements and references online, and to provide search, download, and share functions, or use new social media platforms to organize specific topics and issues, open discussions, and create an academic atmosphere with wide schools of contention. For example, with the "Night Revels of Han Xizai" iPad and Android app produced by the Palace Museum, in addition to providing visitors with a fresh and current interactive experience, it is also an important academic reference with detailed and reliable content featuring 100 content annotations, 18 expert audio tracks and video guide segments, and one postscript, comprehensively integrating research by experts and scholars while incorporating the latest academic opinions. The Forbidden City website has also launched "Famous Paintings in the Forbidden City" (http://minghuaji.dpm.org.cn/) and "Ancient Chinese Painting and Calligraphy Research System" (http://shimage.dpm.org.cn/ggresearchnet/index.jsp), combining the strengths of expert research and new media. Through the combination of high-tech browsing and academic research, users are able to participate in the collation and identification of important information such as inscriptions and seals, stimulating public interest and exploration of traditional Chinese art. Timely publication of collection information and opening scientific research platform resources better serves the public, and also embodies the museum's public social responsibility.

4. *Interactive services in the Big Data environment*

In the era of new media, information devices are constantly generating and exchanging mass data. A Big Data era of mass production, sharing, and application has arrived. In recent years, many museums in China have made great achievements in the development of public services such as navigation, public education, and academic research, accumulating huge data resources. Zhang Lan, director of the Shanghai Museum of History, has stated: "On the one hand, we are contributing data to society, and on the other, social data serves our public culture. We must strengthen our use of the Big Data provided by government, research institutions, and enterprises in the realization of cultural services. Through this data, we are able to discover preferences in service objects, realize public resource sharing, and assume more responsibility for the museum's cultural communication" [5]. In the era of Big Data, digital display must also make full use of the convenience of technological innovation, constantly innovating and developing in order to provide better services to the public. This also offers new development opportunities for the museum [6].

1) Museum services should be oriented towards the needs of the consumer. The public is the service object of the museum as well as the social foundation upon which museums depend for their survival. As public cultural facilities, museums must strengthen their interaction with the public, whether in exhibitions, social education, or academic research, and orient themselves towards public needs. Therefore, before planning digital display projects, it is necessary to conduct investigations to gain in-depth understanding of public needs. The director of the Palace Museum, Dan Jixiang, points out: "For museum visitors, the use of a variety of information channels to conduct a variety of investigations provides a basis for the development of cultural communication, and also brings activities closer to public life. The concept of interaction with society is a requirement for museums in the new era" [7]. In the new media era, museums have relatively stable, loyal followers and fan bases, and social platforms such as websites, smart navigation systems, Weibo, and

WeChat are generally characterized by rapid communication, unlimited time and space, and real-time interaction, meaning that information exchange between museums and the public is no longer one-way, but multi-level and interactive. In order to understand data about a visitor's age, educational background, occupation, behavioral habits, preferences, and suggestions, the Palace Museum has launched ongoing monitoring and evaluation of digital display projects in recent years, making full use of multiple channels to increase public interaction, accumulating a rich visitor database. In the new media era, online interactive coverage is broad, operations are convenient, and a large amount of visitor data and consumer demand can be received in a short time. At this point, various sensor technologies within smart Cloud computing platforms can be used to carefully and effectively analyze and evaluate this Big Data to provide targeted, multivariant, and personalized services, improving the quality of a museum's public service.

2) Big Data in the museum can also provide interactive services for public emergencies. With modern developments and the improvement in living standards, the number of people visiting museums is increasing, especially during the holiday period, when the sudden increase in population density often leads to security risks. Certain large-scale ruin museums, such as the Palace Museum, are often overloaded during the peak tourist season, with the number of visitors reaching 182,000 in a day, putting great pressure on the protection and safety of cultural artefacts. To this end, the Palace Museum is building a dynamic monitoring system, striving to assemble a complete monitoring system for visitor flow in open areas. Smart phones are currently very popular, with high usage rates, and the more mature technology to monitor population density is usually combined with maps to implement geographic positioning through mobile phones. On this basis, supplemented by mobile phone communication, mobile phone push, and multi-latitude Big Data services, the comprehensive analysis, cross-certification, and prediction accuracy will be greatly improved. Therefore, it is necessary for museums to monitor population density growth trends in real time via the mobile internet, as well as the population distribution heating power chart provided by Big Data analysis. When the density reaches a certain level in public spaces, it must be controlled and channeled, using new media such as Weibo and WeChat to disseminate early warning information in a timely manner, reassuring the public and proposing guidance and diversion measures including planning visit routes to prevent stampedes and maximize public safety.

5. Conclusion

The Palace Museum president, Dan Jixiang, has said: "In an increasingly developing modern society, the museum is an excellent community for Chinese cultural undertakings, and must further raise its social service awareness into a mission, continually improving its abilities to serve society and the quality of its public service. Through institutional reforms, we will guide more social resources into museums, and at the same time enhance the use of resources and social service capabilities in order to provide greater and better cultural products and services for the vast general public" [8]. In the new media age, the museum's social functions and roles of education and public service are becoming more prominent. Online museum display must fully utilize new technologies to expand the radiating scope of cultural communication and extend service functions such as tour guides, public education, and academic research. Museums must enhance their reach and visitor experience, thereby breaking through the limitations of the physical building, closing the gap between the public and the museum, and contributing to the overall improvement of the public cultural service system.

References

[1] Song Xiangguang. 'Museums are entering the service era.' *China Cultural Relics*, July 2006, 6.

[2] Can Jian. *New media bluebook: China New Media Development Report No. 5*. Social Science Literature Publishing House, Jun 2014.

[3] Shan Jixiang. 'Social responsibility and social education in the museum'. *Southeast Culture*, 2010, 6.

[4] Ma Xianzhong. 'Playing an important role in the academic research of the museum'. *Journal of Chinese Social Sciences*, May 2013: 452.

[5] Zhang Lan. 'Opportunities and challenges for museums in the Big Data environment'. *Cultural Forces and Museum Challenges*, Shanghai Ancient Books Publishing House, 2013, p. 191.

[6] Wu Ningning. 'Analysis of service innovation and development in museums in the Big Data era'. *China Shrine of Remembrance Research*, 2014, 1.

[7] Shan Jixiang. 'The mission of the museum and protection of cultural public rights'. *Sichuan Cultural Relics*, 2004, 1.

[8] Shan Jixiang. 'Museum management upon entering the service era'. *Dunhuang Cultural Relics*, 2013, 3.

Discussing the Application of the Mobile Internet in Traditional Museum Exhibitions

Zhu Jiwei
(National Palace Museum Department of Data and Information, Beijing 100009, China)

Abstract: Due to the rapid development of the Internet, we have entered the mobile information era, and mobile Internet applications can be seen everywhere. Traditional museum displays deliver multidimensional information. Using the mobile Internet to provide supplementary information is the fastest and most effective mode of transmission, and a significant development for museum exhibition. This paper analyzes the current status of the Internet technology that is used by traditional museums, looking at the impact of mobile Internet development on the traditional exhibition.

Keywords: museum; exhibition; mobile Internet

Preface

Society is rapidly entering the information age. With the promotion of Internet development, the social service functions of mobile Internet technology have been enhanced, infiltrating various areas of public life and work. Cultural and economic developments have brought the museum and public life closer. The management concept of the museum has changed from material-centric to human-centric, adhering to people-oriented service and enhancing its public service functions. Personal mobile Internet terminals have become a new communication mode traditional museum exhibitions based on the fragmented reading, extended reading, knowledge system links, and deep experience of the mobile Internet.

Mobile Internet (MI) is an emerging service that uses mobile wireless communication methods to acquire and provide services through smart mobile communication terminals, involving a terminal layer, software layer, and application layer. The terminal layer refers to smart phones, tablets, e-books, and MID; the software layer refers to an operating system, middleware, a database, and security software; and the application layer involves different applications and services such as for leisure and entertainment, media tools, business, and finance.

In the long run, the realization of diversified mobile Internet technology is the way forward. As of December 2014, the number of Internet users in China reached 649 million, with an Internet penetration rate of 47.9%, and the number of mobile Internet users was 557 million*. Such a large mobile Internet user base is of great significance to the digitization of museums, especially in the function of communication.

1. The importance of the mobile Internet in museum development

1) Mobile Internet technology is constantly growing, and its influence on traditional museum exhibitions will become more prominent and important.

The popularity of mobile phones and mobile terminals has created a bridge between the public and the museum. Therefore, the way in which traditional museum exhibitions incorporate mobile Internet functions to supplement and expound information is an important research topic. The advantages of mobile internet in the development of museums are as follows:

a) Audience-initiated selection. Upon implementing the mobile Internet application layer in a museum exhibition, when visitors walk into exhibits, content will appear automatically, to be introduced in depth through background voices and images.
b) With mobile Internet Big Data, statistics regarding the location of visitors and duration of visits within exhibitions can be counted, understanding audience interest and thereby providing a more powerful scientific basis for the provision of personalized services.
c) Mobile Internet terminals replace earlier touch-screen models in museum exhibitions, making the contents of the exhibition hall even more flexible and convenient to use, realizing exhibition content in one's pocket.

* Published by the China Internet Network Information Center (CNNIC) in Beijing. The 35th Statistical Report on the Development of China's Internet Network.

2) The mobile Internet has strong advantages for museum development.
   a) Mobile Internet terminals can be carried and used at any time to facilitate the explanation and interaction of exhibit information and artefacts.
   b) Via the mobile Internet, visitors are able to interpret various pieces of information regarding the background of collection and compare exhibit information.
   c) Digital technology can be used to virtualize collections, as well as convey sound, sights, and other multidimensional audio-visual experiences to the audience via the mobile Internet and Internet-based terminals.
   d) Internet and mobile Internet terminals can be used to interact with collections on-site for communication and sharing with visitors present at the exhibition as well as off-site visitors.
   e) Mobile Internet terminals can be used to save collection information and become pocket-information.
   f) Mobile Internet can be used to integrate online and offline information in order to enable the OTO (online to offline) mode for exhibitions.
   g) The museum can feed information back through the user application technology layer, obtaining data like user numbers, and cultural artefact preferences in order to provide a reference for the direction of museum services.

2. The status quo of mobile Internet applications in museum exhibitions

Mobile Internet technology is a key to the start of the mobile Internet era, and as the application interface through which users contact the mobile Internet daily, mobile Internet technology has become increasingly significant. It is an important mode in which to improve user experience and stimulate business applications. The mobile Internet transmits collection information that cannot be presented in actual exhibitions to the mobile terminal more intelligently, inspiring visitors to learn about collections.

Traditional museum exhibitions consist mainly of artefacts, and the use of mobile Internet technologies stands to complement the interpretable information that cannot be achieved within an exhibition. So, how can mobile Internet technology be used to serve museum visitors? Let us consider the following analysis.

2.1 The development of the mobile Internet

To realize the network environment of the traditional museum exhibition hall, it is necessary first to realize network coverage within it. In addition to 3G and 4G, the provision of a wireless network environment by the exhibition hall is very important. At present, the market is based around WiFi technology. With the development of technology, Li-Fi technology with a visible spectrum and location services has emerged. Li-Fi transmits weak electrical signals over powerful electric lines, saving the costs of forming new networks and enabling wireless network transmission to cover an exhibition hall. In addition, the use of low-range radio frequency identification technology, Bluetooth technology, and local area network wireless mobile technology (AP) can quickly and effectively push exhibition content to the public's personal information terminals.

2.2 Applications of QR code technology

The development of the mobile Internet has become an important platform for the development of QR codes, as the characteristics of these codes provide a bridge between networks and end users.

The QR code is a key entry point in the mobile Internet, its greatest highlights being high-density coding and large information capacity, accommodating up to 1,580 uppercase letters, 2,710 numbers, 1,108 bytes or more than 500 Chinese characters, several dozen times higher in information capacity than ordinary barcodes. As long as relevant software is installed on a smart phone, or a scanning device is used to scan the QR code, information contained therein can be obtained within a few seconds, and the QR code can simultaneously express information in both horizontal and vertical directions, meaning that a large amount of information can be expressed in a small area. The QR code is a channel by which to guide information flow, but it is the services carried by the QR code and the perfect QR code channel system that are important. With the development of technology, the next important entry point for the mobile Internet will be image recognition and the reduction of intermediate links. As technology matures, image recognition technology will become prominent in traditional museum exhibition halls.

QR codes implement a planar-to-electronic information transformation – an accelerator of information flow that allows people to connect more closely with the network.

2.3 Apps

The app is currently positioned as a mobile Internet terminal preset. WeChat, apps, and QR codes can be combined to directly scan cultural artefacts and read corresponding information, then provide deep information for navigation and linked exhibits. Users are also able to book tours through the museum app. The museum app is an attempt to leverage the current rapidly developing mobile Internet environment, as well as an effort to engage with new technologies and media platforms.

2.4 Multiple technical applications of the WeChat platform

WeChat's "shake", "sweep", and "friend circle" functions guarantee basic user viscosity. With the launch of various new features, WeChat has as gradually expanded from a single communication tool to a multi-business field. As an emerging social media platform, WeChat's public responsibility and guaranteed security have become prominent, and it is now used as a major tool for museum applications. WeChat transmits a large amount of text, in addition to pictures and voice information, using only one-way traffic. It is low cost, and has naturally gained public favor. WeChat offers convenience to public communication, "pocketing" museum information, and also offering enjoyment and communication within "friend circles", allowing the significance of exhibits to have a deeper effect.

In another example, the iBeacon smart sensor uses low-power Bluetooth technology, specifically a low-power Bluetooth broadcast signal function. Such broadcast messages can be perceived across mobile phone Bluetooth, activating a mobile app and enabling it to take in the external environment, as well as pushing relevant information through the mobile app. In other words, the WeChat "shake" function combined with the iBeacon sensor can suddenly become a magical device that perceives and connects all things. Furthermore, by adjusting the power of the iBeacon to control signal range, a transmission area can be chosen within 15 cm to 80 m, realizing transmission where it is pointed. The Disney Family Museum uses the Yunzi iBeacon smart sensor to enable automatic audio navigation of the museum. This are only the tip of the iceberg of a huge number of applications.

Artefacts or works of art can be brought to life using the "shake" function. For example, imagine that you are walking in front of a historical site. You shake your phone, and your WeChat page automatically pushes a voice explanation or video display of the ruin. Or, maybe you have "shaken" at a copy of a certain work of art or historical artefact, which you can now buy directly online via WeChat. When you go into the shop, a bag is already packed and waiting for you to pick it up. These options greatly enhance visitor experience.*

Since the launch of the official WeChat 5.18 platform last year, the Chongqing Three Gorges Museum has continued to upgrade core navigation functions, officially launching a "WeChat Shake and Guide" service during Spring Festival in 2015, providing a more convenient and interactive experience than previous audio guides.

With the development of the mobile Internet and smart mobile terminal technology, visitors only need to use an Internet-based smart terminal such as a smart phone to scan the QR code on an exhibit description card in order to access corresponding web pages, gaining graphic information regarding exhibits, as well as audio and video information by which to understand corresponding exhibits. By saving these web pages, they can read exhibit information multiple times, even when not present at the venue. This information can also be quickly spread to friends via Weibo and WeChat. Data mining and analysis technology can be combined in the era of Big Data, analyzing in-depth consumer preferences and potential market demands. By using the WeChat auto-reply function for mobile terminal "handheld exhibitions", visitors only need to send a specific number

* Introduced by the network SENSORO iBeacon.

or exhibit name to access information and voice explanations. This model breaks the limitations of time and space in past exhibitions.

For example, in April 2015, the upgrade to the Palace Museum's Ceramic Pavillion guide was announced on Weibo and WeChat, with QR codes for the Pavillion and the WeChat tour guide integrated to offer detailed contents. In a sense, the use of Internet terminals in the assistance of exhibitions and displays extends the time and space of traditional museum exhibitions.

3. Prospects for the use of mobile Internet in traditional museums

The Internet is moving in the direction of personalization and mobilization. Using the mobile Internet for the realization of an OTO mode for exhibitions is the future of traditional museum exhibition development. This OTO mode is a new model in which museums use the Internet to connect online and offline space.

The preparation and promotion of museum exhibitions can begin with the mobile Internet, using public, widely-used mobile applications such as Weibo and WeChat to begin publicity, creating a certain butterfly effect before exhibitions open. After mobile Internet promotion has occurred over a certain period, and as the number of views and amount of attention increases, exhibitions can open. Internet publicity can also be used to sell exhibition tickets and reserve them online. After purchasing tickets, visitors are able to use mobile terminal QR codes to redeem or scan them in order to enter an exhibition hall. Meanwhile, through online ticket sales, visitor flow is controlled, with online and offline integration resulting in control of the exhibition site.

Following an exhibition's opening, the mobile Internet enters a critical moment, as Internet-based terminals are used to display museum collection information and exhibition guides. At this stage, the application of mobile Internet terminals in a museum's display must first comply with the principle of functionality. As an auxiliary means of display, mobile Internet terminals cannot replace the dominant position of cultural artefact exhibits themselves, and must choose appropriate technologies based on specific content, objects, and occasions. The primary purpose here is to help deepen exhibit themes and enhance implications, and also highlight cultural artefacts, assisting with the interpretation of exhibition themes and content, therefore providing richer, deeper, and more multi-faceted ways in which to excavate and convey information entailed in the exhibits, once again using technology. With interactive functions, peer-to-peer, point-to-multipoint, multipoint-to-multipoint exhibitions, and audience interactions are made possible, with interactive game experiences and navigation functions available to a participatory audience. In addition, while visitors browse collections, the museum uses the Internet to obtain statistics on visitor traffic and the duration of engagement with exhibits, gaining relevant data with which to analyze and discuss exhibitions.

After enjoying an exhibition, visitors are able to pocket interesting content and participate in further interactivity. This plays a positive role for exhibitions, and also effects the extension of their content. Museum exhibitions match the online realm with the offline realm, which makes it easier for visitors to tour an exhibition, to capture its value, and to provide powerful data for museum research.

In the Internet world of the future, wired and wireless networks will be fused, meaning that networks will be everywhere. If 3G provides users with a highway for wireless communication transmission, then 4G is akin to magnetic levitation. The leap from 3G to 4G is a qualitative one. The mobile Internet is a new stage of development, and 4G will drive it, expanding the application scene. In 2020, we will enter the 5G era, and the speed will increase 1,000 times. Museum exhibitions will be even more multi-dimensionally digitized, more humanized, and more convenient for appreciation and use by visitors of all demographics. The information capacity for propagation will be even greater, the form will be more multi-dimensional, and the services will be further humanized. In addition, wearable devices and the Internet of Things have opened up an era of integrated computing marked by personalization and the Internet of Everything. Wearable devices, combined with mobile Internet terminals, offer instant feedback. Flexible display technology is one of the most important development directions for OLED in the future, having many advantages (such as being ultra-light, ultra-thin, bendable, portable, and impact-resistant), potentially replacing existing flat-panel displays in large quantities. These mobile Internet technologies are hugely important for future applications within traditional museum exhibitions.

An Analysis of "Micro Palace" Service Functions Based on BYOD Technology

Wang Yiyi
(School of Humanities and Social Sciences, Beihang University, Beijing 100191, China)

Abstract: This paper discusses BYOD technology in the case of the WeChat "Micro Palace" public platform designed by the National Palace Museum, analyzing its supporting role in visit services in the form of information push, location services, streaming information, and social sharing, as well as proposing the prospects and challenges for BYOD application within the museum.

Keywords: BYOD technology; Forbidden City, WeChat public platform; visit guide

Preface

Released by Tencent on January 21, 2011, WeChat is an application that provides free instant messaging services. It enables the transmission of text, voice messages, short videos, pictures, and media via network traffic or wireless networks, and can be logged in to from different terminals. It is one of the most popular instant messaging programs in mainland China.

On a public platform service account, four group messages can be sent per month. Messages to subscribers are displayed in a live chat window, and the service account number appears in subscribers' address books, where all subscription account numbers can be seen. You can also apply for customization of up to three navigation bars.

BYOD technology stands for "Bring Your Own Device", i.e. supplying one's own equipment. In the "Horizon Report" released by NMC in 2013, this technology was listed as one of the mainstream technologies to receive attention from museums within the year

[1]. In the museum, BYOD technology is used to replace traditional rental audio guides, providing visitors with more advanced interpretation services. Meanwhile, due to the rich functionality of mobile terminals, more features beneficial to visits and interaction can be added on top of basic interpretation services.

This paper analyzes the support services provided to visitors via the WeChat public account of the Palace Museum as based on BYOD technology.

1. About "Micro Palace" and BYOD Technology

1.1 "Micro Palace"

"Micro Palace" is the public WeChat account for the Palace Museum, which was certified on December 8, 2014 as part of the Palace Museum's digital development. Its basic functional structure is shown in Figure 1.

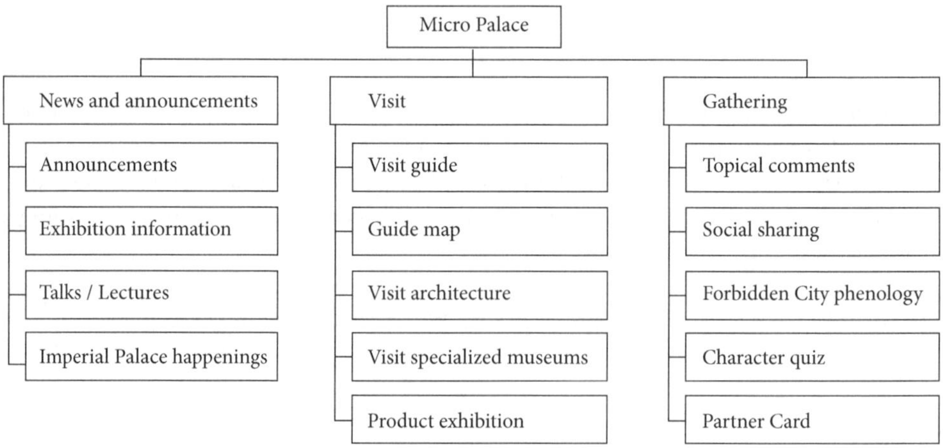

Figure 1 Functional structure of the "Micro Palace"

1.1.1 Take a look

Announcement: Broadcasting information regarding the temporary closure of the Palace Museum's inner exhibition area, and opening information.

Exhibition news: Broadcasting information regarding permanent and temporary exhibitions in the Palace Museum, as well as external exhibitions, including ticket information.

Events and lectures: Broadcasting public welfare lectures of the "Palace Forum" hosted by the Palace Museum, which are held twice a month and require reservation. Notification information includes lecture times, locations, addresses, topics, presenters, and a more detailed explanation of the lecture outline and lecturer.

Palace Museum development: Broadcasting the latest news from the Palace Museum.

1.1.2 Make a visit

Visitor guide: The visitor guide is divided into six sections – opening times, ticketing policy, map, transport links, tour information, and the day's weather. Entering the corresponding number allows visitors to gauge the status of the Forbidden City before, during, and even after visiting, making tours more convenient. Transport links cannot be displayed during geospatial function tests. The day's weather only includes temperature and wind information.

Map: Learn about the location of Forbidden City architecture, exhibitions, and service facilities.

Visit buildings: Interact with the WeChat public platform in three ways – location information, voice, and text that provides building information.

Visit special halls: There are three segments of information within the special section – the Treasure Hall, Ceramics Hall and Clock House. Click on corresponding items to jump to the information interface in which data is presented in a "waterfall flow" form.

Product exhibition: Reporting recent exhibition information in the Palace Museum, as shown in Figure 1. This information overlaps with Exhibition News in the "Take a Look" section.

1.1.3 Get together

Topic comments: Topic comments provide a variety of easy and interesting pieces of cultural knowledge, divided into eight parts – the Imperial classroom, monthly climate, gossip, the fun palace, palace stories, topics, tutorials, and wall covering. Users can comment on articles and view other user comments at the end of each article, but cannot interact in discussions with other users.

Shangsi Flow: This is a small puzzle based on an ancient literary game people played with a floating cup on a winding brook. Users can select different levels of difficulty and adjust the position and orientation of tiles to complete the game.

The Forbidden City phenology: Shows the Forbidden City in four seasons, and the beauty of its festivals. Pull down images to see more images, then tap to zoom in and view the image more closely.

Roleplay: A Q&A-type game, testing a user's roles in the palace by answering questions. Once the test is over, you can choose to share and invite friends to answer questions.

Partner Card: After registering for WeChat membership, users can get a Micro Palace Partner Card, which allows them redeem gifts and view frequently asked questions.

1.2 Analysis of BYOD technology

BYOD technology refers to "Bring Your Own Device", aiming to replace the original voice guide and navigation equipment of museums with the visitor's own smart mobile terminals, adding further functions.

BYOD technology relies on widely-used smart mobile terminals such as the iPad, Galaxy, Nexus, and Surface. Because these devices are getting cheaper and more convenient to carry, visitors are more willing to use their own equipment rather than rent something on-site at the museum, ensuring the basis of BYOD technology.

BYOD technology is a composite technology combining multiple techniques. When using it to improve visit quality, a museum must take into account how to develop mobile terminals apps according to its needs, how to achieve automatic positioning in the venue, how to identify QR codes, how to merge with streaming media technology, and how to navigate.

The museum provides BYOD technology in order to enhance visit quality, as well as reduce the cost of equipment and maintenance. At the same time, BYO equipment is different from traditional audio guides, offering more powerful functionality.

Before a museum faces the challenges of developing BYOD functions, the more important question is whether the venue has the capacity to provide free WiFi for visitors. If not, visitors will face expensive service charges and a lack of high-quality network services, which is detrimental to effective support for the successful implementation of BYOD technology in the visit process.

2. *"Micro Palace" visitor support*

2.1 User requirements

2.1.1 User positioning

Depending on purpose and cultural level, the target population can be roughly divided into three layers: scholars, students, and tourists. Specific divisions are shown in Figure 2.

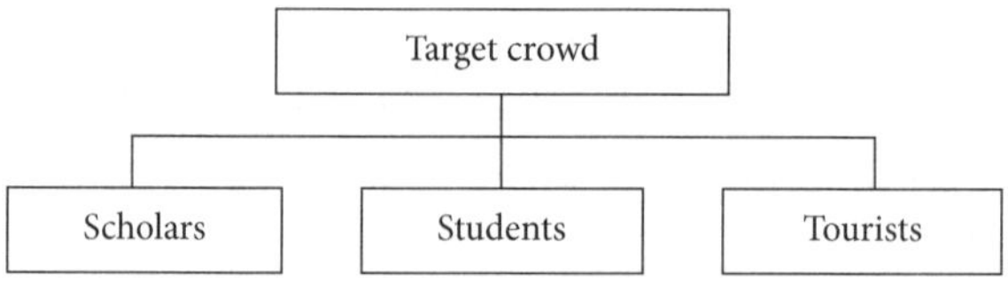

Figure 2 Target crowd classification

The characteristics and descriptions of each target group are as follows.

Scholars:

- Older, with higher cultural levels, and a smaller population;
- Not just intellectuals who are systematically educated; a good understanding of

humanities; often visit major museums and read relevant material;
- Fans of history;
- Seeking opportunities to communicate with people with similar hobbies and knowledge, and are eager to discuss issues with experts;
- A certain ability to appreciate exhibitions and collections.

Students:
- Younger, with middling or high levels of culture, and considerable in number
- Not limited to school students; also includes individuals who are interested in Chinese studies but do not have profound knowledge; this is a broad category;
- Students who have a certain understanding of ancient architecture and history, and hope to deepen their appreciation, not only to improve their education but also as a form of relaxation and satisfaction of their hobbies;
- Although interested in exhibitions, they are not often participatory;
- Very familiar with the network.

Tourists:
- Varying in age, with lower cultural levels and huge numbers'
- Neither students nor scholars;
- Interested in the Forbidden City, but lack detailed knowledge and interest in the humanities'
- Heavily interested in sightseeing;
- Most will not take the initiative to visit more than once, preferring to see many classic collections and ancient buildings in one visit.

2.1.2 User requirements and functions

Different functional targets are set for specific groups of people. See Table 1 for details.

Table 1 Functional Positioning

	Target crowd		
	Scholars	***Students***	***Tourists***
Needs	Communication platform, sense of participation, appreciation	Quick access to knowledge, social contact, sharing, entertainment, appreciation and visit guide	Visit guide, souvenirs and sharing
Functions	Academic lectures, topic reviews, My Card	Visit guide, map, Shangsi Flow, Forbidden City phenology, Roleplay, Partner Card	Visit guide, map, Forbidden City phenology
	Announcement, exhibition information, the Forbidden City happenings; visit architecture, visiting specialized museums, product exhibitions		

2.2 Needs and services

2.2.1 Maps

The map section has a useful feature that provides smart mobile terminals with maps of the Forbidden City's buildings, exhibitions, and service facilities, instead of relying solely on museum signboards or road maps. This eliminates the need for cumbersome audio guides.

When visitors come to the Palace Museum, which covers an area of 720,000 square meters, it is not easy to choose a suitable route or quickly locate ancient buildings or exhibitions of interest. However, by carrying a smart terminal and opening up the "Micro Palace" map, a line drawing of the Forbidden City can be accessed, as shown in Figure 3.

Visitors who want to learn about ancient buildings can click the "Buildings" segment of the map to see locations and information regarding important ancient buildings. For the "Hall of Supreme Harmony", the introductory page includes photos, voice explanations, and text explanations. Because the information is displayed on the visitors' own devices, the museum can provide photos, audio, and text information at the same time – an advantage over traditional navigation services. Text information covers building information and architectural stories and functions, information with which students

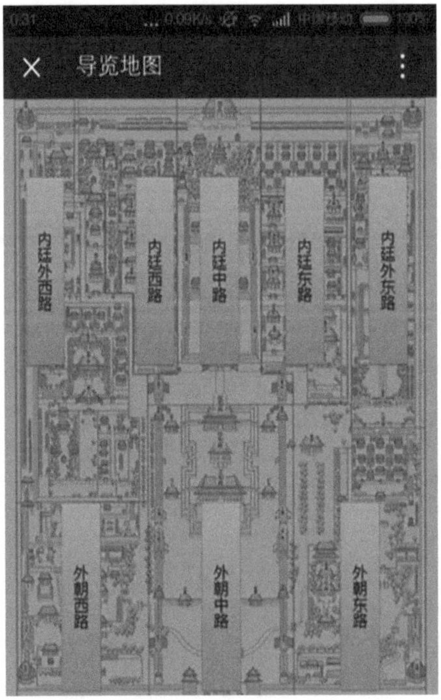

Figure 3 Forbidden City floor plan

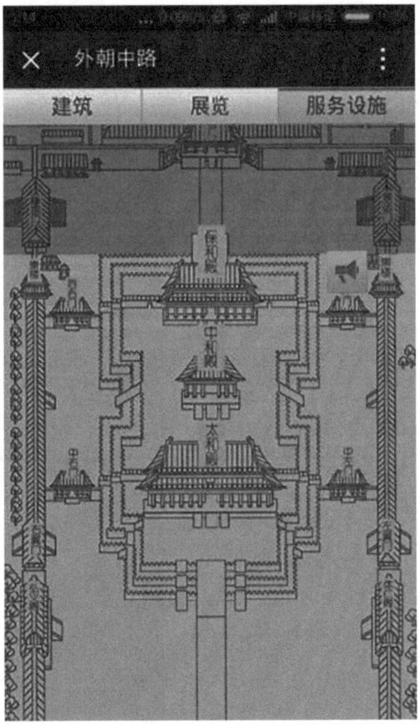

Figure 4 Service facilities floor plan

and visitors can learn about ancient Forbidden City's buildings anytime, anywhere, as well as planning their own routes around the site.

If visitors also want to see the numerous artefacts of the Forbidden City, they can choose the "Exhibition" section of the map, which shows the names of current exhibitions as their location. Most visitors will use the public facilities provided by the Forbidden City, and will want to determine the location of these facilities as quickly as possible. "Micro Palace" deliberately included a "Service Facilities" layer on the map to address visitor needs, providing current positioning for services with one click (including broadcast, consultation, audio guides, explanations, lounges, baggage hold, loans, dining, bathrooms, medical care, ATM, shops, tea houses, and bookstore), satisfying the needs of most visitors, as shown in Figure 4.

2.2.2 Collections and information about ancient buildings

A scholar with a good understanding and prior knowledge of the Forbidden City, or a student interested in Chinese history, will be interested in specific information regarding the Palace Museum's ancient buildings and collections.

In order to meet the needs of the above groups, "Micro Palace" has two important functions; "visiting buildings" and "visiting special halls" within the "Have a Visit" section, which supports those who are currently visiting and yet to visit. These two sections are designed to provide information regarding ancient buildings and the Palace Museum's collections, highlighting the application of BYOD technology in geographic location information and QR code recognition.

1) Ancient building information

Visitors are able to obtain information regarding a building by sending their location to the WeChat platform, reading the name of the building of interest aloud, or entering it. This section uses the geographical location of visitors to display information regarding nearby buildings.

2) Museum collection information

For those who are visiting or are yet to visit, this offers more convenient access to collection information. The Palace Museum has thus included detailed information about the Treasure Hall, the Ceramics Hall, and the Clock House.

Using the Treasure Hall as an example, after entering the corresponding interface, users can view all collections, with images presented in a "waterfall flow" form. Users are also able to search for specific information in the search bar.

2.2.3 Entertainment and leisure

When resting or waiting, visitors like to have some entertainment and interactive activity to pass the time. At this point, they can play "Shangsi Flow" or "Roleplay" in the "Get Together" section of "Micro Palace", lightening the atmosphere and making visits more interesting.

Scholars requiring communication may participate in "Topic Review" in the "Get Together" section and establish a discussion circle with others, while students can register a "My Partner Card" to become a member of the Palace Museum.

2.3 Overview

The application of BYOD technology makes visits more convenient, more profound, and more interesting.

1) More convenient
Visitors can use a personal mobile device to view exhibits and access information at any time without having to browse the website or make a phone call. An electronic map keeps track of their position, and they can choose their own path. Instead of relying on signs to find a route, they can quickly access public facilities based on the electronic map signposting.

2) More profound
Users can select ancient buildings on the map to access voice and text descriptions, as well as recommendations. Exhibit collections are in the palm of their hand, and they can view collections of interest at anytime, anywhere, eliminating the limitations of signs and providing easy access to more in-depth collection information on their own devices.

3) More interesting
Products have been developed to suit the needs of a variety of groups. The addition of social features and sharing can create positive interactions among visitors, as well as between visitors and the museum, turning visits into two-way discussions rather than one-way impositions, and breaking through traditional relationships between visitors and museums.

3. Summary and outlook

The public "Micro Palace" WeChat account developed by the Palace Museum provides users with a wealth of information, including voice, image and text, social and sharing opportunities, and timely exhibition updates during their online visits. It also provides maps and guiding services for offline visitors, relying on user location to explain related exhibits.

1) According to the advantages and requirements of BYOD technology, the following improvements can be carried out in the scope of the National Palace Museum:

a) Providing easy and stable WiFi access
This is the basis for a good BYOD experience. On the basis of free or cheap WiFi, more people will be happy to use BYOD technology to improve their experience. It also helps museums to collect information regarding visits, such as visitor tracking, to analyze the appeal and sustainability of exhibits.

b) Using LBS (Location Based Service) technology, visitor location is automatically obtained within the museum. First, this allows visitors to quickly determine their locations and choose the right visit plan. Secondly, it predicts the exhibits that visitors are viewing or interested in, according to their location, and recommends other resources, providing suggestions for self-exploration.

c) Considering the needs of specific people when pushing information
Amongst target users, teenagers and their family members are the largest demographic visiting museums, and own a high proportion of mobile terminals. Therefore, it is possible to push special topic explanations or exhibitions for adolescents and families of three, providing design concepts in light of the interaction between such families according to the educational background, interests, and communication needs of young people.

2) Specific exhibitions carried out by the Palace Museum are strongly characterized by special topics, focusing on certain collection themes. In this vein, the following improvements can be made:

a) The provision of easy and stable WiFi access
The reasons for this are the same as the institution-wide improvement proposal.

b) Increasing social elements
Using functions such as smart terminal photography and streaming media playback to develop more abundant resources, for example, adding film and music elements to the timeline of the Ceramics Hall in the Palace Museum, and adding an exhibit sharing function for visitors' own photography and paintings.

c) Integrating QR code recognition and augmented reality technology
Upon recognizing QR code information, AR technology is used to present richer and more realistic information via the mobile terminal, which can be collected or shared in order to maintain a visitor's interest for longer.

References

[1] Guo Ji. 'Review of the New Media Alliance Horizon Report (Museum Edition 2013)', [J] *Museum New Technology*, 2014, 2: 2–6.

Discussing the Design and Development of Mobile Museum Applications

Cao Zhiyu
(Department of Research, Military Museum of the Chinese People's Revolution, Beijing 100038, China)

Abstract: The rapid spread of mobile Internet technology has led to more museums attempting to use mobile Internet applications to improve service levels and visitor experience. However, the current development of mobile museum applications is encountering certain misconceptions. This paper draws on mobile application research and practice in museums both in China and abroad, briefly explaining the steps in application design, and referring to the development of application prototypes based on Axure prototyping tools. Due to constraints in time and resources, this paper still has some shortcomings, and requires future improvements.

Keywords: museum; mobile application; design development

Preface

In recent years, the popularity of mobile terminals (represented by smart phones and feature-rich mobile applications) has quickly penetrated public life and work, auguring the arrival of the mobile Internet era. While enjoying a full range of convenient services and new experiences brought by mobile applications, the demand for lifestyle services has also increased.

Meanwhile, Chinese museums have started a wave of free admission, bringing about a surge in popularity that has led to a serious shortage in the capacity of services. In many museums, visitors have to wait in long lines. On entering the museum, they face an outdated form of content, deficient collection explanations, a lack of flexible reservation, and on-site reception staff who are overwhelmed by the numbers and unable to provide adequate services.

In order to change this course of events, improve museum services, and enhance the visitor experience, many museums have leveraged the advantages of predominant terminals, convenient networking, and information expansion brought by the mobile Internet, joining the ranks of exploration and development of museum mobile applications and an improvement to their services.

That being said, the Chinese museum is still in its infancy in this regard. The first wave of mobile applications are more like a transplantation of the museum website into mobile terminals, and are unable to satisfy user demands. Although mobile museum applications have been launched online, visitors only pay brief attention before losing interest.

In the successful development of a mobile museum application to meet visitors needs and offer services, the question is how to balance the practicality and characteristics of the mobile terminal to make it easier for visitors to obtain service applications, and how to reduce the workload of front-line reception staff and improve work efficiency. These have become realistic issues that require serious consideration at the outset of mobile application development in museums.

1. Misconceptions to avoid

Following the development of practical application projects and communication with museum counterparts, the author believes that current mobile museum applications must avoid the following three misconceptions:

1) Failure to pay attention to and understand real visitor needs
Visitors are the ultimate target of mobile museum application services, and without systematic understanding of their needs, even the most advanced technology will be useless.

2) Ignoring the use of mobile terminal features to improve application usability and convenience
Mobile applications are carried by mobile terminals rather than traditional PC devices, and it is difficult to develop practical and user-friendly applications without considering the unique characteristics of mobile terminals in the design and development process in terms of application response, input method, and interface layout.

3) Neglecting the needs of internal staff
Although the mobile museum app ultimately provides services to visitors, it is also used by internal museum staff who serve visitors directly, and their needs cannot be ignored. Mobile applications must be as simple as possible, as museum staff engage in work processes that are highly repetitive. By improving work efficiency, they can in turn provide visitors with faster and higher-quality services.

2. Finding from research and practice in mobile museum apps in China and abroad

In order to avoid the three misconceptions above in application design, the author has looked at literature on museum applications in Chinese periodicals, masters theses, and foreign language journals from 2000 to the present, downloading and testing a large number of apps in an attempt to learn from both Chinese and foreign domains to find answers through field research resources and application practices.

Initially, most of the research in this field began with the introduction of new technologies and the transplantation of PC-based websites. However, with the rapid development of technology and the continuous expansion of application practices, many researchers have returned their focus to visitors, designing applications to address the needs of the public in various environments, as well as interactions between the museum and the public. For example, Galena Kostoska at the University of Treno in Italy looks at the use of mobile applications within a group of elderly and cognitively impaired adults. Firstly, three requirements of elderly people in mobile application services were summed up: large font, high contrast display, and simple sentence structure. Then, using an iPad as the test platform, practical application tests were conducted on 30 volunteers, and suggestions for design improvements were proposed [1].

Some world-famous museums have realized the importance of analyzing user needs and experiences, using a variety of investigative methods. Relying on mature and advanced mobile Internet technology as well as strong economic and cultural strength, they are developing a number of premium application products based on a wide variety of practices that are not only educational and entertaining, but can also subtly enhance a museum's influence.

The implications brought by these research and practices are numerous. Successful application designs go through extensive demand analysis, with initial designs having clear functional positioning and expected goals, and clearly defined requirements that clarify design ideas and principles, creating apps that are beautifully crafted and easy to understand, providing users with positive participation experiences.

3. *How to develop mobile museum apps*

The development of any mobile application is ultimately for the service of users. Therefore, before the actual design and development begins, the needs of users must first be investigated to create a basis. In the application design, utility and mobile terminal-usage and characteristics must be fully considered, designing practical user functions that are convenient in the context of mobile terminals. If necessary, original service processes can be optimized, with a suitable platform selected for prototype production and finally feasibility testing. Application development is carried out in the following steps.

3.1 Demand analysis

In order to solve practical issues and meet user needs (i.e. visitors and museum staff), the design and development of mobile applications must analyze the needs of visitors in three aspects: user groups, processes, and feasibility.

3.1.1 Analysis of user group needs

The analysis of user group needs can be achieved through a random sample questionnaire survey of visitors and interviews with staff in various museum departments. Sun Xiaoye's survey of the National Museum's "Cultural Guide" app at the Chinese Frontier Archaeology Research Center at Jilin University can be used as a reference point for this

Table 1 User Group Analysis

User group	*Main focus*	*Application function expectation*
Audience	What are the matters that need attention before entering the museum? Is it convenient to visit or make an appointment? Will visitors go on a guided tour when they enter the museum? Is it possible to obtain museum information in good time? Are facilities easily accessible and usable?	Visit guide Visit reservation Explanation reservation Audio guide, etc.
Ticket department staff	Mainly concerned with whether audience registration information may be shortened to a certain extent, and whether accessibility services may be provided for the elderly and disabled.	Visit reservation Accessibility reservation Appointment management
Venue maintenance department staff	Mainly concerned with reducing the use and waste of paper tickets, and reducing the workload of maintenance and cleaning staff	Visit reservation Accessible reservation
Interpretation department staff	Mainly concerned with convenience of registration and scientific coordination, as well as the convenience of appointment management	Interpretation appointment Interpretation appointment management
......	......	......

stage [2]. Although the survey was conducted following the app's development, the value of this survey in the demand analysis phase is still worthy of recognition. Under normal circumstances, a user group analysis table can be compiled based on the results of a questionnaire and staff interviews (see Table 1).

3.1.2 Analysis of process demands

Based on the analysis of questionnaires and interviews, and the compilation of user group needs, demands are ranked in descending order according to urgency. The main demands of various processes involved in the mobile application are screened and sorted, to be analyzed and selected in conjunction with development costs and usability.

3.1.3 Analysis of feasibility

In addition to conducting questionnaires and visit surveys, and analyzing and arranging user groups and process requirements, it is also necessary to analyze and demonstrate the feasibility of mobile museum applications. Here, we must consider the number of user groups, the practicality of application design, and help for internal staff.

3.2 Application design

After analyzing the needs of visitors and staff in an on-the-spot investigation, it is necessary to clarify the functional positioning, expected goals, design principles, design ideas, and basic functional framework of the application, so as to complete the design of each individual function in the following steps.

3.2.1 Functional positioning

Functional positioning is the first step in app design. Based on an analysis of actual user needs, a service function that is suited to mobile display is selected from a list of known museum service functions. Simple porting of functions directly from the website is a process that should be avoided, instead taking into account actual user needs, making full use of smart mobile terminal characteristics, and catering to usage habits.

3.2.2 Expected goals

The purpose of designing mobile museum applications is often to remove the limitations of time and space, providing practical and convenient services to visitors anytime and anywhere, while reducing the negative impact of an apparent lack of internal service personnel in relation to audience numbers. With simple and fresh interfaces, and quick, easy operability, with just a touch or click users are able to enjoy services that would ordinarily require long waiting periods and large amounts of manual labor to complete. This is what attracts users.

For visitors, mobile apps should enhance their experience and satisfaction. For museum staff, apps should relieve box office registrars and receptionists of many first-line hospitality duties, reducing their workloads. The use of mobile applications allows museums to leverage their resources and extend their services beyond the constraints of time and space.

3.2.3 Design principles

Although the design of mobile applications is based on existing, mature technologies, the process is by no means a simple case of module integration and functional construction. Design principles must be determined before any functional, practical, easy-to-use, safe, and reliable mobile app is designed. Fu Xiaozhen and partners, who have worked on mobile interaction design and user research for UTStarcom and taobao.com, propose eight principles of mobile application design in their new book *Mobile Design* [3]. Based on these principles, this paper summarizes the process into just six: consider user urgency, minimize text input, maintain a user-friendly and simple interface, cater to finger operation habits, offer easily identifiable text information, and provide stronger security.

3.2.4 Design ideas

Based on the above-mentioned functional positioning and expected goals, according to established design principles in the museum context, the following can be considered: First, figure out user emotions based on actual user needs. Second, enrich the application to gain channels and reduce the threshold for use. Third, consider the characteristics of the mobile terminal to accommodate user habits. Fourth, improve original service processes and utilize limited resources.

3.2.5 Framework function design

With an overview of functional positioning, expected goals, and design principles and ideas, the functional framework of mobile museum application can be depicted, expounding various service function groups. A detailed diagram of the design process offers the most intuitive representation of this phase (see Figure 1).

3.3 Prototyping

With the formulated design, it is necessary to make practical applications for the realization of functions. However, because the actual production of museum apps is mostly achieved by professional companies, the creation of a prototype for verification is an effective means of reducing costs and risks. This can prevent the unfavorable situation in which the design is modified and costs are sharply increased due to functional verification during app development.

3.3.1 Choosing the right production tool

To test and validate different design features, the prototyping tool platform must be intuitive, flexible, scalable, and versatile. To compare production tools, this paper has chosen the latest version of Axure Rapid Prototyping, Axure RP 7.0 (see Figure 2) – a professional rapid prototyping tool developed by the U.S Axure Company. It can quickly create application software prototypes, flexibly expand functions, and support multi-person collaborative design and management control.

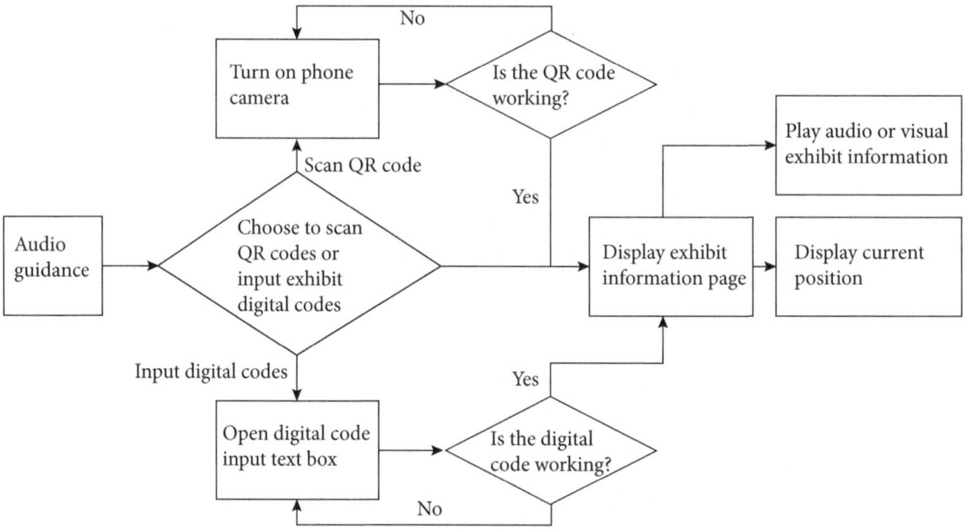

Figure 1 Reservation registration flow chart

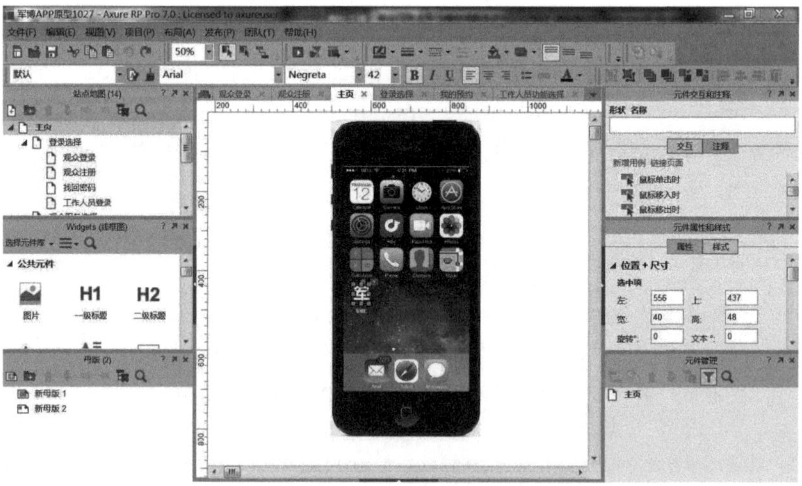

Figure 2 Diagram of the Axure RP 7.0 software development interface

3.3.2 System roles and application scenarios

After selecting a prototyping tool, the permissions and behaviors of the application's roles in each scenario are decided according to a previously designed app function framework (see Table 2).

Table 2 Application System Roles and Scene Assignment Guide

Application scene	*User role*	*Role permissions*	*Main behaviors*
Log in, registration	Audience	According to audience permission allocation, enter corresponding service interface after login	Fill in registration or login info; retrieve login password; enter user service interface
	Staff of various departments	According to audience permission allocation, enter corresponding service interface after login	Fill in login info; perform fingerprint verification
Visit reservation	Audience (including ordinary visitors, special visitors with accessibility needs, etc.)	Schedule appointment according to function settings	Select date of visit; fill in appointment info such as identity card, contact info, etc.
	Box office staff	Manage visitor appointments, assist quick identification for visitors, identify visitor codes	Check audience and visit information; provide appointment contact services for visitors with special accessibility service needs
……	……	……	……

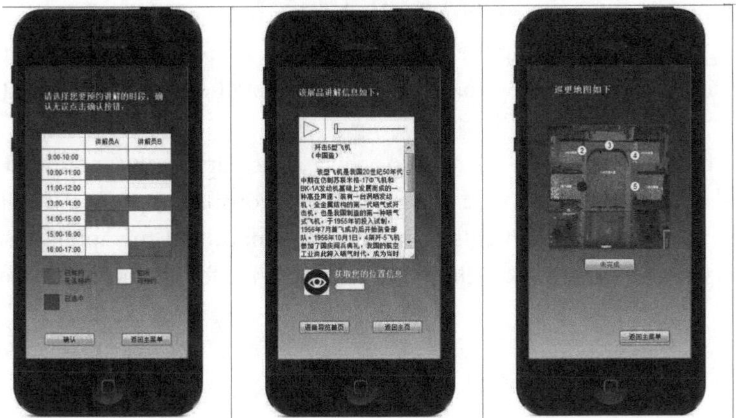

Select appointment time and interpreter	Audio guide to exhibit information	Locate security inspection points

Figure 3 Part of the functional module design

3.3.3 The function and interface design of each module

According to the results of demand analysis and framework function design, after determining system scenarios as well as role permissions and the behaviors of corresponding users, it is necessary to perform specific functions and interaction design for each business function module (see Figure 3), after which a prototype mobile app is born. In order to facilitate subsequent testing and optimization corrections, it is recommended to sort and summarize the functions and design features of each business module (see Table 3).

Table 3 List of Functions and Design Features of Each Business Module

Business module	*Implementation functions*	*Design features*
Login/ registration	User login, registration; password retrieval; secure fingerprint login verification	Multi-role login selection; login and registration error prompts; button function prompts
Visit appointments	Visit appointment; accessible access; visit appointment management	Multi-role appointment selection; graphic appointment selection; QR code quick verification
Appointment explanation	Explain appointment; make appointment	Introduce appointment constraints; consider resource allowance; quick QR code verification

4. Development and release

After fully testing and demonstrating the prototype application, the actual development of the application can be transferred to a professional company. At this stage, the museum must focus on the use of functions and the flexibility of the release format. For example, the application publishing method may flexibly adopt an HTML5 form instead of an app form, significantly reducing the threshold for the use of museum applications. After all, most visitors are not willing to download and install an app each time they visit a museum.

Based on a summary of historical experience, this paper has expounded the development and design steps of mobile museum applications, attempting to meet the needs of users, the practical requirements of mobile terminals, and the application needs of museum service personnel, providing a useful reference for the promotion of museum services.

References

[1] Learning in 140 Characters: The future of museum learning in the digital age | MWA2013: Museums and the Web Asia, 2013. http://mwa2013.museumsandtheweb.com/paper/learning-in-140-characters- the-future- Of-museum-learning-in-a-digital-age/.

[2] Sun Xiaoye. *Applications of prospects for smartphones in museum navigation work* [D]. Jilin University, 2013.

[3] Fu Xiaozhen, Hu Jiachao, Zheng Yuanlong. *Mobile Design*. Electronic Industry Press, 2013, p. 7.

[4] Rousing the Mobile Herd: Apps that encourage real space management [OL]. The annual conference of Museums and the Web, Portland, April 2013, http://mw2013.museumsandtheweb.com/paper/rousing-the-mobile-herd-apps-that-encourage-real-space-engagement/.

[5] Xu Liyuan. *Research on smart phone applications and museum education* [D]. Jilin University, 2013.

[6] Wu Changwen. 'Apps in the age of the mobile Internet and their applications in museums', [J] *Fujian Fair*, 2012, 2: 75–77.

[7] Liu Ningxing. "Research and reflections on the development of childhood mobile museum applications" [A]. *Beijing Digital Science Association, Capital Museum Alliance, China Museum Association Digital Museum Committee, China Cultural Relics Society Photography Committee.* "Integration, Innovation, Development — Digital museums promote the construction of cultural powerhouses" — Proceedings of the 2013 Beijing Digital Museum Symposium [C]. *Beijing Digital Science Association, Capital Museum Alliance, China Museum Association Digital Museum Committee, China Cultural Relics Society Photography Committee*, 2013: 4.

Exploration and Practice of the Wuhan Botanical Garden's NFC-Based Mobile Guide

Han Li and Li Na
(Department of Popular Science, Wuhan Botanical Garden, Chinese Academy of Sciences, Wuhan 430074, China)

Abstract: This paper engages with the innovative scientific communication activities performed at the Wuhan Botanical Garden, which use the advantages of vegetation resources and scientific research personnel within the park to enhance public interaction with science via emerging media. With the introduction of mobile navigation tools such as QR codes and NFC-based guided tours, it has developed a range of functions such as vegetation commentaries and interactive events. This paper analyzes the effects of emerging media in the promotion of scientific communication and public services within popular science venues, hoping to provide a reference point for scientific communication and mobile navigation in the field.

Keywords: Wuhan Botanical Garden science communication mobile guide; NFC (mobile Near-field Communication technology)

Preface

Internet development has enabled offline popular science venues to serve the public across space and time, while the mobile Internet attracts public attention to ever-present devices. Therefore, connecting online content with offline physical venues and organically combining them is a key issue for popular science venues today.

Wuhan Botanical Garden is one of three core botanical gardens under the Chinese Academy of Sciences, with more than 10,000 collected conservation species, more than 10 special plant gardens, and approximately 780,000 visitors per year. Based on the advantages of the park's plant resources and scientific personnel, its missions include the dissemination of information about plants, using the latest means of communication to provide knowledge to the public, and promoting scientific literacy through entertainment within the visitor experience.

Through nearly three years of trials, online carriers such as independent apps, WeChat public platform accounts, and remote cameras have been installed in the Wuhan Botanical Garden. Focus has also been place on offline links, and some exploration and practice have been undertaken to organically combine materials.

1. Plant annotation based on QR codes

With the rapid development of smart technology, the mobile Internet has become a new channel for precise scientific communication. Meanwhile, QR code annotation information has also developed rapidly, and has quickly been accepted by society. QR codes use a certain specific set figures to record symbolic information in a black-and-white pattern distributed across a plane according to certain regulations. They are automatically read by image input devices or photoelectric scanning devices for automatic information processing, with features such as large information capacity, strong error correction function, low cost, easy production, and high durability.

During the Ninth Public Science Day at the Chinese Academy of Sciences in 2013, the Wuhan Botanical Garden launched a series of popular science activities themed around "Science and Technology Innovation and Plant Diversity". For the first time, staff members input QR code "ID Cards" for more than 50 strains of small conserved plant populations and the top ten "Star Plants" within its greenhouses, so that when the public visits the botanical garden, a smart phone can be used to read the QR code on the plant label and access corresponding WAP content to acquire the latest, most comprehensive illustrated information about the plant (see Figure 1). The first launch of the QR "ID

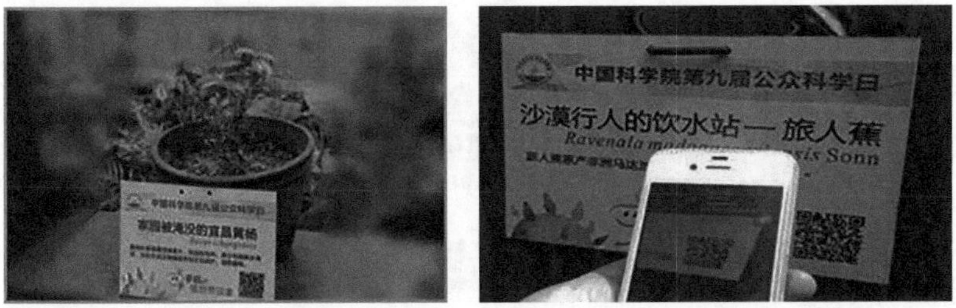

Figure 1 QR code "ID card" loaded on a plant

Cards" has now been publicly recognized. In the past, the nameplate of a small plant would be marked only with its species and origin, with no room for more content. The QR code allows visitors to simply pull out their mobile phone and scan to access detailed information about the plant's morphological characteristics, growth habits, cultivation techniques, breeding methods, key varieties, and plant culture.

In order to promote the use of QR code interpretation within the botanical garden, from 2014 to 2015, nearly 1,000 plants within the park were equipped with QR annotation cards linked to Baidu Encyclopedia and the like. This has greatly enriched the amount of information on the plants' nameplates, while consistent updating of online content also saves the cost of constantly updating physical plant nameplates.

With the use of QR code nameplate interpretation, the disadvantages of QR codes are also highlighted, including the inability to interact in public exchange, as well as the reliance on certain traffic and app software. In order to improve public interaction, Wuhuan Botanical Garden has tried to introduce other models, such as NFC technology.

2. *Plant guides based on NFC*

2.1 About NFC technology

NFC stands for Near-field Communication, or short-range wireless communication in Chinese (see Figure 2). It is a short-range high-frequency wireless communication technology that allows for non-contact point-to-point data transmission between electronic devices for data exchange within 10cm. This technology is derived from contactless radio frequency identification (RFID). Through in-depth study and research into NFC technology by scientists, iPhones and Huawei mobiles have begun to support

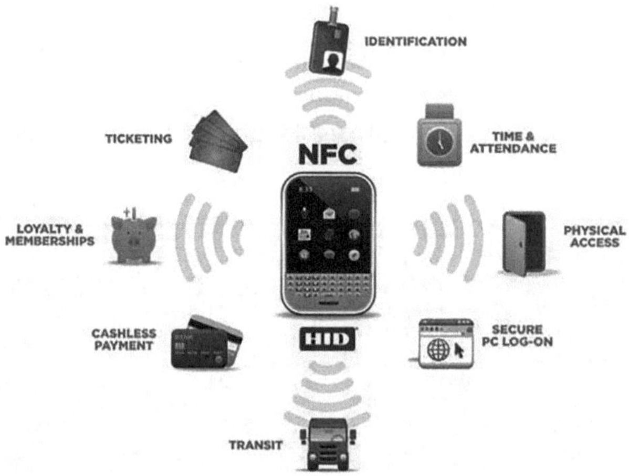

Figure 2 Principles of NFC technology

NFC technology, and Wuhan Botanical Garden staff have also begun to investigate how to exploit its advantages for guided plant commentary.

2.2 Using NFC technology for guides at the Wuhan Botanical Garden

Based on the convenience of NFC technology and the entertainment potential of interaction, Wuhan Botanical Garden has developed a self-guided NFC tour system, relying on existing plant resources. Public visitors with an NFC-enabled mobile phone can access multimedia information regarding the sights and plants when they approach classic commentary signboards or plant nameplates with NFC tags, thus enabling them to enjoy their trip around the park without blind spots. Meanwhile, staff in the park can conduct data analysis according to the time, place, and quantity of public NFC access, to develop personalized and specialized public services.

2.3 Advantages of self-guided NFC tours

1) Near-field communication and low energy consumption
Compared to other technologies, the keyword in NFC is "near", which is especially important in tours of popular science venues. The range of other electromagnetic data transmission technologies is too large, making it impossible to finely locate user position and push information. Although iBeacon, based on Bluetooth technology, can reach minimal levels (tens of centimeters), this has no advantage in large areas with dense plant distribution in the botanical garden. Another great advantage is energy consumption. Power supply is required whether using Bluetooth or other technologies. Even if powered by batteries, they only last two years, so maintenance costs are too high. Therefore, the use of NFC technology for plant navigation and commentary will greatly facilitate museums and botanical gardens in which exhibits or areas have a high concentration of content, greatly reducing energy consumption.

2) Easy to use, large information capacity, and quick interaction
There is a high wastage rate for every one-second increase in the mobile user experience. So, for museums and gardens that receive thousands of visitors per day, the advantages of NFC's rapid interaction are obvious. The average speed to scan a QR code is about six seconds, while NFC needs only 0.1 second, with no need to install an app. Public visitors using NFC-enabled mobile phones need no other hardware. They simply have to stand near to the NFC tag to receive detailed information about nearby plants with high-quality pictures and text, which completes self-guidance and fundamentally solves issues with small and vulnerable physical plant nameplates.

3) Extensive scalability
NFC enables self-guided plant tours, and functions as a membership card, e-ticket, and

souvenir. The Wuhan Botanical Garden conducts corresponding activities each week, and with NFC tags public participants can check in and purchase tickets with one sweep, reducing organizers' workloads and saving participants time and energy.

4) Resistant to wind and sun

NFC tags can be inserted into posters or plastic. They are impervious to outdoor weather, and do not require battery power, thus saving on development costs and making them ideal for outdoor venues such as botanical gardens (see Figure 3).

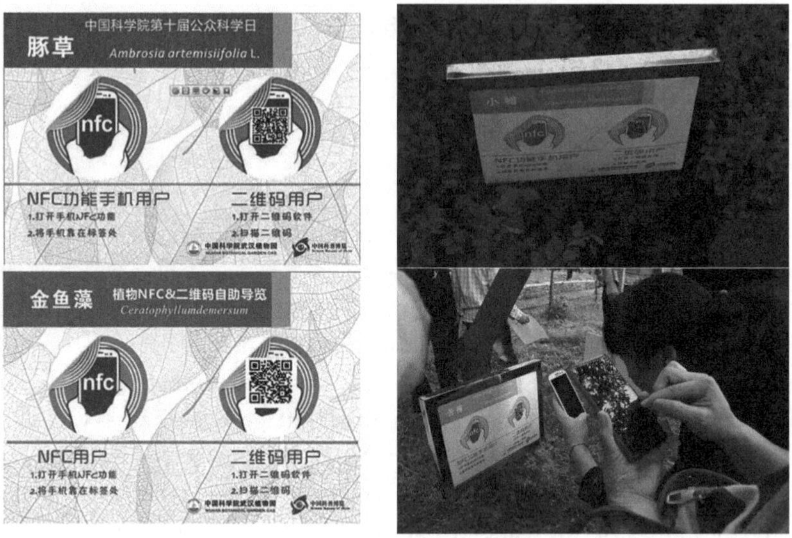

Figure 3 NFC application in the botanical garden

3. *The exploration and practice of popular science at the Wuhan Botanical Garden*

In conjunction with NFC self-guided plant tours, Wuhan Botanical Garden has also carried out a series of interactive activities such as NFC treasure hunts, personalized NFC posters, and NFC event sign-in. In one way, this guides the perceived public interest in plants, satisfying visitors' desire to participate in the scientific process, fundamentally changing the traditional one-way information transmission model of science, and creating a lively and active atmosphere. Meanwhile, the Wuhan Botanical Garden has joined the National Science Resource Grid and Science Cloud Platform, which offers remote observation of plants and convenient public sharing of science resources. This has improved the venue's basic ability for effective information dissemination, and has contributed to the development of informatized science and the need to raise the quality of public science in the new era.

4. *NFC technology in the future development of the Wuhan Botanical Garden*

At present, the application of NFC is still in the trial and optimization stage, but given the rapid development of information technology, there is further space for NFC developments in the future of the Wuhan Botanical Garden.

1) An instant entrance payment
By replacing traditional paper tickets or e-tickets with NFC-enabled smart phones and bank cards, the public will be able to quickly and easily enter a venue simply by swiping a mobile phone or bank card.

2) Popular science activities based on NFC
NFC functions can be used for sign-in, information traceability, and user identification. This can solve issues with online and offline information dissemination, and can also realize the integration of virtual digital content and the real world, simultaneously enhancing public experience and providing a more convenient bridge by which the public can participate in popular science activities, making the effects of these activities more profound.

3) Positioning guides
When a visitor touches a data plate in the park with an NFC tag, their location is pinpointed, realizing NFC-based positioning guidance with no need to turn on GPS positioning or use a paper map to obtain information about surrounding scenery.

4) Analysis of public data
When a user visits a plant nameplate with an NFC tag, he or she registers a series of data such as public access time, order, and number of visits. User demand analysis can be performed based on this acquired data in order to develop personalized services.

References

[1] Song Haiyan. *Design and implementation of the Android-based NFC file transfer system*, 2012.
[2] Zhao Yunyan. *Research on smart poster applications based on NFC technology*. 2012.
[3] Huang Meizhen. *Implementation and application of the Android NFC mobile payment system*. 2012.
[4] Chen Wenzheng. *Implementation of the Android-based NFC system*. 2011.
[5] Yu Chenggang. *Design and implementation of Android-based NFC technology*. 2011.
[6] Xie Weiquan. *Android-based NFC mobile phone introduction systems*. 2013.

Using the Mobile Internet to Share Museum Courses

Liu Jing and Huang Yanjun
(Beijing Natural History Museum, Beijing 100050, China)

Abstract: Museum resource sharing is not a new topic. Its importance has been stressed by industry insiders for many years. In recent years, due to the rapid development of the mobile Internet, museum resources have been spread by new carriers. At present, shared resources are mostly concentrated on collection explanations and exhibition promotion, while the sharing of social education courses remains less of a focus. This paper focuses on the significance of resource sharing in the promotion of social education standardization, the improvement of sustainable results, the strengthening of museum-school integration, the promotion of learning exchanges, and the enhancement of the significance and implementation plans of design levels.

Keywords: mobile Internet; museum courses; sharing

On February 3, 2015, the China Internet Network Information Center (CNNIC) released the 35th *Statistical Report on the Development of China's Internet Network* in Beijing, citing that as of December 2014, the number of Internet users in China reached 649 million, and the number of mobile Internet users reached 557 million, accounting for 85.8% of the total number of Internet users. According to further statistics, 60% of Chinese web users have a positive attitude towards sharing via the Internet, of which 13% are very willing and 47% are more willing. Among 10- to 19-year-old web users, 65.9% are more willing or very willing to share online. Reciprocal sharing of information and resources by web users reduces communication costs and establishes cultural values. With the rapid spread of smart terminals, the mobile Internet has become the new status quo. As the saying goes, the mobile terminal holds the world.

With the advent of the mobile Internet era, under the impetus of Cloud computing, Internet thinking, Internet +, and Big Data, the digital museum is moving from the computer screen to the pocket. The emergence of apps in various forms has transported the original browser-based applications into the palm of one's hand. Digitized museums are developing towards diversified communication techniques, comprehensive information, and technical complexity. This is conducive to the globalized propagation of science and culture, and also to the participation of ordinary visitors [1]. Meanwhile, in the museum industry, the mobile Internet has played a role in the expansion of horizontal communication. This paper proposes how to use the mobile Internet to realize the sharing of museum lesson plans.

1. *The status of the museum's social education curriculum*

1.1 The importance and significance of the museum's social education curriculum

The educational function of the museum is its third key function beyond exhibitions and collections. Museum education has long since broken the traditional singular mode of exhibition viewing, gradually deriving a variety of teaching activities and systematic standards for curriculum design. As a result, museum education has become more scientific and dynamic, attracting the participation of many visitors, especially teenagers.

1.2 The development and use of social education courses at museums

At present, the museum's social education curriculum is designed largely by social workers according to collections and themed exhibitions, including certain teaching objectives, materials, and processes, as well as study plans. Most courses are offered within the venue, while some extend to docking in community venues or schools, with course conveners acting as social workers.

1.3 Issues in the development and use of social education courses at museums

1.3.1 Separation of venues

The social education departments or social sectors in museums offer social education courses based on specific resources and specialties, but they lack sufficient intercommunication with one another, only coming together for industry meetings (such as forums and seminars), occasional competitions (evaluations of outstanding scientific projects in social venues, or evaluations of social education projects), and are limited to a superficial understanding of activity content in brother venues.

1.3.2 Mechanisms to improve imperfect evaluations

Social education in museums lacks a sound evaluation mechanism in contrast to school education. This is reflected in the fact that most activities can only make perceptual judgements based on visitors' enthusiasm; few social educators make scientific assessments of activities via written surveys and study sheets.

1.3.3 Low efficiency

Constrained by a lack of communication between venues and the limited energy of creative staff, the number of executions of each social course within the venue and audience numbers are very limited, while the huge design efforts for teaching can only be applied in a small scope. For example, in the Beijing Museum of Natural History's "Joyful Experiments" event, creative staff designed a theme every month, open to visitors for eight days. Due to workloads, the event was limited by the venue and personnel. It was only able to accommodate 15 people at a time, or 120 people per month – a mere drop in the ocean for a monthly average of 85,800 museum visitors.

1.3.4 Poor integration with school education

Due to the lack of communication channels, school teachers are not clearly briefed on museums' curriculum resources, so they are unable to make use of them. At the Beijing Museum of Natural History, the number of school visitors per day is huge. In 2014, the number of group receptions was 1022, of which 463 were schools. However, only two or three schools had in-depth cooperation.

How can we solve these problems? The "Mobile Internet + museum" model offers the best solution.

2. Using the mobile Internet to share educational resources

The mobile Internet has become a large-scale industry, and the coverage and popularity of mobile apps is now extensive. Within museums, the general direction points towards the extension of digital museums to mobile terminals. Museum apps that use mobile Internet technology are usually based on mass educational resources, with visitors using smart terminals such as mobile phones and tablets to browse information and engage in interactive learning both within and outside the museum. Visitors can post information via interactive third-party social platforms such as Weibo and WeChat. It can be reviewed by "experts" through the system, then published within the mobile Internet system to expand the venue's own education resources. Thanks to the popularity of SNS platforms such as Weibo and WeChat, interaction with and sharing of educational curriculums are realized, with excellent content selected to enrich a museum's educational resources, realizing the self-supporting growth of educational resources and curricula.

In addition to audio services in a variety of languages, the app developed by the Museum of Modern Art in New York allows for the sharing of collections and editing of comments to be sent to social networks, which is undoubtedly the biggest highlight of the app. Because it is so easy to establish hot topics on social networks, with opinion leaders rising through interaction, this kind of educational lesson plan formed by a model similar to crowdsourcing or crowdfunding has the qualities of natural sharing and extensiveness.

The relationships among museums, and between museums and schools, museums and other social groups, and museums and the public, objectively constitute a complex system of two-way exchange and communication between a service and the serviced, education and being educated, and development and being developed in definite social and technical environments, with the social functions of museums reflected in these interactions and mutual influence.

There are over 4,000 museums in China, including more than 100 first-class institutions. Therefore, it can be speculated that the number of social curricula in museums are in the hundreds of thousands. If such huge resources were to be integrated with mobile app platforms, the social benefits would be greatly enhanced.

2.1 The significance of sharing museum curricula via the mobile Internet

2.1.1 Standardization of curriculum design

Standardizing lesson plans on mobile Internet app sharing platforms can guide course design. For example, a lesson plan must include clear knowledge goals, competency goals, and emotional attitude and value goals. It must identify the targeted group and number of people it caters to, as well as the materials needed, detailed teaching steps and procedures, study sheets, evaluation indicators, and PowerPoint presentations. When designers write lesson plans according to requirements, they will learn the standards and basic elements for complete lesson plans while uploading them, thus facilitating the designer to make comprehensive teaching plans based on these factors. This removes the drawbacks of blindness caused by a lack of professional teacher training in the case of many social workers. Meanwhile, the unified display form also lets users understand things at a glance.

2.1.2 The sustainability of resulting applications

There are many benefits of shared platforms for course designers, in line with the sustainability principles advocated by the government, which are highly economically efficient models as a result of smart achievements. For example, the "Stone Tablet Rubbings" activity designed by the Beijing Stone Carving Art Museum can be copied and utilized by other Chinese museums that feature stone monuments. The "Eave Tile Printing" activity at the Summer Palace can be promoted at other museums with ancient buildings, while the "Joyful Experiments" events at the Beijing Natural History Museum can be replicated by other major science and nature venues.

2.1.3 Promoting learning exchange in the industry and improving planning

Social education is an important task for museums. The education curriculum is a core element of social education, embodying the material resources, knowledge resources, and educational expertise of a venue. George Bernard Shaw once said, "You and I are friends. If you have an apple and I have an apple and we exchange apples, then we each still have one apple. But if you have an idea and I have an idea and we exchange ideas, each of us has two ideas." The development of a shared course construction platform helps social and educational staff to break through time and space constraints, gain industry experience, and learn from others' strengths at any time.

2.1.4 Strengthening cooperation between museums and schools

Museum education is an important component of social education, and museum resources are a useful supplement to school education, so the smooth flow of information between museums and schools is undoubtedly the premise of integration. The benefit of using the mobile Internet to share museum education courses is not limited to museum staff. It is also an effective way for school teachers to digest museum resources.

Teachers are able to learn about museum resources through the sharing platform, so as to screen out content to be linked with school curriculum content, and assist in classroom teaching. Meanwhile, teachers can also provide comments and suggestions to the designers, helping them to improve designs and teaching plans.

2.2 Sharing museum education resources via the mobile Internet

The author of this paper has surveyed existing museum apps in China and abroad, finding that the vast majority largely implement functions of museum information release and tour guidance, lacking course sharing functions. Therefore, in the use of the mobile Internet to achieve a course sharing function, corresponding functional modules must be added. To this end, design of functional modules in mobile Internet-based museum app platforms, traditional electronic maps, tour routes, collection information, query reservations, exhibit interaction, personalized recommendations, visual exploration, and search modules must be available alongside the implementation of an interactive museum course sharing community, as well as interactive teaching and e-learning units. The functional module design for mobile Internet-based museum apps is shown in Figure 1.

Interactive community: Provides interactive service functions to visitors, integrated with current mainstream interactive social services (such as Weibo and WeChat). Main functions include:

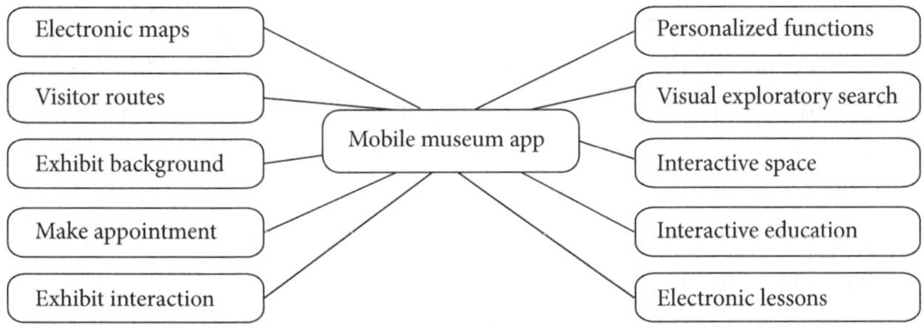

Figure 1 Schematic diagram of the functional modules of the mobile museum app

- A dynamic publishing function for museum information, furnishing the interactive publishing interface of Weibo and WeChat.
- Visitors can use Weibo and WeChat accounts to log in to the museum app.
- Based on Weibo and WeChat group functions, visitors interact with specific topic content, either from a museum course or an audience-initiated discussion.
- Vote online for courses or cases.
- Users can view interactive content of other visitors on the course and participate in interactive reviews.
- The app collects content uploaded by visitors on Weibo or WeChat, and after review and suggestions by online experts, it can be transferred to the course/case resource library of the museum, allowing the resource library to achieve self-growth.

A functional flow chart of the interactive community module is shown in Figure 2.

Interactive teaching: Interactive teaching involves a combination of video, audio, and data communication, supporting rich multimedia teaching methods and courseware production management, with interactive, live, on-demand teaching modules, specifically formatted for the interactive learning platform system for museum audiences.

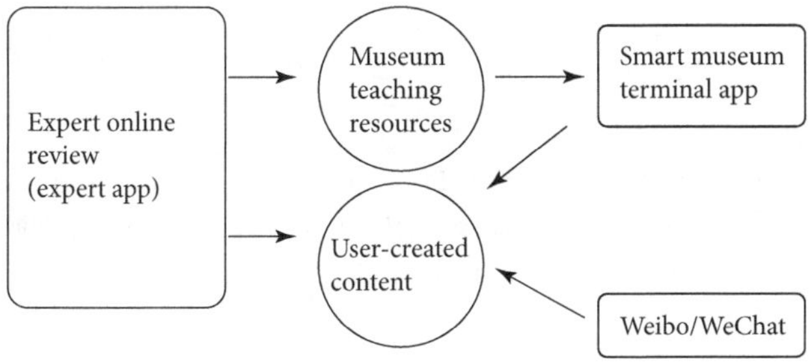

Figure 2 Functional flow chart of the Interactive community module

E-learning units: The mobile Internet-based museum app provides e-learning units for individual use, enabling progress tracking and effective evaluation for on-site teaching and learning activities. This function mainly applies to team activities within the exhibition area, using the tablet terminals provided by the museum. Specific features include:

- Generated e-learning units are stand-alone programs: can be added or deleted according to audience needs.
- Supports question tasks such as multiple choice, fill in the blanks, true or false, question and answer, connection, and drawing.
- Visitors participating in learning tasks receive study forms. After completing all the answers, they can submit them to the system for correction, and return to study points.
- Supports on-site teaching activities, so teachers can see content submitted by students and make comments.
- Supports on-site teaching activities, so teachers can interact online via smart terminals, such as tasks like responding to "race to answer" games.
- Provides two working modes: teacher and student. Teacher mode realizes the management of on-site teaching activities and tasks, as well as the management and assistance of students' progress.

3. Conclusion

With the rapid development of the mobile Internet, the gradual formation of "Internet +" thinking has profoundly affected the development and transformation of museums. The use of the mobile Internet to share museum courses is the embodiment of strengthened sharing of resources between museums, strengthened interaction between museums and people, and strengthened cultural exchange between international museums. While this is required for museum development, it is also an inevitable path for the process of internationalization of China.

References

[1] Hou Feifei. 'Research on the dual-track digital development of digital museums and physical museums'. *School of Art and Design, Beijing Institute of Graphic Communication*, 2009.
[2] Shan Jixiang. *Sichuan Cultural Relics*, 2014, 3.

Research on the Strategic Transformation of Museum Communications and Services in the Mobile Internet Age at the Science and Technology Museum

Hu Jianghua
(Sisgoin Technology Co., Ltd., Beijing 100142, China)

Abstract: In the current mobile Internet era, under the influence of "Internet +", this article focuses on the strategic transformation of communications and services at science and technology museums, and the emergence of new media mobile navigation (micro-Cloud navigation) to assist in this endeavor.

Keywords: Cloud navigation; mobile navigation; Internet +; museum strategy; cultural experience

In 2013, China's mobile Internet market exceeded 100 billion RMB, and it is expected that the total mobile Internet market will reach 600 billion RMB by 2017. So, as a science and technology museum in this environment, we are promoting the development of the cultural science industry and the use of new forms of mobile navigation to chase new growth points.

In the transitional period to a new era, how can science and technology museums meet new development needs? How can they meet the needs of the environment and customers? How can new strategic objectives be implemented?

The answers are to seize the development opportunities of the times, be guided by the general direction, and take advantage of the situation. Museums should also conduct timely research on cultural experience, explore paths for deepening and evolution, and let the new cultural experience lead changes. In this way, they will raise popularization of culture and science and achieve new strategic goals.

1. Introduction

1.1 Issue proposal

1.1.1 The status of the mobile Internet

The mobile Internet is a collective product of the Internet, communication technology, and smart terminals, with the fundamental characteristic of putting the user experience first.

According to statistics, the scale of China's mobile Internet market reached 39.31 billion RMB in 2011. In the first half of 2012, the number of mobile Internet users in China reached 388 million, and the number of users accessing the Internet via mobile phones increased to 72.2%. In 2013, the number of Chinese web users reached 590 million, mobile Internet users 500 million, and the number of smart phones reached 580 million. It is expected that the number of smart phone users will exceed 700 million in 2018.

On the national level, great efforts are being made to promote and guide the healthy development of the mobile Internet as a key platform for the next generation of Internet business. In the 12th Five-Year Plan set out by the Ministry of Industry and Information Technology, information and mobile Internet industry development plans were formulated to promote sound industrial development.

1.1.2 About mobile navigation (micro-Cloud navigation)

Based on iBeacon, micro-Cloud guidance is a smart information management service system application used for deepening museum research and development using various smart terminals such as smart phones.

It relies on the mobile Internet to develop and promote the smart information management service system of the museum. According to research and analysis of museum collection types and quantities, it integrates digital information resources, builds a smart management platform, and establishes a multi-dimensional, scene-based interactive cultural information system to provide visitors with a fully open and comprehensive information interaction platform. The functions of micro-Cloud navigation services are:

- Precise positioning and navigation services;
- Multi-dimensional push messaging services;
- Point-to-point personal marketing services;
- Mobile portable consumption and sharing services.

1.1.3 Subject determination

Science and technology museums excel at honoring history and heritage, as well as disseminating knowledge, but this does not mean that we should be limited to using established methods for the design and display of exhibits. Looking our visitors, we find that the vast majority carry mobile devices.

Therefore, in the current mobile Internet era, this paper discusses the strategic transformation of communications and services, and how mobile navigation can assist in the communication and service of the science and technology museum.

1.2 Research significance

This article is based on thoughts and methods learned from CMPM, combined with modern strategic management theories. It sums up many years of practical experience, reflecting on the use of micro-Cloud navigation in the communications and services of science and technology museums from two dimensions, recognizing what is missing and determining what can be compensated.

2. The era of the mobile Internet and its impact on science and technology museums

2.1 Characteristics of the mobile Internet era

The mobile Internet is a tool that should be used in all industries, as it is part of the information revolution. The integration of the Internet and traditional industry is a further expansion of the new "information energy," and it is defined as a kind of "information energy" today.

Our current position is clear. We do two key things:

1) We are a connector. Through micro-Cloud navigation using the WeChat platform, we are a device that connects human and human, object and object, and human and object. We do not get involved in deep business logic, instead simply working as a connector for the museum.

2) We are part of the content industry, and this is also an open platform, providing intelligent, "decentralized" solutions. We believe that the future "Internet +" model is decentralized, space-oriented, and geographically related, involving thousands of people whose needs can be met. In this way, we maximize connections between museums and

museums, museums and users, and users and other users, optimizing the power of the ecological chain.

Micro-Cloud navigation is a solution that reflects the "Internet +". It is a smart solution that is more long-term and more decentralized.

2.2 The impact of the mobile Internet and its policies on culture and the museum industry

1) From a policy perspective, the State has positioned the cultural industry as a strategic pillar industry.
2) In recent years, cultural consumption and cultural markets have been very active, and the potential of the culture industry has gradually been recognized.
3) The cultural industry has a natural connection to the Internet. The Internet is a carrier of communication and content. It is more interactive than traditional media, and has a more intimate relationship with content.
4) If the cultural industry grows bigger and stronger, it will require a full industrial chain model. For example, international cultural industry giants such as Disney and News Corp carry out a complete industrial chain layout from content production and channeling to derivative development.
5) The consumption momentum of cultural products is getting stronger, and there not many highly practical products are being developed using the cultural attributes of the museum itself. In fact, there are many new ways to potentially interact, such as "shaking out" tickets with WeChat "shake", then forwarding them as gifts. Electronic tickets are capable of containing a large amount of information, and may also be linked to advertisements and coupons for cultural products, which is beneficial for publicity.

3. An analysis of the strategic evolution of science and technology museums

3.1 Progress analysis and the effects of background implementation

China's oldest museums tend to exhibit physical objects and graphic panels. However, domestic museums have developed into the early 2000s, adding electronic touch screens that allow visitors to explore and understand museum information and content. Further development has enriched the means of cultural experience using technology-based effects.

In 1992, UNESCO launched a large-scale "Memory of the World" programme to permanently and digitally store resources of all tangible and intangible cultural heritage of the world, at different levels, in different countries and regions, also realizing resource sharing via the Internet. Now, this concept has become a major direction in the development of the next generation of science and technology museums in the mobile Internet era. On the basis of the mobile Internet today, through Cloud computing and Big Data analysis, smart terminals and the museum's own interactive service system are used to establish the Smart Science and Technology Museum, which facilitates information exchange between object and object, human and object, and human and human.

3.2 The inadequacy of existing strategies in the era of the mobile Internet

The reason why the mobile Internet can be called an era is not because it has created more information, but because it has changed the binary relationship between people and information. It has made people part of the information, changing various relationships and structures of human society, which will also lead to changes in an entire social business model.

Strategic transformation is the process of mutual adjustment and matching between a museum's resources and its environment. Based on environmental changes, new business models are generated through internal resource allocation and external resource integration, so as to achieve a reasonable union and dynamic balance between an enterprise's capabilities, resources, and environment, achieving sustainable competitive advantages.

Passive visits to the museum fall far short of the demands of today's society. Visitors with their own mobile smart devices are only able to use the touch screen or graphic display panels provided by the museum to watch, understand, and interact. Once they have finished reading, visitors have no way to take away or continue their relationship with museum information.

Science and technology museums and corporate exhibition halls have begun to add new elements to traditional planning methods, such as Smart Cities, Smart Homes, burglar alarms, or mobile terminal experiences and WeChat platform interaction. It seems that without paralleling the mobile terminals, it is impossible to keep up with the trend.

4. Choices and goals for the strategic transformation of science and technology museums in the mobile Internet age

Imagine taking out your mobile phone inside a museum and opening WeChat's "shake" function to access explanations of the artefacts around you, as well as coupons for souvenirs and exhibit experience tickets. This is possible thanks to micro-Cloud navigation.

The Micro-Cloud navigation (Big Hand Small Hand Co.) – an O2O service through iBeacon technology – firstly improves efficiency. Secondly, it saves money and brings in more income. Compared to the two common methods of printing coupons and issuing coupons through group buying websites, coupons shaken directly into the hands of consumers can eliminate middlemen and save both material and financial resources.

When it comes to income, Big Hand Small Hand Co. believes that a museum visit have to does not cost money. Space and time can be sold instead. They say, "What is for sale is the time and space one spends inside the venue. This itself can create value. For example, with our iBeacon space, the museum of the future can use the traffic center to gain other income, such as advertising revenue".

Coupons are not the only form of cards that can be shaken, and there are infinite possibilities for customization by individual museums. Put simply, the air is full of information, and full of surprises. It is a case of "information first, material second" [2].

However, there is still no answer regarding how to create other forms of value in traditional museums, and many traditional museum practitioners are anxious, a question that Big Hand Small Hand Co. is considering for the traditional museum.

In addition, unlike group purchase websites where user information is intercepted, Big Hand Small Hand Co. does not want to connect directly with users, but is willing to be a bridge between the museum and the user, leaving the user in the hands of the museum. The idea is to seize the museum's traffic center and bring users back.

According to Big Hand Small Hand Co., "In the future, every museum must have its own user platform. Many will have their own media platforms, and all must have their own traffic center, directly connected to other museums and users in their own industry, with our micro-Cloud navigation becoming a background data platform. We don't intend to intercept users. We want museums to achieve this themselves, so users will naturally remember them. In the current centralized model, the biggest fear is that the museum will be hijacked by large platforms, and this is also not what the museum has planned."

The science and technology museum needs an overall mobile Internet architecture. Big Hand Small Hand Co. says, "There are surprises in the air. Perhaps in the future, users will attend a venue in the hopes of interaction and find there are no surprises, so will not choose that venue again. This encourages the museum to continually improve its business levels and operational abilities, which is actually a two-way interaction. If the museum is well-managed, it will bring in more users, while positive feedback from users will encourage the museum to improve."

5. Conclusions and prospects

5.1 Conclusions

1) The mobile Internet is a brand new product of a new era, and we are in it. As a representative of this product, mobile navigation (micro-Cloud navigation) can be used to create a new, smart information management service system for modern science and technology museums.

2) One of the typical features of the mobile Internet is putting user experience first. Mobile navigation (micro-Cloud navigation) fits with this characteristic, filling the air with information and surprises to meet the needs of the science and technology museum at a technical level. Further, it has abandoned the shortcomings of early mobile Internet services, returning the user base to the museum.
3) Museum development has been a long process. In the era of the mobile Internet, we must gradually abandon old stereotypes, look into the future, and meet actual needs of the current museum community for information interaction.
4) Through the network service platform, mobile navigation (micro-Cloud navigation) integrates all the above content, establishing a complete, new, and smart information management service system that realizes all-round interaction with visitors both online and offline, strengthening audience contact while providing results and efficiency, enhancing the market reputation of both service products.

5.2 Points for innovation

The proponent of micro-Cloud navigation states: "If you shake, you can access magical things. When you touch the air to shake your mobile phone, you can receive greater discounts and benefits. Simply speaking, the air is full of information and surprises. The micro-Cloud guide (Big Hand Small Hand Co.) in the background and acts as a data platform. We do not intend to intercept users, instead seizing the traffic center of the science and technology museum and encouraging users to return."

The science and technology museum needs the support of mobile internet marketing and micro-Cloud navigation.

5.3 Future research prospects

Micro-Cloud navigation is a seed that has sprouted for museum promotion and service. We currently have four service avenues: precise positioning and navigation, multi-dimensional message push, point-to-point close marketing, mobile consumption, and sharing. These are just the foundations. The future development of technology proposes continual expansion to meet the basic needs of science and technology museums. Examples include how to actually turn a museum into a traffic center, and how to guide users to continue their connection with the museum.

References

[1] Xie Huanxu. *Chinese Culture News*, July 4, 2013.
[2] James Gleick. *A Brief History of Information*. People's Posts and Telecommunications Press, Beijing, 2013.

Kickstarting the New "Internet+" Museum Model

Zhang Bin, Ren You, and An Ruixiang
(Department of Culture, Wanda Information Co., Ltd., Shanghai 200040, China)

Abstract: Under the influence of "Internet+", the rise of smart mobile media has provided new opportunities for the development of public museum services, opening a door for imaginative space in which to improve museum service capabilities. This paper explains how museums can use mobile navigation applications to provide new services to visitors in "Internet+" environment in terms of indoor positioning, micro-exchange and operation, and Big Data decision-making.

Keywords: Internet +; museum; mobile navigation; application research

Preface

"Internet +" represents a new form of industry. By harnessing the potential energy of the Internet and integrating deeply with the innovative thinking of "Internet +", traditional industries will undergo a revolution. According to the *China Mobile Internet Data Inventory* and the *2015 Special Forecast Report* from Analysts Think Tank, the number of mobile Internet users in China will reach 790 million in 2015, with the overall scale of the mobile Internet market exceeding 2.3 trillion RMB. Alongside the ubiquitous mobile Internet are new technologies and new models, such as LBS (Location Based Services), O2O (Online to Offline) and mobile payment systems, offering great convenience and excellent experience to users while also encouraging museums to innovate.

The museum is a place for the collection, preservation, display, and study of natural and human cultural heritage, invested with history and future development. In recent years, the levels of public education and services of the museum have increased greatly. Museums have become highlights of urban culture thanks to contributions from the Internet and information technology. The rise of the "Internet+" model has brought changes and opportunities to modern museums, and the question of how to integrate it into innovations has emerged as a key topic in the museum industry. This article explores the "Internet +" museum model, and proposes a smart museum design scheme based on the mobile Internet, providing ideas for modern museum operation.

1. The characteristics of "Internet+" museum information services

The "Internet +" museum is very different from the traditional informatized museum or digital museum. In traditional informatization, we usually think about how information technology can serve the museum, and how to use technology to comprehensively enhance the museum's business capabilities. The "Internet +" model is a reverse, revisiting the museum with Internet thinking and reshaping parts of it. Unlike the digital museum, the "Internet +" museum is itself a process of innovation. There are no mature theories or models, but because of this, the model is full of potential. Next, we will analyze the museum's information service model in terms of three key mobile Internet characteristics.

1.1 Ubiquitous information connection

The greatest contribution of the mobile Internet is its narrowing of space between people and information, positioning smart mobile devices as portable information bridges. It has become an important connector between people and things, people and systems, and people and people, breaking the limits of time and space to provide ubiquitous information services.

Most of a museum's artefacts are stored quietly in showrooms and warehouses, with accompanying digital information generally stored in a computer room, making it challenging for visitors to obtain comprehensive data. The popularity of smart mobile terminals has undoubtedly opened a door for dynamic museum education and publicity. Through a variety of new ways, it passes on information in the fastest and most convenient modes, providing an unprecedented convenient platform in exhibitions, educational events, and audience interaction, achieving "zero distance" interaction with visitors. Meanwhile, this kind of information connection engages with social networks, crowdfunding, and other useful models. It is different from the one-way communication of the past, meaning that visitors are no longer passive information consumers, but producers and communicators of culture, fully activating museum information.

To attain a wide connection of information, the large-scale social opening up of museum resources is indispensable. Foreign museums have made valuable efforts here. For example, the National Museum of the Netherlands and the Metropolitan Museum of Art in New York have released HD image resources on the Internet over the last two years. The Smithsonian built an online service system some years ago, allowing visitors to perform online query of more than nine million items and materials. The institution is still carrying out related digitized management of these resources, making access more open and convenient via the Internet. These successful cases can provide a reference for Chinese museums.

1.2 The circular ecosystem of information services

The era of the mobile Internet is increasingly focused on service ecosystems. In the foreseeable future, isolated services will no longer be competitive. Instead, the direction will be towards closed-loop services that provide connections between services and services, equipment and equipment, and online and offline. This kind of ecosystem provides a "dragon" (or "ouroboros") service model, offering better user experience, developing stronger user commitment, and being more adaptable to long-term operation and development.

For museums, planning an ecosystem of services is feasible. The museum has good upstream and downstream industrial chains — exhibitions, news, transportation, tourism, sociality, publications, and cultural and creative products, as well as a high quality user group — cultural enthusiasts, students, and tourists, with varying consumption characteristics. These elements provide composition space for the "Internet +" museum. For example, the now popular O2O model focuses on how to bring online users to offline consumption and then loop back online through information feedback. For many museums, the largest user portal is tens of thousands of even hundreds of thousands of offline visitors. It is possible to reverse operations and attract offline users to online registration according to categories, guiding them to complete the ecological consumption of visits, explanations, dining, education, souvenir purchase, and social sharing.

1.3 Personalized services and new experiences

With resources becoming extremely rich and the interconnection of a large number of information nodes, the key problem faced by network users is no longer what can be seen, but finding content of interest to obtain higher psychological satisfaction. Faced with the escalation of this psychological need, the classic combination of presenting thousands of people with one face and the traditional "search, list, content" form of display mode, will prove more challenging in meeting the appetite of modern people. Personalized service models offering new information experiences have become important in the developmental direction of the mobile Internet.

As a personal, portable device, the smart mobile terminal provides a great stage for personalized services, easily obtaining user location, hobbies, habits, and social circles, and even attaining physical signs via wearable devices. Through collection and analysis of this basic data, personalized active service models can play an increasingly important role. For example, through the analysis of a user's social network, visit history, and interest settings, the museum can develop a personal social graph and interest map, intelligently calculating the visitor's detailed needs, then providing them with a "private custom" experience including recommendations for events, tours, and souvenirs across different levels of in-depth information, as well as a variety of location-based push services.

Meanwhile, the superior performance of smart mobile terminals also provides a platform for new experiences. With high-definition big screens and high-speed 3G and 4G networks, colorful digital content can be presented to users. Moreover, with high-definition cameras and smart phone sensors, devices can also offer impressive virtual reality performance. For example, when a viewer takes a picture on their smart terminal using augmented reality technology, they can hear the collection "talking" and "moving" on screen, integrating static exhibits with virtual exhibition presentation. In addition, applying the innovation of interactive experience is also very important, such as the "Daily Palace" app, which combines displays, an event calendar, a notebook, and a variety of functions in a light app, receiving high praise from its users.

2. *"Internet +" Museum Design Case Study*

The "Internet +" museum is not a solid model or system; it is more like a process of continuous innovation using Internet thought. In what follows, we will explain the design

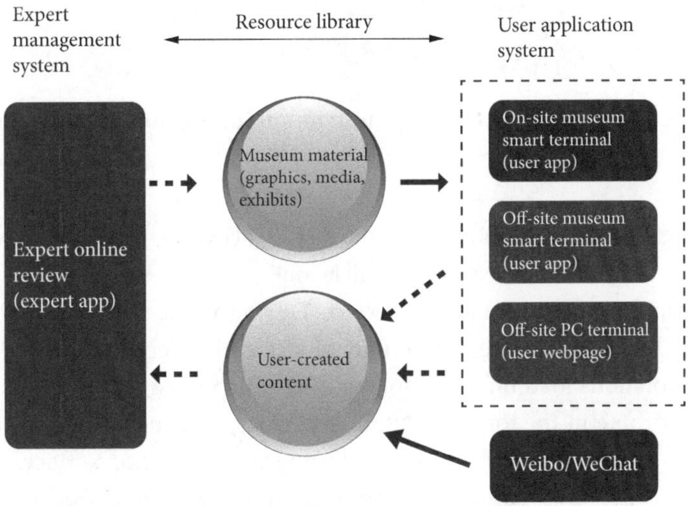

Figure 1 Overall structural plan

of a mobile Internet-based Smart Museum, and try to demonstrate the possibilities of an “Internet +” museum model. The program here uses core technologies like mobile Internet and indoor positioning to set up the backbone of museum information services, supporting the “Museum service + operation + decision” ecosystem services with mobile apps, WeChat, and website channels (see Figure 1).

2.1 Accurate services based on indoor positioning

At the heart of the program is the use of advanced indoor positioning technology within the museum (see Figure 2), enabling the effective positioning of mobile devices with an accuracy of 3 meters, using the WiFi network within the building. Using this technology, users enjoy high-quality and accurate LBS service through mobile phones, gaining positive visitor experiences and deep appreciation.

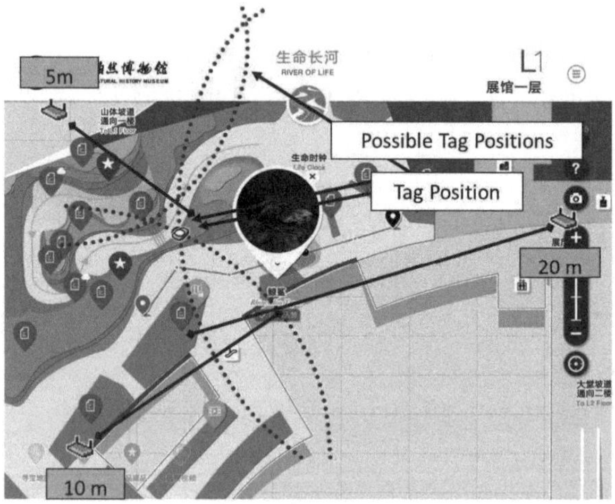

Figure 2 The Shanghai Natural History Museum’s guide app provides WiFi-based indoor location services

Indoor positioning services allow for the display of exhibit information, various public service points, and the entire exhibition hall layout to be loaded on a visitor’s smart mobile terminals, forming a precise guidance service. During a visit, nearby exhibit information is automatically displayed. Users need only glance at the electronic map. Meanwhile, as complete annotations load on the map, the electronic map displays exhibit information along the tour route in the form of text, image, voice, and augmented reality, integrated into the map’s annotation point locations. Furthermore, location services also include a wealth of features, such as reminders regarding visitor flow, recommended routes, treasure hunts, team member searches, instant reminders, and souvenir recommendations.

While visitors experience these services, museum positioning can also collect information in the background, such as user location, path, and visit trajectory, forming visual information resources regarding visit processes. Administrators are able to monitor visitor flow density and guide visitors through a better experience.

2.2 Micro-social contact and operation

Operation is a vital part of the "Internet +" museum model, determining the scope, impact, and effect on an individual user. The first step to improving user conversion is the reversal of O2O (Offline to Online), so we transfer important functions related to events, such as reservations and ticket sales/bookings to apps and WeChat public platforms, leaving only a small number of manual services available. Visitors are thus encouraged to download apps or use the WeChat public platform, resulting in a large number of user conversions from offline to online. We also add a relatively active gift shop to the mobile content, allowing for online purchases of special products or the retrieval of discounts so as to attract user participation, forming a closed-loop linkage with positioning services and personalized push technology that attract visitors to the gift shop.

The maintenance of online activity in museum services is an important issue with online systems and Internet operations. This program innovates online service operations for the museum with the addition of micro-social elements within app and connecting to large social platforms such as WeChat. Besides supporting guide services, the app also provides online columns like audience interest appraisal, expert question and answer, selfies, and social interaction, establishing a micro-community with professional museum characteristics. The museum can then use this self-built micro-community, as well as social platforms such as Weibo and WeChat public accounts, to gather user popularity, realize interaction with exhibition content, select high-quality content with which to continually enrich exhibition resources, and realize the network self-growth of content resources.

Another means of operation is the link with schools and customization of location-guided e-learning guidance through teaching channels within the app. The exhibition hall can thus be transformed into an art or science classroom, becoming a second learning space where students can follow along, complete studies, and answer questions along a specific guided route. Through these linkage operations, many more students and teachers will be attracted to the platform.

2.3 Big Data and decision making

With access to audience location information at any given moment within the museum, as well as visit behavior data, the system is able to generate tens of millions of pieces of log data regarding audience behavior each day, as the basis of Big Data analysis. By modeling such massive amounts of data, analyzing numbers, changes, and correlations between them, information that is of real interest to museum management can be mined.

The system is able to calculate crowd flow distribution according to location data, thereby analyzing the distribution area of visitor flow within the museum and forming a density heat map. The directional flow of visitors within exhibition areas can be discovered, prompting the analysis of time spent in front of key exhibition items. Combined with a comprehensive analysis of personalized user data to explore behavioral characteristics and differences between users, this facilitates the provision of targeted services. Combined with statistics like visitor flow and exhibition, it can unearth patterns of souvenir sales to improve business capabilities. Based on Big Data forecasts, it can act as a warning and guide in special circumstances, such as instances of massive visitor flow. This information allows managers to quantify and analyze the situation, and provides a practical basis for decision-making.

3. Conclusion

"Internet +" stands to have a profound impact on the future development of museums. Museums should open their collection resources more fully, develop their display and dissemination channels, and attract the public on a larger scale. The museum will grow on the basis of its original collections and its display and research functions, becoming a cultural center geared to the needs of society and serving the general public. The "Internet +" museum concept proposed in this paper has been trialed in venues such as the Shanghai Natural History Museum, and has garnered positive responses. Reliance on the "Internet +" museum service model to connect visitors more effectively promises new opportunities for future museum development.

References

[1] Zhang Yongchun. 'On the innovation of means of museum exhibition — The Zhebo "Handheld Museum" Project', [J] *Dongfang Museum*, 2012, 2: 121–125.

[2] Zhu Zhongyi. 'Applications of WiFi-based indoor positioning technology in museums', [J] *Software Industry and Engineering*, 2013, 3: 38–41.

[3] Xiang Kun. 'The use of smartphones in museum navigation', [J] *Cultural Relics Annual*, Dec 2014.

[4] Xiang Na. 'Challenges and countermeasures faced by the museum in the Internet age' [A]. *Modern Industrial Economy and Informatization*, March 2014.

[5] Du Fengyi. 'Research on key Augmented Reality technologies for museum display', [J] *Automation Information*, 2008, 12 (53).

[6] Meng Qingjin. 'Study lists: An effective tool for cooperation between museums and institutional education', [J] *Chinese Museum*, 2004, 4: 15–19.

Analysis of the Importance of Micro-Communication for Cultural Communication in the Palace Museum's Official Weibo Account

Guo Ting
(National Palace Museum Department of Data and Information, Beijing 100009, China)

Abstract: Micro-communication is an emerging and fast-growing means of contacting visitors. The changes it brings to public life are huge and subversive, potentially bringing about a further explosion of online information. In order to achieve more effective communication in today's museums, digitization and informatization are being carried out, and micro-communication offers new opportunities and prospects. Museum development involves making full use of micro-communication, carrying out public education and cultural communication via public social media platforms such as Weibo and WeChat. An analysis of the Palace Museum's official Weibo account proves that micro-communication can be a new component of cultural communication for museums, opening up new terrains.

Keywords: micro-communication, Weibo, cultural communication

Introduction

The core feature of micro-communication is "micro". Its content is "micro-content", the communication experience is a "micro-action", the communication channel is a "micro-platform", and the object of communication is the "micro-audience". As a brand new mode of communication, micro-communication combines characteristics of self-propagation, interpersonal communication, organizational communication, and mass communication. However, "micro" does not mean simple. In the Internet age, it is capable of bolstering a museum's cultural communication. As a micro-communication media platform, Weibo has become a favorite due to its strong immediate release and dissemination capabilities, with extensive propagation, and interactive communication that is both convenient and rapid, along with swift development. There are now more than 500 museums with Sina, Tencent, and Public Weibo accounts. The appropriate use of micro-communication will lead to stronger development of cultural communication for museums in the information age.

1. The advantages of micro-communication

Weibo in Chinese means "microblog" – a broadcast-style social networking platform that shares brief real-time information via attention mechanisms. Users are able to set up personal communities through various clients such as WEB and WAP, and post information in 140 words (including punctuation marks) to be shared instantly. Weibo's attention mechanism is divided into one-way and two-way operations. It is a micro-communication medium. Through mass dissemination, the effect of communication is expanding with its reach, ultimately achieving the same effects as traditional media, and being used to achieve communication purposes.

As a sharing and communication platform, Weibo pays more attention to timeliness and randomness. Microblogging allows for the expression of thoughts and recent developments at every moment. On microblogging platforms, there is no distinction between high and low levels of knowledge. There is no factor of distance, and those who participate have already turned Weibo into a living media circle through interactions, communication, emotional catharsis, and thought records. Meanwhile, it is worth noting that Weibo has become a new means of cultural communication for many museums.

Cultural communication models in the museum must keep pace with the times, and while traditional media can no longer meet the increasingly fast-food style of social development, cooperation with emerging online media is necessary to adapt to new media developments. Weibo has changed the outdated way information is published by museums, and the tedious and lengthy way in which collections have been explained. It brings the public and the museum closer together, taking advantage of the timely and succinct microblogging form to simplify information, with historical and cultural content quickly displayed to users by means of micro-communication, garnering a positive interaction that put the user in touch with the museum. Meanwhile, through the fission-type communication and promotion of microblogs, information is transmitted to the public as quickly as possible. It has also helped museums to achieve the goal of audience expansion and the extension of their culture.

Styles of information dissemination via traditional media are relatively simple, while micro-communication information styles integrate text, pictures, audio, and video with the help of information technology. This satisfies the needs of a variety of groups, transforming communication from a one-way passive acceptance to a multi-media communication. These are the advantages of micro-communication.

Although micro-communication is point-to-point communication like traditional media, thanks to information technology, it is able to interact in a highly convenient two-way mode. The communicator can see the status of the content (such as the number of people following and the number of times forwarded) and form effective interactions based on it, eliminating the impression that the museum is "superior". This provides a basis for planned content by an analysis of commentary and audience data.

2. *The role of Weibo as micro-media in the Palace Museum's cultural communication*

The blueprint of new media set out in the *China New Media Development Report (2014),* edited by the Institute of Journalism and Communication at the Chinese Academy of Social Sciences, points out: "At present, micro-applications based on the mobile Internet, such as Weibo, WeChat, and micro-video platforms, have become popular and mainstream modes of communication, and this is rapidly expanding into areas such as government affairs. The official Weibo account of the Palace Museum (hereinafter referred to as the Palace Museum Weibo) is an official government-certified microblog that has launched on Sina Weibo, Tencent Weibo, and Public Weibo platforms. As of July 24, 2014, the total number of followers across the three official Weibo accounts exceeded 4.46 million. Sina Weibo was launched in March 2010, and has since published 4,012 blog posts, with fan numbers up to 1,367,018. Tencent Weibo launched in January 2011, with 6,342 blog posts and 3,041,958 followers. Public Weibo launched in June 2012, and has published 978 blog posts with 58,932 followers. Ten days after the launch of Tencent Weibo, the total number of followers exceeded 1.58 million, and the audience increased by as much as 400,000 per day. Xinhua News Agency and the *Beijing Morning Post* reported on this phenomenon.

For a museum, microblogging is not easy. Of course, this also illustrates the important position of the Palace Museum in the public perception, which makes its Weibo into a sort of "cultural business card", making micro-communication even more pivotal in its role in cultural communication. When it comes to the simultaneous reporting of major events, important information, in-house exhibitions, and related service information, the Weibo platform has played its due role in guiding public understanding of the Palace Museum in a timely, comprehensive, and accurate manner.

Given the in-depth development of information technology, the cultural communication of museums has already entered the Internet era. The official Palace Museum website has been in operation for more than 10 years, with its various informatized displays growing richer. In recent years, the advantages of micro-communication in the dissemination of information have become more obvious, and it is prime time for the integration of micro-communication into the Palace Museum's cultural communications. This is why microblogging is ideal.

First of all, as an important medium for micro-communication, Weibo provides a good understanding of Internet users. The platform is mature and stable, easily operated by users without additional developments.

Secondly, thanks to the "micro" characteristics of a micro-blog, a small team of five or fewer individuals can handle the publication and interaction of official Weibo information. There is no need to establish a large team or team of experts, which reduces maintenance costs.

Due to the limitations of traditional media itself, how well is published content being transmitted? How widespread is the communication? As a communicator, the Palace Museum simply spreads content, without knowledge of audience preferences. There is a lack of interaction between the communicator and the audience. In terms of ways in which to spread and promote traditional culture more effectively and efficiently, the Weibo platform provides an ideal solution, as the fission-type dissemination and promotion of microblog communication spreads to all interested parties as quickly as possible. The more followers that are interested, published content will be seen by more people. It will be subjectively screened by them, and brought into a further cycle of propagation, meaning that a wider audience will receive the information. This part of the audience may find their way back to the original source and become a long-term followers. Meanwhile, we are also able to attend to follower comments, as well as forwarding, and private messaging, to understand their dynamics and engage in effective interactions. As can be seen from Figure 1, a large number of followers are interested in the original Weibo account, as they see its content being shared by people they care about. So, they pay more attention to it, or recommend it on their own social network, for example, sharing it in their WeChat Moments. This same propagation cycle continues endlessly, growing an highly attentive follower base.

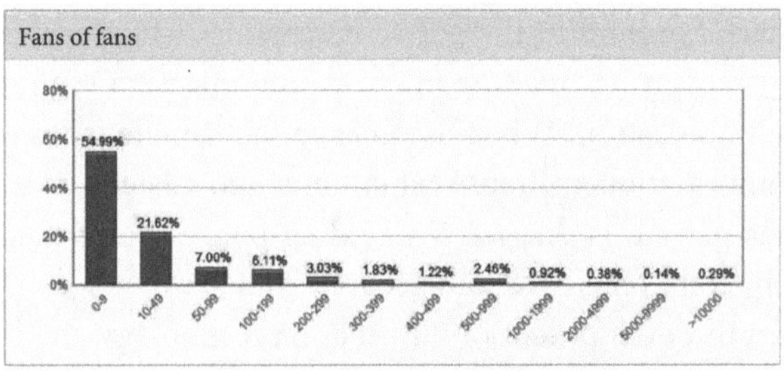

Figure 1 "Fans of fans" analysis

But how can we calculate exactly how many people see a Weibo post? There is a background statistical parameter in Weibo that is able to explain the breadth of this fission-type promotional communication, namely the blog post exposure data (see Figure 2). Blog post exposure refers to the number of times that a post has been seen by Weibo users over a period of time; the higher the content quality, the higher the quality of Weibo followers, and the higher the exposure. After posting a microblog, when followers open their Weibo accounts to the post on their opening homepage screen, one exposure is counted. Exposure data relates to the total number of followers reached by a published post (that is, both the number of actual followers and forwarded followers). When the number of forwards increases, exposure is high. Each microblogging platform has a large number of follower groups, and the force of this communication should not be underestimated. At its most popular, the Palace Museum Weibo reached 18,617,418 individuals, providing its wide reach.

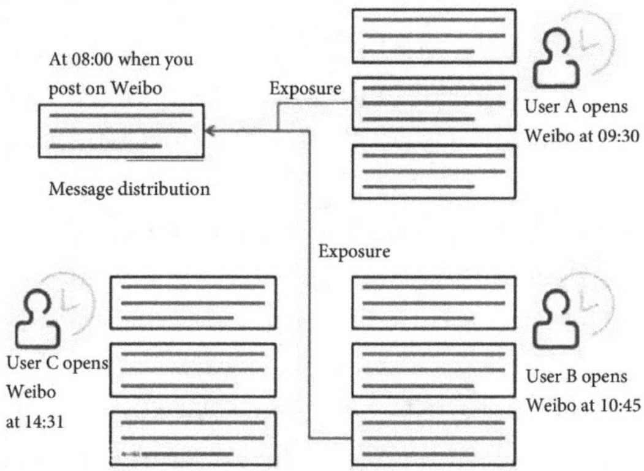

Figure 2 Weibo exposure

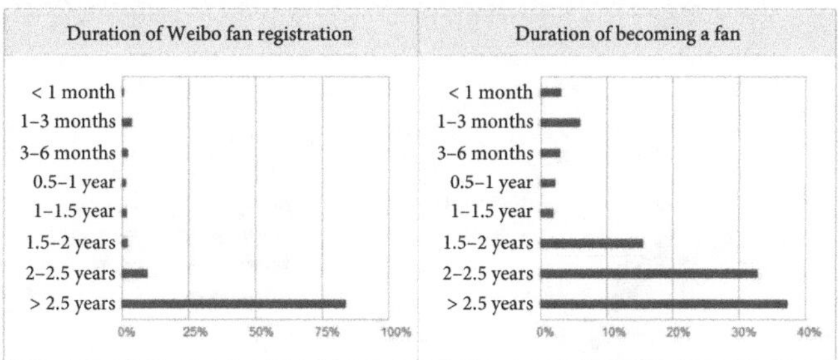

Figure 3 "Duration of Weibo fan registration" and "Duration of becoming a fan" analysis

Based on the analysis of the length of time a user has been registered on Weibo and the length of time a user has been a follower (see Figure 3), users who follow the Palace Museum Weibo have extensive Weibo experience. They are committed followers, paying strong and consistent attention to the Palace Museum. This group plays a vital role in the redistribution of culture.

2.1 Language features of the Palace Museum Weibo

Over the last four years since opening the official Palace Museum Weibo, a complete information dissemination system has been established. From initial content planning to the preliminary examination, first trial, final review, image capture, and utilization of content release, the responsibility of each process is clear, and communication quality is guaranteed. Although the Palace Museum is a public institution, on Weibo it has become a humorous and approachable cultural narrator. The museum's bio on Tencent Weibo reads: "Here is the Palace Museum you can see, and can't see. Here is the Palace Museum you want to know, but don't understand. Here is the Palace Museum you are interested in. Here the freshest, most timely, most reliable, and most powerful Palace Museum." Using this personal, stylistic language shortens the distance between traditional culture and modern society. The institution is affectionately called "Gu Gu" and "Brother Palace" by followers.

The Palace Museum's collection is rich, with far-reaching historical implications. To allow the public to read and understand this content and disseminate it within an extremely limited amount of time, the language used is very important. In Weibo, the language used has to be very brief and to the point. The use of adjectives and technical terms should be cut down as much as possible. The Palace Museum's Weibo content mainly comes from expert research, as well as publications on history, culture, and cultural artefacts. It is rewritten to ensure the professionalism and accuracy of the descriptions.

In terms of the sex and age composition of followers (see Figure 4), the male-female ratio is close, while the age range is largely focused on young and middle-aged people. This group is the main force behind the Internet era, aware of network hotspots, and

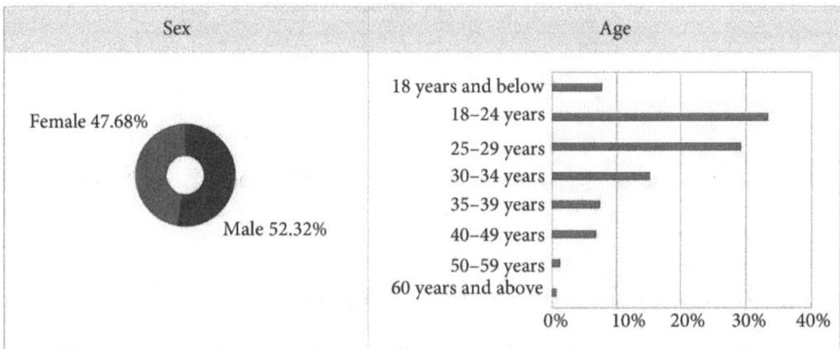

Figure 4 Composition of fans by sex and age

familiar with online language. Therefore, it is essential to use vernacular language. For example "oh", "la", and "ya" are modal particles that evoke a younger tone, and popular network terms like "poke", "pal", "super-cute", and "likes" , have appropriate application in Weibo. In this way, the Palace Museum shakes off it's "old fashioned" image by embracing changing fashions and youth trends, shortening its distance to others within virtual space while creating a change in the impression people have of it.

Refreshing, stylistic language is used mostly in posts with images, such as serious and literary topics like #fallinlovewiththecity#, #forbiddencityphenology#, and #firstsnowintheforbiddencity#. Many young netizens adopt a "literary" persona, willing to cultivate their own sentiments and enhance their disposition and tastes. Many followers like official microblogs to be full of "literary and artistic models", so the Palace Museum Weibo takes this into account when applying text, and when sharing images of buildings, landscapes, flowers, and trees. A picture speaks more than words, so few words are used, explaining the situation with fresh, natural, and immersive language.

2.2 Topic setting on the Palace Museum Weibo

Put simply, "topic" is the name given to keywords used in Weibo searches, placed between two hashtags, and followed by the desired written content, for example: #PalaceMuseum#. In this way, topics can be categorized when publishing content, with scattered single posts gathered up under a single topic. Topics are strong, and can play a role of propagating "micro-exhibitions". When followers open a topic link, all the posts inserted around the topic can be viewed, so there is no need to look at thousands of microblog posts one by one. For followers who browse, continuity and integrity are important, allowing them to discuss topics and thereby retaining their attention. For the Palace Museum Weibo specifically, topic setting allows for division into different categories, gradually revealing its 1.8 million plus artefacts, as well as its resources and research to the public in stages, batches, sub-topics, and multiple formats, cultivating a loyal follower base.

Topics on the Palace Museum Weibo are divided into long-term and temporary categories related to exhibitions, festivals, and events:

1) Long-term topics
   a) Appreciation types
   - #Baskinginthefamilyproperty#: Showing the collections of the Forbidden City, mostly for appreciation.
   - #Fallinlovewiththecity#: With photos as the main theme, pinpointing and reflecting the beauty of the Forbidden City from different angles and details.
   - #Forbiddenelegantcollection#: Synchronized with the release of the "Forbidden Elegant Collection" wallpaper and other electronic gifts on the Palace Museum website.
   - #Forbiddencityphenology#: Based mainly on solar photographs, reflecting the

changes in the creatures and plants of the Forbidden City.

b) Culture type

- #Let'sreadthecalendartogether#: Updated daily with rich content from the "Palace Museum Calendar".
- #Talkingabouttheforbiddencity#: Revealing the 600-year history of Forbidden City, with stories, people, events, and objects.
- #Constructionsecrets#: Explaining the architecture, structure, and historical evolution of the Forbidden City.
- #Fobiddencityhistory#: Major events in the history of the Forbidden City, mainly taken from the "Chronicle of Forbidden City History" column on the Palace Museum website.
- #Culturalcommunity#: Featuring real exhibitions within the museum, and the "Special Cultural Topics" and "Special Edition Home Page" sections on the website, with a single focused explanation of a certain artefact in the special topics and tips.

c) Information type

- #ImperialPalaceinformation#: Synchronized with the "Forbidden City Information" section of the Palace Museum website, broadcasting news about the museum at any time.
- #Littlevisittips#: Tracking and reporting visit information and special event/tour information, and providing information about holidays.
- #ImperialPalaceforum#: Synchronized with the content of lectures released by the "Forbidden City Information" available on the Palace Museum website, explaining lecture content.

d) Activity type

- #Iaskyouanswer#: Small online activities planned for Weibo users to participate in, using surveys or answer forms, implementing reward mechanisms and collaborating with offline promotional activities.
- #DatewiththeImperialPalace#: Website and micro-platform member event previews and reports, combined with micro-activities for online and offline promotion.
- #Forbiddencitylittleclass#: Setting topics and encouraging user participation, re-forwarding expansion of knowledge points for learning through entertainment.
- #Volunteerfocus#: With pictures at the core and supplemented by interviews, introducing Palace Museum volunteers and the current status of volunteer work, presenting them like movie stars to promote positive energy. Appointments can be made with volunteers in conjunction with member activities.

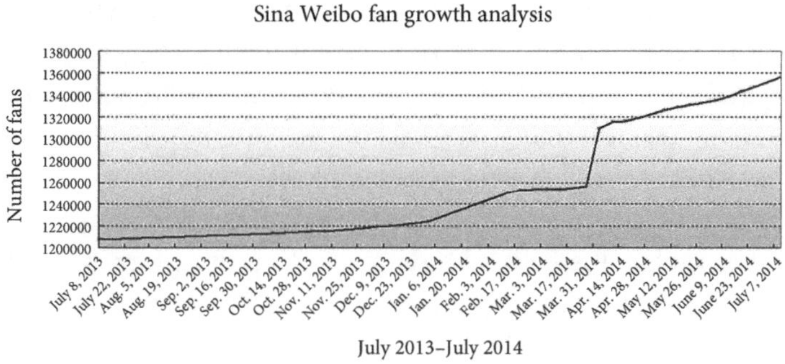

Figure 5 Sina Weibo fan growth analysis

2) Temporary topics

Temporary topics are flexible, time-sensitive, and adapted to the introduction of new ideas and providing followers with a sense of novelty. Topics combine current events with public focus, creating a topical atmosphere that plays a significant role in increased audience attention. According to data (Figure 5), there have been two major increases in the follower growth curve, in the period between the issue of #firstsnowintheForbiddenCity# in February 2014, and #ForbiddenCityspring# in April.

Amongst these topics, #firstsnowintheForbiddenCity# was shortlisted on the Sina Weibo Hot Topic list. Due to the dry weather in the winter of 2013, there had been no snowfall in Beijing, and the desire for snow was particularly urgent. When the first snow finally fell, the Palace Museum Weibo lost no time in releasing a large number of pictures of the snowy Forbidden City, which quickly gathered popularity and forwarding activity. This encourages followers to participate in interactions, "posting" their own pictures of the Forbidden City snow scenes, attracting a lot of attention.

In April, flower season in Beijing, the #ForbiddenCityspring# topic synchronized with the flowering period within the palace. It included historical stories that became very popular with followers. Within a week, follower numbers increased by 50,000. Forwarding by certain online celebrities played a major role in re-transmission. Due to their large fan bases and wide reach, their focused attention and forwarding activity act as great publicity. For a public cultural institution, such a large increase in followers indicates that many of them are deeply interested in the 600-year-old Forbidden City. Many followers have expressed their desire to enter the Forbidden City and enjoy its beautiful scenery, while others have followed in the footsteps of the Palace Museum Weibo, synchronizing their visit and releasing their own pictures, interacting in order to realize the organic combination of online and offline activity.

The Palace Museum Weibo also involves the "24 Solar Terms", seasonal changes, traditional Chinese festivals and modern festivals, along with public interest items. In this way, it explores historical and cultural knowledge, integrates the characteristics of the collections, and offers compelling content, displaying the diversity and richness of the Forbidden City. For example, the serial #ForbiddenCityphenology# topic turns pictures of the changes in the Forbidden City's solar terms into electronic posters and wallpapers that can be used on various terminals. By keeping up with the solar terms and producing beautiful images, the topic now has many loyal followers.

The release of time-sensitive content resonates with a wide audience. For example, when the weather is cold, the Palace Museum Weibo posts information of how people kept warm in the original palace, including hand warmers, clothes and accessories, and brick beds. In summer, it sets topics like #Refreshingsummer#, choosing a palace fan, refrigerator, or artefacts with fresh and elegant colors to explain how the original palace dealt with summer heat.

The Weibo platform also attracts attention with polls and lucky draws (see Figure 6). The Palace Museum Weibo has developed a win-win marketing approach involving event promotion, multi-platform interaction, and award-winning forwarding activity, achieving its own goals of expanded communication, and allowing followers to gain tangible benefits as well.

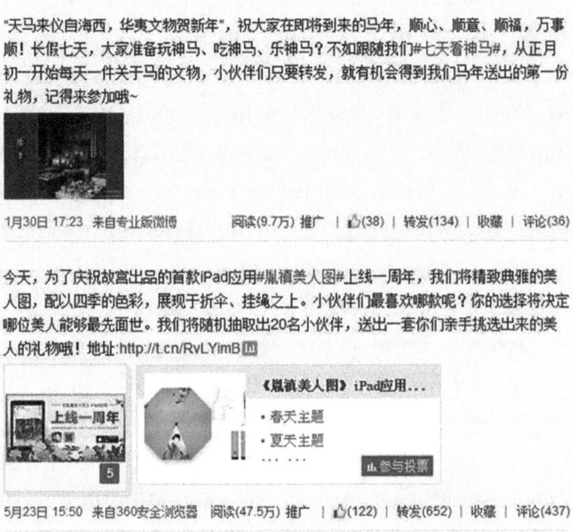

Figure 6 Voting, lucky draws, etc.

2.3 The importance of images on the Palace Museum Weibo

The image has an important position and role on the Palace Museum Weibo, attracting followers through the release of aesthetically attractive pictures. The Internet and smart terminals have brought the public into a "mobile picture reading era". We can now "read pictures" on social platforms anytime, anywhere. Li Peilin defines the era thus: "The era of reading pictures is a state that involves the rapid development of modern society, and the continuous advance of science. Against the background of a relatively perfect communication environment, the use of pictures within information dissemination improves picture quantity and quality. Together with text, it promotes optimized information communication." Yao Zhanxiong believes that the "era of reading pictures" refers to a state in which images are used in large quantity in communication, in the context of modern technological advancement and the acceleration of informatization.

Using micro-communication tools with convenient image publishing functions, the Palace Museum Weibo has released a large number of images of cultural artefacts, buildings, and landscapes, allowing people to "see" the Forbidden City. With the development of science and technology, and the accelerated pace of everyday life, we have entered an era in which words are boring and unattractive. Pictures are needed to constantly stimulate our eyes, inspiring curiosity and touching nerves. Images are easy to understand, so readers and visitors need only accept and appreciate them passively. This makes up for Weibo's 140 word limit that largely prevents deep and comprehensive explanations, and acts as a supplement for the text. It is also an important means of attracting and retaining followers. Figure 7 shows a screenshot of Weibo, a typical case of "the truth is in the picture". Due to a false estimations, previous Weibo content did not include images. Followers were unable to comprehend textual content, because text explanations alone failed to satisfy their curiosity. So, in ck response, photographers took pictures to supplement content. It was a huge success, and content began to spread quickly. This illustrates the importance of images in communication, as a complement for content.

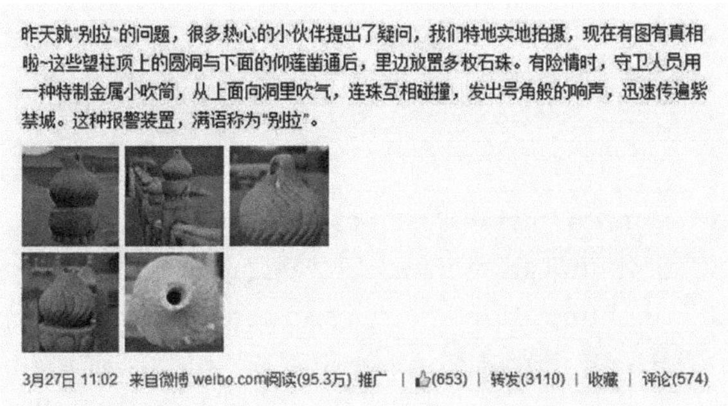

Figure 7 Example of a Weibo screenshot

The two Palace Museum Weibo topics #fallinlovewiththecity# and #ForbiddenCity-phenology# are dedicated to the release of images of palace interiors. The Palace Museum Weibo uses these two topics to attract followers. #FirstsnowintheForbiddenCity# and #ForbiddenCityspring# are two temporary topics that have included a large number of seasonal images, setting a record of an increase of 20,000 followers over two days and 50,000 in a week. In addition to being interested in the history, culture, and specifics of the Forbidden City, it is clear that people are also fascinated by its appearance.

The Palace Museum Weibo has an abundance of carefully crafted illustrations and texts. Without images, it would certainly not be as popular.

2.4 Development prospects for the Palace Museum Weibo

Development prospects include:

- Finding a professional organization to regularly analyze user content and issuance, forming analysis reports (which can be extended to unified user analysis and data comparison across all micro-platforms), adjusting content plans according to these analyses, and monitoring the status of communication.
- Building a micro-platform database connected to the existing management system of the Palace Museum, to share resources across various platforms (websites, mobile terminal sites, and WeChat). An automatic release mechanism is to be established at a later stage in order to reduce labor costs.
- Regular planning of various online and offline promotional activities, striving to share the resources of mainstream operators or platform developers, and successfully use their promotional channels and traffic resources to promote projects.
- Creating a unified micro-platform membership system, issuing a pocket *Micro-Palace* publication (a digital edition or traditional publication) every year or two, alongside other cultural products, to be provided free of charge by subscription or at events for members or enthusiasts; carrying out sharing and forwarding of Palace Museum content, thus attracting more membership and enhancing the commitment of existing members.

3. Conclusion

The function of the museum lies in public education and cultural communication. Museums' means of communication should be engaged with real public life, in order to achieve enhanced communication effects and therefore fulfill the purposes of public education and cultural communication. Means of micro-communication bring the museum closer to public life, even into people's bags and pockets, allowing them to carry the Palace Museum with them. This makes up for the lack in traditional methods of museum communication, offering more comprehensive cultural communication with fewer dead ends. Micro-communication has even better effects than traditional methods, playing an important role with benefits to cultural communication for museums. In the future, there will be even better forms of communication. As long as they are beneficial to cultural communication, we should accept their use by museums.

(The chart data and screenshots in this article are all sourced from Sina Weibo)

Analysis of the Communication Characteristics of Digital Museums in the Big Data Era

Liu Wenhong
(College of Applied Arts and Sciences, Beijing Union University, Beijing 100191, China)

Abstract: As a cultural medium, the museum has always been responsible for the research, education, and dissemination of historical culture. With the rise of digital museums, museums have gained certain mass media attributes, especially in the Big Data age, featuring new communication concepts, functions, platforms, and effects.

Keywords: Big Data era; digital museum; communication

Preface

In the Copenhagen draft constitution, the International Council of Museums (ICOM) stated: "The museum is an open, permanent institution that does not pursue profit, but serves society and social development, standing for the purposes of research, education, and appreciation, collecting, preserving, researching, disseminating, and exhibiting articles of witness to humanity and human environments" [1]. This definition reveals the functional nature of the museum and its relationship with society, and at the same time reveals the fact that although the era of mass communication has not yet arrived, while the museum is not a media organization, it embodies the function of cultural communication and has occupied a position in the history of human interaction.

It is a unique medium, and we must analyze and elaborate on various key elements from a communication science perspective, such as subjects, visitors, content, modes, and effects of communication. With the rapid development of information technology, the application of new technologies in the museum field is also progressing quickly. New technologies such as virtual simulation, physical modeling, Cloud computing, Big Data, the Internet of Things, and the mobile Internet have all offered the museum more space for development. In the current Big Data era, how can the museum reflect its communication functions and realize effective communication? The change in museum communication characteristics within this new era is the subject of this paper.

1. Definition and characteristics of digital museums

In the late 1990s, cultural and artistic circles began the process of digital museum development. On November 28, 2003, the Digital Professional Committee at the Chinese Museum Society was established in Beijing, marking the beginning of museum informatization and digital museum theory. In the process of digitization, with the development of technology, new concepts such as digital museums, museum digitization, electronic museums, virtual museums, online museums, and new media museums emerged. Amongst them, the title of digital museum was most popular. Digital museums are based on museum digitization, while the term "electronic museum" comes from the medium in which the digital museum exists. The term "virtual museum" counters the physical museum, while "online museum" comes with the existence space of the digital museum. All of these terms emphasize a certain aspect of the digital museum [2].

At present, there are multiple definitions of the digital museum. From the perspective of their essence, digital museums stand for the integrated digital reproduction and cognition of physical museums and related phenomena. Specifically, the digital museum is a new model for the management and application of cultural heritage transmission in the context of the information age. This includes digitally collecting, processing, storing, and managing all aspects of natural and cultural heritage information, providing users with services like digital display, cultural communication, education, and research via computer networks – a multi-disciplinary information service system that integrates the sciences of computers, museums, and communication. Digital museum application technology consists of five apparent links to do with data: acquisition, processing, storage, transmission, and output.

In terms of structure, a complete digital museum has four levels:

1) Digital content

The digital data content and data-created works obtained from the digitization of collections are equivalent to physical museum collections. This is the core content of the digital museum.

2) A database
A standardized and regulated collection archive for storing and protecting digital content resources, equivalent to a physical museum's warehouse.

3) A virtual display platform
The presentation of digital museum content, equivalent to the exhibition hall in the physical museum, using multimedia technology, virtual simulation technology, and other means of expression to produce vivid live interpretations of physical exhibits, with virtual communication capabilities across cyberspace.

4) An interactive education platform
Based on the display platform, this involves the collection, analysis, push, and management of various forms of information, deep mining of data, and effective information processing as the basis upon which to realize resource sharing and interaction.

2. *The meaning of the Big Data era*

The concept of Big Data first appeared in the 1980s, and over the following 20 years, it underwent continuous exploration and research in the United States, resulting in the publication of *Digital Planning of the Future: a Report from the Presidential Science and Technology Advisory Committee to the President and Congress* in 2010, and McKinsey & Co.'s *Big Data: The Next Innovation, Frontier of Competition and Productivity*, in 2011, which generated widespread public awareness of Big Data. On March 29, 2012, the Obama administration proposed to invest more than 200 million US dollars for the immediate launch of the "Big Data Research and Development Initiative", which attracted attention from all over the world. Big Data originated as an industry practice, but as the academic world conducted constant research and follow up activities, and with the development of data technology, research from the computer science field infiltrated other disciplines, leading practitioners and scholars in all fields to keep track of the impacts and changes brought about by Big Data while exploring new research methods and ideas.

The era of Big Data is generally considered to have the following four characteristics, referred to as the "four V" features:

1) Large Volume
Data magnitude that has evolved from TB (1012B) to PB and even ZB.

2) Wide Variety
Housing many forms of data, mostly structured and semi-structured data such as web pages, graphics, videos, images, and location information.

3) Fast Velocity
Data streams that are often high-speed, or real-time data streams, often requiring fast, continuous real-time processing; processing tools are also rapidly evolving, perhaps implementing software engineering or artificial intelligence.

4) High Value and Low Density
Amongst the large quantity of information, effective information is like a drop in the ocean.

The greatest impacts of Big Data on the media are twofold:

Firstly, media interactivity, sociality, and intelligence are enhanced, with media integration and human-machine symbiosis being the direction of development, and the role of the public in media content production constantly increasing.

Secondly, traditional media and social media become important data sources. The large amount of data in the media sound field creates considerable social capital, and the mining of data within the database by academia reveals habits of human information dissemination and social activity [3].

The digital museum can be seen as a special form of mass media, and while Big Data has similar effects on it as the media, it is unique. The following section focuses on the characteristics of digital museum communication in the current era.

3. Characteristics of digital museum communication in the Big Data era

3.1 Changes in the concept of communication based on visitors

To a certain extent, differences in communication concepts determine media modes of operation and effects of communication. Further exploration or renewal of the concept of museum communication was carried out first in the context of new museum science in the early 1990s on a world-wide scale, primarily reflecting old concepts of communication.

In the West, the museum as a cultural institution has a long historical background,, with its communication concepts and content having undergone changes over time. In China, museums can be viewed as products of Western influence. In general, its museums focus on research and publicity functions that are closely related to ideology. The relationship between museums and the public, as well as between history and the future, has always lacked opportunities for review from appropriate perspectives and locations [4]. Chinese museums are faced with a lag in their mechanisms, imbalanced development, and a diminishing number of public visitors, which are the main problems faced in traditional museum development. In the current media environment, museum communication is both scientific and cultural. In order to fully mobilize the communication functions of the museum, it is necessary to construct communication systems around scientific communication concepts. We must adhere to an visitor-oriented system that focuses on visitor experience, rather than giving the subject of communication a strong position, treating the audience as a "target" for the inheritance of information. The communication of museum information must be a comprehensive, interactive, two-way transmission activity, involving both sharing and interactive exchange.

3.2 Communication platforms and carrier diversity

Well-known media forms include text, journals, newspapers, magazines, broadcast television, photographs, and books. At the turn of the century, the concept of new media was discussed more in reference to new forms of media emerging alongside the development of satellite communications, digitization, multimedia, and computer network technology [5]. In the digital museum era, the main carrier of information is the website. On a museum's official website, visitors see 2D photos and text regarding museum collections, as well as integrated information based on multimedia technology, and various combinations of information media. This information is realized by hyperlinks and based on database retrieval, enabling visitors to obtain multi-dimensional information, which to some extent exceeds the informational content of physical exhibits.

With developments in media technology, new media such as Weibo, WeChat, and apps will become the main channels of communication. In the Big Data era, public activity has shifted into virtual space. With the progress of digital museums in China and the rapid development of the mobile Internet, more people are expecting to acquire and browse museum-related information on mobile platforms. The effective combination of digital museum application needs with mobile platform characteristics will make use of the advantages of mobile platforms, offering highly convenient display of museum-related information, interaction, and services. Since the launch of the National Museum's "My Museum" app, other museums have followed suit, with the number of companies engaged in the development of museum apps increasing rapidly. Apps have developed from the initial information display and audio navigation, to combining with new technologies such as augmented reality, virtual 3D experience, and image recognition in order to create more innovative and unique interactive experiences for the public. However, the application of domestic museum apps is still at a superficial stage, mainly focused on basic functions such as information display, map guidance, and voice explanation.

3.3 The expansion of communication functions into diversified value orientation

In the study of communication, the mass media has four basic social functions: environmental monitoring, social coordination, cultural inheritance, and entertainment. These four functions reflect the social values of the mass media. On this basis, Wilbur Schramm has expanded upon the economic function of communication. In the era of traditional museums, the physical museum is responsible largely for the crucial mission of cultural inheritance and public entertainment. When discussing the Western museum's showcasing of national images and the entertainment and recreation venues of ordinary people, some people have compared the two functions in a vivid way: the museum is both a country's auditorium and a meeting place for ordinary people.

However, in the era of Big Data, the media attributes of digital museums have been strengthened, and the function of mass communication has fallen on digital museums. Museum websites, now a window for digital museum displays, are no longer restricted by the "walls" of the museum. They are able to collect and disseminate information in a wider boundary, providing multiple ways to collect information for museum developed apps and official museum Weibo and WeChat accounts. This evolutionary information is recorded, and network science and social network analysis are important technologies and methods for Big Data analysis, allowing us to observe the complex behavior patterns of society.

First of all, terminal-based platforms such as websites and apps have accumulated large amounts of activity data on how people apply and learn museum-related information. This data reveals the social codes of human information dissemination, interpersonal communication, and group interaction. In addition, cultural heritage information can be collected and communicated over the network, meeting the public need for dynamic information about the protection and management of cultural heritage through environmental monitoring functions. Furthermore, information conveyed over the website is not just "information"; it also assesses and interprets the value of cultural heritage in detail, providing conservation publicity and education to the public, invoking appropriate reflective responses in the public, and embodying the functions of social guidance and coordination [6].

3.4 Changes in the communication mode: from linear to bidirectional control

In the physical museum, visitors only see the collection within the exhibition hall in a completely "static" form of one-way communication that is mainly based on information display, with most information interaction methods transmitting from "object" to "person", with very limited exchange from "person" to "object". Besides the "intentional attention" of some individuals, effective information dissemination cannot be achieved for the vast majority of visitors. The digital museum, which converts collection-related information to "digital" storage, disseminates and displays it via the Internet, effectively breaking through the space and time constraints attached to exhibiting a physical collection.

The digital museum creatively adopts the information transfer model of "object" → "digit" → "person" [7]. However, it must be noted that "digit" → "person" allows for interactive communication, while "object" → "digit" is still a one-way flow. This kind of information interaction method not only separates close relationships between "objects" and "people", but also lacks management of the synergy between "object" and "object", and "person" and "person", meaning that digital museum is a still a linear, one-way form of interaction. To solve this problem, it is necessary to implement the Internet of Things, acquiring various kinds of information regarding people and objects within the museum in the form of sensors, establishing a two-way connection between "objects", "people", and "digits" using Cloud computing and Big Data analysis technology to achieve "human" control of "objects".

3.5 Monitoring the effects of communication

Similar to all mass media, museums must also pursue communication outcomes and achieve effective communication, or else the propagation process is entirely invalid. Compared with other media, the dissemination effects of museums can be accidental, random, unpredictable, and uncertain time wise. Therefore, research on the dissemination effects of museum has always been a hot topic for discussion, but has yet to enter a practical stage.

In the era of Big Data, communication effects are easily checked, as relevant information can be attained via websites, WeChat, Weibo, and apps. The key to data analysis is not to master large amounts of data, but to extract meaning from it in order to discover hidden patterns and potential relationships [8]. In addition, in the Big Data era, data can be collected and mined, with information integrated to conduct targeted and accurate dissemination and promotional activities.

References

[1] Yan Jianqiang, Liang Xiaoyan. 'Definitions and understandings of museums'. *Chinese Museum*, 2001, 1.

[2] Hu Chui, Zhang Xiaoli. 'A review of digital museum research'. *Digital Museum Research and Practice*, 2009, China Conference (May 2010).

[3] Luo Zhenglin. 'Innovations in the theory of media communications in the Big Data era'. *Young Reporter*, June 2011.

[4] Shen Zhuo. *Research on the concept of communication and display design in contemporary Chinese museums with "people" at the core* [D]. Master's Thesis, Jiangnan University, 2008.

[5] McLuhan, Marshall. *Media is the message — Media and innovative thinking* [M]. Beijing: Tsinghua University Press, 2001, p. 86.

[6] Jin Ruiguo. 'Interpretation of communication in digital museums'. *Museology*, 2010, 8.

[7] Chen Gang. 'Smart Museums — New trends in digital museum development', [J] *Chinese Museum*, 2013, 10.

[8] Yu Guoming, Wang Bin, Li Biao, Yang Ya. 'Communication Studies: The paradigm of the Big Data Era'. *Shanghai Journalism Review*, 2013, 6.

[9] Jia Junzhi, Shi Xuan. 'Research on the construction of digital museum metadata standards', [J] *Journal of Shanxi University (Philosophy and Social Sciences),* Jan 2015.
[10] Li Hengshuang, Wang Guanwei. 'Analysis of the development of digital museums in the new media era', [J] *Main Stage*, 2012, 1.
[11] Zhu Xiaodong. *Research on key technologies in the museum*. PhD thesis, Northwestern University, Nov 2004.
[12] Yang Dandan. 'On the sustainable development of educational museum activities — the example of youth activities at the Capital Museum', [J] *China Museum* 2010, 1.

*Research and Analysis of Indoor Positioning Technology for Museums**

Zhang Yong, Yi Junkai
(Beijing University of Chemical Technology, Beijing 100029, China)

He Kunpeng
(Beijing Gehua Group, Beijing 100007, China)

Abstract: The progress of various indoor positioning technologies has greatly promoted the development of location-based services (LBS). Faced with many available indoor positioning technologies, the museum has begun to use the technology to provide better navigation services for visitors, as well as contribute to daily management. This paper makes a systematic analysis and comparison of current indoor positioning technologies used in the museum, which allow venues to select the appropriate type in order to promote the spread of culture and improve the service experience.

Keywords: museum; indoor positioning; WLAN; Bluetooth

Preface

As digitization progresses, museums are able to provide basic services via mobile terminals, while also focusing on the quality and experience of services. Location-based services (LBS) are the key for mobile terminals to improve service quality. Indoor positioning technology is the core of LBS, so indoor positioning technology has become a popular research topic amongst many scholars, research institutions, and enterprises.

* Funded project: "Research on the Framework of the Digital System for Protecting Cultural Relics", funded by the National Science and Technology Support Program, project number 2014BAK07B01

At present, there is an increasing amount of research on indoor positioning. Common and mature technologies include Radio Frequency Identification (RFID), Ultra-Wideband (UWB), Infrared Red, Ultrasonic, Bluetooth, Cellular Based, ZigBee, geomagnetic positioning, and FM. New indoor positioning technologies and systems include iBeacon positioning [1], Matlab and capacitive sensing indoor positioning [2], MEMS inertial device indoor positioning [3], TC-OFDM physical indoor positioning [4], WLAN and WSN-based heterogeneous network system positioning [5], PowerLine positioning [6], wireless localization techniques based on sets and radial singular functions [7], pressure sensing touch technology indoor floor positioning [8], sound positioning [9, 10], visible light positioning [11-13], and Monocular camera positioning using internal integrated inertial measurement devices (MU and IMU) [15]. This shows the general state of current research. Among the various positioning technologies, each has its own advantages and disadvantages.

Faced with such wide variety in indoor positioning technologies, most analyses of the advantages and disadvantages look directly at a single technology itself or lateral comparisons between various technologies, while industry or field-specific research analysis is still lacking. Given the large quantity of positioning technologies, either technical standards are not uniform, or a choice cannot be made, hindering technological applications and slowing the development of the industry itself. Therefore, this paper performs a comparative analysis of the application of various indoor positioning technologies within the museum field, looking at the shortcomings of these aspects, and hopefully playing a small role in the promotion of future museum development.

1. Factors affecting the choice of indoor positioning technology within museums

The type of indoor positioning technology used in museums is generally determined based on several factors, namely accuracy, usability, coverage, cost, and safety [14, 16].

1) Accuracy

In the Joint Committee for Guides in Meteorology (GCGM), accuracy is defined as the proximity between measured values and true, known values [17]. Usually, positioning objects within the museum include people, collections, and exhibition halls, and the positioning technologies these objects need are not nearly identical due to their differences in height, volume, and size. The accuracy of positioning technology is generally divided into three forms, namely positioning accuracy between 0.1 to 0.5m, known as decimeter level positioning; positioning accuracy between 0.5 and 1m, called sub-meter positioning; and positioning accuracy between 1 and 10m, called meter-level positioning [12].

2) Usability
Usability refers to the percentage of time that is usable under the requirements of accurateness and completeness, and it lies between a random factor (such as communication congestion) and a planned factor (such as routine maintenance) [19]. For museums, usability directly affects the service experience, which in turn affects the strength of a venue's influence. In general, the usability of location technology is divided into three levels: low usability (< 95%), standard usability (> 99%), and high usability (> 99.9%) [14].

3) Coverage
Coverage is the extent to which an indoor positioning technology can provide positioning with high accuracy. In general, the coverage of positioning technology and positioning accuracy are contradictory; the farther the positioning range is, the worse the accuracy, and vice versa. Museums are both large and small, and objects can be positioned far or near. Therefore, the selection of appropriate positioning technology can reduce the costs of corresponding positioning infrastructure to a certain extent, while at the same time ensuring accurate positioning.

4) Cost
Cost is a secondary factor for museums. The museum relies on a definite positioning system to use corresponding positioning technology, and typical positioning systems all include installation and maintenance, as well as infrastructure, components, and equipment. Installation and maintenance costs include initial system setup, testing, and post-maintenance to keep the system up and running; infrastructure and component costs include materials, tools, and instruments required to build infrastructure; position equipment costs refer to purchasing positioning equipment.

5) Security
Security should be the primary consideration for museums, as it directly relates to the safety of collections and personnel. Museums should ensure that positioning technology has as little impact on collections and people as possible, while ensuring that it is not used as a tool for crime. In the choice of technology within the museum, security must be demonstrated to the full.

2. Analysis of indoor positioning technology in museums

Given the current variety of indoor positioning technologies, it is difficult to find one that meets all indoor positioning requirements, and each has its own advantages and disadvantages. Based on these characteristics and specific needs, a museum must select a combination of technologies or a single technology according to its service requirements.

2.1 Classification of positioning technology

There are many ways to classify indoor positioning technology. Collin et al. divide it into two categories according to hardware requirements [20]. Gu et al. have proposed three-way classification based on the network, overall system architecture, and the positioning medium used [21]. Al Nuaimi and Kamel classify according to installed systems within the building [14]. Rainer Mautz classifies according to sensor type [12], while Wei Qing classifies according to the calculating method of positioning, positioning theme, and equipment used [14]. Yu Yang, Zhao Kaifei, and Shen Jia propose a classification based on location technology [22]. Zhang Fan, Chen Diancheng, and Yang Jie also proposed their own classification method [23]. Table 1 shows a detailed explanation of the above-mentioned classifications.

Table 1 Classification of Existing Indoor Positioning Technologies

Author/Year	*Basis*	*Technical categorization*
Collin et al., 2003	Hardware within the building	Existing hardware and stand-alone hardware within the building
Gu et al., 2009	Network, system architecture and location media	1) Based on the existing network, divide into network-based and non-network based technology; 2) Based on the overall system architecture, divide into autonomous positioning architecture, infrastructure positioning architecture and self-directed infrastructure-supported architecture; 3) Base on the positioning medium used, divide into ultrasonic, RFID, vision-based and sound-based localization technology
Al Nuaimi & Kamel, 2011	Mounted system	Fixed device indoor object and personnel positioning
Rainer Mautz, 2012	Sensor system	Camera, infrared, magnetic induction, sound, WLAN, RFID, UWB, Agile GNSS, pseudo-satellite, other radio, internal navigation, magnetic system and other infrastructure systems
Wei Qing, 2014	Positioning method, theme and equipment	Network-based, mobile-based terminal and terminal-network based hybrid positioning technology
Yu Yang et al., 2014	Proximal relationships	Proximity relationship positioning, geometric relationship based positioning and scene analysis positioning
Zhang Fan et al., 2012		AGPS positioning technology, wireless positioning technology and other positioning technologies

The above classifications of indoor positioning technology are based on the whole technology, not analyzing or classifying a specific industry or field. What follows is a comparative analyses of current indoor positioning technology from the museum perspective.

Based on objects within the museum, indoor positioning technology is divided into three categories: personnel positioning, collection positioning, and exhibition hall positioning.

1) Personnel positioning refers to the location of visitors, staff, and personnel within the museum. As this includes people of all ages, especially children, positioning technology must be at least sub-meter level, including Bluetooth, WLAN, and MU/IMU.
2) According to the size of the collection, positioning is divided into: WLAN, ultrasonic, and RFID for large-scale collections (1 m or more); Bluetooth and light-based for medium-sized collections (0.5–1 m); RFID and UWB for small collections (0.5 to 0.15 m); and UWB and pressure touch floor positioning for micro-collections (less than 0.15 m).
3) Positioning technology used for positioning within the museum uses WLAN. Figure 1 is a graphic representation of these technical categories.

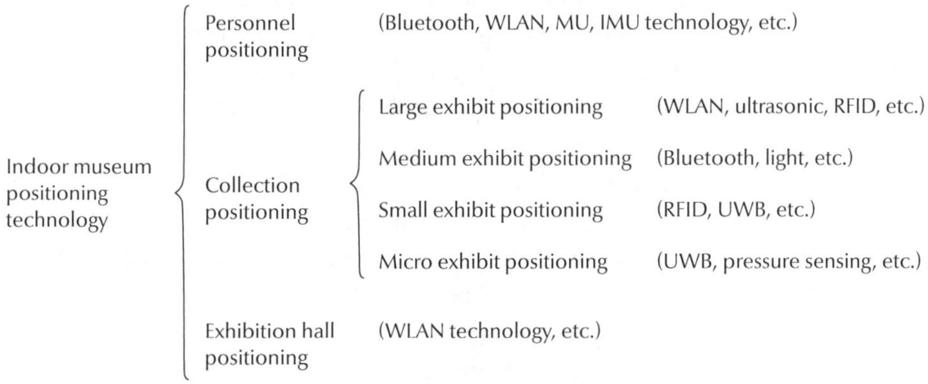

Figure 1 Classification of indoor positioning technology in museums

2.2 The application of positioning technology in museums

1) WLAN positioning

WLAN positioning is also known as WiFi positioning. At present, most museums already have WLAN networks, so this technology is the most commonly used [5, 19, 16]. Visitors are able to access the museum's WLAN network anytime through mobile devices, such as smart phones, tablets, PDAs, and e-readers. According to the accuracy and scope of the positioning technology, it is suitable for people, exhibition halls, and large museum collections. Sizeable artefacts include large-scale porcelain, pottery, stoneware, bronzes, woodwork, and carvings [24]. However, WLAN is affected by non-line-of-sight and multipath propagation [25], resulting in signal transmission distortion, which causes reduced positioning accuracy. There are many mobile phone apps available for museums using WLAN positioning, such as the Zhongshan Warship mobile phone app, the Shaanxi History Museum app, and the Suzhou Museum app.

2) Bluetooth positioning

Bluetooth positioning is also one of the more popular positioning technologies used in museums, the reason being that it is easily integrated into mobile devices, and is very low cost [23, 18, 26]. Meanwhile, with the introduction of Bluetooth 4.0 [27] and iBeacon technology [1], its positioning accuracy, positioning range, and energy consumption have all increased. For museums, this is suited for positioning medium-sized items, such as porcelain, pottery, ironware, stone carvings, bamboo, and woodwork larger than 0.5m [24]. Furthermore, user location positioning is also possible through mobile terminals with integrated Bluetooth, providing navigation services and the like.

3) Light positioning

Light positioning is one of the newer positioning technologies, usually utilizing a beacon generated by an LED light source for positioning [12, 13]. As lighting devices are generally used within the museum for lighting purposes, coverage is extremely extensive, and its use in museum positioning can both meet lighting requirements and reduce costs. When the height of the beacon is 1m, the circular probability of error (CEP) is about 20cm [11], so the technique is more suited to positioning small objects (small artefacts and miniature collections) within display cabinets. However, as a visible light positioning method, it is susceptible to occlusion, affecting the accuracy of positioning results. Indoor positioning using visible light has no electromagnetic pollution, and is easy to deploy [11, 12], so it appears to be one of the most promising technologies for the future.

4) Power line and fingerprint location positioning

Power line positioning refers to the input of high frequency broadband signals on power lines [6]. As with light positioning, power lines are found in almost every corner of museums due to the large amount of lighting and electronics. The use of power line positioning allows the museum to save the costs of deploying a great deal of positioning infrastructure. Due to the widespread availability of power lines and high positioning accuracy (such as positioning accuracy within 1m achieved by integrating the KNN positioning or Naive Bayes algorithm), small, medium, and large collections can all be accurately located within the museum. However, the security of this positioning method is still unclear and requires verification, though the technology is extremely promising.

5) Indoor positioning floor based on pressure sensor touch technology

Pressure sensor touch technology carries out positioning by setting up a plurality of pressure sensors on the corners of a solid plane. The technology has advantages of environmental robustness, strong persistence, low strain, and three-dimensional information perception and positioning accuracy (the absolute error of positioning can reach up to 2cm) [8]. Since this positioning technology requires a positioning floor with specific sensing devices, deployment costs are relatively high. It is mostly suited to the positioning of a few particularly expensive or miniature collections, such as town hall treasures.

6) RFID positioning
This technology is one of the most mature in comparison to WLAN and Bluetooth. Due to simplicity of deployment and low cost [19, 26], it is especially suitable for the management of museum collections [16], enabling effective tracking, checking, and counting of information, and offering great convenience to collection management. However, this technology is unable to provide direct indoor navigation services via mobile terminals, and is not easily integrated with other existing positioning systems.

7) Ultrasonic positioning
Ultrasonic positioning has the lowest adoption rate of positioning technologies in museums. For museums, the costs for ultrasonic positioning are high due to the need for professional positioning equipment. However, due to its high positioning accuracy, it is well-suited to the positioning of both people and collections in museums, particularly for small and micro collections.

8) UWB positioning
UWB, or ultra-wideband, is a more specialized positioning technology than ultrasonic. Therefore, there are relatively few application examples in the context of museum

Table 2 Comparison of Indoor Positioning Technologies in Museums

Positioning technology \ *Influencing factors*	*Accuracy*	*Applicability*	*Coverage*	*Cost*	*Security*
WLAN	Meter	Standard	> 100 m	Lowest	High
Bluetooth	Sub-meter	Standard	< 100 m	Relatively low	High
Light positioning	Decimeter			Low	High
Power lines	Sub-meter			Low	
Pressure sensing	Decimeter			High	High
RFID	Sub-meter	Low	50 m	Higher	Higher
Ultrasonic	Decimeter	Low	10 m	High	Low
UWB	Decimeter	Low	20 m	Highest	Low

positioning technology. Due to its highly accurate positioning, UWB is suited to the positioning of micro-collections in museums, but the main obstacle faced in museum application is its high cost.

Table 2 shows a comparison of the various positioning techniques described above.

3. Indoor positioning plan

Judging by the above analysis of the various technologies within the museum, it is a challenge to completely satisfy all positioning requirements with one single technology, and therefore, it is useful to consider the combination of several technologies [28]. Here are a few combination options.

1) WLAN + Bluetooth

This solution makes full use of WLAN infrastructure while simply installing the basic Bluetooth positioning module, which is very cheap, making the cost of this solution extremely low. Meanwhile, this solution requires no user adaptation, so is very practical. However, due to the positioning accuracy limitations of the two positioning technologies themselves, the positioning of some small and micro collection artefacts is not entirely accurate. However, with improvements to various positioning algorithms and software quality, the positioning accuracy of this solution can be improved in the future. The system structure is shown in Figure 2.

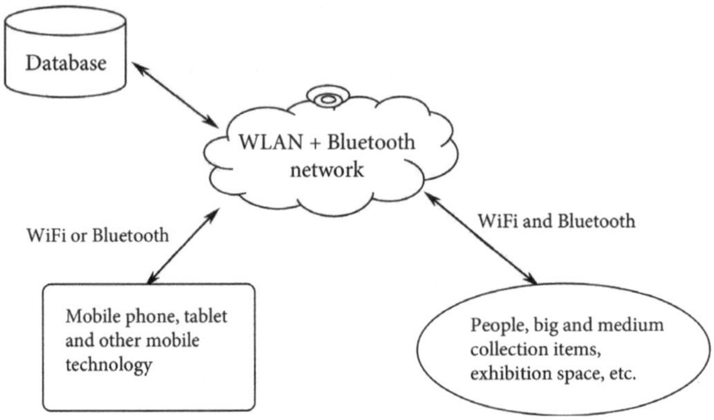

Figure 2 WLAN + Bluetooth positioning

2) WLAN + Bluetooth + UWB

This positioning scheme makes it possible to realize comprehensive positioning of personnel, collections, and exhibitions within the museum, and the research and application work done with these three technologies is relatively mature. Greater investment is needed in the establishment of this kind of positioning system, as Bluetooth and UWB require additional equipment installation, and the operation requirements of UWB are very sophisticated. Generally, this scheme is suited to museums with relatively complete collections, large scales, and relatively high budgets. The system structure is shown in Figure 3.

3) WLAN + Light + Power Line

This solution makes full use of museum infrastructure. Neither the museum nor visitors need install any further positioning equipment. With the most simple and convenient operation and maintenance prospects, this is the lowest cost positioning solution. However, since both the light and power positioning technologies are still in experimental stages, there is still time before this solution can be realized. However, with the advancement of technology, it appears to be the most promising for the future. Its system structure is shown in Figure 4.

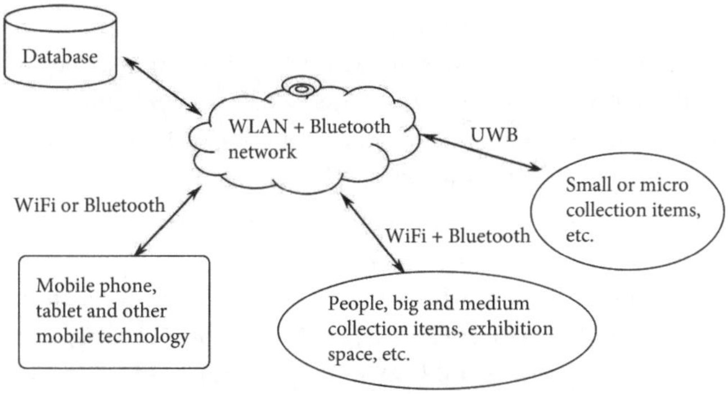

Figure 3 WLAN + Bluetooth + UWB positioning

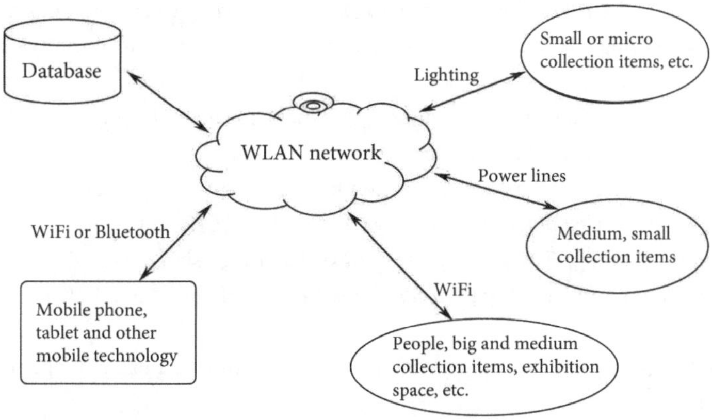

Figure 4 WLAN + light + power line positioning

4. Summary

This paper has looked at several key factors that influence the museum's choice of indoor positioning technology. Then, based on both more mature indoor positioning technologies and newer positioning technologies, suitable techniques have been selected and classified accordingly, analyzing the application prospects of these technologies within the museum at the same time. Finally, three potential solutions have been proposed.

In summary, the museum must adopt appropriate positioning techniques based on its actual state, in order to enhance service levels and promote its influence.

References

[1] Zhang Zhuosheng, Ai Haojun, Ma Fangfang, Xue Jingyuan and Zhang Yin Shengyuan. 'Smart Museums using iBeacon positioning technology', [J] *Electronic Engineering & Product World*, 2015, 1: 31–35.

[2] Zheng Chunhua. 'Indoor positioning systems using Matlab and capacitative sensing', [J] *Measurement & Control Technology*, 2014, 10: 143–146.

[3] Li Jinfeng, Wang Qinghui, Liu Xiaomei and Cao Shun. 'Pedestrian indoor positioning systems using MEMS inertial devices', [J] *Computer Measurement & Control*, 2014, 11: 3761–3763.

[4] Deng Zhongliang, Yin Lu. 'Indoor positioning systems based on the TC-OFDM system', [J] *Telecommunication Network Technology*, 2015, 3: 32–35.

[5] Yu Qin, Xiao Zhihui, Jiang Wei. 'Co-location of heterogeneous networks based on WLAN and WSN', [J] *Telecommunications Science*, 2013, 7: 83–88.

[6] He Jian, Wan Zhijiang, Liu Jinwei. 'Indoor positioning technology based on power lines and location fingerprinting', [J] *Journal of Electronics & Information Technology*, 2014, 12: 2902–2908.

[7] Gu Zhaojun, Wang Zhongpeng, Liu Fei. 'Wireless positioning technology using sets and radial basis functions', [J] *Computer Applications & Software*, 2015, 1: 255–258.

[8] Wei Hengxi, Chang Xiaoming, Liu Weiling, Duan Jinjun. 'Indoor positioning based on pressure sensing touch techology', [J] *Microcontrollers & Embedded Systems*, 2014, 10: 22–25.

[9] Zhao Xiaoyan, Zhou Lin, Wu Zhenyang. 'Sound source localization algorithms based on compression sensing', [J] *Journal of Southeast University (Natural Science Edition)*, 2015, 2: 203–207.

[10] Xu Bin, Xia Xiaohu. 'Research on indoor sound location method using microphone arrays', [J] *Journal of Hefei University (Natural Science)*, 2015, 1: 30–34.

[11] Chen Canfeng, Tan Wei. 'An indoor positioning system based on visible light beacons and its implementation,' [J] *Computer Engineering & Applications*, 2015, 5: 82–87.

[12] Wu Nan, Wang Xudong, Hu Qingqing, He Rongxi. 'High-precision indoor visible light localization methods based on multi-LED,' [J] *Journal of Electronics & Information Technology*, 2015, 3: 727–732.

[13] Moon, Myoung-geun and Su-il Choi. 'Indoor position estimation using image sensors based on VLC'. *IEEE*, 2014: 11–14.

[14] Zhang, Y., J. Tan, Z. Zeng, et al. 'Monocular camera and IMU integration for indoor position estimation', [J] *IEEE Engineering in Medicine and Biology Society Annual Conference Proceedings*, 2014: 1198–1201.

[15] Al Nuaimi, K. and H. Kamel. 'A survey of indoor positioning systems and algorithms'. *Innovations in Information Technology (IIT)*, IEEE Convention, 2011: 185–190.

[16] Kuflik, T., J. Lanir, E. Dim, et al. 'Indoor positioning in cultural heritage: Challenges and a solution' [C]. *Electrical & Electronics Engineers in Israel*, IEEE Convention, 2012: 1–5.

[17] Mautz, R. *Indoor positioning technologies*. PhD thesis, *Habil.ETH*, Zurich, 2012.

[18] Wei Qing, Peng Rui. 'Discussion of positioning technology and indoor positioning schemes', [J] *Guangdong Communication Technology*, 2014, 5: 57–63

[19] Al-Ammar, M.A., S. Alhadhrami, A. Al-Salman, A. Alarifi, H.S. Al-Khalifa, A. Alnafessah, and M. Alsaleh. 'Comparative Survey of Indoor Positioning Technologies, Techniques, and Algorithms. *Cyberworlds* (CW), International Conference on IEEE, 2014: 245–252.

[20] Collin, J., O. Mezentsev, G. Lachapelle, et al. 'Indoor positioning system using accelerometry and high accuracy heading sensors'. In *Proceedings of ION GPS/GNSS 2003 Conference*, 2003: 9–12.

[21] Y. Gu, A. Lo, and I. Niemegeers. 'A survey of indoor positioning systems for personal wireless networks'. *Communications Surveys & Tutorials*, IEEE 2009, 11(1): 13–32.

[22] Yu Yang, Zhao Kaifei, Shen Jia. 'Application, research status and prospects of indoor positioning technology', [J] *Telecommunications Network Technology*, 2014, 5: 46–49.

[23] Zhang Fan, Chen Diancheng, Yang Jie. 'Comparative study of indoor positioning technologies and systems', [J] *Guangdong Communication Technology*, 2012, 11: 73–79.

[24] Gao He. 'Classification of museum collections', [J] *Chinese Museum*, 1988, 4: 18–20, 96.

[25] Zeng Wei, Huang Liang. 'A WLAN-based indoor location algorithm using sparse representation', [J] *Computer Applications & Software*, 2014, 12: 175–177, 244.

[26] Curran, Kevin, Eoghan Furey, Tom Lunney, Jose Santos, Derek Woods, and Aiden McCaughey. 'An evaluation of indoor location determination technologies'. *Journal of Location Based Service*, 2011, 5: 61–78.

[27] Mo Qian, Xiong Shuo. 'An approach to indoor positioning algorithms based on Bluetooth 4.0', [J] *Journal of Astronautic Meteorology & Measurement*, 2014, 6: 66–70.

[28] Lymberopoulos, D., Liu J., Yang X., et al. 'A realistic evaluation and comparison of indoor location technologies: experiences and lessons learned' [C]. *Proceedings of the 14th International Conference on Information Processing in Sensor Networks* ACM, 2015: 178–189.

CHAPTER 5

Entering the Museum

— Excavating and Utilizing the Cultural Tourism Value of the Museum

Discussing the Development of the Cultural Tourism Value of Museums in Historical Buildings – The Beijing Ancient Architecture Museum

Huang Xiao
(Beijing Ancient Architecture Museum, Ancient Building Communication Department, Beijing 100050, China)

Abstract: The Beijing Ancient Architecture Museum uses ancient buildings as its site, so elements like the architecture, overall layout, and environment offer great advantages in the development of cultural tourism value. Combining practice and theory in this work, the author suggests that through ways and means like extensive external publicity, the maintenance of its own resource advantages, and the enhancement of service levels, the museum's cultural implications can be explored and the value of tourism enhanced, transforming the museum into a leisure entertainment venue that educates through entertainment in order to attract more visitors. If visitors are willing to enter, the museum can better and more fully utilize its functions of social education.

Keywords: museum; tourism; ancient building conservation; publicity and service

With the continuous development of the tourism industry and the consistent improvement of public cultural literacy, tourism has become an indispensable leisure activity for many. People are less satisfied with the "box ticking" sort of tour where they just glance at things hurriedly. Instead, they want to feel and understand the history and cultural characteristics of a place in depth. Therefore, as venues that disseminate scientific and cultural knowledge to the public via collections of physical materials, regional museums are especially and increasingly being included on travel agendas. Faced with this phenomenon, on the basis of their functions of publicizing education and disseminating scientific and cultural knowledge, museums must fully exploit their cultural implications, enhancing the value of tourism while meeting visitors' needs for leisure and entertainment. Amidst this, due to their architecture, layout, and environment, museums within ancient and historical buildings possess greater advantages for the development of tourist value. We must exploit these advantages to the full, turning the museum into a cultural and recreational place for entertainment and education, as well as a preferred tourist destination.

1. Extensive external publicity to attract visitors

One of the most important tasks in the development of tourism value is building a reputation in order to attract visitors, which is the driving force and goal of continued development. However, as far as Beijing museums based in ancient buildings are concerned, the majority are still widely unknown, with generally small audience numbers. In the author's opinion, the reasons for this are as follows: First, these are mostly small-scale topical museums, such as the Beijing Ancient Architecture Museum and the Beijing Stone Carving Art Museum. Due to strong specificity, most of the visitors to these museums are engaged in learning and research in a particular topic; ordinary visitors may not be so keen. Second, while most museums are open to the public free of charge, museums in ancient buildings conserve their historical architecture by controlling the flow of people while ensuring maintenance funds. These museums are unable to implement free admission, so the public must consider the price of entry when choosing whether to visit. Third, the ancient buildings that some museums rely on have important historical status in regards to architecture, history, and religion, but they are not as well known to the general public as, for example, the Forbidden City, the Temple of Heaven, or the Lama Temple. Moreover, their locations are less easy to find. With the development of the city, ancient buildings have become hidden in a jungle of high-rise buildings, and cannot be found unless visitors are in close proximity. One example is the Temple of Agriculture – the grounds on which the Beijing Museum of Ancient Architecture is located. Although it has been encroached upon by urban development, it still occupies about 20,000 square meters. There are only 500 meters between the gates of the museum and the Central Axis facing the Temple of Heaven, but it is difficult to find because the museum gate is not on the main road. Many tourists are unable to find it even when they are nearby. Even people who work or live in the area do not know that there is a museum there. Another example, the Zhenjue Temple, where the Beijing Stone Carving Art Museum is located, is less than 200 meters from the northwest gate of the Beijing Zoo, which is jammed with visitors, but it is difficult to find as it is also far from the main road. Therefore, in order to enhance touristic value and attract visitors to such museums, the first step is letting people know that the museums exist and what they contain.

1.1 Promotional publicity using emerging media

With the development of technology, people are increasingly dependent on networks and mobile terminals. Internet promotion has become crucial, and we must continue to innovate while keeping pace with development, using new things to bring visitors closer. Weibo and WeChat have assumed a positive role in bridging the museum and the audience, with the ability to offer details about exhibitions or related background knowledge. Its main positive features are strong timeliness, allowing visitors keep abreast of museum developments, as well as high interactivity, meaning that visitors are able to ask questions and report their needs directly to staff through a museum's official Weibo and WeChat. They can also query the museum exhibits and services, and make other comments, which the museum can respond to immediately. In addition, current publicity via Weibo and WeChat is largely dominated by young staff. Abiding by strict conditions of public opinion regulations and museum standards, they are able to use specific network language and adhere to certain Internet trends to publish Weibo posts or interact with web users. In this way, the museum becomes closer to the people – less old-fashioned and unfathomable. In close communication, it is beneficial for museums to listen to audience opinions and suggestions more directly, then make corresponding adjustments.

Amidst the development of smart phones, mobile apps have gained strong momentum. One example is the AppStore, which Apple launched in 2008, initially with fewer than 500 apps. Over the following three years, numbers increased to 500,000 apps, with the cumulative number of downloads reaching up to 15 million, and these numbers continue to grow exponentially. Therefore, the launch of the museum's own app is a great means of publicity and promotion. Over the last two years, some Chinese museums have begun to cooperate with companies to develop apps. The most common app types are as follows. One is an app based on the museum itself, with content including information about exhibitions and exhibits, the most important feature being a museum tour guide and explanation feature. There are also thematic exhibition-based apps. For example, "Zhang Daqian Painting Exhibition" was a joint venture between the Taipei History Museum, the Sichuan Museum, the Jilin Provincial Museum, and the Shenzhen Museum. A Shenzhen cultural company then developed an exhibition app, not only for appreciation and display of exhibits, but also updating advanced notices about tour times, and detailed geographical information regarding the exhibition hall, allowing for convenient visits for all. There are also popular science or game apps based on cultural artefact collections, such as the Palace Museum's "Yong Zheng's Beauties" and "Emperor for a Day". These apps facilitate visits, and enhance interaction between visitors and the museum, providing successful cases that we can learn from.

1.2 Enhancing attention by organizing cultural events

Ancient architecture reflects social production, quality of life, science and technology, craftsmanship, artistic style, customs, and habits of the time. It is also a carrier of culture. The first construction of buildings used for sacrifice and religion was carried out for cultural activities. For example, the Temple of Agriculture, where the Beijing Museum of Ancient Architecture is located, was an important place of worship in the Qing and Ming Dynasties. In addition, such buildings often have a garden-like environment and open spaces, which may serve as platforms for events related to classical culture.

The museum recognizes its own cultural implications and unique advantages, and carries out various cultural activities to showcase them. Firstly, it can attract more visitors based on this. Secondly, it can provide opportunities by inviting the media to carry out publicity and promotion. For example, the Beijing Ancient Architecture Museum used its advantage as the site of the Temple of Agriculture to fully re-enact the ancestral worship ritual of the royal family in the Qing Dynasty, with an interview report featured on TV programs "Special Concern" and "Here It Is Beijing", as well as other popular programs on Beijing Satellite TV. Thirdly, museums must cooperate with other institutions and organizations to assist one another. For example, museum can be used as venues to launch new products, promote or publicize the works of famous writers, painters, and musicians, and raise public interest with the influence of famous brands or celebrities. For example, in 2010, the Confucius Temple and Guozijian Museum invited the ceremonial performance artist Ha Hui to perform at the opening ceremony of the first "Beijing Guozijian Confucius Temple Chinese Culture Festival" at the Confucius Temple, at the same time holding a media conference at this first stop of the "Ha Hui: Sixteen Flower Sounds — New Ceremonial Music World Tour Concert" in Beit Zion, Israel. These performances have received a lot of attention, and have been widely reported by major media, with performance videos widely circulated.

2. Maintaining resources and achieving sustainable development

A museum at an ancient site has unique advantages in terms of site resources, and its architecture, overall layout, and environment can be important factors to attract visitors. The ancient architecture itself is the most important "collection", so the conservation of ancient buildings is a responsibility that the museum must shoulder. Ancient buildings carry historical memory. They are not simply wooden or brick structures, but witnesses of changing dynasties and a nation's rise and fall. They reflect social production, lifestyles, science and technology, craftsmanship, artistic style, customs and habits. Once they are destroyed, we can recreate their outward appearance, but the historical and cultural meaning contained in the building is difficult to recover. On the basis of protecting ancient buildings and maintaining resources, we must explore their cultural implications and enhance touristic value.

Many museums at ancient sites were originally designed to protect the ancient buildings. The Temple of Agriculture, for example, was an important place of worship for the Ming and Qing Dynasties. After the demise of the Qing Dynasty it was nationalized by the National Government, turned into a park and reserved for commemorating revolutionary martyrs, and later abandoned due to various internal and external government issues, with the outer part of the first agricultural altar constantly being encroached upon. After the foundation of New China in 1949, the inner hall of the Temple of Agriculture was occupied by a school, with the main halls serving as student dormitories, conference rooms, libraries, and school-run factories. During the Cultural Revolution, factories and large numbers of residents moved in. A lot of the land in the inner and outer altars was divided and used by units belonging to various departments, with modern houses such as bungalows being built. Some of the original altars were destroyed, and while it was fortunate that the main group of buildings is still standing, its status remained at risk due to unprotected use throughout the year. In 1989, the Beijing Ancient Architecture Museum was established in the Temple of Agriculture, allowing for the beginning of scientific conservation and repair work. In the 10 years directly following the establishment of the museum, repair work on all the inner altar buildings was completed. After that, on the basis of overall completed repair, rehabilitation, maintenance, and repair work are carried out as and when they are required, according to the state of the building. Meanwhile, new technologies and means have been used for ancient building conservation. One example is the 3D laser scanning virtual construction project undertaken in cooperation with the Beijing University of Architecture in 2010. Advanced "3D laser scanning" technology was used to accurately map the shape of the building, with this "digitization" enabling building materials to be recorded and preserved more realistically and accurately, providing a basis for future conservation and repair.

We must also pay attention to the governance and maintenance of the surrounding environment in which ancient buildings are located, trying to avoid the development of new buildings within the historical courtyard. When construction is necessary, special attention must be paid to the unity of overall style. There are also usually old trees in the courtyards of ancient buildings, so it is necessary to protect and rejuvenate them while performing overall green work in order to provide a pleasing environment and attraction for museum visitors.

3. Improving awareness of public services and enhancing service levels

As an public education institution, the museum can intuitively disseminate varied scientific and cultural knowledge to its visitors. By presenting physical illustrations directly to the audience, the museum is more influential than other written or image-based materials, and visitors are more likely to gain a vivid and concrete impression. From perspective of tourism development, when people choose to visit a place that is unfamiliar, its historical and cultural characteristics will be the area they are most interested in, meaning that the museum is one of the best ways in which visitors can learn about this content, and therefore a must-see on the agenda. While the museum cannot position itself as a high-level "teacher", it must attempt to develop a closeness with its audience through services, allowing visitors to gain pleasure.

3.1 Hosting good exhibitions is the basis for all museum services

The exhibition is one of the direct carriers of the museum's cultural public services, one of the core cultural products of the museum, and the basis upon which the museum carries out social education functions. Therefore, creating boutique and popular exhibitions must be the focus and goal of every museum.

Exhibitions, and particularly the basic exhibit display, reflect the nature and mission of a museum, embodying its system and direction, and reflecting the level of basic functions such as collection, research, and display. This is the greatest difference between museums based in ancient sites and general ancient construction management offices or parks. A museum must run high-quality exhibitions based on its own characteristics and nature, and must conduct full research on collections and related discipline systems, attracting tourists for a worthwhile experience.

3.2 Improving service quality and supporting facilities

Exhibit explanation is one of the most important ways in which the museum serves visitors, allowing them to deepen their understanding of exhibits and displays, master key points, increase interest in their visit, and broaden their horizons. Lecturers and guides are the most direct contact between museum staff and visitors, and can also be said to represent the image of the museum. Therefore, the level of the explanation provided and professionalism of the lecturer/guide are important for the improvement of overall museum service levels and quality. Nowadays, modern explanation tools have been widely used in various museums to provide visitors with convenient explanation services. Conditions permitting, museums are employing more people to engage in explaining and communicating with visitors via two-way communication, offering a cordial atmosphere, raising the museum's reputation, and assisting in the understanding of audience perspectives.

With the growing awareness of museums regarding public services, facilities have been set up to provide convenient services, including barrier-free access, umbrella rentals, large luggage storage, and seats for resting. Beyond maintaining these facilities, we must work hard on the details in accordance with visitors' needs, such as the provision of mobile phone charging and reading glasses rental.

3.3 Developing souvenirs to meet visitors' cultural consumption needs

Due to a fondness for a certain element or memory, or as a gift for relatives and friends, visitors will most likely purchase some type of souvenir. Thinking from this perspective, a museum's cultural and creative products may act souvenirs. So, when developing these products, the museum must highlight its own unique characteristics, creating products that embody its exhibits and architectural features, which cannot be bought or even seen anywhere else. This will increase visitors' willingness to shop, and the products will serve as another form of publicity for the museum. The Beijing Ancient Architecture Museum has launched notebooks and key chains depicting colorful images of various ancient buildings, in a total of eight categories. There are 40 sets of rulers based on various window lattices, which are very popular among visitors.

With the development of the cultural and creative industry, more museums have begun to develop their own cultural and creative products. The creative elements are different yet the forms are relatively monotonous, such as books, pens and other stationery, key chains, fridge magnets, and postcards accounting for the vast majority. Objectively speaking, it is extremely difficult to innovate, so in the development of cultural creative products, digital technology can be introduced to inspire breakthroughs. For example, AR (augmented reality) technology can be used to create postcards in cooperation with corresponding mobile apps that allow items on the postcard to move on the paper when the angles are changed and flipped by the user – a novel concept that also inspires visitors' appreciation and understanding of the exhibits.

References

[1] Beijing Ancient Architecture Museum. *Beijing Ancient Architecture Museum Collection*. Beijing: China Legal Publishing House, 2012.

[2] Wang Hongjun. *Foundations of Chinese Museum Studies*. Shanghai: Shanghai Ancient Books Publishing House, 2001.

[3] Chen Xu, Li Xiaotao. *Research, Preservation and Repair of Beijing Xiannongtan*. Beijing: Tsingua University Press, 2009.

Thoughts on the Cooperation Between Industry-specific Museums and the Tourism Industry

Liu Jia
(General Administration Department of the China Railway Museum, Beijing 100055, China)

Abstract: In recent years, the development of museum tourism has been accelerating, yet many industry-specific museums have been neglected by the tourism industry due to various factors, affecting their survival and development. Therefore, this paper attempts to provide feasible strategies and suggestions for promoting the cooperation between industry-specific museums and tourism, so that they may enter and integrate into the increasingly prosperous and promising cultural tourism market and achieve their own sustainable development.

Keywords: museum; tourism; management; cooperation

In recent years, cultural tourism has developed rapidly, and has gradually become the focus of tourism. An important reason for this is that tourists' demands have changed. People are no longer satisfied with the visual impact of landscapes, but hope to gain deeper emotional enjoyment and spiritual satisfaction through their own cultural literacy and aesthetic approaches. As such, the museum has become an important element of cultural tourism, and the most active and promising sector of the tourism industry.

Museums have signed cooperation agreements with travel agencies to increase numbers of visitors. They are also cooperating with tourism agencies to create cultural tourism brands, design and develop cultural products, and provide diverse experiences and services, in order to achieve mutual benefits. On the whole, while this kind of cooperation is basically self-governing, without the establishment of a unified organization or channels with corresponding rules and regulations, it lacks effective supervision and restraint, causing elements of chaos and problems for the growth of museum tourism. It is important to note that museum tourism presents a significant imbalance when it comes to differing sizes, natures, and geographies of museums. Generally speaking, large-scale and well-known museums have an advantage in collections, displays, facilities, capital, management, and services, and are able to cooperate easily with the tourism industry for mutual benefit. Meanwhile, small-scale, low-profile industry-specific museums cannot compare to the big, comprehensive museums in terms of hardware and software, and are inevitably neglected by the tourism industry. The reasons for this situation are, of course, multifaceted: due to an over-emphasis on professionalism, the industry-specific museum's operation mentality have become relatively closed-off, lacking promotion, openness, and inter-museum communication, with unitary collection content and exhibition display that cannot be updated on time, limiting audiences. Furthermore, from the perspective of profit, the tourism industry instinctively prefers large-scale, comprehensive museums with high incomes. It does not tend to pay attention to the investment or development of industry-specific museums, and therefore does not encourage necessary guidance to tourists. Many visitors are eager to visit larger museums due to the influence of inherent consumer ideology or perception, disregarding industry-specific museums. Despite the policies of free opening for public benefit, the tourism industry has not seen the importance of industry-specific museums, nor has there been a fundamental change in audience consumption concepts. On the contrary, the imbalance between museums has intensified. Such situations and reasons require us to carefully examine, explore, and reflect, seeking answers to the important and urgent issue of how industry-specific museums can enter and integrate into the increasingly prosperous and promising cultural tourism market and realize their own development.

First of all, industry-specific museums must recognize the situational environment and learn to operate within it. In the past, there was a large gap between museums and tourism. The museum lacked a deep understanding of and identification with the tourism market, and habitually rejected tourism. Nowadays, museums and tourism are developing together, with booming cooperation between the two. More museums and tourism industries are becoming aware that cooperation is the general trend and inevitable path, and both parties are able to break through respective bottlenecks. Therefore, industry-specific museums must adhere to the development direction and path of cooperation with the tourism industry, understanding and grasping the dynamic and trends of cultural tourism, and participating in the cultural tourism market with an open attitude.

Although museums are non-profit public cultural institutions that do not seek profit as their goal, they are not guaranteed to be "insulated" from businesses, as they are both institutions that serve the public [1]. Under the premise of adherence to its basic nature, and putting social benefit first, the museum can achieve economic benefit with reasonable and compliant business activities, relying on its own economic benefits to achieve sustainable development, thereby creating greater social benefit and realizing a double harvest of social and economic gain. Industry-specific museums should not miss out on such opportunities and stagnate due to their own shortcomings or disadvantages. Instead, they must apply relevant laws and regulations in bold innovations of business concepts, models, and methods. At the same time, they should learn from the successful experiences and practices of Chinese and overseas museums regarding cultural tourism. Industry-specific museums must also maintain a clear and persistent understanding of their operations, avoiding adverse tendencies to deviate from or violate the social nature and basic functions of the museum, and not making the mistake of seeking nothing but profit.

Secondly, industry-specific museums must have clear positioning, and must project their unique features. While there are many differences between individual venues, amidst the universal commonality of the museum, each has its own unique personality. Unlike large-scale and comprehensive museums, industry-specific institutions have the distinct characteristics of their culture, and are more likely to relate directly with the public, offering a higher sense of affinity. Therefore, to carry out accurate and clear positioning in order to integrate and serve public cultural needs is a key factor in the sustainable development of industry-specific museums. This requires the industry-specific museum to strive to become a window for the display and popularization of characteristic cultures, and to develop social education in order to become an important place for leisure and cultural services. This must be based on a complete understanding of its own advantages, external environment, and audience needs, establishing various channels and platforms to close its distance with the public in order to become a cultural service institution. Some industry-specific museums must also consider their identities and responsibilities within their respective industries, and must shoulder the important mission of promoting industrial development. Therefore, the positioning of the museum must be considered comprehensively from a multi-dimensional perspective. It must be based on reality yet tailored, and have a certain forward-thinking nature without being too high and far-reaching, as this will be counterproductive. In addition, the industry-specific museum must make the most of its uniqueness and work hard to make flexible use of its cultural advantages, leveraging its cultural characteristics and attraction. This will trigger attention from the public and media, allowing the museum to find its direction and path into the cultural tourism market. In short, clear positioning and outstanding features are important guarantees by which industry-specific museums may find their position and role on the broad stage of cultural tourism, and achieve close cooperation with the tourism industry.

Thirdly, industry-specific museums must improve displays and focus on public participation. Due to a long-term lack of funds, primitive technology, and lack of professional staff, the exhibition methods of many industry-specific museums possess varying degrees of obsolescence or rigidity, as well as a lack of timely, necessary improvements or innovation in the context, form, and means of display. This means that they are unable to attract attention, spark interest, and make an impression. Therefore, for the industry-specific museum to align with cultural tourism, it must be people-oriented and culture-focused, using cultural collection resources, keeping up with the times, and promoting innovation, especially for exhibition concepts, forms, and means. Increasingly popular digital and multimedia technologies must be applied to exhibitions, with high integration and unification of artistic display and modern technology realized through modern sound and illumination, electronic publication, electronic screen display, and audio navigation, all in order to make displays more modern, popular, interesting, and enlightening, as well as to attract more visitors. Meanwhile, realistic exhibition improvement must avoid blind innovation seeking. It cannot simply rely on advanced technology to achieve goals, otherwise the museum falls into the trap of putting technology first, resulting in exhibitions being flashy but soulless. On the other hand, the industry-specific museum must pay attention to participation, interaction, and experience in improved displays, so that visitors can use own hands to operate and experience the exhibits during their visit, while gaining knowledge and entertainment value [2]. This will enhance the museum's recreational features to create a relaxing and pleasurable browsing environment that guides the visitor through an immersive experience.

Finally, industry-specific museums must increase publicity and expand visitor numbers. Today, with the rapid development and wide spread of information technology, media and channels of information exchange are abundant, convenient, and timely. Under these circumstances, there are more opportunities for the development of the museum, and also more challenges. This means that the museum must pay greater attention to strengthening publicity and promotions, otherwise it is likely to be cast aside by the increasingly strong waves of informatization.

The strengthening of social publicity is even more important for industry-specific museums. Through active and effective publicity, both government authorities and the community will be able to fully recognize the important position and role of the museum in the development of cultural tourism, so that the museum can obtain policy and financial support from the government and all sectors of society. This will help the industry-specific museum to establish an image, expand its influence, create an atmosphere, and raise awareness in order to attract visitors. Therefore, industry-specific museums should make all employees aware of the importance of publicity as a matter of urgency. They can use print materials, websites, email, blogs, WeChat, and other modern media, as well as going deep into communities, enterprises, and schools to hold exhibitions. In this way, they will attract visitors to attain audience diversity, so that publicity measures can have tangible effects.

In addition to the above suggestions, while industry-specific museums strive to achieve close cooperation with cultural tourism, they must focus on improving service levels and quality, understanding and considering varied audience needs in the quest for accurate service positioning. They must embrace the "people first, service-oriented" philosophy, respecting each visitor and providing warm and thoughtful service in order to promote an intimate relationship. Meanwhile, inter-museum exchange and cooperation must not be ignored. It is important to strengthen the links between museum societies and associations in order to obtain necessary support and assistance, and also to work collaboratively with other museums, including the sharing of collections, experts, and even visitors, achieving common development though cooperative models and mechanisms to form close strategic partnerships.

In short, the development of industry-specific museums is not only an expression of the social division of labor, but also a mirror that reflects an open and flourishing society. Only by possessing the belief, determination, and courage to innovate and keep up with the times can industry-specific museums seize opportunities, keep a foothold in reality, and implement feasible measures and methods in order to inject new vitality and create new situations for the museum tourism industry, while realizing healthy development in their cooperation of mutual benefit with the tourism industry.

References

[1] Zhang Min. 'Museums and Tourism', [J] *Chinese Museum*, 2004, 1: 25.

[2] Wang Hongjun (ed.). *Foundations of Chinese Museum Studies* [M]. Shanghai: Shanghai Ancient Books Publishing House, 2001, p. 287.

/ 3 /

The Creation of Smart Scenic Spots and the Development of Museum Information

Song Meili
(Department of Social Education, Beijing Xuannan Culture Museum Management Office, Beijing 100053, China)

Abstract: With the fast development of cultural undertakings at museums, and the wide application of multimedia technology, informatization has become a key task in museums. This refers to the application of information technology in museum in order to improve efficiency and raise the utilization rate of cultural resources, as well as to realize intelligent, scientific, and informative management, thus providing better cultural services in order to promote sustainable business development. In 2012, the Beijing Tourism Development Committee proposed the idea of building a "smart landscape", emphasizing the use of smart networks to realize smart services and smart management of scenic areas, having the same development concepts and service groups as the museum when it makes use of informatization to improve the quality of cultural services. This article focuses on how to use the development of smart scenic spots to promote the development of museum informatization, achieve the organic integration of the two, and serve large numbers of visitors more efficiently.

Keywords: smart scenic spot; informatized cultural service

With the rapid development of the times, informatization and digitization have penetrated various fields within society and public life. On January 12, 2015, the Central Office and the State Council issued the *Opinions on Accelerating the Creation of a Modern Cultural Public Service System*, proposing to accelerate the digital development of cultural public services. This was the first time that China had officially proposed the concept of digital development of cultural public services. It is thus clear that strengthening museum informatization meets the need for social development, and also satisfies the unlimited public pursuit of cultural resources. Tourism is closely related to cultural public services, and the newly promulgated *Museum Regulations* point out that encouraging strengthened cooperation between museums and tourism will enhance museum development, and that the integration of tourism and museums will be enhanced through policy orientation and the need to serve the public. In 2012, the Beijing Tourism Development Committee proposed the idea of using smart networks and other technologies to build a "smart landscape", promoting the development of the tourism industry while improving the quality of cultural services. This takes on the same development philosophy and common service groups as museum informatization.

1. The concept of smart scenic spots and museum informatization

Smart scenic spots* use smart networks to provide comprehensive, thorough, and timely perception of geographic elements, natural resources, tourist behavior, staff tracking, scenic area infrastructure and service facilities. It also involves the visual management of visitors and staff; the optimization of the re-engineered scenic business process and smart operation management; the formation of strategic alliances amongst upstream and downstream enterprises in the tourism industry to protect the authenticity and integrity of heritage resources and improve the quality of tourist services; and the achievement of comprehensive, coordinated, and sustainable development of scenic areas. It presents the characteristics of market networking, service functionalization, management informatization, and investment benefits.

With regard to the informatization of museums, the author believes that the means of informatization must be applied to all museum undertakings in order to improve efficiency and fully utilize cultural resources, as well as realize intelligent, scientific, and informatization of management, thus providing improved cultural services to visitors and promoting sustainable development.

Looking at the concept of the smart scenic spot, its development, means of application, and service objects are all basically consistent with museum informatization, both ideas coinciding in terms of content and work. If we create smart scenic spots, promote museum informatization, and encourage the organic combination of the two, visitors will benefit.

* The Beijing Tourism Development Committee formulated a *Code for the Construction of Smart Scenic Spots in Beijing* in 2012.

2. *Using the development of smart scenic spots to develop museum informatization*

2.1 Improving network infrastructure to create the basic conditions for museum informatization

Creating network infrastructure is an important step in the development of smart scenic spots, and is also a necessary condition for museum informatization, as the exchange of museum data materials relies on a basic network. A resource sharing platform can be built on the basic network for the purpose of resource sharing amongst various museum departments, promoting overall museum operations. Database technology may also be introduced to effectively manage museum collection information and improve information management. Meanwhile, mobile terminal products that are rich in content and form can be developed to explain museum cultural resources and make visits more lively and interesting.

At present, the connection mode of the basic network mainly involves wired network and wireless network forms. LAN technology based on the wired mode has been very popular, but in recent years, with the development of network technology, more wireless network applications and related products have arisen. For example, the mobile phones we use frequently in life have basic WiFi functions, and user demand for wireless networks is rising. The wireless network replaces part of the transmission medium of wired LAN with radio waves, supplementing and extending the wired network. Not limited by the location of cables and ports, one can connect to the network using wireless Internet access devices in a given space, saving the costs of wiring between terminals and switching devices, allowing greater flexibility, convenience, and efficiency for the creation of museum networks and information exchange. Through wireless network technology, museum staff can remove the restrictions of network cables, accessing the Internet for work anytime, anywhere. Visitors can also enjoy the convenience brought by museum informatization any time in the exhibition hall.

In 2011, the Beijing Tourism Commission launched a wireless network coverage project for the city's A-list scenic spots, providing financial and technical support for the development of hardware facilities. The Beijing Xuannan Cultural Museum (the AAA National Tourism Scenic Spot where the author is located) has participated in the project, adding wireless network equipment to each exhibition hall so that visitors can use mobile phones, computers, and other terminals to connect to the Internet wirelessly during museum tours, while also providing the basic conditions for further informatization.

2.2 Establishing a smart security system to enhance safe museum operations

The museum is a collector of precious cultural artefacts as well as a common space open to the public. Safeguarding the people, property, and objects within the museum is therefore an important foundational task. With the progress of the times, relying solely on “human defense” can no longer meet developmental and operational needs. Modern information technology, electronic technology, and network linkages are gradually being applied to museum security, effectively reducing manpower and raising safety precaution factors. Smart security systems are an important part of the development of smart scenic spots, including video surveillance, traffic monitoring, electronic access control, patrol systems, and emergency management. Video surveillance is a basic component with high popularity, offering 24-hour monitoring, with a control panel that can be used to zoom in and rotate cameras in order to reduce and avoid security risks. The recording and data saving functions of video surveillance systems allow for image data to be saved within a certain period of time, providing a reliable basis for the handling and resolution of unexpected safety issues. Flow monitoring systems produce real-time statistics regarding the number of visitors, and also statistically monitor the heat distribution of tourist attractions and the length of time visitors spend there, setting up automatic alarms for flow overruns so as to facilitate timely guiding of crowds and avoid accidents. Patrol systems are flexible applications of the access control system, facilitating the supervision and management of control personnel, and comparing pre-set patrol plans with actual patrol records to track missed inspections, delays. This makes personnel management more accurate. Based on the museum’s carrying capacity and specific needs, other security systems may also be developed, such as anti-intrusion alarm systems, emergency smart broadcasts, and fire protection systems. The focus of smart security systems is not only the security functions of every system, but the scientific and effective linkage mechanisms between various systems. For example, when a smoke alarm system detects a fire, it begins an emergency smart broadcast and guides evacuation, with automatic linkages to an electronic access control system to open access control, a linked fire protection system that engages a fire extinguishing spray, and a linked video monitoring system to monitor and record the accident area. A series of technical means are used to achieve effective prevention, timely processing, and scientific linkages for sudden situations in order to minimize losses to the highest degree.

2.3 Using an electronic tourist service system to broaden the service model for museum informatization

Providing diversified services to meet visitors’ cultural needs is a long-term value goal of the museum, with the electronic tourist service system mentioned as part of smart scenic spots, including the establishment of an electronic ticketing system, a self-guided tour system, interactive experiences, and multimedia presentations.

The electronic ticketing system works through computer networks, information encryption, and recognition technology. An electronic ticketing system is able to realize all-round, real-time monitoring and management functions, such as computerized ticket sales, automatic ticket checking, ticketing data statistics, and basic analysis of visitor information. For visitors, the electronic ticketing system provides more ways to purchase tickets as well as convenient security clearance, saving time. For museums, it can help to avoid the shortcomings of manual ticket checking such as slow statistical analysis, frequent loopholes, high error rates, and high labor intensity. Daily tasks such as ticket sales, ticket inspection, inquiry, summary, statistics, reports, and marketing are managed via a highly information-based, smart system platform, which efficiently and accurately provides detailed data for management and decision-making, thus improving efficiency while saving manpower. Through the functions of real-time statistics and analysis of the smart ticketing system, instantaneous visit counts can be grasped in a timely manner, and split-flow can be carried out during peak periods, thereby reducing reception pressure and raising visitor satisfaction.

The scenic self-guided tour system is a modern tour system based on technologies such as wireless communication, global positioning, the mobile Internet, and the Internet of Things. Visitors are able to realize self-guided tours via mobile phones or navigation terminals, as well as set tour progression according to factors such as personal preferences and time requirements in order to meet individual needs. With continual technological updates, the self-guided tour system provides multi-lingual audio navigation, and also combines content with sound effects, animation, and pictures to form a 3D narration that meets visitor needs with different languages and levels. Through radio frequency identification technology (RFID) and geographic positioning systems (GPS), the navigation terminal can automatically identify the labels of nearby cultural artefacts, prompting and playing related content to make visits more interesting.

In 2013, the Beijing Tourism Commission launched the development of a self-guided tour software system for A-list scenic spots, with 110 scenic spots participating in the development project, and The Beijing Xuannan Cultural Museum was one of them. The A-list scenic self-guided tour software system exists in Android and Apple (iOS) versions. The viewer uses a mobile phone with a GPS location function to download and install the self-guided tour software, which offers complete self-help services within the scenic spot, including navigation and tour services, self-guided tour guides, public service facility location query, self-location services, and picture browsing. The project has been officially launched via software system development and expert appraisal, and is free for visitors to download and use on the Beijing Tourism Network. Research and development into the system provides the conditions upon which the museum can improve the self-guided tour system. On this basis it can increase functions such as in-depth multimedia tour guides, interactive visitor comments, and electronic ticket sales as per needs, providing further convenience for self-service tours.

With the development of the cultural service concept, both the museum and tourism industries are paying more attention to the dominant status of visitors, advocating participatory, inspirational, and interesting tour modes. Interactive experience and multimedia display are a product of this, with applications including technologies such as virtual reality (VR), augmented reality (AR), dynamic 3D imaging, and touch-screens. In museums, interactive experiences and multimedia display technologies are used widely, with some qualified museums also offering special visitor experience areas or interactive experience areas for young people. Using the appropriate application of information technology and multimedia technology to design dynamic displays and interactive experiences will lead to breakthroughs in the museum's education model, from single, static displays to visual, auditory, tactile, and other sensory experiences, presenting cultural artefacts and information to the audience in more comprehensive, in-depth, and vivid exhibitions. It will also enhance the effects of displays while increasing interactive participation, guiding visitors from passive acceptance to active participation, discovery, and contemplation, so as to develop the museum's educational function. The appearance of electronic service systems has brought huge changes to museum services. The application of modern technical means has satisfied the cultural needs of a variety of groups, broadening the forms of a museum's cultural services, making them smarter and more usable.

2.4 Building a publishing platform for tourism information

There are many ways to create an information publishing platform. Smart scenic spots mainly use Internet-based portals that provide information services, as well as mobile information dissemination channels composed of modern mobile communication technology, the Internet, and the mobile Internet.

Portals are the most basic platforms for publishing tourism information. Smart scenic spots require portal sites that have tourism services as the core content and operate normally online. Through the portal site, visitors can: browse information; make queries; use panoramic 3D live action hybrid reality to create virtual tours online; carry out travel route recommendations and trip planning; read introductions to scenic services; obtain traffic navigation; download service; offer opinions and feedback; and access electronic business marketing functions. The appropriate use of these portal resources complements the physical museum. For example, with virtual reality (VR), people can visit the museum via network simulation, using ultra-high precision scanning technology to zoom in and appreciate cultural artefacts, offering more sensory experiences and allowing visitors to access cultural knowledge without restrictions of space and time. In the design of the portal, it is also possible to increase the academic forum section, strengthening communication with visitors and promoting a "secondary fermentation" of information in the dissemination process.

Mobile information dissemination is a new method of disseminating information that has developed rapidly in recent years, using modern mobile communication technology, the Internet, and mobile Internet technology to realize information transmission and business promotion via handheld terminals such as mobile phones and PDAs. At present, China has a large mobile phone user base utilizing widely popularized mobile apps, Weibo, and WeChat. Information dissemination is no longer peer-to-peer, point-to-point communication, but more of a fission-type spread. The promotion of mobile information dissemination is the pursuit of larger museum audiences, and an inevitable trend in the development of museum informatization. From a technical perspective, the mobile information publicity platform is able to realize almost all the functions provided by the portal, but at the same time, it can develop more interactive experience forms. Mobile app projects, such as "Forbidden City in the Palm" launched by the Palace Museum, is a good exploration and attempt to use mobile information dissemination to interactively communicate between museums and visitors, realizing information push, GPS automatic tour guides, voice explanations, and satisfaction surveys while speeding up the dissemination of information and opening up new channels.

3. *The relationship between the development of smart scenic spots and museum informatization*

3.1 Consistent goals and mutual promotion

Museums and tourism have a common goal to meet the cultural needs of the public. The museum provides essential resource conditions for the development of tourism, while tourists are among the museum's main service targets. The development of smart scenic spots is conducive to the sustainable development of tourism, and also plays an encouraging role in boosting informatization in museums. The rich cultural resources of the museum and research advantages of professional domains provide the necessary resource conditions for the development of smart scenic spots. At the same time, the development direction of museum informatization, multimedia technology, and smart security are widely used in museums, providing valuable practical experience for the development of smart scenic spots.

For museums, participation in the development of smart scenic spots accumulates relevant experience and puts forth practical suggestions for informatization, as well enjoying more support from policies, funds, and technology within the tourism industry, guaranteeing the implementation and operation of projects. Therefore, finding the right combination of work, logical arrangements, and organic integration furthers the win-win development of both the tourism and museum industries, and also benefits the public.

3.2 Further considerations brought about by the creation of smart scenic spots

There are coincidences and cross-overs between smart scenic spots and museum informatization, but the two are not exactly the same. Collection, research, and education are the three basic functions of the museum, and informatization must be deepened and expanded around them. The development of smart scenic spots can provide references for museum education, services, and operation, but there is little focus on key business areas involved in museum collection and research. This is an area we must consider in museum informatization, such as the use of databases to carry out informatization management of collections, the use of environmental temperature and humidity monitoring systems to enhance protection, and establishing a database of scientific research to protect and utilize cultural research findings. In addition, some museums in ancient buildings must also focus on the use of modern technology to monitor and protect the building structures themselves.

Due to the application of various programs, technologies, and software devices in the creation of smart scenic spots and museum informatization, a large amount of data is generated. Through the strengthening of system management and unified application of Big Data, the formation of a complete informatization plan to avoid duplicate development and data inconsistency is an area to which museums must pay special attention. Especially when it comes to small and medium-sized museums, the overall planning of informatization must be done well, with sufficient understanding and positioning of their own situations beforehand, and never blindly following or copying.

Museum informatization should be based on characteristics of a museum's own cultural resources, and must not remain on the surface of technology. At present, the technical means of museum informatization is rich, with many excellent professional teams both in China and abroad engaging in research, development, and application in the field, providing technical support. In the face of ever-changing technical means, choosing appropriate display methods to give prominence to the museum's own characteristics is a question worthy of close consideration in developing informatization. This requires us to pay more attention to resources themselves, and undertake research and data collation in a down-to-earth manner. We must find the appropriate combination of sites and corresponding technical forms based on deep mining of collection resources. This way, the means of digital technology can serve the content itself, helping visitors to interpret its cultural implications. This the chief significance of informatization and digitization in museums.

3.3 Possible issues with various management systems

Museums and tourism belong to two different industries. Museums are public welfare undertakings, while tourism is the pillar industry among the national "tertiary industries", and a business-oriented enterprise. Therefore, in real life, there can be situations in which social and economic benefits contradict one another. As a developing museum, we must keep in mind the purposes of service to the public and social development, putting social benefits first in our development, and encouraging the appropriate development of cultural and creative industries, enhancing the museum's capacity for self-propagation. Under the premise of empowering social benefits, we must improve economic benefits, thus balancing the relationship between the two industries.

In addition, since museums and tourism belong to two different industries, there can be differences in work methods and processes across different fields. It is necessary that we find appropriate cooperative models. At present, the development of smart scenic spots is in its initial stage, with some projects having yet to be tested or requiring further improvements. This means that we must forge reasonable expectations for the future, and prepare to face and solve any issues that may arise.

*Research on Design Models for Digital Museum Display**

Yang Kun
(Beijing Union University, Beijing 100101, China)

Abstract: With the development of digital technology and its expanded application in the domains of cultural artefact display and research in museums, many new models have appeared in digital display design. This paper will classify and discuss these models in detail, proposing a new direction for the future.

Keywords: museum; display method; display technology

Recent years have seen the development of digital technology and the expansion of virtual reality technology, digital interactive technology, and multimedia within cultural artefact exhibitions and research domains within museums. As a result, the modern display concept has broken through the walls of the traditional museum, and showcases, panels, models, and text are no longer the only means of display. Digital display models that emphasize human-orientation are the new direction for the development of exhibitions.

In the museum, digital display utilizes interactive art to carry out information stimulation of the visual, auditory, and tactile senses, so that visitors can be immersed in the environment while at the same time autonomously choosing navigation routes and theme according to personal preferences. Using interactive methods while realizing the educational function of exhibits, this method can enhance the appeal of the exhibit and the awareness of a visitor in active exploration.

* Fund Project: A project supported by the National Science and Technology – "Exploration and Application of Smart Exploration and Display Technology for Improved Cultural Tourism Value for Museums" (2014BAK08B02)

Digital museum exhibitions have multi-faceted characteristics in terms of content interactivity, design dynamics, and technical multi-dimensionality, causing differences to appear in digital design models.

There are four main types of digital design model, divided into physical museum exhibitions (two types) and virtual digital exhibitions (two types).

1. *Digital exhibition design for physical museums*

From the perspective of museum development, the safekeeping, protection, display, and education and entertainment value of cultural artefacts are the main functions of the museum. Due to the limitations of cultural artefacts themselves and the exhibition space of the museum, the display function of exhibits cannot be fully exposed to visitors. Even if exhibits are on show, it is difficult to display relevant information attached to the exhibits and the environment in which cultural artefacts are located, so the display effect does not meet expected requirements. Digital exhibition display in physical museums is a positive supplement to traditional forms of display.*

Digital exhibitions in physical museums largely combine traditional exhibition methods with modern digital technology, enhancing the concrete means of display in the physical museum while solving the issue of insufficient exhibit information that arises in the traditional model. In this way, information generated as a result of the integration of physical exhibits and digital, virtual exhibits is the information obtained by visitors within the physical museum, making its transmission more effective, thus playing an important role in information enhancement.

Digital exhibitions in the physical museum now consist of the "Augmented Display Space" and "Virtual Display Space" exhibition models.

1.1 "Augmented Display Space"

Due to a lack of flexibility in the museum's physical display, and a lack of emotion in its virtual display, the Augmented Display Space came into being. It integrates virtual display into the museum's physical display space, using digital technology to: enhance visitors' experience of physical museum exhibitions; expand the display space of the museum; enhance audience interest in exhibits; and provide comprehensive exhibit information, making up for the shortcomings of traditional display methods while providing a new way to develop physical display methods [1].

The museum's Augmented Display Space integrates virtual information space within the physical museum, giving the display space interactivity and navigability. In the real exhibition space of the museum, once visitors are close to exhibits, they enter a virtual

* Huang Qiuye, Ye Ping. 'New conceptions of digital museum exhibition design', [J] *Art Theory*, 2008, 12.

information space corresponding to the exhibits via smart digital positioning systems that combine physical and virtual information space. Through smart digital positioning systems, visitors can select self-navigated information through the system's virtual information space, adjusting the tour route according to their needs, thus highlighting personalized display design. Visitors are able to compare information in the real space of the museum with that in virtual space, enhancing their interest in exhibits and providing a good platform for the realization of museum functionality. The digital smart navigation system that Augmented Display Space relies upon can provide visitors with personalized display and interactive panels, meanwhile helping them to find places of interest in real space and optimizing routes accordingly.

In the Augmented Display Space of the museum, the way in which visitors obtain information, and how much information they obtain, depends on the determination of their route. Since the virtual information space was added to museum display, the visit route has been doubled in order to provide information. Within Augmented Display Space, the mobile digital smart navigation system presents a navigation path according to visitor interests and a real space planning scheme, presenting a tour route in both physical and virtual space. The smart digital navigation system of the physical museum's Augmented Display Space shows the trajectory of visitors within the physical space at any time and in any place, providing information that visitors require and that is of interest. If they do not submit explicit requests, the system provides default information. Visitors are free to roam within the space. The system will automatically calculate exhibit information from information storage according to a visitor's location, guiding them to attend to and remain in locations of interest as the system automatically plays multimedia descriptions. If visitors want to go to a specific exhibit area, the system will set up an appropriate route and display corresponding to guide them there. On the way, the system will not provide any other information along the way, to ensure that visitors arrive at designated exhibition areas in a timely and efficient manner. Visitors who wish to enhance their understanding

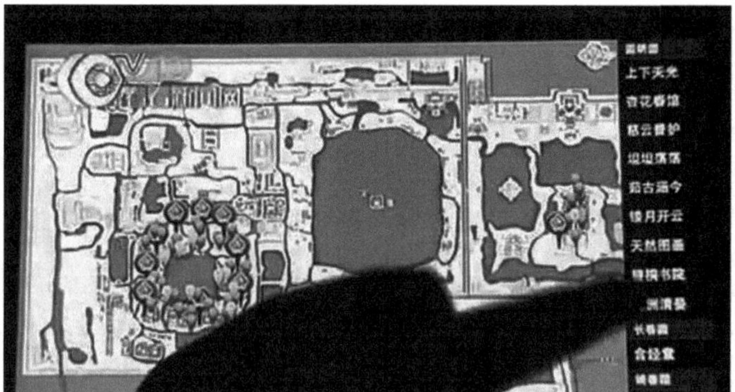

Figure 1 Yuanmingyuan mobile navigation interface

of exhibits are invited to interact with the system and obtain more information by clicking the appropriate virtual system button (see Figure 1).

1.2 "Virtual Display Space"

The physical space of the museum was originally used only as a space for displaying entities, but with the development of digital technology, exhibit information also has special display space within the physical venue, forming an exhibition model in which physical exhibit information is separated from virtual information, dubbed "Virtual Display Space".

At present, Virtual Display Space is represented in the interactive digital exhibit design within the physical display space of the physical museum, and interactive display means have been applied more within the physical museum. Visitors can now interact with live scenes and exhibits within physical display space. With the development of digital technology and the expanded use of digital technology within the fields of artefact display and research, the humanization of interactive digital exhibitions stands to gain prominence. This actual experience environment will make up for the monotony of physical display space, enriching the content of the physical museum and attracting more visitors. It offers a traversing of personal experience, stimulating personal interest, enhancing visitors' understanding of exhibits, and fully empowering the purposes of the physical museum.

Virtual Display Space relies on modern digital display devices to create physical display space via digital technologies, such as virtual reality, digital interactive technology, and multimedia (see Figure 2). All aspects of information regarding the exhibits are created by humans, through coordination between various interactive display devices and connections between display content, forming a full range of interpretations. Meanwhile, some interactive devices are also provided for visitors to use if they wish. After browsing exhibits, visitors can learn more about them through inquiry equipment, using digital application software to re-arrange the exhibition according to the knowledge they have gained, which may then be uploaded onto the museum's website to form a personalized digital museum.

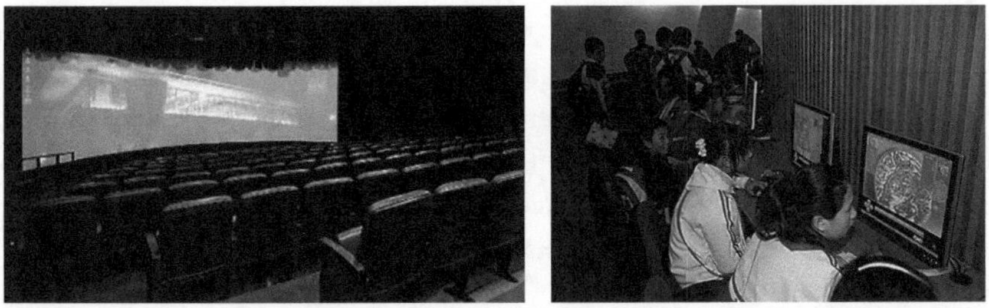

Figure 2 The Capital Museum's digital auditorium and interactive multimedia zone

2. Virtual Digital Display

The digital exhibition breaks through traditional exhibition modes, realizing the combination of physical display and digital technology, and promoting the research and display of museum collections. However, the limitations of physical space, language, and time on the digital exhibition are not fully reconciled, giving virtual digital display an opportunity to develop.

Virtual digital display uses digital technology to collect and compress digital information, forming a virtual exhibition space via networks and computer technology. It allows visitors to view the digital exhibits within the virtual information space, separate from the real exhibits. The use of digital display technology to achieve digital exhibit research, display, and information dissemination provides visitors with more personalized services. Virtual display mainly involves aspects of entertainment, interactivity, extensibility, and accuracy. The key is interactivity, as visitors gain a high degree of autonomy within the virtual museum.

Virtual display design is the integrated design of multiple technologies, effectively mobilizing the senses, and thereby enhancing interest and offering strong appeal and sensory expression. Research into digital display design of the virtual museum is mainly based on the crucial role of the visitor in the exhibition, namely the way in which information is accepted, and the visitor experience.

Virtual digital display is divided into virtual information digital display and physical information digital display designs.

2.1 Virtual information digital display design

At present, the digital display of virtual information takes the form of the online museum, in which a collection management system is established via digital technology. Museum visits, browsing, information inquiry, and retrieval are provided through network site forms, so visitors accept information and services via the museum network. This is a passive mode of virtual digital display, and although the museum is able to provide some exhibit information, there is a lack of information exchange and feedback between visitors and exhibits. With the rapid development of digital technology, virtual museums have added humanized exhibition models based on passive digital display, incorporating the concept of personalized visits, so visitors have a certain degree of independence within the virtual museum, according to their own needs. This means that each visitor to the same virtual museum obtains different display information, highlighting individual characteristics even more strongly. With the continuing expansion of digital technology, the virtual museum proposes a multi-faceted display system framework for digital display design in terms of needs like personalization and expanded transmission of exhibit information.

The virtual museum display system framework is divided into three parts, as shown in Figure 3. From top to bottom, the top is the user end, referring to the personnel who use the system, connected via the Internet and server. It is an interactive user interface with personnel contact, providing options like querying, editing, browsing, viewing, and entertainment. The middle is the content management server, including planning, membership, scripts, multimedia objects, directories, and queries. All functional and operational user needs are here. The bottom is the database, where the data generated or used in categorical operations is concentrated. Due to differing storage methods, the database is divided into an information database and an archive database, with the former storing data corresponding to different categories, and the latter storing complete or linked content in the form of archives and files. All of this content is mutually integrated and reorganized with website architecture in order to present complete and diverse content that meets visitors' requirements [2].

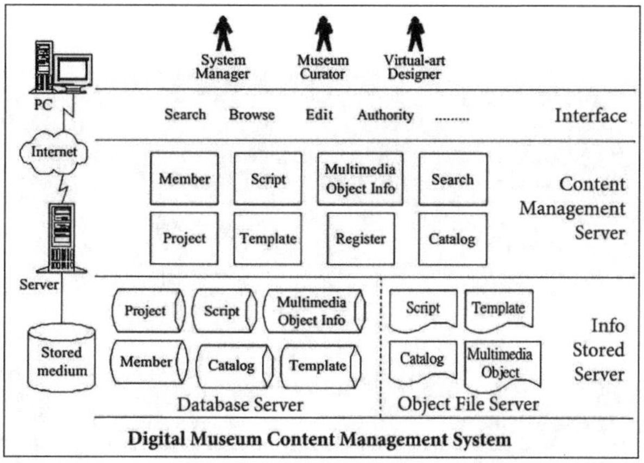

Figure 3 Virtual museum display system framework [2]

2.2 Virtual display design of physical information digitization

The digital virtual display of physical information refers to digitization of physical museum and exhibits, forming a virtual museum environment similar to that of the physical museum, which visitors can visit anytime and from anywhere, making visits faster and more convenient. This form of virtual display brings a sense of immersion through digital technology such as virtual reality, allowing people to integrate into the virtual scene, enhancing display effects and allowing interaction with exhibits in the scene at any time in order to realize the free interaction of man and machine. It arouses audience interest, mobilizes sensory excitement, and maximizes audience enthusiasm in participation, attracting visitors to exhibitions, and thus better reflecting the value of cultural artefacts. This is of great significance for the development, management, research, and maintenance of cultural artefacts.

Digital virtual display of physical information provides visitors with a space for exhibit display. In the virtual museum visitors can interact with the virtual environment, selecting exhibits they wish to visit, formulating appropriate tour routes, participating in display activities, and even communicating with other visitors or playing games within the virtual system to achieve virtual roaming.

At present, the types of virtual technology for the display of digital information are divided into virtual reality technology and laser holographic imaging technology. Virtual reality technology is divided into desktop, immersive, distributed, active, and passive modes, based on graphics and graphic schools. By cooperating and integrating with one another, these digital technologies are able to move the physical museum into virtual space, enhancing display effects, and realizing humanized multi-dimensional information display. During the interaction between humans and computers, interaction models are realized in which the senses of vision, hearing, and touch are fused into one body. (See Figures 4 and 5).

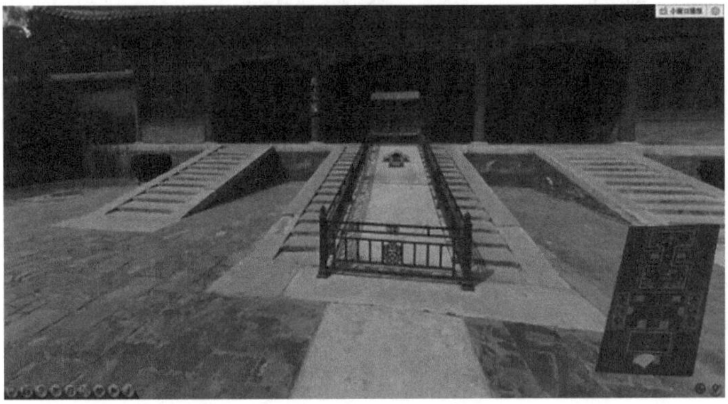

Figure 4 360° panoramic virtual roaming of the Palace Museum

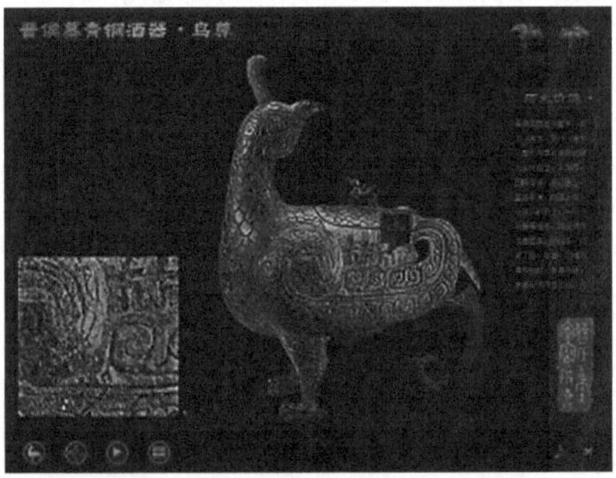

Figure 5 A bronze vessel from the Jinhou Tomb – virtual display of a bird statue

With the development and integration of digital technology and network technology, digital museum display will develop in the direction of combined physical and virtual museums, namely an online and offline digital display model. This model requires the perfection of physical museum hardware and software, and rich virtual museum content, with physical and virtual museums seamlessly connected and coordinated. This creates a social relationship between the visitors and the museum in which visitors can access information and services at all times, freely transitioning between virtual and real space, with improved exhibit display effects, enhanced attraction, and attendance to their needs, inspiring them to explore the museum on their own initiative while fully developing the basic service function of the museum.

In the fast-developing information age, digital museum display is inseparable from digital technology, and the use of digital technology for information dissemination has become the mainstream in museum digital display design. In digital display design, digital technology is an important technical means and support for development, and the rapid development of digital technology is sure to revolutionize the business of museum display.

References

[1] Huang Qiuye. *Research on digital exhibition and display technology in museums* [D]. Masters thesis, Jiangnan University, July 2008.

[2] Chen Baixun, Hong Zhengxin. *Models, concepts and systems for digital museum virtual display.* Taiwan New Century Digital Library and Digital Museum Seminar.

[3] Chen Gang. 'Analysis of basic digital display characteristics in museums', [J] *Southeast Culture*, 2009.

[4] Li Min. *Design and implementation of Jinhou Tomb bronzeware display based on mobile technology* [D]. Master's Thesis, Beijing University of Technology, June 2014.

The Digitization of Cultural Heritage – Research and Exploration of Digital Technology in the Han and Tang Emperors' Tombs

Xie Haoyi
(China Stamp Museum Exhibition Department, Beijing 100005, China)

Abstract: For a long time, culture has gone hand in hand with science and technology, with every major advancement in the latter bringing about revolutionary changes to the development style, mode of communication, and expression of the former. The integration of culture with science and technology profoundly affects the progress of society as a whole. In the same way, the methods of protecting cultural heritage are also developing with changing times. The advancement of science and technology, as well as continuous improvements and development of information technology, have offered digitization as a new and important way to protect cultural heritage.

Since the 1990s, developed Western countries have been converting their own cultural heritage into digital forms on a large scale, at once better preserving culture and promoting the continued inheritance, advancement, and development of the country and its own culture. Exploration of the digitization of China's cultural heritage began in the late 1990s. However, there are few examples of digital cultural heritage projects with complete architectures that use key technologies and clear processes.

This paper hopes to assess the space, scope, and method of applications and specific related digital media technologies for the protection of cultural heritage, using the digital research project of the Han and Tang Dynasty Emperor Tombs as an example – a key research project of the Shaanxi Provincial Education Department.

Keywords: cultural heritage; Han and Tang emperors' tombs; digital technology, virtual reality

Preface

Cultural heritage is a precious and non-renewable resource. With the progress of economic globalization and the acceleration of modernization processes, China's cultural ecology is undergoing tremendous changes, and cultural heritage and its living environment are seriously threatened. Many famous historical and cultural cities, streets, villages, ancient buildings, and ancient sites have been destroyed, and countless pieces of important cultural heritage have disappeared or been lost, while national and regional cultural characteristics are reducing quickly. A lot of cultural heritage carrying historical information and value is rapidly disappearing, never to return. Strengthening the protection of cultural heritage is thus an urgent task.

In the early 1990s, UNESCO launched the "Memory of the World" project to promote the digitization of worldwide cultural heritage. Meanwhile, developed countries such as European nations, the United States, and Japan have carried out digitization of cultural heritage. In the United States, digitization in social sciences, life sciences, and archeological discoveries has comprehensive and higher-level applications, which have digitized the nation's cultural heritage while developing other ancient heritages, such as reproducing the Roman Colosseum and a virtual Pompeii. Japan's digital virtualization of social and life sciences has developed more rapidly, and is widely used. Japan is committed to research into network and digital technology, such as color treatment in the restoration of lost cultural sites such as *Qi Bu County's Achievements and Progress in the Establishment of Digital Archives*. It is also cooperating with cultural institutions in other nations to seek effective preservation of world heritage via digital network technology. In China, the 1997 National 863 International Cooperation Project, "Research on the Auxiliary Reproduction and Protection of Ancient Murals" formed a standardized auxiliary method based on digital technology for the protection research and reproduction of murals through actual conservation and reproductions of the artwork in Dunhuang's Mogao Grottoes. In 1998, the proposal and implementation of the "Digital Palace" concept took another important step in the digitization process.

1. The digitization of cultural heritage

1.1 An inevitable direction in the protection of cultural heritage

Digital technology surpasses the original methods of cultural heritage conservation, assisting in all aspects of protection including the exploration, protection, research, and display of cultural heritage. It uses digital means to record cultural heritage information more comprehensively, and permanently records the current state of cultural heritage, providing support and new means of research and display.

At present, China sees the importance of the use of digital technology and information technology in strengthening the conservation and development of cultural heritage. "The digitization of cultural heritage refers to the use of digital technologies such as digital acquisition, storage, processing, display, and communication, to transform, reproduce, and restore cultural heritage into shareable and renewable digital forms, interpreting information from new perspectives, saving it in a new way, and using it to meet new needs" [1].

1.2 The need to analyze case studies of cultural heritage digitization

In looking for a case study to inspire China's digital cultural heritage project, most information available is limited to general, overarching summaries of the project, or simply discussions of only a certain technology, while overall conceptions, preliminary preparation processes, and the application of related technologies are rarely explained in a systematic fashion.

Cultural heritage itself is composed of a large number of objects, sites, cultural artefacts, literature, manuscripts, pictures, and books. Using digital technology to restore these complex, tangible or intangible materials to historical scenes or facts requires coherent thinking and reasonable and certified artistic creation that is holistic and comprehensive in its conception. The overall conception determines each part within. Therefore, to perform the overall planning and structured management well is of great significance to every segment in the implementation of labor division that will promote the progress of the project.

2. Digitization of the Han and Tang Dynasty tombs

The number of imperial tombs in Shaanxi is unparalleled, including nine Western Han Dynasty tombs and 18 from the Tang Dynasty. These imperial tombs have profound cultural implications. They are a source of Chinese pride, and major tourist attractions for global visitors.

2.1 Research objectives

1) Digital combination and integration of materials related to these emperors' tombs.
2) Digital repair of damaged cultural artefacts such as buildings and carvings.
3) Use of digital technology to carry out spatial digital restoration of imperial mausoleum sites that have cultural and artistic value.
4) Combining generated virtual space scenes with actual shots and animation footage to create a documentary that presents historical facts.

2.2 *Elements of the project*

The project was carried out in accordance with a process of expert research — discussion and planning — comprehensive investigation — collection classification — sorting and classifying — multi-faceted research — hierarchical production — final integration — visual display, involving field visits to the mausoleum sites of nine Western Han Dynasty emperors and 18 Tang Dynasty emperors, studying the site and surrounding life production in depth. A large number of research materials were prepared, with various levels of research conducted, offering a history of the tomb combined with expert opinions to set out a proposal and research focus suited to the project requirements.

2.2.1 *Expert research and debate*

Zhang Jianlin and several other archaeological experts visited the tombs several times with the project team in order to study and investigate the status of the mausoleums, focusing in recent years on archaeological sites at the Qianling, Chongling, and Jianling mausoleums, then discussing plans and objectives.

2.2.2 *Establishing a complete and effective work team*

The project was divided among an expert group, a data collection group, a data classification and compilation group, a two-dimensional image data integration processing group, a virtual reality space group, an animation group, and a post-production and editing group - a total of seven groups.

2.2.3 *Data collection*

The project participants were divided into six batches according to divisions of labor, carrying out an inch-by-inch inspection of the mausoleums of the Han and Tang Dynasties. A detailed survey of culture and production in the villages surrounding the mausoleum was carried out, with nearly 2,000 GB worth of photographs taken (more than 10,000 images). The film and television group was divided into four batches, focusing on site remains in the Qianling, Chongling, Qiaoling, and Jianling mausoleums, providing basic materials for documentary production and animation simulation, shooting 500 GB, or 800 minutes, of material. The animation group was divided into four batches to survey the Qianling, Chongling, and Qiaoling mausoleums, and a large number of sketches were completed.

2.2.4 *Field research and imaging practices*

Geography and images were the two major components. The geography component was based on history and humanities, while the image part related to videos created by students based on field research.

2.2.5 Research into the design, architecture, and carvings of the mausoleums

This research considered the general outline, shape, layout, geomantic omens, and topographic features of the nine Han Dynasty and 18 Tang Dynasty tombs, categorically comparing and classifying the distribution structure of "mountain tombs" and "underground tombs" to draw up the investigation routes. Meanwhile, each tomb was drawn separately, with existing remains presented in accordance with current layout, selecting the mausoleum systems of the early Tang, middle Tang, and late Tang Dynasty in Qianling, Chongling, and Jianling on which to expand animated virtual display, with green mountains and blue waters used as the basic style of animation in restoring the buildings on the mausoleum grounds.

1) Architectural style

Combing historical materials to understand the architectural features and styles of Han and Tang Dynasty imperial mausoleums, exploring and surveying the ruins, looking for architectural forms that correspond to historical materials.

Characteristics of Han Emperor tombs: The Western Han Dynasty followed the characteristics of Qin mausoleums, with the grave built with an earth mound as a tomb, shaped like a pyramid with the upper part cut off, called a Fangshang (cross spread); next to the tomb are sleeping chambers, streets going east and west, and throne rooms and rest halls; smaller scale than the Qin, with relatively low soil mounds, the outside of the cemetery is surrounded by ancestral temples and a mausoleum city. Major changes in the mausoleum system in the Eastern Han Dynasty involved the relocation of the ancestral temple from outside the cemetery to inside, and adding a memorial hall with a large bell hanging next to it [2].

There is little known data regarding Tang Dynasty mausoleum systems, so research depends on archaeological excavation and reading between the lines of related literature. Due to the special relationship between mausoleum architecture and palace architecture in the Tang Dynasty, palace architecture can be used as a reference, often displaying its appearance in accordance with archaeological excavations.

2) Stone carvings

The data and images collected from the field were summarized in high-quality pictures and text, with location recovery, display of characteristics, and drawings carried out on carvings on Han and Tang Dynasty imperial tombs, focusing key analysis on the Qianling, Qiaoling, Chongling, Tailing, and Zhenling mausoleums, which have better preserved stone carvings in their present state with high artistic standards. A summary comparison of stone lions, winged horses, ostriches, stone figures, ornamental pillars, and battle horses is combined with on-site measurements and archaeological data, to analyze changes in characteristics of early, middle and late Tang dynasties through the carvings' characteristics, accessories, and volume, specifically volume, form changes, and carving techniques.

2.2.6 Establishing a multimedia database as comprehensive data architecture for the project

Digital data collection in the project formed a large number of information sets, such as text, image, sound, video, soundtrack, and 2D/ 3D model data, requiring the application of a multimedia database to manage it permanently and efficiently. Metadata was used to catalogue the collected digital information, setting up a catalogue template, and realizing automatic cataloging functions as well as audiovisual key frame detection and sorting functions. The management and storage of digital data and various data formats were carried out.

2.2.7 Integration and production

The two-dimensional image data integration processing group, the virtual reality group, the animation group, and the post-production and editing group undertook comprehensive and arduous production in this stage, conducting meetings and discussions at irregular hours. The data collection, data classification, and editing groups were required to supplement data in circumstances of insufficient production data, with guidance from experts and professors.

2.3 Application analysis of digital technology in the project

1) 2D software was used largely for cropping and splicing images, graphics, color correction and patching, image size and format setting and adjustment, and plane and elevation drawing of the overall mausoleum layout using implementation software such as Photoshop, Illustrator, and AutoCAD.
2) Virtual Reality technology is an important part of cultural heritage digitization, reproducing the original appearance of cultural artefacts through their textual descriptions, and using CT scanning or other 3D techniques on existing artefacts for restructuring. For above-ground buildings that have been destroyed, virtual reconstruction is possible by applying virtual technology to historical data. Remote sensing and dynamic capture of the mausoleum sites can be used to realize 3D display and real-time 3D graphic generation technology.
3) 3D software is applied in Sketchup and 3DSMAX for digital scene sketch modelling. This project mainly used polygon modelling methods. For the modelling of large scenes and distant objects, overall and rough modelling methods are adopted, that is, modelling methods reducing the level of detail; for key objects and close-up scenes, subdivision level modelling is used to increase object authenticity. 3D application software such as Maya and Zbrush are used for the reconstruction of stone sculptures.

4) Augmented reality (AR) technology developed through virtual reality (VR) technology, and virtual 3D constructions are integrated into actual shooting environments to overcome the visual bluntness of digitized images.
5) Rendering technology is performed largely in Photoshop and 3DSMAX within the project, focusing on materials and textures, ray tracing, rendering, and overall ambient lighting.
6) Animated roaming technology will traverse real scenes of the main body of existing imperial tombs, and their surrounding natural and man-made landscapes, using modelling and rendering methods to virtualize the original features of Han and Tang tombs and carry out roaming design.
7) Clipping, editing, and post-production of video footage uses software such as After Effects, Nuke, Adobe Premiere and Sony Vegas, the software integrating each part of the project, smoothly combining video material and audio data into a whole according to the project's established narrative sequences.

3. Conclusion

Digital technology is constantly innovating and improving, and is being used increasingly for the protection of Chia's cultural heritage, although there are still issues that must be seriously considered. First of all, digital technology does not simply involve technology or software alone, but is based on the needs of project objectives, combining knowledge of a variety of subjects and using cross-production across a variety of technical software – operations that require a complete system, which in its turn requires comprehensive consideration, careful labor division, and multi-faceted cooperation.

Secondly, a comprehensive and in-depth deconstructive analysis of successful cases can be used to find key points and borrow from successful experiences. The deconstructive analysis and self-assessment of a project can also helps find solutions to unsolved problems, find root causes for existing shortcomings, and continually improve through analysis and summary.

Thirdly, China's digital application level is still far behind that of other countries, requiring project operators with comprehensive understanding of digitization and how to structure digital cultural heritage projects.

Finally, the fundamental purpose of cultural heritage digitization is the use of digital technology to better protect cultural heritage. Therefore, at this fundamental point, China must avoid projects that are purely for technology, instead making full use of this great developmental opportunity by learning from domestic and international success stories, acquiring technologies and experiences in order to explore new in-depth methods and measures for cultural heritage protection, and striving to promote the digitization of cultural heritage.

References

[1] Jiang Shen, Lu Xiaobo. Interactivity and applications of display communications in the digitization of cultural heritage — the example of the contemporary communication of Dunhuang culture.

[2] *The Underground Palace of Beautiful Ancient Chinese Architecture: The construction of the imperial mausoleum*. Beijing: China Architecture & Building Press, 2010.

[3] Zhu Xiaoyun. *The digitization of French culture makes a big move* [N]. Chinese Culture News, 30 Sept 2010.

[4] Zhou Jianlin. 'China has established an initial intangible cultural heritage protection system' [EBDL]. <http://www.chinanews.com/ cul/news/2010/06-14/2344560.shtml> (accessed 02 May 2011).

[5] Xu Xiaohan, Hu Qingfang, Feng Lin. 'Let cultural heritage live forever in digits'. Chutian Metropolis Daily, 23 Aug 2012.

[6] Li Fang, 'Cultural heritage – how to enter the "digital age"'. *People's Daily*, 10 Dec 2008, 5.

[7] Shan Jixiang. 'China's plans for the protection of cultural heritage over the next five years'. *Urban and rural construction*, Dec 2005.

[8] Wu Xiaochuan. 'Tradition in reality — Thoughts on the field research and image practice lessons of the Tang Emperor's Mausoleum'. *Art Education*, 2012, 4.

[9] Jiao Nanfeng. 'New developments in the archaeological field work for the Western Han Dynasty'. *Archeological Relics*, 2011, 3.

Mining and Utilizing the Cultural Tourism Value of Zhengyang Gate

Li Qing
(Business Department, Zhengyang Gate Management Office, Beijing 100051, China)

Abstract: The continuous growth of cultural tourism has brought museums and tourism closer together. The number and variety of museums in Beijing are great, and museums on ancient architectural sites are an important component. Zhengyang Gate is an ancient construction dating back 600 years, having played a role in military, ceremonial, and transportation history. Now, as key national protection units, the Zhengyang Gate and Archery Tower are located separately in Tiananmen Square and Qianmen Commercial Street, serving society through developing museums and other businesses. By examining the historical and cultural background, geographic location, and visitor characteristics of Zhengyang Gate, this paper analyzes the possibility of using it to mine cultural tourism value, proposing the exploration and utilization of such value in the aspects of enhanced visitor experience, souvenir development, and strengthened cooperation with other institutions.

Keywords: Zhengyang Gate; cultural tourism; museum; exhibition

Tourism is one of the fastest growing and most influential industries in the world. In recent years, cultural tourism has gained increasing favor from practitioners and travelers, even becoming a form of fashion. In 2009, the Ministry of Culture and the National Tourism Administration issued the *Guiding Opinions on the Promotion of Combined Culture and Tourism*, proposing that "culture is the soul of tourism and tourism is an important carrier of culture. Strengthening the deep integration of culture and tourism will help promote reform of the cultural system, accelerate the development of the culture industry, promote transformation and upgrades to the tourism industry, and meet public consumption needs, helping to promote the inheritance and protection of Chinese cultural heritage, expand the influence of Chinese culture, enhance China's soft power, and promote cohesive social development".

The process of cultural tourism is behavioral, realizing the purposes of perception, understanding, and observation of specific cultural content through tourism. The emergence of touristic activities using cultural resources as their main content, including historical sites, architecture, national art and folklore, as well as religion, is closely tied to the needs of travelers. Some scholars in the field have divided cultural tourism into four types, namely, historical and cultural forms represented by cultural artefacts, historical sites, ruins, and ancient buildings; modern cultural forms represented by modern culture, art, and technological achievements; folk customs represented by daily life and customs, festivals, sacrifices, wedding and funerals, sports activities, and clothing; and moral and ethical culture characterized by interpersonal communication. It is clear that the development and popularity of cultural tourism has linked tourism closely with the museum industry – a relatively quiet industry with rich historical culture – as museums are becoming destinations for tourist groups and individuals.

According to the regulations of the International Museum Association, the museum is a non-profit, permanent public institution that serves society and its development. It collects, protects, researches, disseminates, and displays the physical and intangible heritage of humanity and human environments for the purposes of research, education, and appreciation.

1. Conditions and possibilities for the development of cultural tourism at Zhengyang Gate

1.1 A long history and an ideal location

Lizheng Gate, the predecessor of Zhengyang Gate, was built in the Yuan Dynasty, and formed the south gate of the Great Capital of the Yuan City Wall. In the 19th year of the reign of the Ming Emperor Yongle (1421), Zhu Di moved his capital to Beijing. In the process of rebuilding the city, the south wall of Lizheng Gate was moved one kilometer south, that is, from today's Chang'an Street to the location of the three other gates. As a result, this brick and stone part of Zhengyang Gate dates back 600 years. The main parts of the gate and archery tower above the city base were rebuilt after being destroyed by the Eight Nation Alliance war in the late Qing Dynasty, and are over 100 years old. During the Ming and Qing Dynasties and up to the Republic of China era, Zhengyang Gate played multiple roles as a part of military defense, ceremonies, and transportation, and it has rich cultural and historical implications. In 1988, it was listed as a key national cultural artefact protection unit, and in 1990 and 1991, the gatehouse and archery tower were opened to the public. At present, Zhengyang Gate has a fixed exhibition and temporary exhibition hall, with the permanent *Lofty Zhengyang —The History and Culture of Beijing's Zhengyang Gate* being launched in 2014. For more than 20 years, the Zhengyang Gate Management Office has overcome the weaknesses of immovable cultural artefacts, launching nearly 100 exhibitions, publishing books, conducting lectures, and exploring social education activities as a national key cultural artefact protection unit.

Today, Zhengyang Gate stands on the central axis of Beijing's old city, with the gate tower located at Tiananmen Square and the south end of the Chairman Mao Memorial Hall. The archery tower is located at the northern end of Qianmen Street, which has been an important commercial district in Beijing since ancient times, as well as a gathering place for merchants and travelers. Many time-honored brands operate here today as illustrations of Old Beijing business culture. In this way, Zhengyang Gate is a museum surrounded by tourist attractions. If "the museum is destined to earn objective profits through ever-developing cultural tourism", then Zhengyang's superior geographical location and ancient architectural features enhance its potential and possibilities for raising cultural touristic value in comparison to other museums.

1.2 Advantageous demographics

The visitor is the main proponent of cultural tourism. Researchers McKercher and Cros have divided cultural visitors into five categories in their book *Cultural Tourism and Cultural Heritage Management*, listing accidental-discovery cultural tourists, conscious-purpose cultural tourists, accidental cultural tourists, casual tourists, and sightseeing cultural tourists. These categories are not mutually exclusive, but the depth of experience, pursuit, and regard for cultural tourism in the decision to visit the destination are different. According to surveys, visitors to Zhengyang Gate generally cover all five types.

In August 2014, Zhengyang Gate's management office launched a permanent exhibition, *Lofty Zhengyang —The History and Culture of Beijing's Zhengyang Gate*. At the time of the exhibition, the social education department conducted a detailed audience survey.

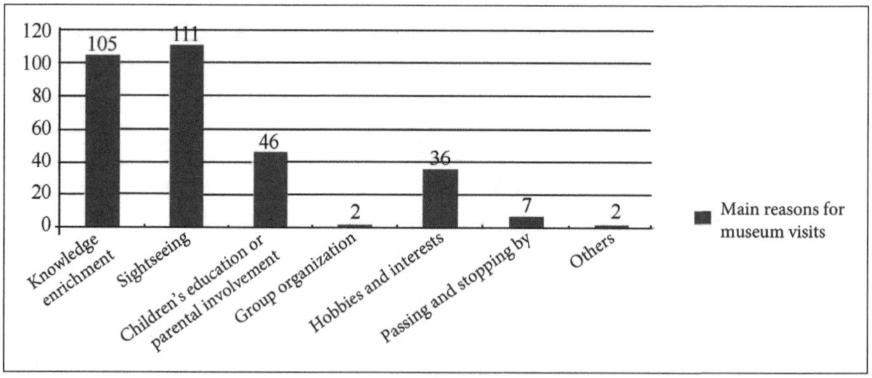

Figure 1 Main reasons for museum visits

The results indicated that visitors with the purpose of sightseeing and knowledge development accounted for the vast majority, with the number of sightseeing visitors ranking first (see Figure 1).

Amongst the survey's participants, knowledge of Zhengyang Gate came from various channels, with "television networks" and "passing through" being the main sources (see Figure 2). Most visitors were interested in "the history of Zhengyang Gate" (see Figure 3). After the visit, the greatest impression on visitors was "the unique architectural style of Zhengyang Gate" (see Figure 4). These results illustrate that Zhengyang Gate is an attractive location for all types of cultural tourists.

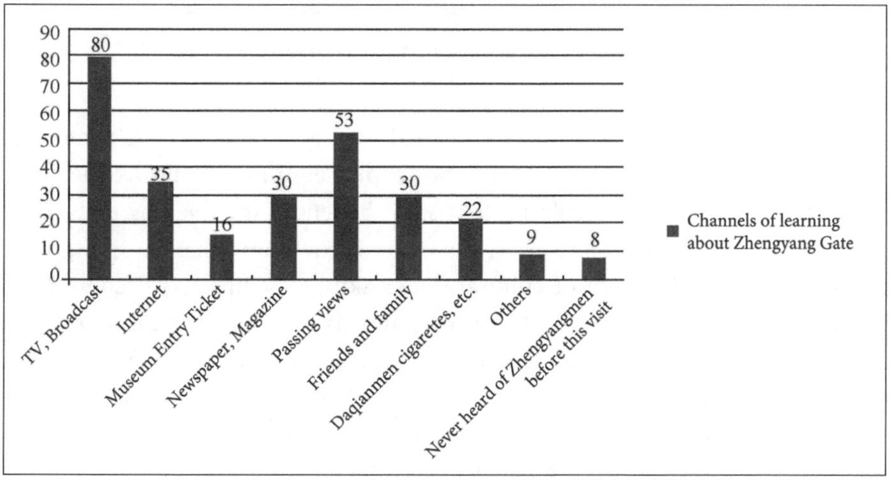

Figure 2 Channels of learning about Zhengyang Gate

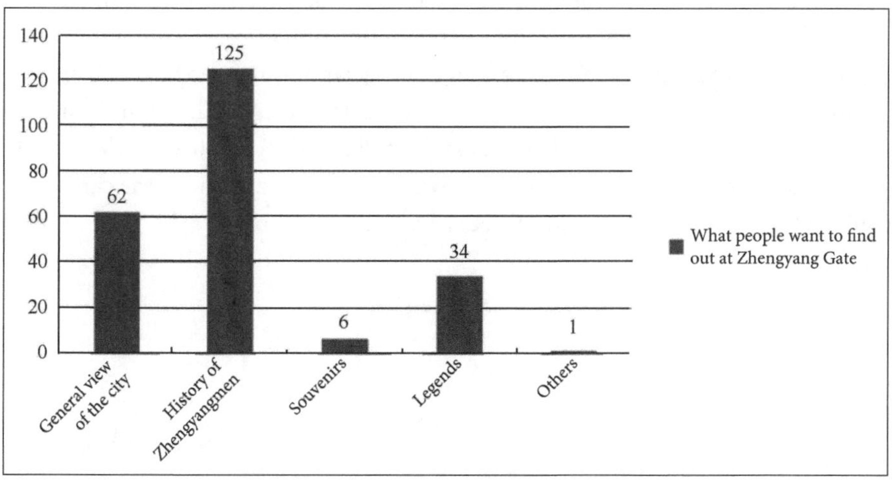

Figure 3 What people want to find out at Zhengyang Gate

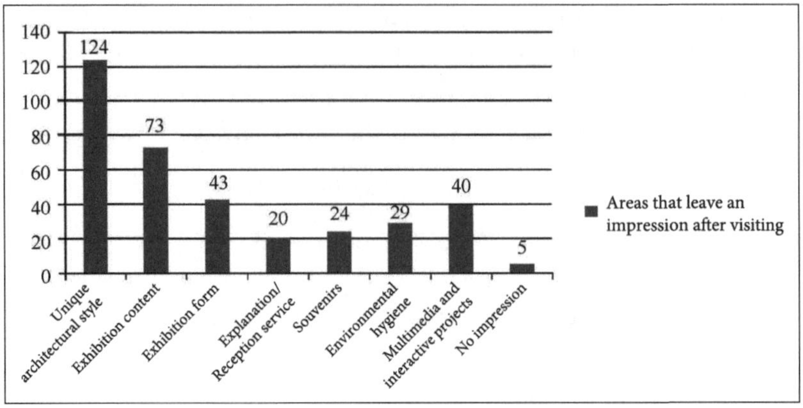

Figure 4 Areas that leave an impression after visiting

Visitors to Zhengyang Gate can be divided into local and foreign demographics according to the scope of their activities, and the museum must consider their varying needs. Enhancing the value of cultural tourism is beneficial to satisfying the visit needs of both local residents and foreign visitors. In recent years, thanks to the national emphasis on the development of cultural undertakings, Beijing has become home to increasing numbers of museums. Due to its status as a national political and cultural center, museums on ancient architectural sites are particularly prevalent. For foreign cultural tourists, museums on ancient architectural sites such as Zhengyang Gate are important destinations; for foreign tourists who are not interested in cultural tourism, such museums are also attractions thanks to their local characteristics.

Meanwhile, Beijing has been an immigrant city since ancient times. Since the latter part of the 20th century, its area and population have undergone rapid expansion. Among the current 21.148 million people living in Beijing, the permanent resident migrant population is about 8.027 million, with a registered population of 13.163 million including a large number of urban immigrants.* Residents, especially young people, are familiar with the history and culture of their city. Museums based on ancient architectural ruins represent the history and culture of different periods in Beijing. The 600-year-old Zhengyang Gate has witnessed a series of important historical events, and is one of the carriers of the capital's culture. Exploring its cultural tourist value stands to meet the visiting needs of local residents.

* Data from the Beijing Municipal Bureau of Statistics website.

2. *Mining and utilizing the cultural tourism value of Zhengyang Gate*

In order to develop, a museum must first find the right location and identify a key audience to serve. Zhengyang Gate's location, the characteristics of ancient sites, the absence of movable cultural artefacts, and visitor analysis indicate that the mining and utilizing of cultural tourism value is a direction with strong developmental potential. Given the conditions of Zhengyang Gate, this can be carried out in the following ways.

2.1 Improved visitor experience

Fixed exhibitions have a long-term period. The permanent exhibition at Zhengyang Gate lacks cultural artefacts and exhibits, so it can only achieve effective knowledge dissemination through accurate information, rigorous logic, clear display and steady style, with exhibitions, exhibits, and interactive events fitting seamlessly with the overall atmosphere of the ancient buildings. However, this alone is not enough. In order to enhance the value of cultural tourism, the visitor experience must increase within core exhibitions and services (see Figure 5). "Visitors pay attention to the pursuit of special experiences during the visit and have high expectations. In addition, tourists are very attracted to the so-called 'blockbuster' exhibition". This means that the creation of visitor experience requires accurate content for information display, digital display methods, and the application of modern technology.

In 2014, the *Lofty Zhengyang — The History and Culture of Beijing's Zhengyang Gate* was launched, with an interactive experience project designed in the *Full Map of Qianlong Jingcheng*, based on detailed research on modern and ancient maps of Beijing both inside and outside the city. Meanwhile, by setting up two high-resolution 1.5-meter tall, 46-inch

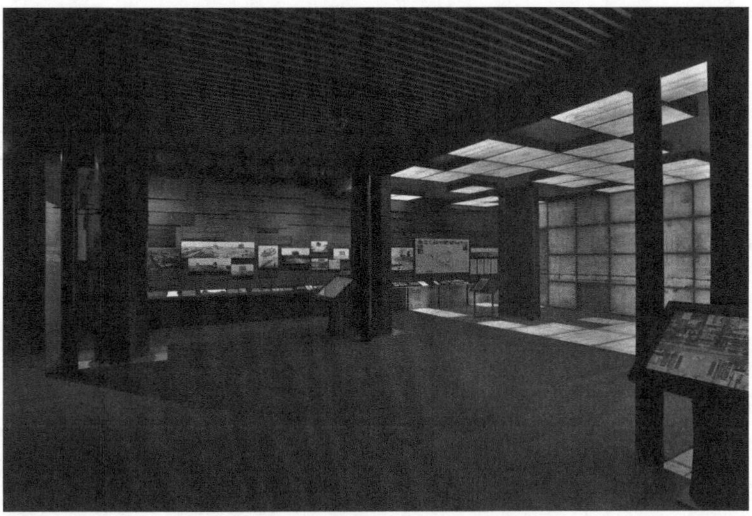

Figure 5 Visitor experience hall

touch screens, the Beijing city map drawn in the 15th year of Emperor Qianlong's reign (1750) is contrasted closely with the modern map. Through touch selection, visitors can compare the same location separated by more than 260 years, locating attractions and areas of interest that may still exist, as if alternating between ancient and modern time. This project is very popular among visitors. Flying through the central axis is another interactive experience project launched within the exhibition, in which visitors stand under the camera and raise their arms to the side. Combined with film techniques in a movie taken of the old Beijing central axis, the visitor appears to be flying through the area on the screen. This project has received wide acclaim by visitors from underdeveloped areas (see Figure 6) during normal opening hours.

In visitors surveys, the majority of respondents believe that the focus of future exhibition design should be "scene props" (see Figure 7), illustrating the need for compelling displays. "Any display, as long as there is movement, will attract the attention of the audience". Therefore, the proper use of digital multimedia display methods to increase visitor experience is the future direction of exhibitions held at Zhengyang Gate,

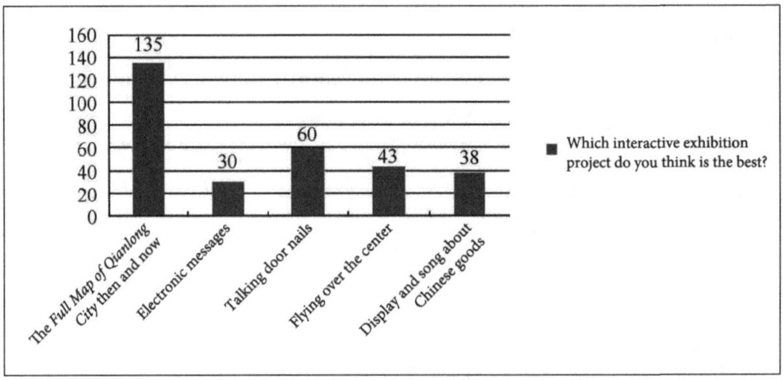

Figure 6 Which interactive exhibition project is the best?

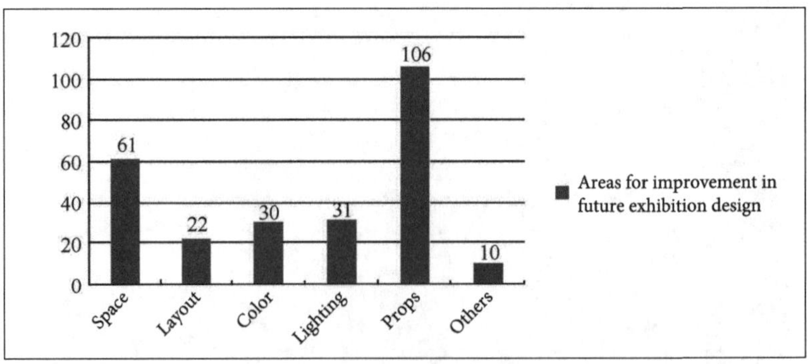

Figure 7 Areas for improvement in future exhibition design

especially historical and cultural exhibitions. In this regard, we have made active attempts, including application through various channels for the overall digital restoration and interpretation of the Zhengyang Gate complex, including working with digital media companies on the basis of historical data collection and research, using digital display means, and integrating rich historical information into relative 3D image space and transforming it into multiple forms. It also involves the use of computer-aided design such as AUTOCAD, 3DMAXS, and UDK to create perfect exhibition display effects, and presenting the tourist community with historical and cultural implications through vivid imagery and related methods. Meanwhile, we are attempting to provide a digital service method that is rich in content and also convenient, so that visitors are able to understand ancient buildings and exhibition content through the WeChat platform and audio guides.

2.2 Souvenir development

Generally speaking, tourists buy souvenirs for themselves or their families and friends during a trip. Visitor surveys show that 68% of respondents would purchase souvenirs, and 1% would decide whether or not to buy depending on the situation (see Figure 8). At present, most tourists suggest that Zhengyang Gate souvenirs are "not prominent". Therefore, the development of souvenirs must first overcome a lack of movable cultural artefacts and elements for reference. This can be achieved by identifying the characteristics of the city gate and their cultural meanings. The other aspect is to create a souvenir to reflect the visitor experience and spark memories.

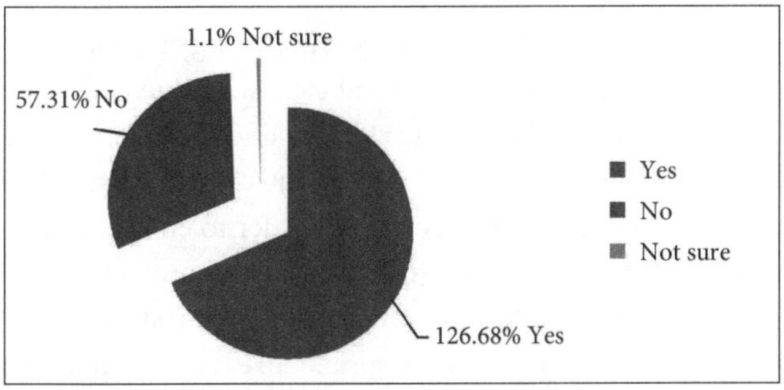

Figure 8 Will visitors buy souvenirs related to Zhengyang Gate or the exhibition?

2.3 Strengthening cooperation with other institutions

First, cooperation should be made with tourist institutions. Zhengyang Gate is located in Tiananmen Square. Each day, many people go into to the square to participate in a flag-raising ceremony at the Monument of the People's Heroes, while endless groups enter the Mausoleum of Chairman Mao. These activities are concentrated largely in the morning. Zhengyang Gate faces the south gate of the Mausoleum of Chairman Mao, making it a good combined activity for visitors to the attractions mentioned above. In the off-season, when visitors to Zhengyang Gate are fewer, attempts to cooperate with travel agencies may increase ticket sales and the possibility of souvenir sales.

Second, cooperation should be made with similar museums to achieve a 1 + 1 > 2 result. Small and medium-sized museums have had success in strengthening exchanges and cooperation, and holding exhibitions and events. The cooperation between Beijing Former Residences of Famous People and museums is a typical case. Beijing's old city gates were originally a whole, interconnected structure. The parts that have survived to this day are the Zhengyang Gate, the Archery Tower, the Deshengmen Archery Tower, and the Dongnancheng Corner Tower. From a national perspective, the Xi'an City Wall, Nanjing City Wall, Pingyao Ancient City Wall, and Jingzhou City Wall are all better preserved city gates, as well as famous tourist attractions. Some have already achieving significant success in the conservation of cultural artefacts and the development of tourist resources, strengthening contact with relevant units, cooperating in exhibition organization and exchange, and the exchange of cultural and creative products. This is a good way to form a brand in order to promote tourism.

Third, museums can strengthen their cooperation with multimedia companies. In the era of Big Data, and digital and Internet + museums, digital means should be used to explore modes of exhibition presentation, service audience models and museum management, and develop innovative working methods, in order to meet visitors' needs and carry out museum work more efficiently and conveniently.

In summary, tapping into the value of cultural tourism has positive significance for the improvement of a museum's visibility, the exertion of its functions and influence, as well as a key developmental direction. Whether it be Zhengyang Gate or other museums, the importance of mining and utilizing cultural tourism value is the deep excavation of immovable and movable museum cultural artefacts, and the core lies in identifying each museum's position and characteristics. Museums must develop strategies for exploring and utilizing the value of their cultural tourism in order to enhance their appeal through ensuring quality. For Zhengyang Gate, this includes the protection of immovable key cultural artefacts and ancient buildings, detailed and accurate exhibition content in a variety of display methods, and the appropriate use of scientific and technological exhibition means to provide warm and thoughtful service that includes digital display and education methods.

References

[1] [US] Kotler, Neil and Philip Kotler. *Museum Strategy and Marketing*. Beijing: Beijing Yanshan Publishing House, Dec 2006.

[2] [US] Hall, Margaret *Exhibition Theory — 21 Issues in Museum Exhibition*. Beijing: Beijing Yanshan Publishing House, Dec 2006.

[3] Liu Weimin. *Report on the Integration of Culture and Tourism*. Shandong: Shandong University Press, Oct 2011.

[4] Song Zhenchun and Li Qiu, *Research on the Cultural Tourism Industry and Urban Development*, Beijing: Economic Management Press, Feb 2012.

[5] Wei Yansheng (ed). *Chinese Cultural Tourism. Beijing*: Tourism Education Press, Aug 2014.

[6] Beijing Zhengyang Gate Management Office audience questionnaire survey statistics (Aug 2014).

Analysis of the Cultural Tourism Value of the Great Wall Museum

Zhu Dongfen
(China Great Wall Museum Business Department, Beijing 102112, China)

Abstract: The Great Wall has had a close relationship with tourism for a long time. The cultural implications of the Wall are an important resource for tourism, and the Great Wall Museum is a good provider and disseminator of this resource. This paper provides an analysis of the engineering and construction, military defense, and cultural tourism value of the Great Wall in order to impart, protect, develop, and utilize this valuable resource.

Keywords: Great Wall; tourism; value; impart

With the development of the economy, living standards are constantly improving, and leisure time is more abundant, increasing the popularity of the tourism industry. When people venture outside, they are no longer simply glancing over things hurriedly. Through touring, they hope to discover the history and culture behind the places they visit. The historical and cultural implications carried by museums can satisfy this demand. The Great Wall Museum of China completed and opened to the public in 1994 as an auxiliary tourism facility for the Great Wall of Badaling, enriching the content and expanding the scope of tours, and allowing people to learn more about the Wall.

Faced with the rich tourism resources of the Great Wall, and exploring its unique value in terms of engineering and construction, military defense, and cultural tourism, the Great Wall Museum offers history and culture, and also effectively develops Great Wall resources for its sustainable development as a public resource.

1. Millennial Engineering

The Great Wall of China is a wonder in the history of global architecture. Recognized as a World Heritage Site by UNESCO, it is the greatest architectural project in the history of human civilization, embodying the perseverance and great wisdom of ancient Chinese people, as well as showcasing the extraordinary achievements and value of Ancient China's engineering technology.

The Great Wall Museum offers information about the Wall's construction across the Spring and Autumn Period and the Warring States Period more than 2,000 years ago, at which point the Chu State built the "Fangcheng". After the Qin Dynasty unified China, the Great Walls of Yan, Zhao, and Qin were connected. The Great Wall of China stretched thousands of miles, from Lintao in the West to Liaoning in the East. It was built by more than 20 successive vassal states and feudal dynasties. Among them the Jin, Han, and Ming Dynasties built walls on a large scale, more than ten thousand *li* in length. In the history of the world, no construction has involved the efforts of so many dynasties, with such a long construction cycle and length. Furthermore, the architectural style of the Great Wall is also highly diverse.

This high-rise wall made of brick, stone, and earthwork runs over land and mountains, shores of rivers, lakes, and seas, and desert grasslands. It covers complex and steep geological terrain, entailing particularly difficult construction work. The building of the Great Wall required a profound understanding of geology, surveying, applied mathematics, mechanics, geometry, architecture, organized engineering, material transport, scientific and technical information, and strict construction management. The ancient working class of China fully exerted their wisdom and great creative spirit, building not only a huge and complicated construction, but one that has stood for thousands of years. It is a hugely impressive feat, and one that has kept the world in thrall.

The architectural artistry of the Great Wall is grand indeed. First of all, it is rich in form. The walls contain hundreds of impregnable passes, and thousands of enemy lookouts and beacon towers. In these constructions there are exquisite architectural structures. At the narrowest portion of the Simatai Great Wall, it is only one brick wide. There are also beautiful brick carvings, such as those on the Zhangjiakou and Datong enemy lookouts, which break up the monotony of the Great Wall. At the impregnable passes, there are a wide variety of constructions of different grades, such as in Badaling, Chadaocheng, featuring not only walls, towers and beacons, but also temples, government offices, and stations.

The Great Wall traverses grassland and desert, crosses river gullies, and leaps between hills and mountains, winding endlessly. It is not simply a construction project, but a great work of architectural art. With its huge size, magnificent body, majestic posture, and its dominating spirit, the Great Wall has attracted countless admirers and has become a symbol of the Chinese nation.

2. *Surviving conflict*

As Ancient China's great defense project, the Great Wall was originally built for military protection purposes. It was initially begun by the Chu State began in order to defend itself from northern nomads and enemies. Later, the Qi, Yan, Wei, Zhao, and Qin began to build their own portions for the same purpose. After the Qin unified the Six Kingdoms, Qin Shi Huang sent generals to attack the Xiongnu in the north to connect the walls of various states to form the Great Wall. It then played an active role in preventing Xiongnu invasion, protecting the economic and cultural development of the central plains. Sun Yat Sen once commented: "Although the First Emperor was brutal, the Great Wall has merit in later generations, combating major incursions."

The Han Dynasty continued construction of the Great Wall for over 20,000 miles in length, as a part of the ancient Silk Road. In order to defend against the intrusion of Tartar and Oirat Mongols in the Ming Dynasty, the construction of the Great Wall continued uninterrupted between the reigns of Hongwu and Wanli. Measuring 8,851.8 km, it stretched from Jiayuguan Pass and Gansu Province in the west, to Liaodong Hushan in the East.

The Great Wall was not only the front line of defense for rural people, but also a frontier base for counterattacks on nomads. An observation post was deployed on the front of the wall by which to investigate the enemy; the depths of the wall were equipped with armies to be mobilized in event of war. In ancient times, nomadic cavalry possessed excellent maneuverability and strong attack power, but they had no clout against the walled defense system, meaning that the Great Wall was able to resist intrusions by northern people.

The main parts of the Great Wall include city walls, defensive forts, ramparts, and beacon towers. These constructions are extremely elaborate in design, with sound defense functions that are as solid as iron walls. Once addressed by an enemy, these facilities demonstrated their full capability, serving a powerful blow to adversaries.

The Badaling Wall is located at the north opening of the Jundushan Guangou Ancient Road in Yanqing, Beijing, in the midst of the 40 miles of Guangou. In the Ming Dynasty when there were four lines of defense at Nankou, Juyongguan, Shangguan, and Badaling, the latter was an important military barrier, and a source of defense for the capital of Beijing due to its high, steep terrain. It was equipped with defensive forts, enemy lookouts, and beacon towers, with military facilities such as battlements, lookout holes, and gun perforations on the walls. Watchtower structures were built strategically in the dangerous terrain of the city wall, varying in size. Some are one story, others are two or three, with an arched gate, arched window, and staircase, and space for soldiers to temporarily rest or store weapons and food.

In times of war, the watchtower occupied a commanding position on the Wall, using the most advanced weapons of the time — artillery, Franc guns and cluster guns — to take turns to attack the enemy. If the enemy was close, soldiers would prepare bows and arrows to shoot them. If the enemy attacked the Wall below, thousands of stones of different sizes could be thrown at them. If the enemy broke through the gates into the city, defenders from both sides of the watchtower would rush in from all directions, forming a dense cross-fire network, encircling the enemy and trapping them. This is why the watchtowers are also known as barbicans.

There are also beacon towers with wide vision built on high-posts or dangerous spots on both sides of the Great Wall — another important part of the military defense system. These were a special facility for the transmission of military and smoke alarms. In the event of an enemy attack, smoke signals were burned during the day, and a fire was lit at night. Later, gunshot alarms were also added, with the number of enemies determined by the number of gunshots. As long as the enemy was spotted in front of the beacon tower, smoke and fire could be lit down the wall post by post. In an instant the command post could be seen hundreds of thousands of miles away, and the army would be mobilized for war.

Therefore, the main purpose of the Great Wall was military defense, especially in a time of cold weapons, resisting foreign aggression effectively and having important military value.

3. *Inheriting culture*

At the Great Wall Museum there is a special exhibit featuring the certificate issued to the Wall by the World Heritage Committee. The Committee's evaluation of the Great Wall of China is as follows: "Its value in culture and art is equivalent to its historical and strategic importance. From this, it can be seen that the Great Wall is an important carrier of Chinese culture, inspiring countless poems and legends around its architectural facilities, historicity, character, and anecdotes. The unique nature and style of the Great Wall has written an immortal chapter in the treasure chest of Chinese culture."

The many poems written about the Great Wall are rich in praise for its strategic location and magnificent landscape, offering reflections of the times and emotions, such as Li Bai's "From tens of thousands of miles, a long travelling wind blows through Yumen Guan"; Wang Changling's "The moon is the same as that of the Qin Dynasty, and the border pass the same as that of the Han Dynasty; the men defending the border thousands of miles away have not returned"; Wang Wei's "We drink one more cup, no comrades remain at the West gate of Yangguan"; Wang Zhihuan's "The flute need not blame the willows; the spring breeze does not cross Yumen Guan"; and Mao Zedong's "The sky is high and the clouds are pale; the wild geese fly southward. If we fail to reach the Great Wall we are not men, we who have walked 20,000 miles..."

Amongst the many gates of the Great Wall of China, there are few like Badaling, which has been walked by so many military strategists, politicians, travelers, writers, traveling merchants, peddlers, garrison soldiers, and wandering officials. When passing through Badaling, some have been amazed by its natural dangers, some have admired the beautiful scenery, some have been worried about state affairs, some have mourned ancient events, and a lot have left poems behind. In the Jin Dynasty, Liu Ying was the first to propose the name "Badaling", writing in the poem *Late to Badaling*: "Moving through two mountains, hundreds of miles up and down, a constant steady stream, like running water, shaped like the lumbar spine of a whale, like a snake from head to tail…"

Among the countless Great Wall folk stories, one of the most influential to be passed down was *Lady Meng Jiang*, which accused Qin Shihuang of tyranny, revealing the hardships and suffering of laborers building the Great Wall, and praising loyal love until death.

The Yanqing folk collection *Legends of the Badaling Great Wall* was selected for the second list of national intangible cultural heritage, covering a wide range of topics. *The Origin of the Three Great Wall Passes, Yuan Renzong Born in Yanqing,* and *Li Zicheng Goes to Beijing* are legends with distinct local characteristics, unique expressions, rich mythology, and legendary vibrance, and are a crucial part of the culture of the Great Wall of China.

In addition to playing an important role in military defense, the Great Wall also provided a spiritual garrison, inspiring people to defend China and build her up. In modern times especially, the Chinese people use the Great Wall as a symbol of their nation. The Chinese national anthem *March of the Volunteers* contains the lyrics "Arise, ye who refuse to be slaves. With our flesh and blood, let us build a new Great Wall", calling on the people to resist invasion at the most critical time. A line in the *Great Wall Ballad* has a unique lyrical appeal that has inspired patriotism in hundreds of millions of Chinese children: "The Great Wall is ten thousand miles long. Outside and inside it is my home town… Forty thousand (四万万) compatriots with one heart; the new Great Wall is ten thousand miles long".

Today, the Great Wall is regarded as a rock-solid and invincible symbol of victory. The public hailed the People's Liberation Army for defending national sovereignty, guaranteeing the smooth progress of a "Great Wall of Steel". The Three-North Shelter Forest System Program across the western northeast, north, and most of northwest China is known as the "Green Great Wall".

4. Tourist attractions – A bridge of friendship

The Great Wall of China is world famous, and was one of the first sites to be included on the Word Cultural Heritage List, enjoying a high reputation around the globe. The Wall at Badaling is an outstanding section thanks to its exquisite materials, technology, and imposing manner. It has become a link for friendly exchange between people of the world, having long occupied an irreplaceable position in the hearts of people and leaders.

Like a "long rainbow of friendship", it acts as a tunnel through ancient and modern times, setting up a bridge that connects China with foreign countries.

Since its public opening in 1954, the Great Wall at Badaling has received more than 160 million Chinese and overseas tourists, hosting more than 500 heads of state and government from all over the world. Photographs and materials from these visits are in the Great Wall Museum, which is unique amongst similar Chinese museums, planting seeds of friendship and peace. On November 1, 1998, the Belgian Prime Minister Jean-Luc Dehaene left a message about the Great Wall after touring Badaling, saying "The Great Wall is not a border, but a symbol of friendship amongst the people of the world"; on September 13, 2000, President Fernando de la Rúa commented: "To visit the greatest miracle of mankind — The Great Wall — one's mood cannot be expressed in words. The Great Wall is a bond of human friendship and peace"; on February 22, 2002, United States President George Bush wrote: "I hope that our people enjoy peace and good fortune forever"; on September 7, 2004, Saint Lucia's Governor-General, Pearlette Louisy, commented: "Very stirring. I will always treasure the memory of my trip to the Great Wall. The Great Wall stands for peace and harmony — a symbol of the world's good wishes."

Long years cannot sunder the Great Wall, nor can geographical distance lessen its attraction. The Great Wall Museum will strive to tap into and make full use of important Wall resources to bring greater cultural enlightenment to its visitors while protecting the sustained development of its tourism resources.

References

[1] Badaling District Office. *Badaling Great Wall.* Beijing: Beijing Tourism Press, Oct 1988.

[2] Xu Cheng. *Ningxia Archaeological History Research Paper.* Ningxia People's Publishing House, 1989.

[3] Luo Zhewen. *Luo Zhewen's Great Wall Collection.* Foreign Language Press, 1996.

[4] Yanqing Cultural Committee. *Legends of the Badaling Great Wall.* Beijing Publishing House, April 2010.

[5] Meng Xianli. *Discussions of Badaling and the Great Wall.* People's Posts and Telecommunications Press, Oct 2014.

Mining and Utilizing the Value of Cultural Tourism for Sustainable Museum Development – The Great Wall Museum of China

Qu Tianxiao
(Business Unit of the Great Wall Museum of China, Beijing 102100, China)

Abstract: China's history of over five millennia created a remarkable ancient civilization, leaving behind a large number of historical artefacts that are rich in art, history, culture, and scientific value. As a place in which to protect, collect, study, and display these historical artefacts, the museum carries the history and culture of a country and region. With economic development and social progress, the public is paying more attention to personal cultural needs. With the advantage of its rich cultural resources, the museum plays an important role in modern tourism, providing visitors with high-quality cultural experiences and exerting the functions of social education. Therefore, mining and utilizing the cultural tourism value of the museum to meet the public need for multi-faceted cultural experience has a positive effect on the promotion of sustainable museum development.

Keywords: museum; cultural tourism value; mining and utilizing

In recent years, with the development and progress of society, the rise of tourism, and the increase of tourist demands, the museum has become increasingly prominent in its function as an important cultural tourism resource, becoming a key carrier for unique historical culture and the enhanced attraction of cultural tourism. This paper explores the relationship between cultural tourism value and museums, combining certain achievements of the Great Wall Museum of China in mining and utilizing cultural tourism value, then putting forward some opinions and suggestions regarding sustainable museum development.

1. Relationships between cultural tourism value and the museum

1.1 The concept and implication of cultural tourism value

Cultural tourism is a behavioral process realizing the purposes of perception, understanding, and observation of cultural content specifically through tourism. It is a touristic activity with cultural resources as its main content, involving historical sites, architecture, national art, folklore, and religion, and its emergence is closely related to changing tourist demand.

Cultural tourism value is first reflected in meeting the public's growing touristic demands. With the improvement of living standards, tourism products such as sightseeing and amusement activities can no longer satisfy public tourism needs, and the pursuit of cultural enjoyment has become popular amongst modern tourists. Cultural tourism enables people to understand cultural implications whilst pursuing leisure and enjoyment.

The value of cultural tourism is also reflected in the conservation and inheritance of history and culture. Alongside the development of urbanization, many areas have destroyed the old to make way for the new, demolishing houses for road repairs. This has led to the disappearance of many local characteristics, and the remains of history and culture have been destroyed. However, by tapping into the value of cultural tourism and its development, the architecture and culture of different regions can be preserved, so that people from all over the world can experience it.

1.2 The relationship between cultural tourism value and the museum

Through the protection and collection of cultural heritage and natural specimens with inheritance value from the development of human history, museums have designed unique exhibitions and cultural products through comprehensive and systematic research. They do this in order to provide scientific research, cultural education, and appreciation-based recreation and entertainment services to the public, meeting the desire of tourists to cross time and space barriers to dialogue with history, and experience the cultural tourism value behind cultural artefacts through the artefacts themselves. Therefore, the museum plays the roles of an excavator, processor, and communicator of cultural tourism value in the development of cultural tourism, combining market research to produce a variety of cultural tourism products to tourists. Visiting the museum allows people to audio-visually understand the political, economic, cultural, and social development of a country or region.

2. *The significance of mining and utilizing cultural tourism value*

2.1 The museum as a national school

In a recent study of the museum, Xi Jinping pointed out that the museum is a big school, letting history and cultural artefacts speak, passing down the achievements and glory of the past, and enhancing China's self-esteem and pride. Xi presented the museum with a new historical mission and modern role, pointing it in a direction of sustainable development.

Thus, the museum has a responsibility to systematically organize its own cultural resources, allow collections of cultural artefacts to live, fully exploit its cultural tourism value, absorb research from various fields, and use modern technology to present rich and colorful social education activities to develop vivid and unique cultural and creative products. In this way, the public will be edified by culture and enlightened by wisdom, establishing a strong national identity thorough cultural confidence. This is the significance of tapping into the cultural tourism value of the museum.

2.2 Promoting the shared development of museums and tourism

Culture is the source of life and the inexhaustible motive behind tourism. In order to inherit and develop non-renewable cultural resources, the museum collects, studies, and protects cultural artefacts such as objects, ancient buildings, historical materials, and works of art, realizing the sustainable development of cultural resources. Meanwhile, a high concentration of cultural resources gives the museum a strong tourist attraction, making it a unique, high-quality tourist resource. Public interest and concern will be sparked by the appropriate mining and utilization of the museum's cultural tourism value, protection, and inheritance of local culture; integration with local tourism resources; and the launch of featured brand events, offering educational and interactive experience-based cultural tourism, creating new economic value and promotion through a variety of media, promoting the shared development of museums and tourism.

3. *Achievements of the Great Wall Museum in exploiting cultural tourism value*

Located in the Badaling Great Wall Scenic Area, the Great Wall Museum of China is a thematic museum based on the Great Wall, comprehensively reflecting its historical, military, architectural, economic, cultural, and artistic status. In recent years, in order serve visitors more efficiently, the Great Wall Museum has undergone comprehensive renovations, improving the content of its exhibitions with new, breakthrough designs. Secondly, by exploring its own characteristic cultural tourism value, the museum has also planned and launched a series of brand activities, accruing great social benefits.

3.1 Mining the value of cultural tourism by combining various means of display to create exhibitions that satisfy the public

In order to enhance its appeal, the Great Wall Museum met the 2008 Olympic Games with a brand new appearance, addressing visitors' needs and improving their understanding of exhibitions. From 2007, a large-scale transformation of basic displays was carried out, from the preparation of outlines to expert demonstration, modification, and on-site construction, re-opening in May 2008. Following the change, the theme of the museum became "World Wonder — A Historical Monument", with basic displays in four parts: "Over 2000 Years — Continuous Construction", "Grand Accomplishment — The Endless Great Wall", "The Great Wall Inside Out — One Family" and "Long-lasting Spirit Carried Forward". Compared with the original exhibition, the new content is more abundant, covering more than 2,000 years including the military defense function and systems of the Great Wall, the architectural structure and layout, and the economic, artistic and cultural developments along its length, as well as demonstrating the Great Wall as a symbol of the Chinese nation, and the protection of global cultural heritage.

The range of the exhibitions concentrates on cultural artefacts and the essence of specimens unearthed along the Great Wall, accompanied by full and accurate historical documents and photo models. It showcases the grandeur and history of the Great Wall, and the wisdom and dedication of the working people of Ancient China, giving a detailed formulation of the Wall as a symbol of the Chinese nation, and the bonds of unity and integration between all ethnic groups. In addition, multimedia displays and interactive technology have been introduced, increasing interest in the exhibitions and audience participation. In all eight exhibition halls, four LCD screens and three projection playback devices have been installed to play loops of commentary videos and images related to exhibition content. Five touch screens have also been set up, on which visitors can click to browse content and cultural artefacts, Great Wall poetry, legends, historical photos, and messages of world-wide heads of state who have visited Badaling, and watch a feature film about the Great Wall's military defense system. A custom-made virtual imaging display device is used to illustrate how fire and smoke were used to transmit enemy signals inside and outside the wall, and project scenes of warfare such as rolling stones and logs, arrows, and artillery strikes against enemy, creating a historical picture of the Great Wall while providing visitors with a lot of relevant information.

From the audience angle, more attention is paid to the details of the exhibit display. For example, a mirror has been installed under the exhibit of Yanqing Zhou seals, so that visitors can see the words engraved on them. Rubbings are shown on the top of bricks and the title monuments for visitors to read. To create a comfortable environment, the entrance and exit of the museum and the entrance and exit of each exhibition hall are equipped with disability ramps. Audio guides are offered in seven languages including Chinese, English, Japanese, and Korean. During holiday periods, a post is set up to provide wheelchairs for tourists who need them, along with free sewing kits and boiling water.

3.2 Promoting the Great Wall and carrying out various forms of social education

The Great Wall Museum is an important window through which the history and culture of the Wall can be showcased to an international audience, and an important foundation for the promotion of Chinese national spirit. The question is, how can we ensure that it fulfils both of these functions? We must walk in both directions, inviting visitors to come in, and allowing the museum to go out.

Entry is encouraged through a series of rich and colorful cultural activities, using a variety of media to promote the museum and attract visitors. On "5.18", International Museum Day, each year a series of Great Wall Knowledge Lectures are held. Using festivals like Patriotism Education and Publicity Day, Children's Day, and Military Day, museums will host student groups for enrolment ceremonies, education contests, and speech competitions. Folk exhibitions and entertainment activities are held during traditional festivals such as Spring Festival, Qing Ming, and Dragon Boat Festival.

Going out involves visiting schools, communities, and military camps via mobile exhibitions, lectures, knowledge contests, and regular activity courses in primary and secondary schools, achieving positive social benefits. Since March 2014, the staff at the Great Wall Museum have visited Qianjiadian Schools, National Primary Schools, and 12 primary and secondary schools, giving the "World Wonder Historical Monument" lecture series, using exhibition panel displays and slide presentations to educate students on the history of the Great Wall and its construction. Museum staff also went to five primary and secondary schools including Badaling Elementary School, Kangzhuang Middle School, and Xiatun Elementary School, to run an interactive teaching program called "Popular Science Base — Great Wall Rubbings and Inscriptions", during which students mastered inscription and rubbing skills in actual operation processes, and enriched their knowledge of the Great Wall.

4. Mining and utilizing cultural tourism value to achieve sustainable museum development

In recent years, although the Great Wall Museum of China has achieved clear results in exploiting the cultural tourism value of museums, the advantages of cultural resources have yet to be transformed into economic advantages. Further efforts are required to achieve sustainable museum development, with an emphasis on the development of cultural derivatives and the cultivation of museum personnel to improve economic benefits and competitiveness.

4.1 Paying attention to the expansion of derivative products and the improvement of economic benefit

Museums have an advantage in the development of souvenirs, as they can create and produce items with collectible value based on their rich collection of historical cultural artefacts. Editing and publishing various text-based books and producing audio-visual products can provide visitors with a more comprehensive and clear understanding of a museum's history, as well as the cultural background of its exhibitions and artefacts. The Great Wall Museum has printed and published a picture album, providing visitors with a tool for a deeper understanding of the Wall and museum. It includes cultural artefacts along the wall, the history of the Great Wall and its maintenance and protection, and the various activities the museum has developed. The next step is to publish a collection of stories related to the Badaling Great Wall, including legends about locations and historical figures.

In addition, souvenirs developed by a museum have typical, local characteristics, demonstrating local customs and culture. They are designed to meet the potential psychological needs of visitors, while promoting the development of cultural tourism and improving the economic benefits of museums. At present, the Great Wall Museum deals in various cultural products related to the Wall according to exhibition themes, including models, keychains, works of art, and postcards. The next step is to enrich the product types, making products that are more practical and interesting, drawing on the design concepts of the Forbidden City and the Capital Museum to develop a souvenir series featuring the Great Wall Museum. A variety of operating activities are also being planned for development according to specific museum characteristics, including special holiday events with cultural themes, expert-led lectures, and parent-child activities.

4.2 Cultivating professional museum management personnel and establishing a high-quality presentation team

A lack of professional knowledge is a common problem amongst museum staff in China, and museum management personnel with professional knowledge are rare. Sustainable development of the museum requires professional managers to motivate the trajectory. Museums must not only engage grassroots staff training programs, but also train professional management personnel based on the museum's specific characteristics; strengthen collection knowledge and professional knowledge learning; conduct assessments; and expand personnel exchange across same and relevant industries, learning from other museums and tourist attractions.

Although modern technology has a high rate of development, and museum cultural artefacts can be explained to visitors in a comprehensive and systematic manner via multimedia and other means, having a guide explain things is irreplaceable. A guide has the advantages of warmth, interaction, timely understanding of visitors' moods, and immediate answering of questions. The more visitors a museum has, the more important the role of the guide. Therefore, with the rise of tourism, museums must adapt to current conditions and strengthen guide training, establishing a team of guides with high cultural level and a love for their job. In this way they will tap further into the cultural tourism value of the Great Wall, providing high quality explanation services to tourists, and promoting sustainable museum development.

The Great Wall, containing the crystallization of ancestral wisdom, waits for us to take up the responsibility and mission of the times, calling us to care for it, interpret it, and pass it along. The author believes that we have the responsibility and ability to turn the Great Wall Museum into a grand educational institution that upholds the spirit of the Great Wall and flies dreams of revival, cultivating public sentiments of national responsibility, inspiring their consciousness of protecting the Great Wall and their belief in promoting its spirit.

References

[1] Wang Hongjun. *Foundations of Chinese Museum Studies (revised)*. Shanghai: Shanghai Ancient Books Publishing House, Dec 2001.

[2] Beijing Museum Society (ed.). *Policy and Exhibition — Diversified dimensions of museum exhibition construction*. Beijing: China Book Publishing House, Nov 2012.

[3] *Museum Development Forum Committee. Museum Development Theory* (2012). Beijing: Beijing Yanshen Publishing House, June 2013.

[4] Zhang Yinglan. *Protecting civilization, inheriting culture, creating the future together — Important discussions on the work of the museum by General Secretary Xi*. Beijing: China Cultural Relics News Agency, Feb 2015.

[5] Liu Wei. *Tourism Science*. Beijing: Higher Education Press, Jan 2014.

Mining and Utilizing the Cultural Tourism of the Museum – Experience from London, and a Case Study from Beijing

Zhang Jingyu and Zhang Jingqiu*
(College of Applied Arts and Sciences, Beijing Union University, Beijing 100191, China)

Abstract: With advancements in society, the role of museums in cities has become increasingly important. The ability to explore and utilize the cultural tourism value of the museum corresponds to the depth and breadth of the urban landscape and its cultural implications. This paper provides an analysis of the number, distribution, and ways of realizing the value of tourism at London museums, and suggests how the cultural tourism value of Beijing museums can be improved through the establishment of museum resources. Drawing lessons from the successful cases of the London museums, the paper clarifies the value of museums in cultural tourism and the significance of digital museums in this field.

Keywords: museum; cultural tourism; value mining; London experience; Beijing case study

Preface

As a place for soliciting, collecting, exhibiting, and studying natural and human cultural heritage, the museum reflects various aspects and changes in the history of human civilization, cutting-edge science and technology, and the artistic and cultural domains,

* Communications author jingqiu@buu.edu.cn.

providing the public with an educational institution, building, or public place to acquire knowledge and education, and to appreciate culture. In this way, museums play an irreplaceable role in the recording of human civilization and the development of culture [1].

In the 1970s, strategic thinking about the museum's capacity to serve society was on the rise. In Europe and the United States, the "three E functions" of the museum were developed — Educate, Entertain, Enrich — referring to its roles in cultural education, recreational entertainment, and life enrichment [2]. With urban development, the museum's role in cultural education and ideological education came to the fore, with exhibition themes growing closer to public life and more attention being paid to social value. As a vehicle to bring museum development closer to public life, cultural tourism is an effective method to realize this positioning.

Cultural tourism is an important component of urban tourism. Compared to urban tourism generally, it focuses on a high-quality, deep-reaching change in the proportion of cultural implications in tourism activities. American scholar Robert McIntosh defined cultural tourism as follows: "Cultural tourism includes all aspects of travel, whereby travelers become familiar with the heritage and history of others and their contemporary ways of life or thought" [3]. Therefore, cultural tourism provides more choices for urban tourism, including visits to the former residences of famous figures, museum visits, and trips to institutions of higher education. It expands the breadth of urban tourism, enabling travelers to accumulate knowledge while sightseeing, and therefore enhancing perception and thinking, tapping into urban tourism.

Some Chinese scholars have conducted research into the museum as the main body of cultural tourism, summarized as follows.

1) Integrating museums and cultural tourism: The three basic functions of collection, display, and scientific research must be developed in a balanced manner [4], adapting a new model of museum operation to the modern market [5].
2) Increasing the attraction of cultural tourism via museums: Establish a unique and positive image [6], unearth the cultural implication of exhibits [7], improve interpretation systems [8], improve perceptions of authenticity [9], enrich exhibit forms [10], and expand service functions [11–12].
3) Summarize the value of cultural tourism in museums: 1. To demonstrate local customs and unique features, shaping the uniqueness and individuality of the city [13]; 2. To understand cultural excellence in depth, popularizing cultural knowledge and passing on cultural heritage [14]; 3. To improve the scientific and cultural literacy of the public, and the ideological and moral quality [15].

The cultural tourism value of the museum resides in the attraction of the museum itself, the value and significance of its existence, and the guarantee of the quality and depth of the knowledge obtained there by visitors. Therefore, the exploitation and utilization of cultural tourism value is the focus of future museum development.

1. London

London's museums lead the world in terms of visitor numbers and the quality of their collections. London has the largest proportion of museums in the world, ranking among the best in terms of development. With this pedigree, London bears many successful case studies that are worth studying, and can be used as a reference for the exploration and utilization of cultural tourism value in Beijing.

1.1 An overview of London museums

As one of the birthplaces of the museum, the United Kingdom has many resources in this field, and its museums are among the best in the world. To date, there are about 2,500 museums in the UK, and more than 40% of residents visit museums at least once a year. More than 80% of overseas tourists visit museums as travel destinations, and museums and exhibition halls have an annual turnover of 9 billion British pounds.

In terms of London specifically, there are 250 large and small museums across 1,605 square kilometers of land. It is the city with the largest proportion of museums in the world (see Figure 1).

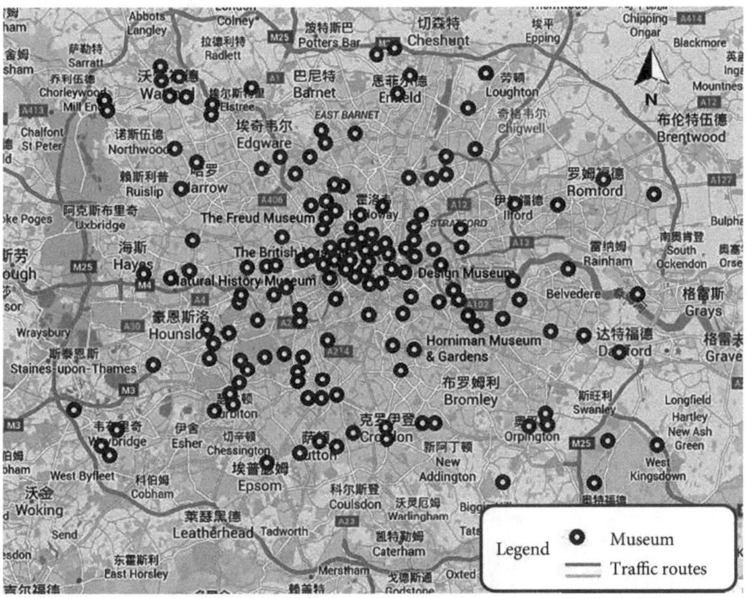

Figure 1 The distribution of major museums in London

1.2 Methods for realizing cultural tourism value

1.2.1 Cooperation between museums and higher education institutions

London museums have established cooperative relationships with universities by borrowing staff and technology to improve development and then feed museum resources back into university education. The 250 museums in London have established sharing and cooperation projects with higher education institutions, and museums employ outstanding personnel from these institutions in the form of volunteers and lecturers with advanced knowledge and expertise on collections, conservation, and curation. While in its own process of development, the museum also feeds back to university education, providing new teaching ideas and experimental environments for the university using collection-related resources and experience. This positive cooperation between universities and museums is an important method for the realization of cultural tourism value.

1.2.2 The design and sale of museum souvenirs

Museums in London combine individual characteristics with exhibition themes to launch a variety of unique souvenirs. The Design Museum is a unique self-positioning establishment with a design focus. Every year it hosts a "Designs of the Year" exhibition to showcase exhibit types from seven major design domains, embodying an emphasis on modernity and design. Its souvenirs stand out among other museums in the field.

As the best-known museum in London, the British Museum is known for its rich and beautiful collections. Souvenir designs use reputed characteristics in a flexible way, integrating elements of collections into daily necessities, and launching a batch of souvenirs such as badges with Chinese porcelain patterns (see Figure 2), Victorian coin-shaped chocolates (see Figure 3), and bearded Viking fridge magnets.

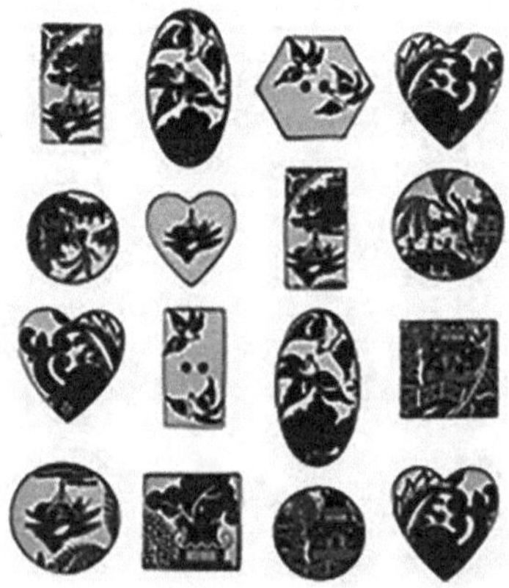

Figure 2 Chinese porcelain patterned buttons

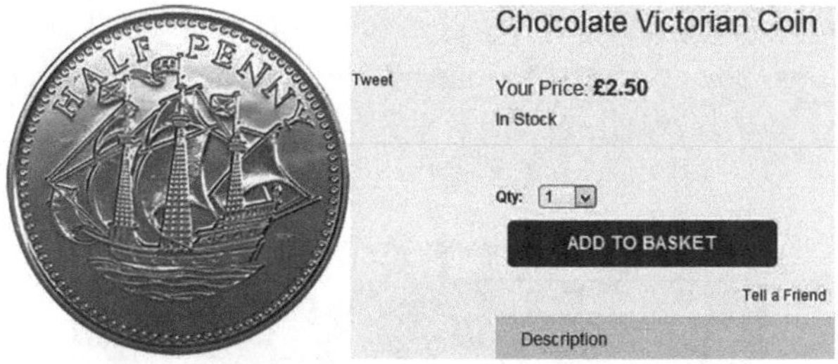

Figure 3 Victorian coin-shaped chocolates

Source: (Museum of London website http://www.museumoflondon.org.uk/london-wall/)

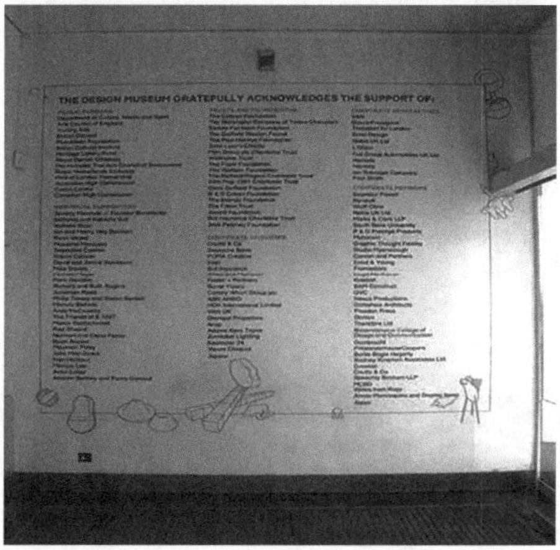

Figure 4 Wall of acknowledgement at the Design Museum

(Source: Wang Xiaoyu. *Go to London, visit museums*. Henan Science and Technology Press, 2013)

1.2.3 Multiple forms of promotion

1) Acknowledgement wall

At the Design Museum there is a wall of acknowledgements (see Figure 4), on which the museum expresses gratitude to its various funders. Sponsors also leave supportive words and signatures. Messages of gratitude on the wall draw attention to the museum as an institution.

2) Network media

London Museums pay significant attention to online media as a promotional window (see Figures 5 and 6), with pages designed to reflect their orientation. Feature information is updated in real time, providing business hours and exhibit information to those who have yet to visit, and offering a viewing platform to enrich the cultural tourism experience. Networks also offer online sales, previews of upcoming exhibitions, volunteer recruitment opportunities, current key products, and detailed introductions to the museum's collections. The creation of museum network media has become an important way of promoting cultural tourism value.

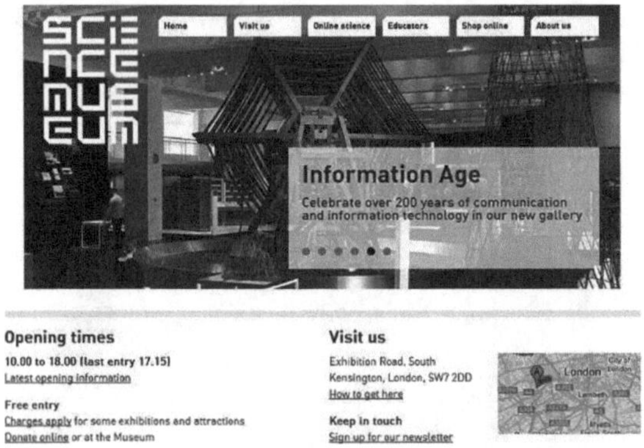

Figure 5 Science Museum promotional page

(Source: Science Museum official website http://www.sciencemuseum.org.uk/)

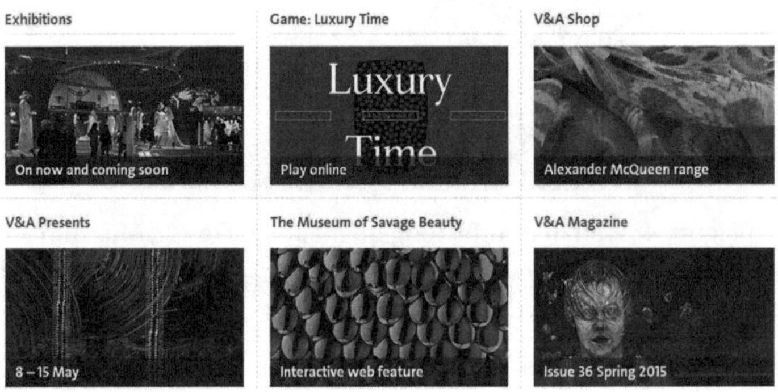

Figure 6 Design Museum promotional page

(Source: Design Museum official website http://designmuseum.org/exhibitions/design-museum-tank)

3) Continuing integration of technology and culture

The British Museum recently partnered with Sketchpad to launch a new public service – a 3D printing service where visitors can print 3D versions of the museum's signature collections. Continued integration of cultural exhibits and new technologies is an important factor in ensuring a museum's vitality, while at the same time constantly replacing the old with the new and keeping pace with the times, giving a new image to a museum's seemingly old buildings, and thereby increasing attention and recognition.

2. *A Beijing Case Study*

2.1 Status quo of museum resources

Beijing has a total of 151 categories of museum resources, divided into six major groups: historical and cultural, personal memorial, cultural and artistic, ethnic and religious, natural science, and science and technology (see Figure 7).

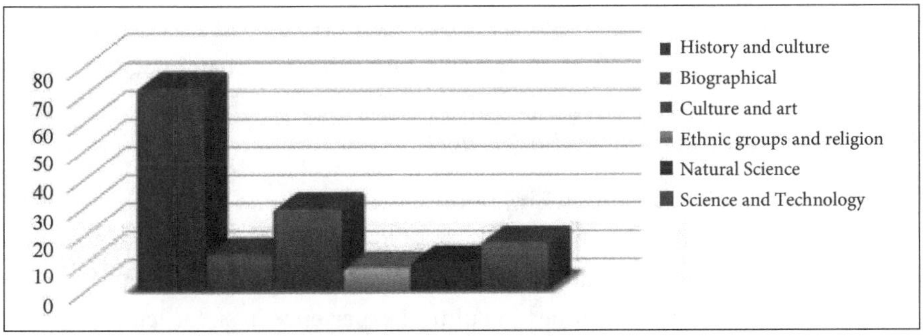

Figure 7 Composition of museum types in Beijing
(Source: Data compiled by the Beijing Municipal Bureau of Cultural Relics http://www.bjww.gov.cn/)

Museum resources are unevenly distributed throughout districts and counties. Overall, the urban areas have more and the suburbs have fewer, with more in the east and fewer in the west. Museum resource categories are more abundant in city centers (see Figure 8).

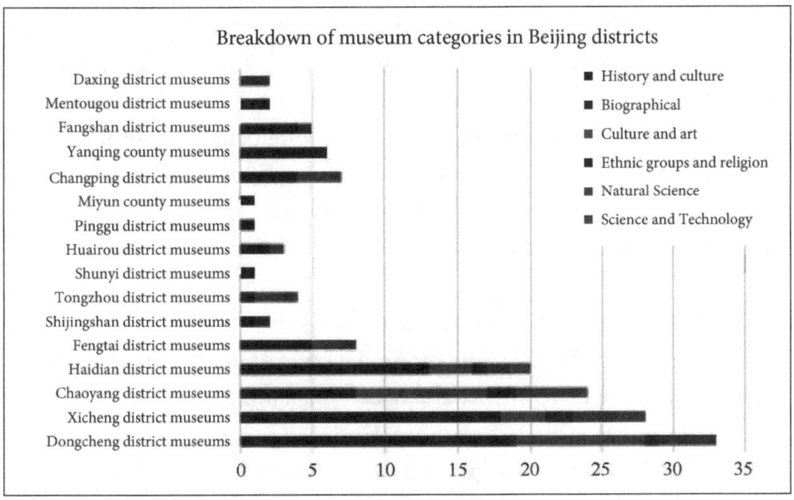

Figure 8 Composition of museum types in Beijing
(Source: Data compiled by the Beijing Municipal Bureau of Cultural Relics http://www.bjww.gov.cn/)

2.2 Summary of the cultural tourism value of museums

The museum not only serves the citizens of a region, but also attracts visitors from other areas and the rest of the world. For example, in 2014, the Beijing National Palace Museum's visitor count exceeded 100,000 in three consecutive days over the holiday period for the National Day of the People's Republic of China. The arrival of a large number of visitors increases a museum's income, and drives the operation of various related industries such as urban catering, accommodation, and transport, which increases the city's overall income. Meanwhile, cultural tourism serves urban development by enriching tourism products, exerting educational functions, and establishing an urban image.

2.2.1 Enriching tour routes and tourism products

As the main bodies of cultural tourism, hundreds of museums in Beijing are enriching the diversity of tourism products, thereby increasing the competitiveness of tourism products in Beijing. The museum's functions of displaying collection and promoting cultural make up for the shortcomings in cultural experience in sightseeing. Scenic spots and museums work together to create tour routes, using museums as a node for the start and/or end of the tourist experience, combining cultural education and the physical act of sightseeing into a mature tour route that is high in cultural implications.

2.2.2 Cultural communication and communication platforms: empowering educational functions

1) Exhibition space for intangible cultural heritage

The museum bears a heavy responsibility in protecting intangible cultural heritage in the city. In doing so, an important element is planning and designing museum exhibitions to provide a physical space and display platform. The museum broadcasts videos of intangible cultural heritage, introduces inheritors, and exhibits achievements via multi-dimensional and three-dimensional displays, informing people about the status quo and value of this heritage, which is conducive to its promotion and conservation. Meanwhile, this also expands the forms and cultural value of the museum, providing more space for urban cultural education and tourism functions.

2) The physical embodiment of a second classroom

Physical collections, scientific research, and social education are the three basic functions of the museum. The former two are processes of a museum's own improvement, while the social education function is the core purpose. Knowledge acquired by young people from textbooks is limited in terms of length and transmission method. Museums containing different pieces of knowledge in various categories support young people's education with their rich collections and comprehensive, systematic, and intuitive scientific research explanations. Thus, the museum is regarded as a second classroom, with museum learning as an extension of classroom knowledge. It is an urban cultural landscape that serves the youth as the physical embodiment of a second classroom.

3) A cultural exchange platform
Collection resources are a window through which people understand the world. By visiting museums, they broaden their horizons, gain knowledge, and become more accomplished. With modern progress, knowledge is moving towards "citizenization", and the developmental direction of the museum involves increased integration with social life. Visiting a museum will become a routine cultural and leisure activity for the public. The various kinds of museums provide a platform for urban residents to interact with people who enjoy the same hobbies, and to explore professional matters.

2.2.3 Establishing a city's image

A city's image is divided in two parts: a physical image and an urban implication. The physical image refers to the outward image of the museum building as a city landmark, while the urban implication refers to urban culture made up of the collection of various fields such as architecture, citizen culture, and science and technology. The cultural tourism value of a museum is established by both parts of the city's image.

1) Physical image — Establishing urban landmarks
Museum buildings have typical, commemorative, and permanent features, and their unique architecture can embody a city's cultural symbology and image. For example, the National Museum located on the south end of Beijing's Chang'an Street features a roof with traditional northern elements such as cornices and brackets. It is a modern building that fully reflects the essence of Chinese architecture, thus becoming an important part of Beijing's cultural landscape, demonstrating national spirit and culture. Therefore, the National Museum can be seen as a landmark, distinguishing Beijing from other cities. Unique, high-quality museum buildings are the cultural business cards of a city, demonstrating its style, carrying its unique culture, and shaping the urban environment as a landmark construction.

2) Urban implication — Passing down the cultural context of the city and enhancing cultural taste
Museums in cities perform the early stages of collection work by collecting and protecting historical objects. These collections reflect rich cultural heritage accumulated over long-term urban historical development, including the current style of the city, and future development directions. People visit museums to observe history, examine the present, envision the future, and inherit urban context. Greater cultural identity enhances national cohesiveness, with the improvement of public culture driving the rise in urban cultural taste and creating a positive atmosphere.

3. The cultural tourism value of Beijing museums

In observing London museums in terms of cultural tourism value, and analyzing the number, type, and efficiency of existing Beijing museums, lessons can be learned.

3.1 Achievement pathways

3.1.1 Broadening communication to attract more tourists

According to the successful development of London museums, there are three key ways to improve museum publicity.

1) The celebrity effect
For Beijing's three major categories (history and culture, culture and art, ethnicity and religion), we can learn from the wall of acknowledgements at the London Design Museum, featuring the handwriting of historical celebrities, public figures, artistic pioneers, and religious leaders. This improves both the cultural atmosphere of the museum and its sense of history, while also playing a financial role. With the celebrity effect, the museum will attract more visitors.

2) Creating a perfect network platform
Network media has become the most favorable publicity tool for museums. With the current popularity of smart phones, the promotion of museum apps has become an increasingly important component in the development of network platforms.

3) Replacing the old with new
With continual new advancements in exhibition content and improved publicity, more people will be attracted to museums for cultural tourism.

3.1.2 The scientific display of cultural artefacts: three-dimensional presentation

When it comes to the display of cultural artefacts, scientific concerns (including the spatial location of museum collections, the design of museum lighting, and the density and layout of exhibits) affect way collections are shown. For representative, high-value and top quality collections, there should be a combination of digital technology, with video displays to show materials, uses, and manufacturing processes behind the exhibits. Virtual reality technology can be combined with sound, light, and electricity to display exhibits three-dimensionally for large-scale cultural sites and historical environments. The use of appropriate display methods and three-dimensional presentation technology are the best ways of exhibiting collections.

3.1.3 Refining exhibition descriptions and exploring cultural artefacts

History- and culture-based museums largely display exhibits in the form of cultural artefacts accompanied by text descriptions. At present, the text descriptions at Beijing museums are relatively simple, listing tag information such as place of discovery and artefact name. Higher-quality text descriptions offer an outline of the historical background of the collection, as well as its origin and development. Reading and perceiving historical, cultural, and scientific value through text, and in regards to complex collection experience, objects that were lost overseas, engaged in high-priced bidding, and donated privately must be reflected in text descriptions to increase the sense of history. With the increase of international tourists, text descriptions with standard incisive English translations are another crucial way to raise cultural tourism value. Strict yet scientific, detailed and in-depth exhibit descriptions enable a better understanding, allowing visitors to maximize their knowledge. For the historical and cultural category, this speaks to a key role of the museum, and is of great significance in enhancing the value of cultural tourism.

3.1.4 Cooperation with educational institutions for mutual gain and development

As in London, Beijing museums also see students as their main attendees. The museum is a preferred place for off-campus activities for educational institutions. However, in London, the museum is not simply a place to visit. Museums and institutions have established cooperative relations, meaning that museums recruit student volunteers and guides, and employ highly educated students as reserve personnel, combining museum resources with daily teaching, and providing education-related resources such as experimental sites and cultural artefacts. This cooperation is most crucial in museums that focus on natural sciences and science and technology, and is also an important way to establish cultural tourism value.

3.1.5 Creating unique products and expanding the significance of souvenirs

Chocolates shaped like Victorian coins, badges with Chinese porcelain patterns, and Viking fridge magnets at the British Museum are all daily necessities that feature themes of the museum. High-quality souvenirs allow visitors to remember the cultural achievements exhibited in the museum, as well as the museum itself. Beijing museums must draw on this success, aiming to sell special souvenirs and launch experiential activities that align with their themes, as this offers economic benefit in the form of crucial funding while reflecting museum culture.

3.2 Digital museum application

Digital museums uses electronic databases, digital imaging technology, and three-dimensional modelling technology to push the exhibits through the limitations of physical space, displaying the historical background, material composition, and graphic symbolism of physical exhibits via image data. They also integrate existing resources with sound and optoelectronic technology to simulate specific scenes (such as war scenes) with stereoscopic rendering, and simulate site remains through three-dimensional modelling technology.

These new technologies are applied in the mining of cultural value in museums. They are able to extend the exhibition space of the physical museum via the coexistence of cyberspace and multidimensional physical space, reducing the pressure on traditional museums as a carrier, and extending the visiting time of tourists within museums. By interacting online with visitors and citizens, we can also dig deeper and fully utilize the cultural value of the museum in order to expand its influence and its cultural tourism value.

References

[1] Wang Hongjun. *Foundations of Chinese Museum Studies* [M]. Shanghai: Shanghai Ancient Books Publishing House, 2006, pp. 61–66.

[2] Guan Xin. 'Digital Museums and Public Education', [J] *Museum Research*, 2006, 1: 3–7.

[3] Tang Youbo, Xu Zhe, Guo Qingsheng, Gao Menghe. 'Tourism Culture and Chemistry — A vision of cultural chemistry', [J] *Journal of Shanghai University (Social Science Edition)*, 1991, 4: 4–12.

[4] Guo Fuchun. 'Preliminary study on the construction of museums and urban cultural tourism' [N]. *China Cultural Relics*, 2006, 6.

[5] Chen Yan. 'On the ecological museum model of national cultural tourism developmen' , [J] *Journal of Yunnan Minzu University (Social Sciences)*, 2009, 2: 52–55.

[6] Liu Xi. 'The museum strings together highlights of Yangzhou cultural tourism' [N]. *China Travel News*, 2 April 2007, 6.

[7] Ge Hua, Tang Chunsong. 'Guilin Museum produces a new trump card of cultural tourism with Meiping' [N]. *China Cultural Relics*, 22 Dec 2006, 6.

[8] Zhao Xia, Yun Hong. 'Cultural communication in the translation of Salt Culture tourist attractions — the example of the Zizong Salt Industry History Museum', [J] *Journal of Sichuan University of Science and Technology (Social Science Edition)*, 2013, 6: 38–41.

[9] Zheng Qin, Zhou Chunyan. 'Research on the development of cultural tourism products from the perspective of authenticity — the example of the Three Gorges Museum of Chongqing', [J] *Journal of Neijiang Normal University*, 2013, 2: 69–74.

[10] Feng Xiaomei, Zhao Dandan, Cao Qing, Liu Qingyou. 'Research on Wuyue cultural tourism development based on the concept of the "ecological museum"', [J] *Hunan Agricultural Sciences*, 2011, 2: 32–33, 43.

[11] Xie Peihong. 'On the experiential design of museum cultural tourism products', [J] *Business Times*, 2009, 19: 126–127.

[12] Liu Xi. 'The museum strings together highlights of Yangzhou cultural tourism' [N]. *China Travel News*, 2 April 2007, 6.

[13] Yi Xiaofa. 'Making the museum into the new center of urban cultural tourism' [N]. *China Travel News*, 17 Feb 2006, 6.

[14] Sun Mingqiang, Tang Fujing. 'Cultural tourism is the best option — Analysing the ecological museum experience', [J] *Contemporary Guizhou*, 2009, 6: 42–43.

[15] Zhang Lei. 'Creating cultural tourism products to improve the city's soft power' [N]. Jiujiang Daily, 08 Jun 2007, 1.

[16] Lin Ziping. 'Several basic issues in understanding the social benefits of museums', [J] *Chinese Museum*, 1986, 4: 32–37.

Making the Visit a Pleasure – The Application of Smart Navigation and Positioning Systems in the Museum

Shi Yi
(Beijing Stone Carving Art Museum, Beijing 100081, China)

Abstract: The smart near-field geographic positioning system is an application developed specifically for museum scenes. Based on iBeacon technology, M-Beacon is able to realize smart museum navigation, real-time visitor interaction, as well as the analysis and application of collected data in real-time.

The museum will quickly and efficiently turn existing background exhibit information into audio and video data, simply by creating a base station, server, and pre-production content. The system provides visitors with a more humanized service and enhances audience perception, and also becomes a platform for real-time communication and interaction between museums and their visitors, collecting relevant data from visitors for museum use, then providing information that is better suited to personal needs through data sorting, analysis, and database updates. The application of the M-Beacon system has played a positive assisting role in the improvement of museum service levels and development.

Keywords: intelligence; navigation; system; museum; application

As a custodian of the relics of human culture, the museum has four main functions: collection, display, education, and research, playing an active role in the development of social culture. Since the beginning of the current century, traditional museums have been transformed into institutions of the new era, becoming the "spiritual homeland", "cultural oasis" and "knowledge temple" for the public. They are richly endowed by nature to play a role in the promotion of popular science.

The realization of all museum functions requires the exhibits to be carriers. Museum exhibits are a product of cultural accumulation. The information and value contained within them are often concentrated in the characteristics of a specific region or period, so it is important to present exhibits to visitors in a variety of forms. Good museum design and planning incorporate the perceptions and feelings of visitors, allowing them to be immersed and inspired the unique cultural implications of exhibits.

The Beijing Stone Carving Art Museum is a good example. Its exhibits and collections contain more than 5,000 ancient stone inscriptions, as well as precious statues from the Northern Dynasties, highly artistic Jinyuan stone carvings, and epitaphs of successive dynasties after the Tang. It is home to some of Beijing's earliest stone carvings, such as *Tomb Signs of the late Qin from Youzhou in the Han Dynasty* on stone columns and tower components, as well as the exquisitely carved Qing Dynasty Stone Worship Halls, the unique epitaphs of Nalan Xingde's wife Lu, as well as *Zhijinzhai, Jingdetang,* and other famous stone calligraphy carvings. There is also a collection of stone carvings known as the Beijing Forest of Steles. In addition to offering artistic enjoyment to visitors, this rich collection faces the new challenge of how to convey the cultural information contained in its exhibits while stimulating interest and exploration.

1. The developmental status of the museum

The role of the museum in public life is becoming increasingly important. By the end of 2014, the number of registered museums in China reached over 4,100, increasing at a rate of about a 100 each year. Since introducing free entry in 2008, museums across China have received 500 million visitors each year, with an average growth rate of between 50 and 70%.

However, current museum operations still face certain issues and challenges.

1) Increased utilization
With free entry and the growth of public awareness about exhibitions, audience numbers are increasing. The improvement of effective museum utilization rates is an important issue in achieving the functions of a museum.

2) Humanization of museum services
Current museum exhibit explanations mainly consist of manual explanations, audio guides, and mobile phone QR code scanning. With visitor explanation services, it is necessary to further enhance humanized and targeted design to provide a richer perception.

3) Insufficient working capital
Most museums rely on funding allocation. Daily operations and special exhibition planning require a lot of manpower and material resources, the cost of which is high.

With continued development, it is particularly important to fully empower the social functions of cultural communication and public education. In the past, museum improvements tended to focus on hardware equipment – a "backstage" improvement. Meanwhile, the front-end experience of direct contact with visitors was often overlooked. Studies have shown that the sensory experience element of interactive display is the easiest way to stimulate subjective audience initiative. This is because sensory forms of interaction pay greater attention to audience individuality, providing a more valuable experience that realizes public participation in the truest sense, allowing visitors to make better use of museum resources. With the advent of the Internet age and the mobile Internet era, mobile terminals such as smart phones have become indispensable, fostering a usage habit among a high number of users. Using mobile terminals such as smart phones to enrich visit experience is one of the ways in which museums can develop. The realization of museum study tours via mobile phones is a current trend of modern times.

2. About iBeacon

In June 2013, Apple released iBeacon at WWDC (Apple's World Wide Developers Conference) – a technology offering precise micro-positioning through low-power Bluetooth technology. With this technology, devices are able to receive signals sent by other iBeacons within a certain range. Information from one can be passed to other users within a certain range. The characteristics of the technology include long transmission distance, low cost, fast transmission rate, multiple supported devices, strong base station endurance, and easy promotion, and it is able to realize accurate indoor positioning and smart push functions.

iBeacon includes:

A server: Used for background data storage. Content pushed by the base station and information fed back through mobile devices are stored here;

My Beacon: A Bluetooth transmission base station using iBeacon protocol. With a built-in Bluetooth 4.0 chip, it can transmit and receive 2.4GHz Bluetooth signals and interact with smart phones within a range of 60 meters.

Mobile phones and other mobile devices: Smart phones with relevant apps and Bluetooth 4.0 or above.

Using smart access control as an example, the working principle is as follows: A mobile phone carrying MyBeacon software comes close to the entrance of a building using MyBeacon access control. MyBeacon identifies personnel information and sends it to the server. The console then displays relevant information such as the name and photo of the entering personnel. Once this technology is implemented, it can be applied in museums, exhibitions, shops, stadiums, and areas with large flows of people, where exhibits and crowds need to interact with one another.

The Belgian technology company Prophets installed the iBeacon receiver in the Rubens House in Antwerp. With the app, visitors are able to open different presentations depending on the location of each piece of art, with mini games and features making the viewing experience three-dimensional and vivid.

The Southern Buddhist Millennium Classics — Qibao Ruiguang China Southern Buddhist Art Exhibition took place in Taiwan, introducing the Beacon navigation system within the exhibition. When visitors walked toward the cultural artefacts, their mobile phones immediately displayed a related introduction featuring high-resolution pictures, allowing users to operate the interface and appreciate the cultural artefacts more easily.

Macy's in the United States began deploying iBeacon sensors in 2013, under the name shopBeacon. Customers entering various floors or shops received different offers. Macy's also worked with local US app Shopkick – a shopping service that rewards customers in store based on their location. Customers can first look at a certain product at home, then when they enter Macy's, ShopBeacon technology reminds the customer to buy it.

The US Major League Baseball Stadium has an iBeacon sensor that can connect to a user's mobile phone via low-power Bluetooth technology. Most devices are placed at fan access points, so that when followers are detected entering the stadium from all directions, the iBeacon base station will push app notifications on user iOS devices, including maps, venue information, promotional videos, and seat information.

3. *Applications of iBeacon in the museum field*

Some countries have begun to apply this technology to museums. Belgian technology company Prophets installed iBeacon in the Rubens House in Antwerp, developing related applications and allowing visitors to receive information about pieces of work at different locations they stop at, enriching their experience.

There are three main applications for this technology in the museum field:

1) Realizing smart museum tours through iBeacon technology

Rich and vivid background information regarding exhibits can be stored on the server and pushed to visitors' mobile phones, who receive this information via iBeacon base stations located near the exhibits. This allows them to learn more and appreciate the displays.

2) Realizing interaction between the museum and visitors

a) Visitor feedback: suggestions and opinions regarding the museum can be pushed to the server via the app, creating an effective channel by which the museum can collect suggestions.
b) Accurate information push relying on data mining: based on historical visitor

data, judging visitor preferences, and pushing exhibitions or events that can be of personal interest.

3) Organizing more dynamic cultural exhibition fairs with iBeacon technology
An iBeacon base station is placed in front of each booth, with information pertaining to the exhibitor of each booth made into new forms of information such as video, audio, and text. Visitors download the app at the entrance and turn on Bluetooth. When they visit each booth, the app automatically pushes exhibitor introductions, and the organizer can also push other relevant information to visitors.

4. About M-Beacon

The "Smart Near-field Positioning System Development and Demonstration" project for museums, announced by the Beijing Stone Carving Art Museum, was approved by the National Science and Technology Commission. The project designed the "M-Beacon Smart Navigation and Positioning System" specifically for the museum, for applications in the museum hall. Based on iBeacon technology, M-Beacon is able to realize smart navigation within museums, as well as real-time museum-visitor interaction, and the analysis and application of data collected in real time.

The three aspects of hardware parameters, app functions, and background data analysis modules are described below:

1) Hardware parameters
Hardware supports light, temperature, and acceleration sensors, which monitor the environment around exhibits with low cost, long battery life, stable signal, and easy installation features.

2) App functions
Navigation: After turning on Bluetooth, visitors can receive background information about exhibits in the current location by shaking their device.

Recommendations: The app recommends exhibitions or events that can be of interest based on a visitor's history and preferences.

Indoor navigation: Using iBeacon base station positioning technology to achieve accurate indoor navigation in each functional areas, visitors will be able to find their way around.

Interaction: Accurately information can be pushed to visitors based on historical visit data. Visitors can leave direct messages and interact through app software.

3) Background data analysis
The background server saves and analyzes terminal service data connected to the base station, automatically generating reports for management personnel to view, and interpreting the behavioral habits of visitors and optimizing future exhibitions. Real-time understanding of the operating status of base station equipment at each location, and timely warnings of base station abnormality are also performed.

A multi-dimensional portrait of visitors shows popular exhibits and recent traffic cycles. This includes statistics and analysis of the sex, educational background, age, and geographical location of visitors, as well as audience volume in the exhibition hall. Meanwhile, it is possible to count and analyze the number and length of stops in front of exhibits, in order to judge the most popular ones.

5. System advantages

5.1 For museums

1) Realizing the intelligent push of exhibit information, saving manpower and material resources, improving explanation efficiency, and exerting the role of the museum.
2) Indoor intelligent positioning can enrich the museum's navigation service and establish a more accurate museum map.
3) Through the behavioral analysis system and statistical analysis of visitor browsing data, a database of visitor preferences can be established, recommending exhibitions/museums that individuals can be interested in, as well as building a model of visitor interests and hobbies.
4) Establishing an interactive platform for the museum and visitors via the information feedback system, realizing real-time dialogue and interactive education.

5.2 For visitors

1) Intelligent push information and diversified exhibit information (text, pictures, video, and audio) provide visitors with richer, more diverse, and more convenient "immersive" experiences.
2) Intelligent indoor positioning functions provide visitors with detailed and accurate indoor positioning and navigation, also providing push information for surrounding services.
3) The use of personal mobile devices enables visitors to become the main museum subject, mastering and preserving information, and establishing long-term learning and visiting habits.
4) The information feedback system enables visitors to engage in dialogue with the museum, establishing a community in which to share and exchange information.

In contrast to dedicated museum apps and WeChat navigation, we find that the M-Beacon system is more convenient to use, because after opening the software and Bluetooth, no additional operations are required. Promotional content is rich, including text, images, audio and video. Two-way, real-time information interaction is realized, and the app can be applied to any museum at a low cost.

6. Six key characteristics of the system

1) Initiative
For the first time, an intelligent active museum public service system for mobile terminals has been established in the museum, combining hardware and software. The museum intelligently and spontaneously offers explanations and guides to visitors, fully liberating them in the process. This is an innovative and breakthrough concept for museum visits.

2) Three-dimensionality
The use of Bluetooth wireless transmission is flexible, real-time, and mobile terminal friendly, providing a more immersive 3D experience for visitors.

3) Stability
The push of exhibit information is not affected by visitor flow or the environment, and is therefore more convenient. The system is able to integrate exhibits from multiple museums, and visitors can easily carry information home after the visit to deepen their understanding.

4) Universality
Visitors can easily achieve the same viewing experience on different mobile terminals, and the system can operate normally in both outdoor and indoor environments. It is highly adaptable to outdoor venues and cultural artefacts.

5) Intelligence
The system habitually analyses visitors' browsing habits, judging their preferences and intelligently recommending potential exhibits of interest through reasonable collection of arrival times, length of stay, browsing history, and movement path.

6) Interactivity
The system's real-time information feedback scheme establishes a dialogue between visitors and the museum, providing a platform for interactive learning.

The base station and server equipment for this system are reasonably priced. As the museum has background cultural resources for exhibits, it is simple and fast to generate audio and video materials. By simply implementing base stations, servers, and preliminary exhibit content making, audience perception can be increased once the system is installed. It also becomes a platform for real-time communication with visitors, providing effective user contact entry for the promotion of subsequent cultural and creative derivatives.

The application of the smart navigation and positioning system within the museum has created a brand new platform by which visitors can understand and interact with the it. The system has advantages such as convenient operation, rich information resources, diversified interaction, data collection and analysis, and customized promotional information for the audience, providing users with personalized services, as well as offering reliable data to understand audience needs and improve consistently. The application of this new technology plays a positive role in the improvement of museum service levels and overall development.

Urban Information Culture and Tourist Destination Art

Zhang Haoda
(Department of Information Management, Peking University, Beijing 100871, China)

Abstract: A city's information culture ecosystem involves not only its natural environment and physical locality, but also the human environment and virtual network space. In addition, the historical inheritance, customs, and linguistic features of cities and communities are essential components of the urban information culture ecology. So, how can natural and human landscapes, historical allusions, social sites, and commercial shopping activities be organically combined? It is important to find, exploit, and construct destination art spaces. Destination art always has a story, and with the story comes an opportunity to connect the various links mentioned above. The author has experienced the attraction of destination art spaces through various foreign travels, and has seen the economic value created through cultural creative products. If handled well, such interdisciplinary cross-field work offers great rewards with little effort.

Keywords: information culture; tourism economy; destination art; public art; urban image

The acceleration of urbanization in China in the 21st century has vastly altered the structure of the original urban cultural ecosystem, directly affecting the function of the city and the quality of urban life. With the transformation of traditional urban functions, the income generated by the tourism economy has been a great attraction for urban administrators, not only due to being a green economy, but because it offers great opportunities for sustainable development.

In recent years, the author has visited many of the so-called "cultural and creative industrial parks" that have been established in cities in order to attract tourists, from the philosophies upheld by administrators to large-scale "song and dance drama" projects like "Impression Lijiang". It is not difficult to see a lack of awareness among many urban managers regarding their own urban information culture, as they swarm to produce "Impression XXX" dance and musical dramas. This reflects a lack of creativity and confidence, and a limited understanding of their own resources. What's more, old forms rich in historical and cultural information are destroyed and replaced with cheap ideas and materials, described euphemistically as "cultural industry."

The author hopes that city administrators will view urban information and tourism culture in the digital age from a three-dimensional perspective, for example:

1) Looking at the relationship between people and cities from the perspective of cultural anthropology;
2) Looking at the development of modern cities from the perspective of informatization;
3) Looking at the implication and extension of the concept of urban public art from the perspective of digital communication;
4) Looking at urban tourism and destination art from the perspective of natural history and popular culture.

Cities in China are undergoing major demolition and construction, and while people are dismantling what is considered "waste", we cannot disregard its role as a symbol or imprint in residents' minds. Whether an office worker rushing to work or a citizen relaxing at home, there is a feeling of dependence on the old surrounding environment. This is apparently unfelt ordinarily, but the disappearance of the buildings and facilities they see every day can trigger disappointment and loss. In most cases, waste is not readily identifiable as cultural artefact, but pieces must be identified and treated as art in terms of their spiritual significance.

Meanwhile, although some cities and facilities have been retained, native populations have disappeared, or have been driven out due to high land prices, replaced by traveling salesmen or traders from surrounding areas. One consequence of this is that tourists no longer have the chance to exchange information with the original city, and no chance to experience the original form of urban life. What's left is a noisy, homogenous tourism market.

1. The culture and design of urban information

Throughout the process of social development, commercialization has led to the rapid expansion of urbanization. The acceleration of urbanization has greatly changed the structure of the urban cultural ecosystem, directly affecting the functions of the city and the quality of urban life.

The city's information culture ecosystem includes not only natural environments and physical localities within the city, but also the human environment and virtual network space, while the historical inheritance, customs, and linguistic features of cities and communities are essential parts of urban information culture ecology. As the tourism economy plays an increasingly important role in the promotion of urban development, the integration of typical urban landscapes and landmarks (which are inherent in urban information culture) into artistic concepts of destination tourism is an overlapping project that involves interdisciplinary action. If handled properly, it will have broad-ranging effects.

Urban information design refers to the design of space within and outside the city, social structure, and civic living environments, in which people achieve political, economic, cultural, scientific, technological, and aesthetic goals. An important foundation for its existence is a city's level of commercialization. The main forms of urban information design are presented in visual aspects including urban navigation systems, commercial advertisements, urban sculptures, architectural sculptures, environmental art, landscaping, landmark architecture, public art, and destination art, together forming the external image of the city.

2. Destination art

In the Chinese art world, people often confuse "destination art" with "public art".

In recent years, economic reform has accelerated the urbanization process, partly out of a demand for large-scale urban development and construction, and partly out of a demand for self-expression from interest groups. The terms "city image" and "public art" are therefore linked.

Perhaps "destination art" should be used to describe many of China's current works of "public art", as the basic concepts of public art are not often found in such works. However, their existence is of great significance to the tourism economy, making them into destination art. However, they may also not be destination art, as they do not necessarily coexist uniquely with the local environment.

What is destination art? It can be argued that it is a specific concept, only existing in a specific environment (on a site), as it is part of a whole. In other words, it is not a work of art that can be moved freely and can be viewed independently by people in any location. Such works have a unique coexistence with the local environment.

People can be used to roaming in beautiful natural landscapes, but they also value the sight of man-made scenes such as bridges, pavilions, pagodas, and other poetic and artistic edifices. From the perspective of tourism, both natural and human landscapes can be included in the concept of "destination art", measured by their relationship to the tourism economy.

At present, the Chinese government is advocating for the transformation of labor-intensive, process-oriented cities into creative and cultural service cities – a general trend within deepening economic reform. Alongside transformation in urban economic forms, some previous material forms related to human life and production must be abandoned. However, as far as the tourism economy is concerned, the outdated remains of human activities have precisely the potential to become "new" cultural landscapes, becoming "destination art" on popular tourist routes.

In places that have been forgotten or abandoned, beautiful scenery still exist. The design strategy of tourism culture must not be limited to the act of getting rid of the old to make way for the new, but must also utilize so-called "waste". One example is Beijing's 798 industrial art zone and its follow-up creations. In addition to ancient capital tourism, these have become the most famous and important destination art landscapes within "new" Beijing tour routes, retaining and making full use of the original terrain plant structures. Striving to reduce costs while minimizing transformations, they create a distinctive artistic landscape. By changing the functions of original buildings, facilities, and venues, abandoned factories, equipment, pipes, and gantries are repackaged and artistically modified whilst preserving original structural styles. New and old structures coexist and complement one another, becoming a place for leisure with ultra-modern consciousness.

Some of the renovated factories have become parks and dining areas, and others have become exhibition spaces, galleries, and art sales venues. This reproduces part of the history of the industrial area, and also creates a large-scale destination art space for citizens and tourists to relax in and enjoy. The formation of a spatial focus within a complicated urban space attracts public attention, strengthens the centrality and cohesiveness of the space, and at the same time showcases a high-quality creative cultural life. Here, the author is focusing on rendering art in a "reserved" destination, while of course it is also feasible to make "newly created" destination art according to guidelines, despite the fact that the associated costs and risks are relatively high. Notably, as a destination art space for urban information culture and tourism, 798 is a successful case. However, it cannot be blindly imitated, as it involves a variety of intersections of time, space, background, and politics, which cannot simply be copied.

3. *Urban public art*

Public art exists in the urban cultural and ecological realm, involving material and virtual space. Art work that allows for public continued participation is completed by artists under the guidance of certain civic awareness, using public cultural resources as media within the public environment.*

The "public" in public art has three meanings:

1) It is addressed to the public: i.e. targeted at the public;
2) It is for the public: from public position and for public interests, it is not concerned with private positions or personal interests;
3) It is about public affairs: it is usually related to a public affair or a major issue in society.

Urban public art is of great significance in the overall construction of a city. In order to build a modern and emerging city, we must improve the cultural and artistic atmosphere, and enhance the overall image of the city on the basis of improving urban infrastructure. Urban public art aims to shape a space composed of free and equal citizens, the key elements of this space being openness and interactivity. Public art speaks to the public and is closely related to public affairs. Therefore, it is able to stimulate public thinking about social affairs and living spaces.

If public art works were simply to attend to visual aesthetic appreciation, ignoring the "participation" aspect and simply satisfying visual enjoyment without universal humanistic concern, they would show only weak "public" consciousness. Public art intends to beautify the city. It does not exist for its own construction, and depends largely on the revelations that citizens gain from it.

4. *The dissemination of tourism information and destination art design*

The dissemination of tourism information refers to the communication of information regarding tourism-related scenic spots, travel dynamics, and urban planning systems to consumers via service media. Tourism information dissemination is a key component of tourism marketing and promotion, and is an important condition for promoting the industrialization of tourism culture.

Tourism products are different from other general commodities. With the development of information technology and the gradual maturity of the tourism industry, consumers

* Zhang Haoda. *Visual Communication: Information, cognition, interpretation* [M]. Beijing: Peking University Press, Dec 2010, 1st edition, p. 139.

are increasingly dependent on and demanding of information services. Before they travel, tourists generally use a variety of media to research their destinations, and design their own travel routes and accommodation arrangements by means of Internet-based services. Their decision-making process involves input, management, output, and feedback of information. Here, the author wishes to emphasize that only when various pieces of tourist information are expressed in a systematic form (to be effectively accepted and recognized by consumers) is a potential tourist destination likely to actually be chosen by a customer.

How can we organically integrate natural or manmade landmarks, historical signs, human and social artefacts, and commercial shopping spaces? Finding, excavating, and building destination art becomes very important. With it comes a story, and with a story there is an opportunity to link the various areas mentioned above. The author has experienced the attraction of destination art spaces through various foreign travels, and has seen the economic value created through cultural creative products.

Comprehensive cultural information within tourism information must be treated as part of the integrated marketing communication, customizing a "cross-media linkage" key point based on submitting to the overall marketing strategy. Especially with smart phones today, it is characterized by a data service center that is capable of collecting, storing, and transmitting multimedia information in the form of an app (referring to third-party smart phone applications), using destination art as a thematic thread throughout. Meanwhile, the development of scenic location websites is an important part of the scenic spot information project. It is integrated with e-government and e-commerce systems, and combined with "scenic spot" natural resources, cultural resources, biodiversity, digital topographic maps, and other databases to provide cultural tourist information in the form of Big Data. Administrators must realize that in contrast to professional commercial travel websites, the portals of tourism channels are relatively neutral, displaying images of scenic spots more objectively and comprehensively so that tourists may form a more comprehensive understanding.

5. *Digital tourism information design*

As the propagation of online travel information is not limited by time and space, as long as tourists are offered WiFi access, users with handheld mobile terminals can enter the network to view travel information transmitted through the network. Interactivity in this kind of information exchange transcends the interactivity of interpersonal communication in a sense. Moreover, network communication provides a number of social platforms on which to engage in interactive communication. For example, the current "WeChat Moments" involve text, images, audio, video, and animation in one, bringing great opportunities for editing and disseminating stories based on destination art. Potential consumers have the opportunity to be attracted to and moved by online stories related to destination art before being a visitor themselves, becoming "followers" even before they make a trip, then inspiring a mental drive to complete the process from "desire" to "action" as quickly as possible.

The following are specific design recommendations.

5.1 Scenic websites

There are two kinds of scenic website. The main type is the professional tourism website provided by management bureaus or administrative companies regarding scenic spots. These websites play a unique role in the dissemination of tourism information about scenic spots. The owners and builders of these websites include management personnel from tourist attractions and facilities. Their duties include the development and conservation of tourist attractions, the improvement of infrastructure and related services for tourist attractions, the promotion of scenic spots and regional culture, and the publication of images of scenic spots. The second kind is the business-type travel website. Some of these websites stem only from considerations of content development and integrity, while others are used for the pursuit of commercial interests, such as online marketing for travel agencies. The information provided by these websites plays a certain role in inducing tourism consumption.

5.2 Design navigation equipment

To the majority of ordinary tourists, sophisticated digital technology broadcasts may simply be operator gimmicks. The development of genuine, easy-to-operate digital handheld navigation devices allows visitors to pick up content and master travel more easily. These devices are cross-industry cross-technology products.

5.3 Designing a guide system

Developing a visual guidance system with strong visual effects allows tourists to view things under the direction of a scientific navigation system release, improving the utilization rate of scenic spot resources, promoting the flow of tourists, and enabling visitors to get the most out of their visit.

5.4 Designing QR codes and a WeChat public platform

To make full use of terminal equipment such as smart phones, a QR code scanning system can be used to release timely information with a substantial amount of content, as well as a real-time WeChat public platform for more effective on-the-spot learning.

The information provided by online media advertising and network information dissemination is monitored by the public in real time. In the era of Big Data, online media outlets have the opportunity to break through limitations of their own storage, allowing any mobile terminal devices to receive the same amount of information.

5.5 Virtual reality and augmented reality

Destination art involves scattered scenes and forms belonging to different places, and visitors often see partial links without fully understanding the whole picture at once. The appropriate use of virtual reality and augmented reality technologies by means of apps can provide visitors with scalable information – an effective means to overcome this obstacle.

6. Realistic issues

In many cities in China, information related to destination art in tourism tends to indulge in false positivity, glossing over reality. Such "image projects" are simply a waste of manpower and money. A cohesive society not only involves coordination within society, and also between humans and nature. True destination art can only be called such if it publicizes human concerns and integrates art into the vast landscape.

We must remember:

Nostalgia — What is the source of enthusiastic public participation?
Nostalgia — How can we sustain public interaction?
Nostalgia — Digital technology must support content creation.
Nostalgia — We must not use digital technology just to show off.

Destination art originating from urban information culture is not simply a matter of building or setting up a few independent works of art in local areas, nor is it a matter of finding a few representative works of art in historical sites that belong to the local cultural atmosphere. In fact, the most important thing is that it conveys a clear and convincing concept belonging to the local community. This concept must have unique local characteristics while acting as a common human pursuit. No matter where tourists come from, the language they use, the color of their skin or their appearance, they can be touched by these "objects" there and then. They travel a great distance to experience this feeling.

Finally, the author would like use visuals to demonstrate these concerns. The following are 38 specially selected image frames. Please read and consider the captions while browsing the pictures.

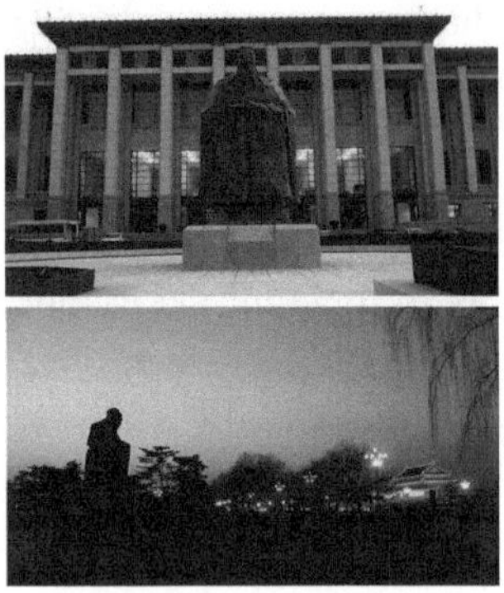

001 | Public Art—Confucius Statue on the east side of Tiananmen Square, Beijing

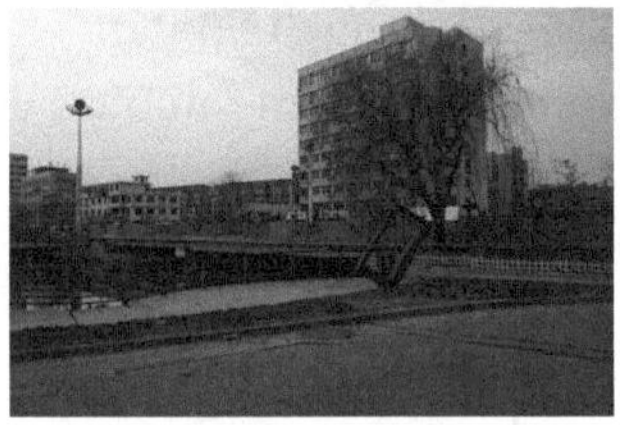

002 | Public Art — Uncoordinated Decoration, Yancheng

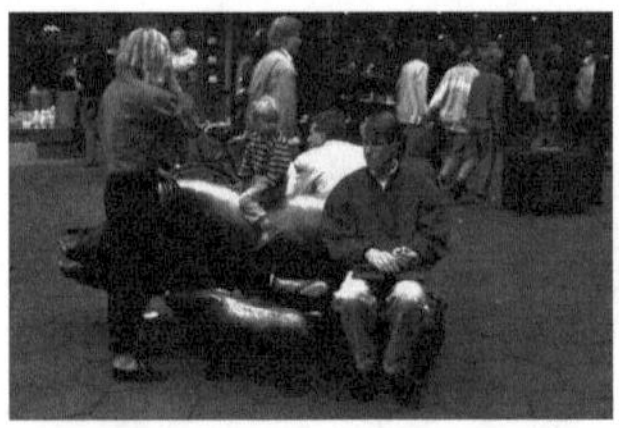

003 | Public Art — Pedestrian Street Sculpture, Copenhagen, Denmark

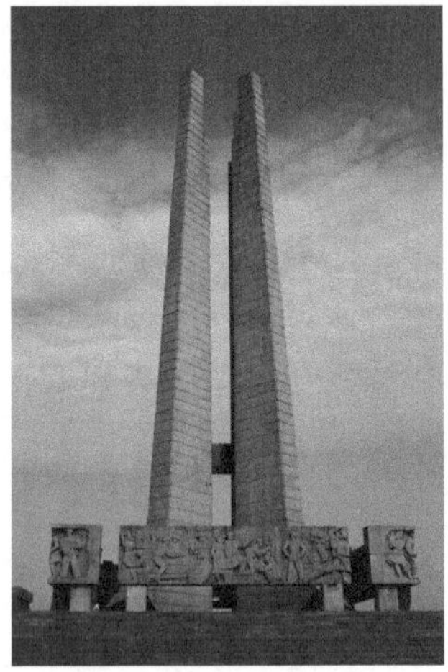

004 | Public Art — Earthquake Monument, Tangshan

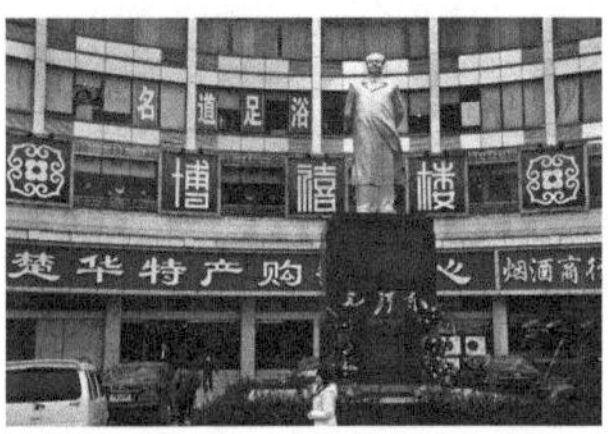

005 | Public Art — Mao Zedong Standing Statue, Changsha City, Hunan

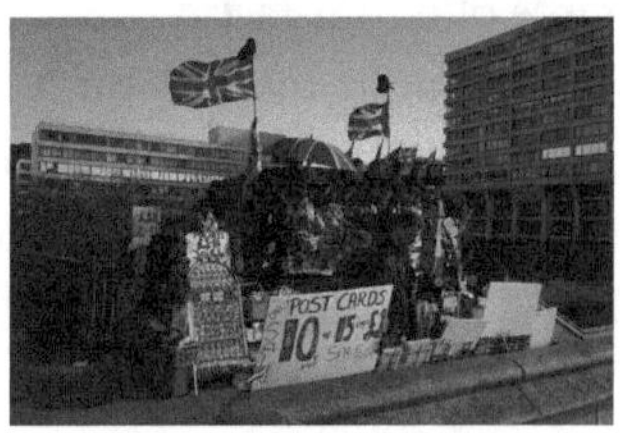

006 | Public Art — Souvenir Stand, London

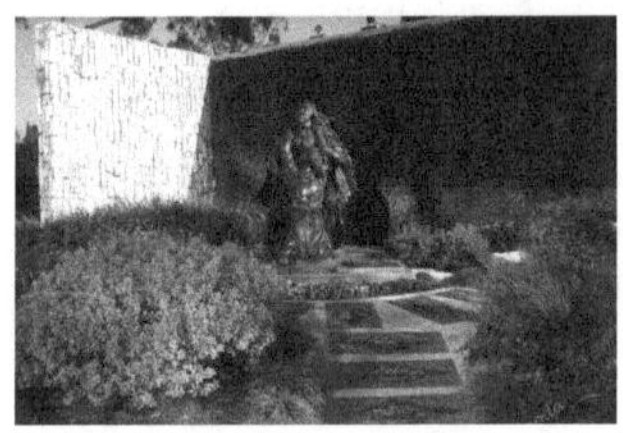

007 | Public Art — The Prodigal Son Returns, Los Angeles

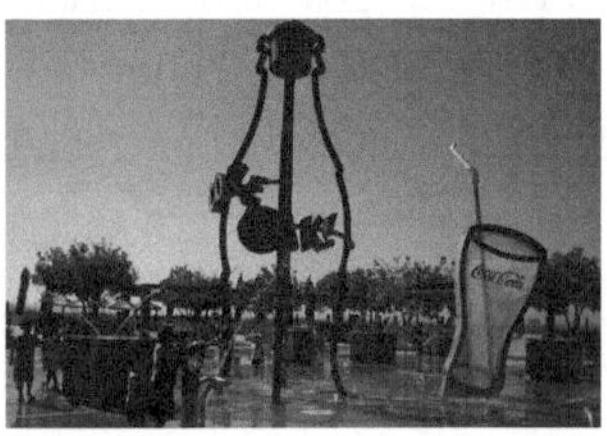

008 | Public Art — Los Angeles, USA

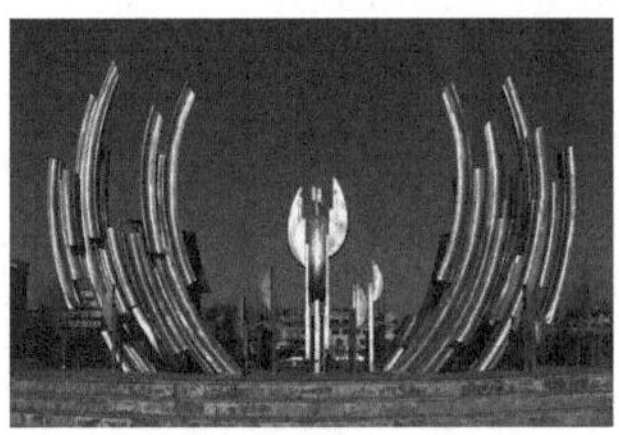

009 | Public Art — Minxin River, Shijiazhuang

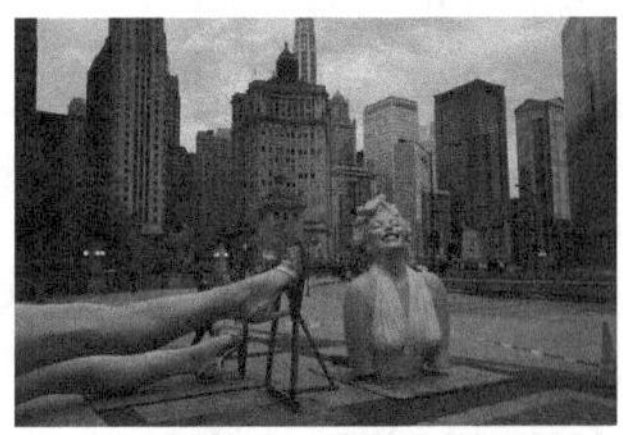

010 | Public Art — Eight-meter-tall Marilyn Monroe Sculpture, Chicago, USA

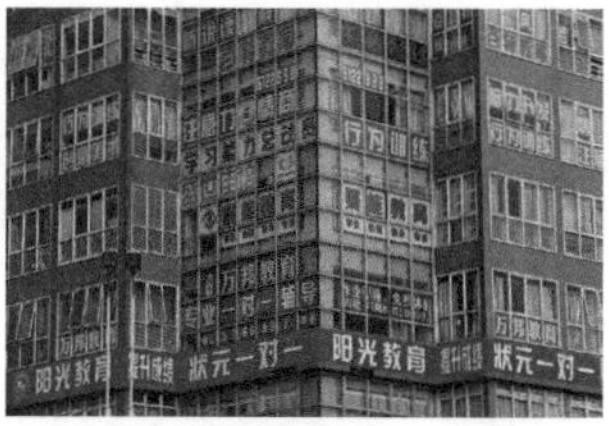

011 Destination Art — Anyang Cram School Building, Henan

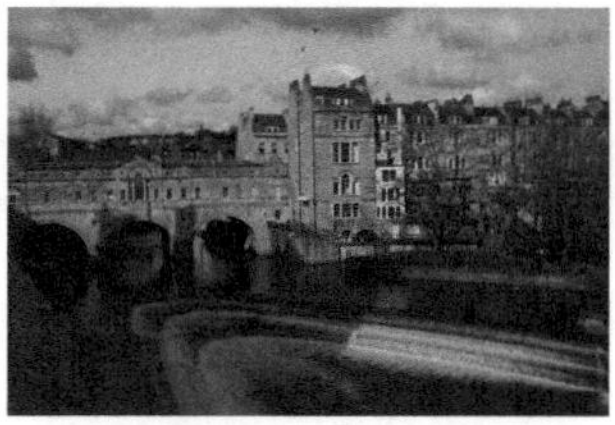

012 Destination Art — Bath Old Town, UK

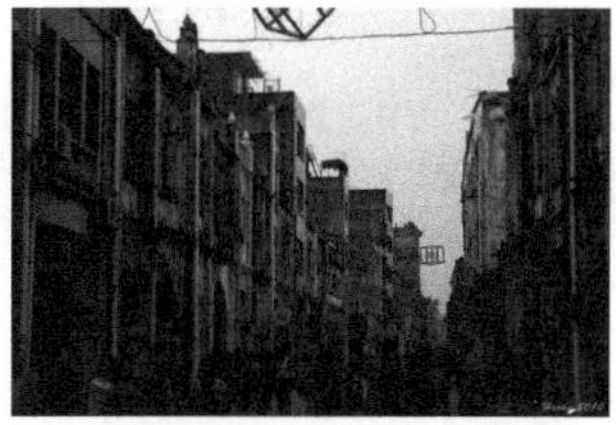

013 Destination Art — Beihai Old Street, Guangxi

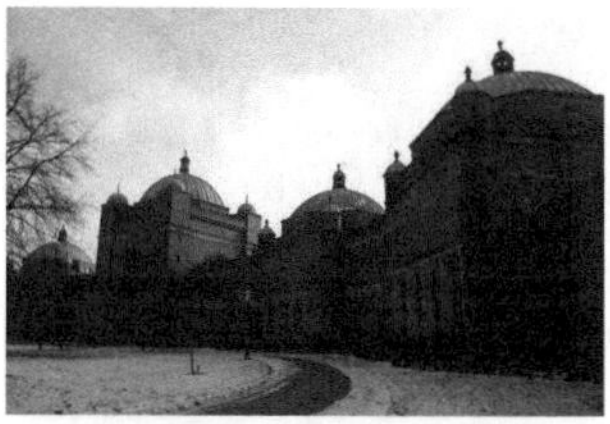

014 Destination Art — University of Birmingham, UK

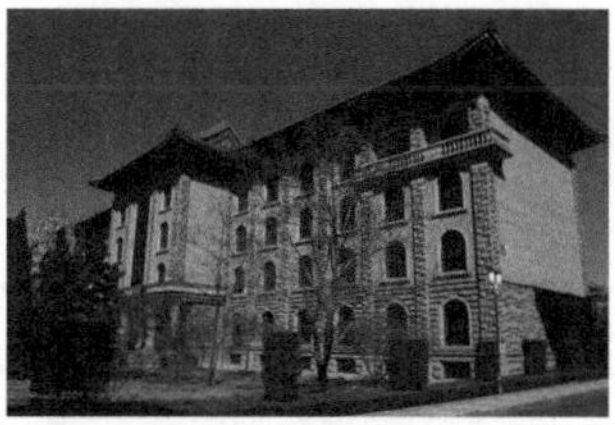

015 Destination Art — Central Party School

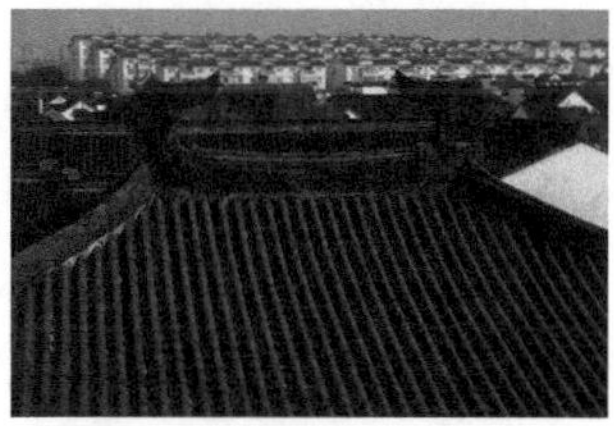

016 Destination Art — Hanshan Temple, Suzhou

017 Destination Art — Helan Mountains Rock Art, Ningxia

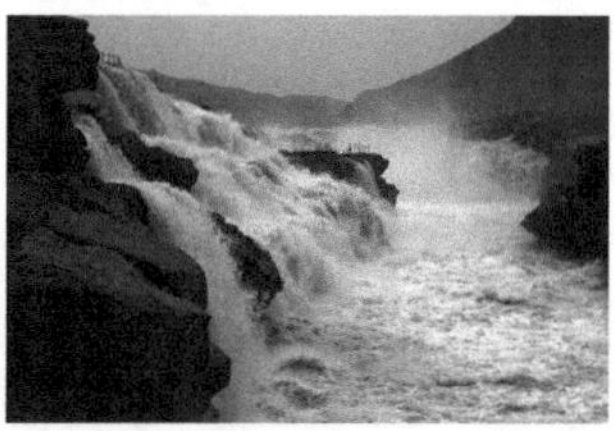

018 Destination Art — Hukou Waterfall, Shaanxi

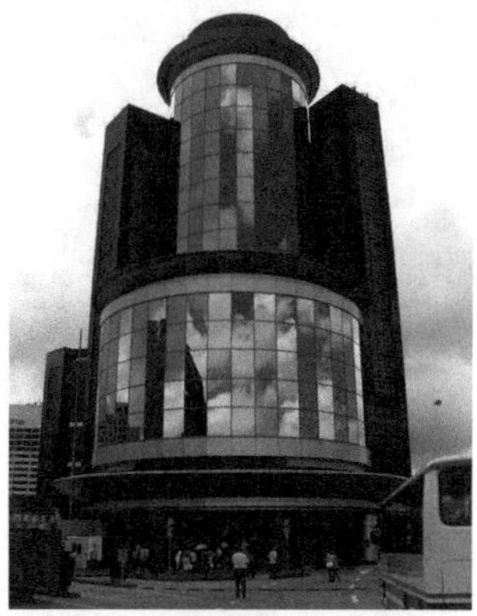

019 Destination Art — Marina Bay Sands Casino, Macau

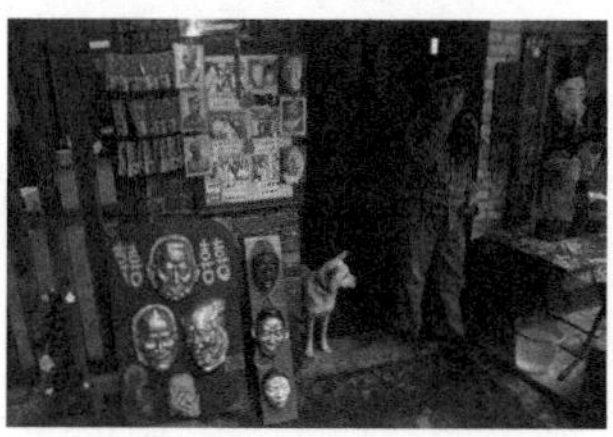

020 Destination Art — Jiufen, Taiwan

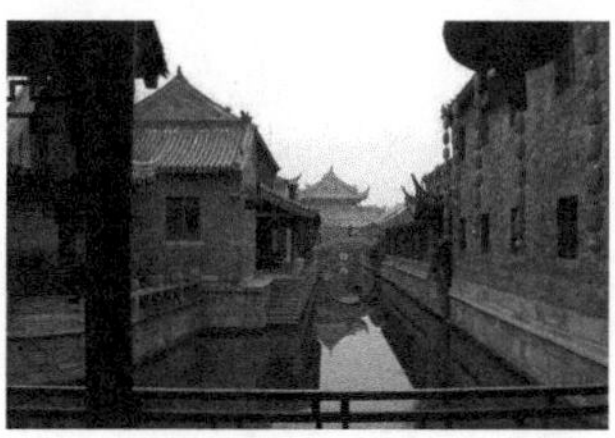

021 Destination Art — Looking at a Lonely Space, Taierzhuang

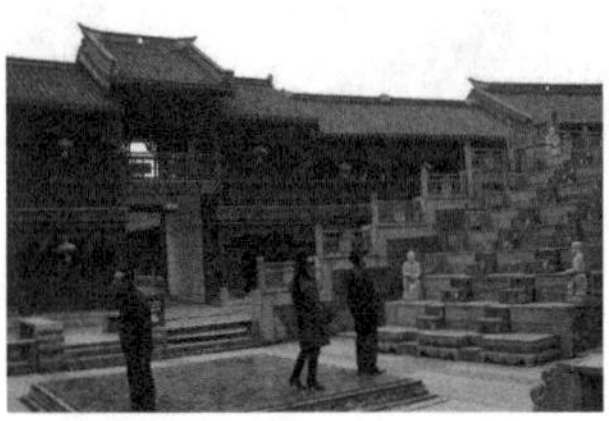

022 | Destination Art — Far Away, Taierzhuang

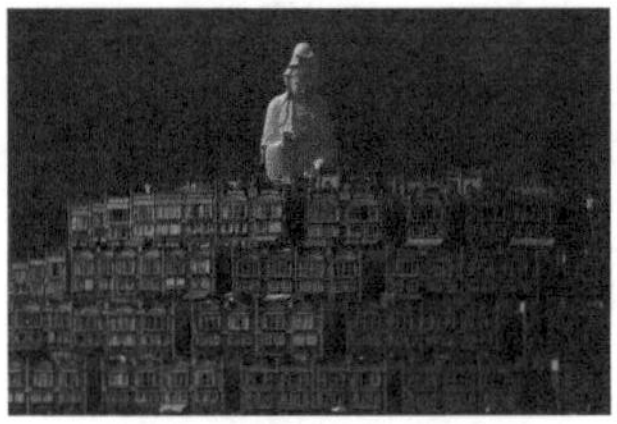

023 | Destination Art — Tsz Mountain Monastery, 76-meter-tall Guanyin Statue, Tai Po, Hong Kong

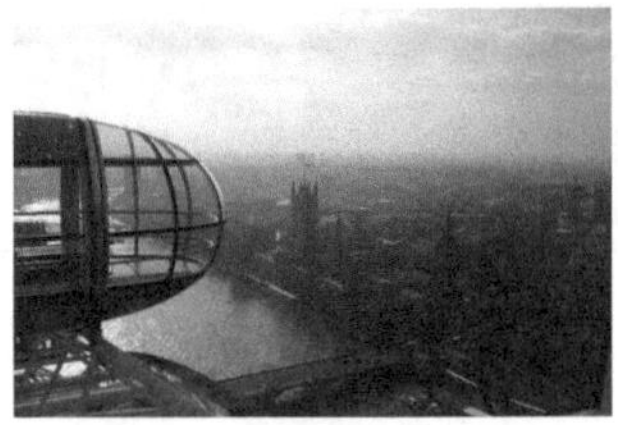

024 | Destination Art — View of the Thames from the London Eye

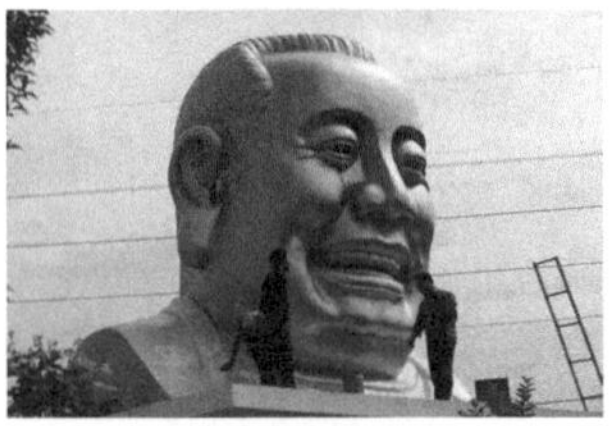

025 | Destination Art — Longhua Fun Water Park Buddha Statue, Yanshi, Luoyang, Henan

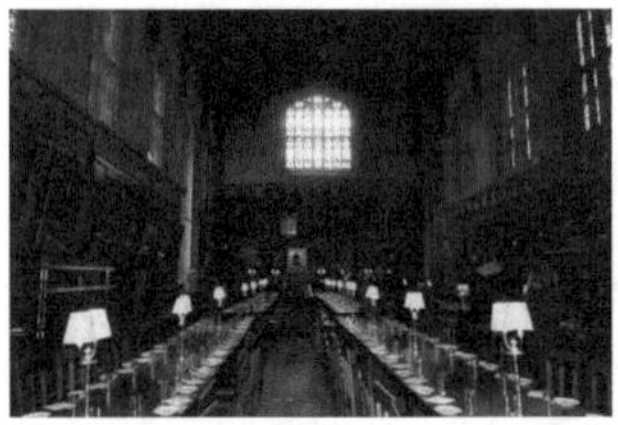

026 | Destination Art — Christ Church College, Oxford University, Oxford, England

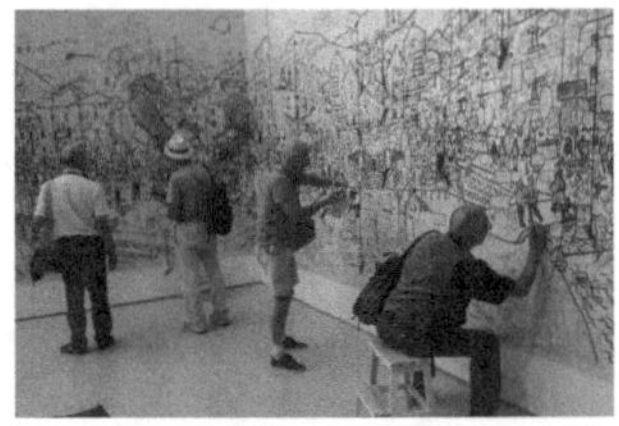

027 | Destination Art — Participatory Art at 798 Beijing

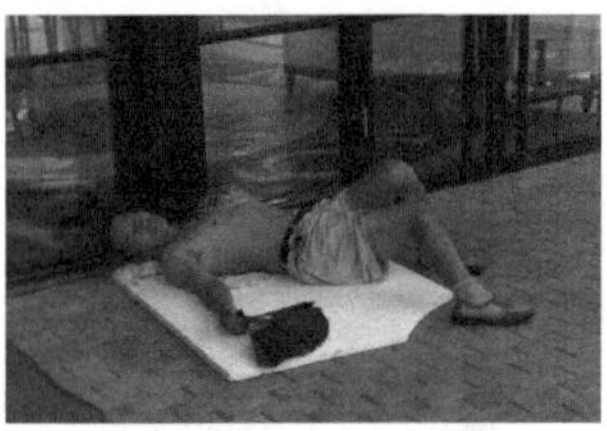

028 | Destination Art — Performance Art at 798 Beijing

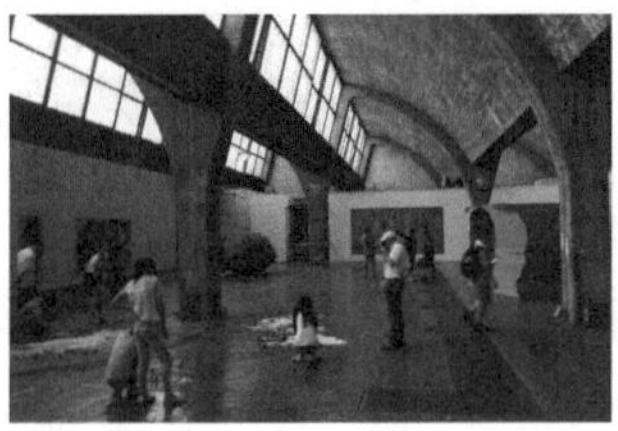

029 | Destination Art — Installation Art at 798 Beijing

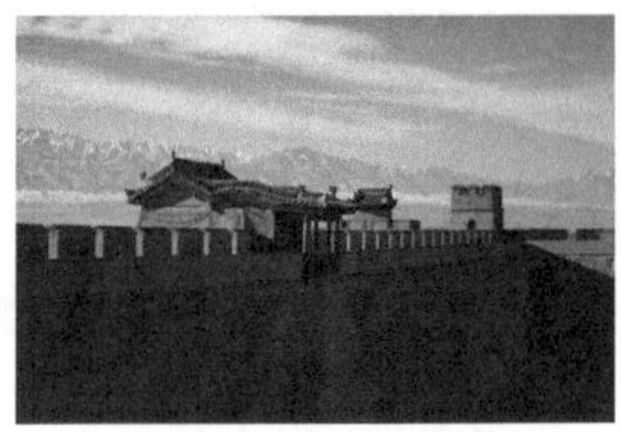

030 | Destination Art — Qilian Mountains, Jiayuguan

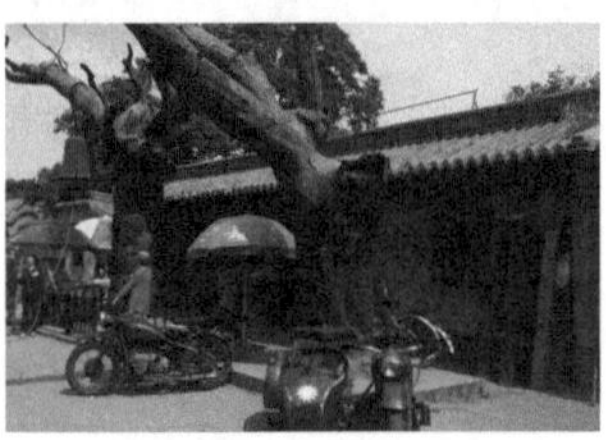

031 | Destination Art — Ranzhuang Tunnel Warfare Memorial, Hebei

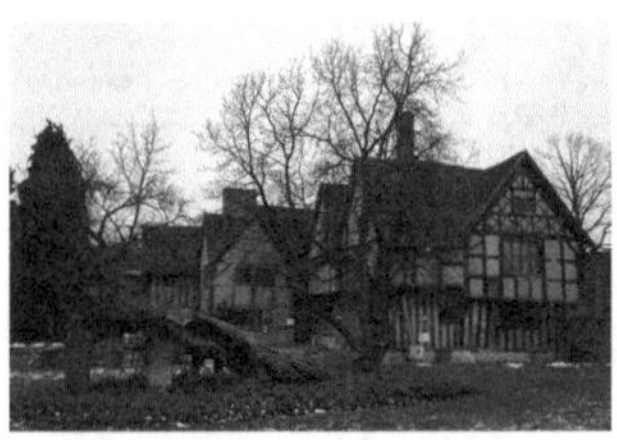

032 | Destination Art — Shakespeare's House, UK

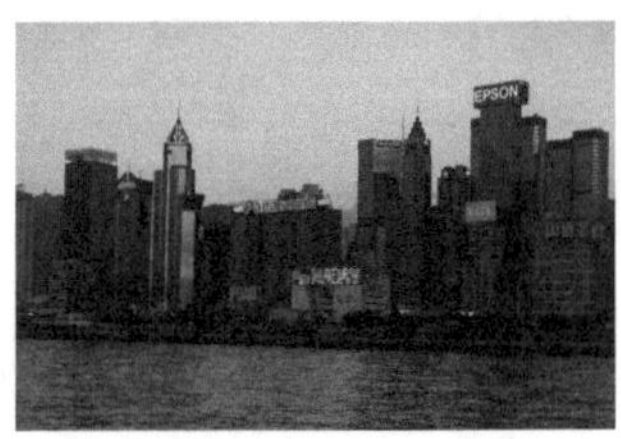

033 | Destination Art — Commercial Advertising, Hong Kong

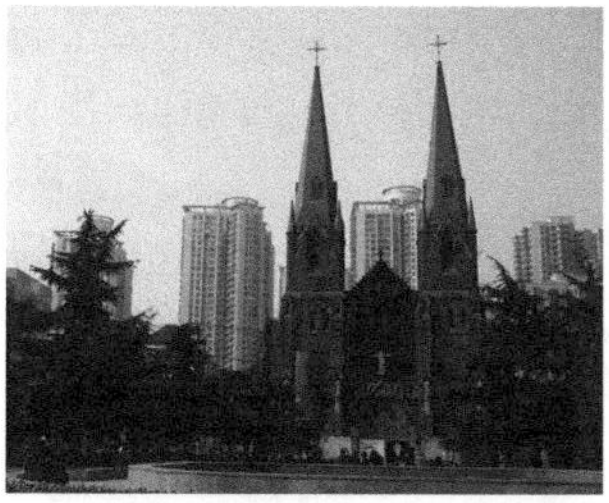

034 | Destination Art — Xujiahui Church, Shanghai

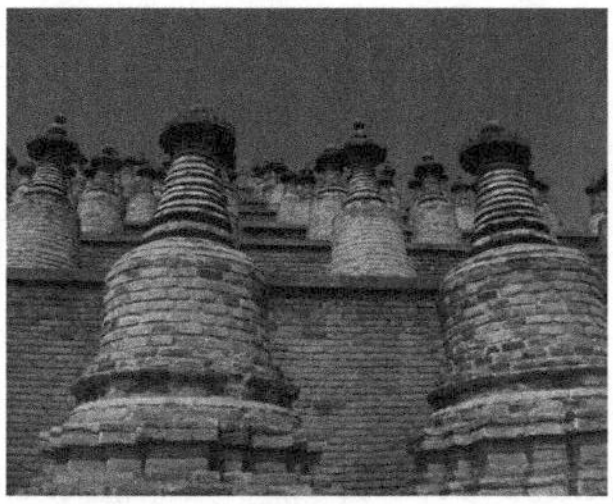

035 | Destination Art — One Hundred and Eight Stupas, Qingtongxia

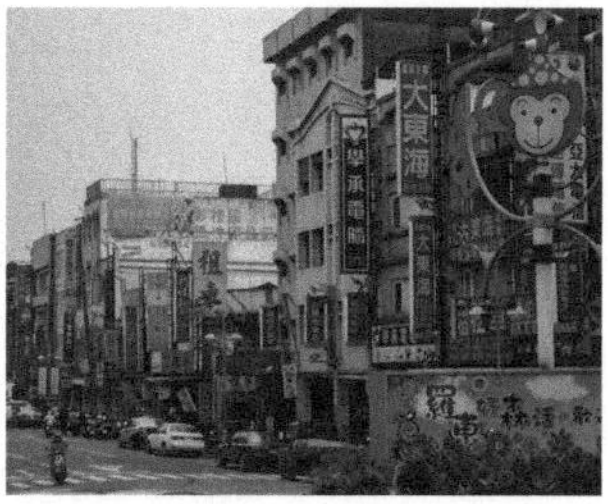

036 | Destination Art — Luodong Town Image Advertising, Yilan, Taiwan

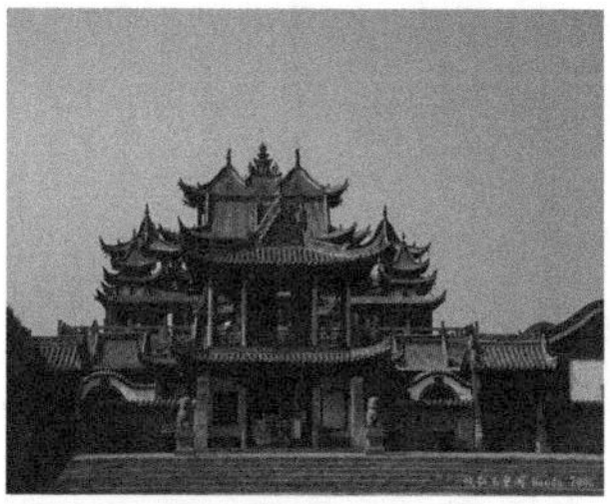

037 | Destination Art — Jade Emperor Pavillion, Pingluo

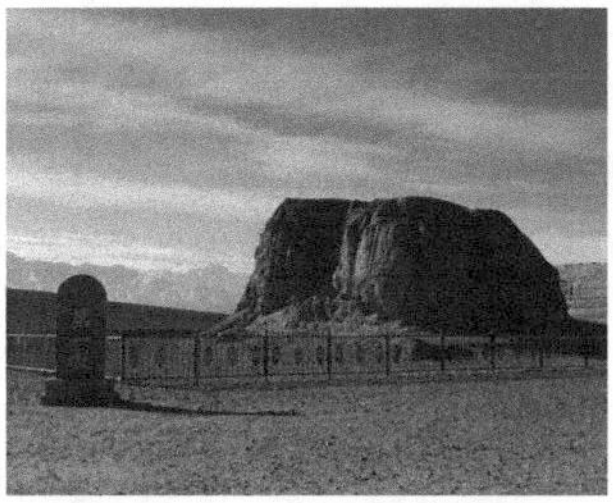

038 | Destination Art — History, Gansu

001—Public Art—On January 12, 2011, on the east side of Tiananmen Square in Beijing, a bronze statue of Confucius, totaling 9.5 meters in height, was completed in front of the National Museum. However, after standing for 90 days, it was moved from its original position.

002—Public Art—A piece of work in Yancheng City, Jiangsu Province; at one in appearance with the urban style, but divided within.

003—Public Art—A sculpture that provides rest on a pedestrian street in Copenhagen, Denmark.

004—Public Art—Earthquake Monument in Tangshan City, Hebei. In the early morning of July 28, 1976, a massive earthquake of 7.8 on the Richter scale occurred in Tangshan. A city of one million people was razed to the ground in an instant.

005—Public Art—A standing statue of Mao Zedong in a housing project area in Changsha, Hunan Province.

006—Public Art—A souvenir stand in London, England, showing the city's style.

007—Public Art—"The Prodigal Son Returns" sculpture in a street park in Los Angeles, USA, based on Luke 15:11-32 from the Christian Bible.

008—Public Art—A public art work in Los Angeles, USA, which visitors may interact with using water guns.

009—Public Art—A group of sculptures located on the bank of the Minxin River in Shijiazhuang, Hebei Province. There did not used to be a river in Shijiazhuang City. This river was excavated by hand, and water was channeled from the Shijin irrigation canal into the city on September 29, 1997. Officials call it a "civil river", while the public calls it an "image project".

010—Public Art—On May 7, 2012, an eight-meter-tall statue of Marilyn Monroe by artist Seward Johnson was dismantled. This model was developed from the classic image of Monroe's 1955 film *The Seven-Year Itch*.

011—Destination Art—Located on the famous remedial class building in the center of Anyang, Henan Province, this advertisement showcases it as a place where people love to learn.

012—Destination Art—Old Town of Bath, England. In the 1st century AD, the Romans built a large number of bathing communities in this hot springs town, which became a holiday resort. It is the only city in the UK to be included on the World Cultural Heritage List.

013—Destination Art—Built in 1883, Guangxi Beihai Old Street - Zhuhai Road is more than 100 years old. In the past, "sailor culture" promoted the development of the street, and after the Reform and Opening, one word from the writer Shu Yi ensured the Old Street's survival.

014—Destination Art—Founded in 1825, the University of Birmingham is one of the top 100 universities in the world, and the famous Chinese geologist Li Siguang is among its most outstanding alumni. To date, the University of Birmingham has produced eight Nobel Prize winners and two British Prime Ministers.

015—Destination Art—The Central Party School of the Communist Party of China in the western suburbs of Beijing. Its architectural style includes both traditional Chinese roof elements and Yan'an cave features, as well as Romanesque stone walls and Rococo-style balconies, as an expression of inclusive thinking.

016—Destination Art—Standing on the top floor of the bell tower of Hanshan Temple outside Gusu, townhouses are visible through the temple roofs and ornaments.

017—Destination Art—Rock paintings of Helan Mountain, Ningxia. Helan Mountain is a place where northern ethnic minorities such as the ancient Xiongnu, Xianbei, Turks, Uighur, Tubo, and Tangut raised animals, hunted, lived, and reproduced, leaving behind many beautiful rock paintings.

018—Destination Art—The Hukou Waterfall on the Yellow River in Yichuan, Shaanxi Province - the largest waterfall in the upper reaches of the Yellow River, with a nine-meter drop.

019—Destination Art—Located in the Marina Bay Sands Casino in Macao, it is said that this architectural shape is a metaphor for male genitalia.

020—Destination Art—Jiufen, a mountain city in Taipei, where decades of gold mining made a small town flourish. Miners' gold made Jiufen into a city that never slept. It is a major cultural tourism destination within Taiwan's successful economic transformation, and is the setting for the film *A City of Sadness*.

021—Destination Art—Located in Taierzhuang New Town, Zaozhuang City, Shandong Province; a brand new "Old City" where no one really lives. It is a stage with no players. The buildings are just props for tourists to walk around.

022—Destination Art—In Taierzhuang – a brand new "Old City" – tourists are "passing travelers" while "citizens" come to work.

023—Destination Art—A 76-meter-high white Guanyin statue stands behind a group of buildings at the Tsz Mountain Monastery in Tai Po, Hong Kong.

024—Destination Art—From the London Eye one can see the Thames, Westminster Palace (the House of Parliament, listed as a World Heritage Site in 1987), Westminster Abbey (the church of the royal family) and Big Ben.

025—Destination Art—In April 2013, a large Buddha Head was built at the Longhua Water Park in Luoyang City, Henan Province. Due to its suspicious shape and demeanor, it was ordered to be changed, and in June 2013, it was replaced by a dragon head.

026—Destination Art—Christ Church College was the first college built at Oxford University. It is the only church-style college in the world, and the largest in Oxford, England. Its land occupation and architecture belong to the college itself, and the government cannot interfere. Christ Church was built 800 years ago. It has trained 13 British Prime Ministers over 200 years, and is an important location in the *Harry Potter* films.

027—Destination Art—An exhibition at the Beijing 798 Art District, where visitors sign a participatory art work.

028—Destination Art—A tourist appears to sleep in the scenic area of the 798 Art District in Beijing – a piece of performance art reflecting an understanding of artistic freedom and respect for people.

029—Destination Art—An installation art by a Korean artist at the 798 Art District in Beijing.

030—Destination Art—The Qilian Mountains in the distance in Jiayuguan, Gansu, evoking memories of Wang Chingling's frontier poems.

031—Destination Art—Ranzhuang, Hebei Province, is the setting for the film *Tunnel Warfare*. In the commemoration of the 70th anniversary of the Second World War, the author captures the participation of tourists in a play — the gamification of history by spoof movies has finally paid off.

032—Destination Art—Shakespeare's House, Stratford, 180 kilometers from London, where Shakespeare was born and died.

033—Destination Art—Hong Kong's Victoria Bay. Some buildings appear as landmarks during the day. When the evening lights are lit, they are carriers of commercial advertising, clearly demonstrating the city's location to visitors.

034—Destination Art—A catholic church in Xujiahui, Shanghai. In the late Ming Dynasty, Matteo Ricci converted university scholar Xu Guangqi to Catholicism and passed on his scientific knowledge. The descendants of Xu were born here, hence the name “Xujiahui”. In 1847, the French Catholic Jesuits built this church, and the Xujiahui area became a window for the import of Western culture.

035—Destination Art—The One Hundred and Eight Stupas, located in Qingtongxia City, Ningxia, were built during the middle and late Western Xia Dynasty. The stupas are a commemorative Buddhist construction, and are some of the few remaining cultural artefacts from the Western Xia.

036—Destination Art—Image advertising in Luodong Town, Yilan, Taiwan. There are many forests in Luodong Town, creating a beautiful natural environment. Although it is the smallest administrative unit in Taiwan, it is a very famous tourist area.

037—Destination Art—Jade Emperor Pavillion in Pingluo, Ningxia, is a group of structured wooden buildings surrounding a two-story pavilion with gable and hip roofs. It consists of the Chenghuang Temple, Guanyin Temple, Niangniang Temple, Sanqing Hall, Doumu Palace, and the Jade Emperor Hall, and is the largest ancient building complex in the northwest of China.

038—Destination Art—Jiayuguan’s Great Wall “First River Pier” is a pier at the western end of the Ming Dynasty Great Wall, built on a cliff nearly 56 meters high on the edge of Taolai river. The “First River Pier” at Jiayuguan and the “Old Dragon’s Head” at Shanhaiguan guard each end of the Great Wall.

Note: Thirty-two of the pictures marked with the following numbers belong to Zhang Haoda: 002, 003, 004, 005, 006, 007, 008, 009, 012, 013, 014, 015, 016, 017, 018, 019, 020, 021, 022, 023, 026, 027, 028, 029, 030, 031, 032, 033, 034, 035, 036, 037, 038. The remainder are sourced from the Internet.